PRECONDITIONING METHODS: ANALYSIS AND APPLICATIONS

TOPICS IN COMPUTER MATHEMATICS

A series edited by David J. Evans, Loughborough University of Technology

Volume 1 PRECONDITIONING METHODS: ANALYSIS AND APPLICATIONS
Edited by David J. Evans

Additional volumes in preparation

ISSN: 0275 5815

This book is part of a series. The publisher will accept continuation orders which may be cancelled at any time and which provide for automatic billing and shipping of each title in the series upon publication. Please write for details.

Contents

Introduction to the Series vii

Contributors ix

Preface xi

On Preconditioned Iterative Methods for Partial Differential Equations (D.J. EVANS) 1

The Extrapolation Technique as a Preconditioning Strategy (A. HADJIDIMOS) 47

Further Preconditioning Strategies for the Numerical Solution of Certain Sparse Linear Systems (D.J. EVANS AND A. BENSON) 69

Alternating Direction Preconditioning Methods for Partial Differential Equations (D.J. EVANS AND C.R. GANE) 81

Preconditioned Iterative Methods for the Numerical Solution of Elliptic Partial Differential Equations (D.J. EVANS AND N.M. MISSIRLIS) 115

SSOR Preconditioning of Toeplitz Matrices (L. ANDERSSON) 179

On the Convergence Theory and Application of Symmetric Compact Preconditioned Iterative Methods (D.J. EVANS AND E.A. LIPITAKIS) 189

Analysis of Incomplete Factorizations with Fixed Storage Allocation (O. AXELSSON AND N. MUNKSGAARD) 219

Stabilized Incomplete LU-Decompositions as Preconditionings for the Tchebycheff Iteration (H.A. VAN DER VORST) 243

Modified Incomplete Cholesky (MIC) Methods (I. GUSTAFSSON) 265

Preconditioning Symmetric Indefinite Matrices (P.E. SAYLOR) 295

The Preconditioned Chebyshev Iterative Method for Unsymmetric Linear Systems of Equations (D.J. EVANS AND M.A. KAMMOONAH) ... 321

The Preconditioning by Direct Factorization Methods for Solving Self-Adjoint Partial Differential Equations (D.J. EVANS AND I.C. DEMETRIOU) 355

Preconditioned Iterative Methods for the Generalized Eigenvalue Problem with Large Sparse Matrices (D.J. EVANS AND J. SHANEHCHI) 379

On the Application of Preconditioned Direct Methods (D.J. EVANS AND M. HATZOPOULOS) 401

Development of an ICCG Algorithm for Large Sparse Systems (A. JENNINGS) 425

Preconditioned Conjugate Gradient Methods for Transonic Flow Calculations (Y.S. WONG AND M.M. HAFEZ) 439

An Efficient Preconditioned Conjugate Gradient Method. Applications to the Solution of Non-Linear Problems in Fluid Dynamics (R. GLOWINSKI, J. PERIAUX, AND O. PIRONNEAU) 463

Preconditioning Conjugate Gradient Algorithms for Solving Finite Difference Systems (D.A.H. JACOBS) 509

An Approximate Q.I.F. Method for Parallel Computers (D.J. EVANS AND E.A. LIPITAKIS) 537

Index 553

Introduction to the Series

In recent times there has been a broadly based and vigorous pursuit of mathematics and the need for its interaction with other sciences and disciplines, particularly the computer, is of paramount importance.

The rapid development and widespread use of computers in the academic, scientific and industrial fields has stimulated the need to move away from the traditional areas of mathematics to a more computer-orientated approach to the subject.

For some time now the emergence of a new form of mathematics has been evident and can most clearly be seen in the research papers submitted for publication to academic journals such as the *International Journal of Computer Mathematics*. This has led not only to inevitable increase in its frequency of publication but has also prompted the publishers to introduce a new series of concise monographs, each concerned with one particular aspect of Computer Mathematics. There is a need to acquaint readers with the advances in these subjects which are rapidly achieving tremendous importance and which are likely to impinge upon many aspects of our daily lives. Future volumes are planned at irregular intervals as dictated by need or the development of new topics.

David J. Evans

Contributors

L. ANDERSSON	University of Lulea, Lulea, Sweden
O. AXELSSON	University of Nijmegen Toernooiveld, Nijmegen, The Netherlands
A. BENSON	University of Technology, Loughborough, Leicestershire, UK
I.C. DEMETRIOU	University of Technology, Loughborough, Leicestershire, UK
D.J. EVANS	University of Technology, Loughborough, Leicestershire, UK
C.R. GANE	Central Electricity Research Laboratories, Leatherhead, Surrey, UK
R. GLOWINSKI	Paris VI University, Paris, France
I. GUSTAFSSON	Catholic University, Nijmegen, The Netherlands
A. HADJIDIMOS	University of Ioannina, Ioannina, Greece
M.M. HAFEZ	Joint Institute for Advancement of Flight Sciences, The George Washington University, Washington, D.C., 20006, USA
M. HATZOPOULOS	Michigan Technological University, Houghton, Michigan 49931, USA
D.A.H. JACOBS	Central Electricity Research Laboratories, Leatherhead, Surrey, UK
A. JENNINGS	Queens University of Belfast, Belfast, Northern Ireland
M.A. KAMMOONAH	University of Technology, Loughborough, Leicestershire, UK
E.A. LIPITAKIS	University of Technology, Loughborough, Leicestershire, UK
N.M. MISSIRLIS	University of Athens, Athens, Greece
N. MUNKSGAARD	CE-DATA, Virum, Denmark
J. PERIAUX	AMD/BA, St. Cloud, France
O. PIRONNEAU	Paris Nord University, St. Denis, France

P.E. SAYLOR — University of Illinois, Urbana, Illinois 61801, USA

J. SHANEHCHI — University of Technology, Loughborough, Leicestershire, UK

H.A. VAN DER VORST — Academisch Computer Centrum Utrecht, Utrecht – de Uithof, The Netherlands

Y.S. WONG — Institute for Computer Applications in Science and Engineering, NASA Langley Research Center, Hampton, Virginia, 23365 USA

Preface

This volume provides an up-to-date account of an interesting new research topic concerned with the numerical solution of partial differential equations, i.e., the analysis and application of preconditioning strategies.

The main purposes of the book are to present sufficient theoretical and practical details to enable the reader to implement the presented methods in order to solve practical problems and/or to pursue theoretical studies to advance the knowledge of preconditioning theories. Also included in the presentation are practical applications involving elliptic and parabolic partial differential equations, eigenvalue problems, least squares and variational methods, optimal control, engineering structures, transonic flows and parallel computation.

I wish to thank the invited specialists who have contributed to this volume, and my graduate students who, over the years, have helped enormously in progressing my knowledge, understanding and development of the subject.

Finally, I wish to express my deep gratitude to my secretary, Judith Briers, for her excellent typing of the manuscript.

David J. Evans

ON PRECONDITIONED ITERATIVE METHODS FOR PARTIAL DIFFERENTIAL EQUATIONS

D.J. Evans

Dept. of Computer Studies, University of Technology, Loughborough, Leicestershire, U.K.

ABSTRACT

This paper surveys the preconditioning techniques, introduced by Evans (1968) and defines 2 classes of preconditioned iterative methods for the numerical solution of the sparse linear system $A\underline{u}=\underline{d}$. The main difference between the methods is shown to depend on whether the conditioning matrix R consists of components derived from a splitting or incomplete elimination/factorization of A. Some theoretical and experimental results for the iterative schemes are given when A has been derived from the finite difference/element discretisation of 2nd order self-adjoint elliptic and parabolic partial differential equations in 2 space dimensions. Finally, the application of both forms of preconditioning to stationary and non-stationary iterative methods are presented and computational results compared.

1. INTRODUCTION

This paper is concerned with the use of iterative methods for the solution of N linear algebraic equations of the form

$$A\underline{u} = \underline{d} \ , \qquad (1.1)$$

where the matrix A is positive definite, large, non-singular with non-vanishing diagonal elements. We may, without loss of generality, set all the diagonal coefficients $a_{ii}=1$, then $A=I-L-U$ where L and U are respectively lower and upper triangular matrices with null diagonals. We note that if we premultiply (1.1) by A^{-1}, then we obtain immediately its solution $\underline{u}=A^{-1}\underline{d}$. However, there are certain difficulties for computing A^{-1}, especially when (1.1) is derived from the approximation of elliptic partial differential equations, so instead we consider the case where the above system is premultiplied by a non-singular matrix R^{-1}, where R^{-1} is an approximate inverse of A. Thus, we transform the original system (1.1) into the following preconditioned form,

$$R^{-1}A\underline{u} = R^{-1}\underline{d}. \tag{1.2}$$

The matrix R will be referred to as the conditioning matrix and according to our previous observation we first require this matrix to approach A so that the spectral condition number of $R^{-1}A$ is very much less than the spectral condition number of A where the spectral condition number of a non-singular matrix M is defined to be $\kappa(M)=||M||\ ||M^{-1}||$.

After we form system (1.2), we can define the following general iterative scheme

$$\underline{u}^{(n+1)} = \underline{u}^{(n)} + \alpha R^{-1}(\underline{d}-A\underline{u}^{(n)}) \quad , \tag{1.3}$$

where α is a real parameter to be defined later. It can be easily seen that the constructed iterative scheme (1.3) is completely consistent with (1.2) if and only if R is non-singular and $\alpha \neq 0$.

We shall be interested in the possible forms which the matrix R can possess within the contents of the above discussion, e.g. when R=I and $\alpha=1$ we can formulate many of the fundamental iterative methods based on the splitting principle which have been used extensively to date and show that their convergence performance inversely depends on the spectral condition number.

The simple Jacobi method which is based on the splitting A=I-B can be written in the form

$$\underline{u}^{(n+1)} = (I-A)\underline{u}^{(n)} + \underline{d} = \underline{u}^{(n)} + \underline{r}^{(n)} \quad , \tag{1.4}$$

showing that the change in each component is equal to the corresponding component of the residual vector $\underline{r}^{(n)}$.

Two forms of the above equation (1.4) in which a constant factor α or a different choice α_n for each iteration is multiplied by each component of the residual vector, and then added to each component of the present iterate $\underline{u}^{(n)}$ give rise to the equations

$$\underline{u}^{(n+1)} = \underline{u}^{(n)} + \alpha\underline{r}^{(n)} = (I-\alpha A)\underline{u}^{(n)} + \alpha\underline{d}, \tag{1.5}$$

and

$$\underline{u}^{(n+1)} = \underline{u}^{(n)} + \alpha_n\underline{r}^{(n)} = (I-\alpha_n A)\underline{u}^{(n)} + \alpha_n\underline{d}, \tag{1.6}$$

which are the well known Simultaneous Displacement and Richardson's method respectively. (Young, 1954).

Briefly, in the method of Simultaneous Displacement, i.e. equation (1.5), if we denote the error vector $e^{(n)}$ by

$$\underline{e}^{(n)} = \underline{u}^{(n)} - A^{-1}\underline{d},$$

then the error vector can be shown to satisfy the equation

$$\underline{e}^{(n+1)} = (I-\alpha A)\underline{e}^{(n)} = (I-\alpha A)^{n+1}\underline{e}^{(0)},$$

and, since the error operator $(I-\alpha A)$ is kept constant throughout the iteration, then the method is classified as a stationary linear iterative process. For the iteration to converge, the criteria necessary is that the modulus of the largest eigenvalue or spectral radius, of $(I-\alpha A)$ is less than unity.

If we assume that the spectrum of real eigenvalues λ_i of A are bounded by the values a and b such that

$$0<a\leq\lambda_i\leq b<\infty, \quad (i=1,2,\ldots,N), \tag{1.7}$$

then the criteria for convergence makes it necessary for

$$|1 - \alpha\lambda_i| \leq 1, \tag{1.8}$$

which gives the permissible range for values of α to be

$$0<\alpha<\frac{2}{b}. \tag{1.9}$$

The fastest convergence rate is obtained by choosing α so that the spectral radius of $(I-\alpha A)$ is minimized. Clearly the best choice of α is the one for which

$$1-\alpha a = -(1-\alpha b),$$

i.e.

$$\alpha = \frac{2}{(a+b)}. \tag{1.10}$$

With this choice of α, the convergence factor or spectral radius is, for both i=1 and N,

$$|1-\alpha\lambda_i| \leq \frac{b-a}{b+a} = \frac{P-1}{P+1}, \tag{1.11}$$

where P=b/a is the P-condition number of A and is defined as the ratio of the maximum eigenvalue to minimum eigenvalue of a positive definite matrix.

We further define the asymptotic rate of convergence R_∞ by the formula

$$R_\infty = -\log_n|\gamma| \,, \tag{1.12}$$

where γ is the spectral radius of the error operator $(I-\alpha A)$ and is given by (1.11). For the method of Simultaneous Displacement we obtain the rate of convergence from (1.11) and (1.12) and it is given by

$$R_\infty \simeq \frac{2}{P} \,. \tag{1.13}$$

Similarly, for Richardson's method, we have

$$\underline{e}^{(n+1)} = \underline{u}^{(n+1)} - \underline{u} = (I-\alpha_n A)\underline{e}^{(n)} = \prod_{i=0}^{n}(I-\alpha_i A)\underline{e}^{(0)} = Q_{n+1}(A)\underline{e}^{(0)} \,, \tag{1.14}$$

and since the error operator changes for each iteration, this is a non-stationary linear iterative process.

If the N eigenvalues of A, λ_i with corresponding eigen-vectors v_i form a basis for the space, then since the eigen-values and eigenvectors of $Q_{n+1}(A)$ are $Q_{n+1}(\lambda_i)$ and $\underline{v}_i$, respectively, we obtain the result

$$\underline{e}^{(n+1)} = Q_{n+1}(A)\underline{e}^{(0)} = \sum_{i=1}^{N} a_i Q_{n+1}(A)\underline{v}_i = \sum_{i=1}^{N} a_i Q_{n+1}(\lambda_i)\underline{v}_i \,.$$

Clearly for the error vector $\underline{e}^{(n+1)}$ to be small we need the value of $|Q_{n+1}(x)|$ to be small over the entire interval $[a,b]$ under the constraint $Q_{n+1}(0)=1$. Such a polynomial has been given by

$$|Q_{n+1}(x)| = \frac{T_{n+1}\left(\frac{b+a-2x}{b-a}\right)}{T_{n+1}\left(\frac{b+a}{b-a}\right)} \,, \tag{1.15}$$

where $T_{n+1}(x)=\cos[(n+1)\cos^{-1}(x)]$, is the Chebyshev polynomial of degree $(n+1)$ over the interval $[-1,1]$. The α_i are chosen so that the roots of $\prod_{i=0}^{n}(1-\alpha_i x)$ are coincident with the roots of $T_{n+1}\{(b+a-2x)/(b-a)\}$. The details concerning the choice of α_i are given by Forsythe & Wasow (1960).

The maximum value of $Q_{n+1}(x)$ as given by (1.15) for $n\to\infty$ (since the maximum absolute value of the numerator is unity), is

$$\max_{a\leq x\leq b} |Q_{n+1}(x)| = [T_{n+1}(z)]^{-1}, \text{ where } z=\frac{b+a}{b-a}, \tag{1.16}$$

and using the relationship for $T_{n+1}(z)$, $z>1$ given by

$$2T_{n+1}(z) = [z+\sqrt{(z^2-1)}]^{n+1}+[z-\sqrt{(z^2-1)}]^{n+1},$$

we see that (1.16) simplifies to

$$\max_{a\leq x\leq b} |Q_{n+1}(x)| \leq \frac{2}{[z+\sqrt{(z^2-1)}]^{n+1}},$$

and it follows that the eigenvalues $|Q_{n+1}(\lambda_i)|$ are uniformly bounded as $n\to\infty$. The asymptotic bound to the convergence factor is

$$[z-\sqrt{(z^2-1)}],$$

and hence the rate of convergence as $n\to\infty$ is

$$R_\infty = \log[z+\sqrt{(z^2-1)}]. \tag{1.17}$$

For substantially large problems, we have

$$z \simeq 1 + \frac{2}{P},$$

and the final result for the rate of convergence is

$$R_\infty \simeq \frac{2}{\sqrt{P}} . \tag{1.18}$$

Methods involving two previous iterates by way of a three term recurrence relationship are well known. The second order Richardson's method, for example, is defined by the formula

$$\underline{u}^{(n+1)} = \underline{u}^{(n)}+\alpha(\underline{d}-A\underline{u}^{(n)})+\beta(\underline{u}^{(n)}-\underline{u}^{(n-1)}), \tag{1.19}$$

where α and β remain constant throughout the iteration and are chosen to provide maximum convergence to the solution.

The theoretical justification for the choice of α and β has been given by Frankel (1950). For brevity, we state the final result. The optimum values are

$$\alpha = \left(\frac{2}{\sqrt{a}+\sqrt{b}}\right)^2 \quad \text{and} \quad \beta = \left(\frac{\sqrt{a}-\sqrt{b}}{\sqrt{a}+\sqrt{b}}\right)^2, \tag{1.20}$$

the spectral radius is $\sqrt{\beta}$ and the rate of convergence is again

$$R_\infty \simeq \frac{2}{\sqrt{P}} . \tag{1.21}$$

The Chebyshev acceleration of (1.6) has the form

$$\underline{u}^{(n+1)} = \underline{u}^{(n)}+\alpha_n(\underline{d}-A\underline{u}^{(n)})+\beta_n(\underline{u}^{(n)}-\underline{u}^{(n-1)}), \tag{1.22}$$

where the parameters α_n and β_n vary with each iteration and are defined by Stiefel (1958) to be of the form

$$\alpha_n = \frac{4T_n(z)}{(b-a)T_{n+1}(z)} \text{ and } \beta_n = \frac{T_{n-1}(z)}{T_{n+1}(z)} . \quad (1.23)$$

Since the coefficients α_n and β_n are less than unity, round off difficulties do not arise as they do in (1.6) due to the later values of α_i (i=1,2,...,m) when the number of parameters m is large and hence the method given by equations (1.22) is to be preferred over equation (1.6). Other advantages such as the elimination of the need to preselect m, the number of parameters such that after m applications of (1.15) the desired accuracy is attained and the precalculation of the α_i outweigh some of the disadvantages of the necessity to retain two iterates. Since the minimizing polynomial for the iteration (1.22) is identical to that given by (1.15), it follows immediately that the rate of convergence of this method is equal to the result obtained in (1.18). (Stiefel, 1958).

Finally, the Chebyshev semi-iterative method (Golub & Varga, 1961) can be derived from (1.22) if we assume that a and b have values $1-\mu$, $1+\mu$, respectively, where μ is the spectral radius of the error operator of the Jacobi method. The recurrence relationship for the Chebyshev polynomials then becomes

$$\alpha_n = 1+\beta_n , \quad (1.24)$$

and the iteration takes the particularly simple form

$$\underline{u}^{(n+1)} = \alpha_n[\underline{d}+(L+U)\underline{u}^{(n)}-\underline{u}^{(n-1)}]+\underline{u}^{(n-1)} ,$$

or in standard form

$$\underline{u}^{(n+1)} = \underline{u}^{(n)}+\alpha_n(\underline{d}-A\underline{u}^{(n)})+(\alpha_n-1)(\underline{u}^{(n)}-\underline{u}^{(n-1)}) . \quad (1.25)$$

The parameters α_n are calculated from the formulae

$$\alpha_1 = 1,\ \alpha_2 = \frac{2}{2-\mu^2} ,\ \alpha_{n+1} = \frac{1}{(1-\mu^2\alpha_n/4)} , \quad (1.26)$$

for n=2,3,..., whilst the rate of convergence is generally the same.

If we now consider a different splitting technique to that given by (1.3) and write

$$(I-L)\underline{u} = U\underline{u}+\underline{d} \ , \tag{1.27}$$

then, since the matrix (I-L) is non-singular and strictly triangular, the left-hand side vector $\underline{u}$ can be derived in a simple way by the well known process of back substitution. Consequently, the equations given by (1.27) are what we term *"easily solvable"*.

Multiplying both sides of (1.27) by a real parameter ω introduces the principle of overrelaxation and then adding u to both sides yields, after rearrangement,

$$(I-\omega L)\underline{u} = [(1-\omega)I+\omega U]\underline{u}+\omega\underline{d} \ . \tag{1.28}$$

Simple insertion of iteration superscripts serves to define the Successive Overrelaxation iterative method S.O.R. (Young, 1954) given by

$$\underline{u}^{(n+1)} = (I-\omega L)^{-1}[(1-\omega)I+\omega U]\underline{u}^{(n)}+\omega(I-\omega L)^{-1}\underline{d} \ , \ n\geq 0. \tag{1.29}$$

The purpose of the iteration becomes clear and sensible when we see that the formula uses the most recently computed components of $\underline{u}$ at each stage of the iteration, i.e., to compute $\underline{u}^{(n+1)}$, it uses the values $u_1^{(n+1)}, u_2^{(n)}$ and $u_3^{(n)}$, etc. This principle results in a minimum of computer storage requirements as well as speeding up the iteration process.

From (1.29) we see that the iteration matrix $L_{\sigma,\omega}$ has the form

$$L_{\sigma,\omega} = (I-\omega L)^{-1}[(1-\omega)I+\omega L^T] \ , \tag{1.30}$$

where σ is a certain ordering of the points and ω the successive overrelaxation parameter. Then, for certain matrices possessing a special property (Property A) it has been shown that there exists a fundamental relationship between the eigenvalues of $L_{\sigma,\omega}$ and B. (Young, 1954).

The real significance of this relationship is that it can be used to explicitly determine the real parameter ω which minimises the spectral radius.

The value of ω_b which gives the least spectral radius of

$L_{\sigma,\omega}$ for any consistent ordering is then

$$\omega_b = \frac{2}{1+\sqrt{1-\rho^2(B)}} , \qquad (1.31)$$

and the corresponding spectral radius is thus

$$\rho(\mathcal{L}_{\sigma,\omega_b}) = \omega_b - 1 ,$$

or in terms of the P-condition number of A,

$$\rho(\mathcal{L}_{\sigma,\omega_b}) = \left[\frac{\sqrt{P}-1}{\sqrt{P}+1}\right], \qquad (1.32)$$

by using equation (1.11).

Consequently, the convergence rate of the S.O.R. method can be shown to be

$$R_\infty = \log\frac{(1-1/\sqrt{P})^2}{(1+1/\sqrt{P})^2} \approx 4/\sqrt{P} , \qquad (1.33)$$

for $P \gtrsim 1$.

Thus, an order of magnitude improvement in convergence rate can be obtained simply by the correct choice of ω and the small changes in the iteration process by going from equations (1.4) to (1.29), i.e. by using the most recent values of the iterate wherever possible and a consistent ordering of points on the grid.

By extending the ideas derived from established accelerated methods such as the S.O.R. method given by equation (1.29), the S.S.O.R. method (Sheldon, 1955) given by the equations:

$$\begin{aligned} (I-\omega L)\underline{u}^{(2n+1)} &= [(1-\omega)I+\omega U]\underline{u}^{(2n)}+\omega\underline{d} , \\ (I-\omega U)\underline{u}^{(2n+2)} &= [(1-\omega)I+\omega L]\underline{u}^{(2n+1)}+\omega\underline{d} , \end{aligned} \qquad (1.34)$$

and the Extrapolated Modified Aitken method (Evans, 1963) given by

$$(I-\omega L)(I-\omega U)\underline{u}^{(n+1)} = [(1-\omega)I+\omega^2 LU]\underline{u}^{(n)}+\omega\underline{d}, \qquad (1.35)$$

can be obtained.

The strive to obtain left hand sides such as (1.34) and (1.35) is essential for higher order iteration methods for although the equations appear to be implicit in form, the solution at each iteration is obtained from simple algorithms which permit forward and back substitution processes on a right hand side vector. These operations involve the known triangular matrices of the original matrices A in which the sparseness is retained. These methods can also be shown to have convergence properties dependent on the spectral condition number of the coefficient matrix.

A comprehensive list of all the iterative methods based on the splitting principle is given in Evans and Missirlis, (1982).

2. AN ALTERNATIVE PRECONDITIONING STRATEGY

In addition to the principle of splitting which forms the basis of many iterative methods we can also allow the form of R in (1.3) to be based on the factorisation of A into easily invertible matrix factors. Methods for such factorisations are well known and are summarised in Table 2.1.

However, since the direct factorisation process of a sparse matrix A results in a large number of the zero entries of the matrix being replaced or 'filled in' by non-zero values. Severe disadvantages arise from these entries which appear in the numerical process for only are they a source of rounding errors in any computation in which they are involved but they greatly increase the amount of work and computer storage involved in the algorithm.

Exact Method	R conditioning matrix	Iterative Method (1.3)
1 Gaussian Elimination $A=\mathcal{L}^{-1}\mathcal{U}$	$\mathcal{L}^{-1}_s\mathcal{U}_s$	Evans (1974)
2 Triangular Factorisation $A=\mathcal{L}\mathcal{U}$	$\mathcal{L}_s\mathcal{U}_s$	Stone (1968), Evans & Lipitakis, (1979)
3 Choleski Square Root $A=QQ^T$	$Q_sQ^T_s$	Dupont (1968), Meijerink & Van der Vorst (1977), Gustafsson (1978)
4 Root Free Factorisation $A=\mathcal{L}D\mathcal{U}$	$\mathcal{L}_sD_s\mathcal{U}_s$	Evans & Lipitakis (1982)
5 Root Free Choleski $A=\mathcal{L}D\mathcal{L}^T$	$\mathcal{L}_sD_s\mathcal{L}^T_s$	Kershaw (1978)
6 Normalised Symmetric Factorisation $A=DTT^TD$	$D_sT_sT^T_sD_s$	Varga (1960), Evans & Lipitakis (1980)

TABLE 2.1: Approximate Factorisation Methods

Thus, research in recent years has been directed towards a number of indirect or iterative methods in which the strategy is to reduce the fill-in terms of the factors $\mathcal{L},\mathcal{U}$ etc. to a minimum whilst still retaining the property that $A\approx\mathcal{L}_s\mathcal{U}_s$, etc. where $\mathcal{L}_s,\mathcal{U}_s$ etc. denotes the corresponding sparse matrix factors. Of course, the strategies by which each method retains its sparseness varies differently and the reader is referred to the literature for further details. In the paper, only the strategy for the approximate or incomplete Gaussian elimination method will be discussed.

3. MINIMISATION OF THE CONDITION NUMBER

To obtain improved convergence rates for the methods of successive approximations previously considered, we require the coefficient matrix of the system (1.1) to have a P-condition number as small as possible. If this criteria is not satisfied, then it is advisable to prepare the system or pre-condition it beforehand. This preparation consists in passing from the given system (1.1) to the equivalent system

$$QA\underline{u} = Q\underline{d} \quad , \tag{3.1}$$

where Q is a certain non-singular matrix chosen so that any application involving its use is one which forms an easily solvable system and the "condition" of the resulting system is less than its original value.

If we pursue the 'splitting' approach, one such form of the matrix Q is $(I-\omega L)^{-1}$ and the system (3.1) can then be re-written as $(I-\omega L)^{-1}A(I-\omega U)^{-1}[(1-\omega U)]\underline{u} = (I-\omega L)^{-1}\underline{d}$, which with the aid of an intermediate transformation vector $\underline{x}$ given by

$$\underline{x} = (I-\omega U)\underline{u} \quad , \tag{3.2}$$

and an ω, which is an acceleration parameter to be defined later, simplifies the process to its final form,

$$(I-\omega L)^{-1}A(I-\omega U)^{-1}\underline{x} = \underline{b} \quad , \tag{3.3}$$

or

$$GAG^T\underline{x} = \underline{b} \quad ,$$

where $G = (I-\omega L)^{-1}$ and $\underline{d}_\omega = (I-\omega L)^{-1}\underline{d}$.

It can be shown by further matrix manipulation that (3.3) can be expressed in the iterative form,

$$(I-\omega L)(I-\omega U)\underline{x}^{(n+1)} = [(1-\omega)(L+U)+\omega^2 LL^T]\underline{x}^{(n)} + \underline{d}_\omega \, , \tag{3.4}$$

from which we can deduce that ω is some acceleration parameter for when $\omega=0$, we obtain the Simultaneous Displacement method (1.5) and when $\omega=1$, we derive a method equivalent to a forward Gauss-Seidel process followed by a backward Gauss-Seidel process or the Aitken iteration method as it is widely known, (Evans, 1963).

Unfortunately unlike the S.O.R. method with its

consistent ordering property we are unable to find a suitable ordering of the points on our network in which the eigenvalues of the error operator of (3.4) are related in a simple way to the eigenvalue of the Jacobi matrix given by (1.4), except for the simple case of the σ_1 ordering, previously discussed in Evans (1968).

We now develop a general theory for the condition of the derived system (3.3), i.e.,

$$(I-\omega L)^{-1}A(I-\omega U)^{-1}\underline{x} = B_\omega \underline{x} = \underline{d}_\omega . \tag{3.5}$$

So we have to solve the system (3.5) where B_ω is a function of ω and we are free to choose ω such that the P-condition number of B_ω is smaller than that of A (if it is possible).

Assume that B_ω possesses the eigenvalues:

$$\lambda_1, \lambda_2, \ldots, \lambda_N ,$$

where λ_1, λ_N are the eigenvalues of largest and smallest modulus respectively, and let $\underline{y}_1, \underline{y}_N$ and $\underline{z}_1, \underline{z}_N$ be the corresponding right and left eigenvectors of B_ω.

Then, by definition, for every eigenvector and its corresponding eigenvector we have the relationships:

$$B_\omega \underline{y}_i = \lambda_i \underline{y}_i , \quad i=1,\ldots,N, \tag{3.6}$$

$$\underline{z}^T B_\omega = \lambda_i \underline{z}^T , \quad i=1,\ldots,N. \tag{3.7}$$

From (3.6) we have:

$$(I-\omega L)^{-1}A(I-\omega U)^{-1}\underline{y}_i = \lambda_i \underline{y}_i , \quad i=1,\ldots,N,$$

and premultiplying this by $(I-\omega L)$ we obtain:

$$A(I-\omega U)^{-1}\underline{y}_i = \lambda_i (I-\omega L)\underline{y}_i . \tag{3.8}$$

Let

$$\tilde{\underline{y}}_i = (I-\omega U)^{-1}\underline{y}_i , \tag{3.9}$$

then

$$(I-\omega U)\tilde{\underline{y}}_i = \underline{y}_i , \tag{3.10}$$

substituting (3.10) into (3.8) we obtain:

$$(I-L-U)\tilde{\underline{y}}_i = \lambda_i (I-\omega L)(I-\omega U)\tilde{\underline{y}}_i . \tag{3.11}$$

Similarly from (3.7) we can show the result

$$\underline{z}_i^T (I-\omega L)^{-1}(I-L-U)(I-\omega U)^{-1} = \lambda_i \underline{z}_i^T .$$

Post-multiplying by $(I-\omega U)$, we obtain:

$$\underline{z}_i^T (I-\omega L)^{-1}(I-L-U) = \lambda_i \underline{z}_i^T (I-\omega U) \quad , \tag{3.12}$$

and if we let

$$\underline{\tilde{z}}_i^T = \underline{z}_i^T (I-\omega L)^{-1} \quad , \tag{3.13}$$

then

$$\underline{\tilde{z}}_i^T (I-\omega L) = \underline{z}_i^T , \tag{3.14}$$

which on substituting (3.13) into (3.12) gives the result

$$\underline{\tilde{z}}_i^T (I-L-U) = \lambda_i \underline{\tilde{z}}_i^T (I-\omega L)(I-\omega U) . \tag{3.15}$$

Post-multiplying (3.15) by $\underline{\tilde{z}}_i$ to get

$$\underline{\tilde{z}}_i^T A \underline{\tilde{z}}_i = \lambda_i \underline{\tilde{z}}_i^T (I-\omega L)(I-\omega U) \underline{\tilde{z}}_i \quad , \tag{3.16}$$

from which we can obtain the following result for λ_i, i.e.,

$$\lambda_i = \frac{\underline{\tilde{z}}_i^T A \underline{\tilde{z}}_i}{\underline{\tilde{z}}_i^T \underline{\tilde{z}}_i + \omega \underline{\tilde{z}}_i^T (A-I) \underline{\tilde{z}}_i + \omega^2 \underline{\tilde{z}}_i^T LU \underline{\tilde{z}}_i} \quad . \tag{3.17}$$

If we now denote the quantities:

$$\underline{\tilde{z}}_i^T A \underline{\tilde{z}}_i = \tau_i \quad , \quad i=1,\ldots,N, \tag{3.18}$$

$$\underline{\tilde{z}}_i^T \underline{\tilde{z}}_i = \rho_i \quad , \quad i=1,\ldots,N, \tag{3.19}$$

$$\underline{\tilde{z}}_i^T LU \underline{\tilde{z}}_i = k_i , \quad i=1,\ldots,N, \tag{3.20}$$

and substitute (3.18), (3.19) and (3.20) into (3.17) we obtain for each eigenvalue λ_i of B_ω the result,

$$\lambda_i = \frac{\tau_i}{\rho_i (1-\omega) + \omega\tau_i + \omega^2 k_i} \quad . \tag{3.21}$$

By definition, the P-condition number of B_ω can be written as

$$P_\omega = \frac{|\lambda_1|}{|\lambda_N|} \quad ,$$

i.e.,

$$P_\omega = \left| \frac{\tau_1}{\rho_1(1-\omega)+\omega\tau_1+\omega^2 k_1} \right| \Big/ \left| \frac{\tau_N}{\rho_N(1-\omega)+\omega\tau_N+\omega^2 k_N} \right| ,$$

$$P_\omega = \frac{|\tau_1(\rho_N(1-\omega)+\omega\tau_N+\omega^2 k_N)|}{|\tau_N(\rho_1(1-\omega)+\omega\tau_1+\omega^2 k_1)|} \quad . \tag{3.22}$$

In order to minimise the P-condition number P we attempt

to find a value ω that satisfies the relationship:

$$\frac{dP}{d\omega} = 0 ,$$

which on differentiating (3.22) with respect to ω, we obtain a polynomial expression of the form:

$$a\omega^4+b\omega^3+c\omega^2+d\omega+f = 0,$$

where:

$$a = k_1k_N(\tau_1'\tau_N-\tau_N'\tau_1)+\tau_1\tau_N(k_N'k_1-k_1'k_N),$$

$$b = \tau_N^2\tau_1'k_1-k_N\tau_1^2\tau_N'+\tau_N\tau_1(-\rho_1'k_N+\rho_N'k_1+k_N'\tau_1-k_1'\tau_N-$$
$$-\rho_N'k_1+\rho_1'k_N)+(\tau_1'\tau_N-\tau_N'\tau_1)(-\rho_1k_N-\rho_Nk_1),$$

$$c = (\tau_1'\tau_N-\tau_N'\tau_1)(\rho_1k_N+\rho_Nk_1+\rho_1\rho_N)+(\rho_1k_N-\rho_Nk_1)\tau_1\tau_N+$$
$$+\tau_1^2\rho_N\tau_N'-\tau_N^2\tau_1'\rho_1+\tau_N\tau_1(\rho_N'\rho_1-\rho_1'\rho_N-2\rho_1k_N+2\rho_Nk_1+$$
$$+\rho_1k_N'-\rho_Nk_1'+2k_N\tau_1-2k_1\tau_N+\rho_N'k_1-\rho_1'k_N+\tau_Nk_1-\tau_1k_N+$$
$$+\rho_1'\tau_N-\rho_N'\tau_1),$$

$$d = -(\tau_1'\tau_N-\tau_N'\tau_1)2\rho_1\rho_N+\tau_1'\tau_N^2\rho_1-\tau_N'\tau_1^2\rho_N+\tau_N\tau_1(2\rho_1'\rho_N-2\rho_N'\rho_1+$$
$$+2\rho_1k_N-2\rho_Nk_1+\rho_N'\tau_1-\rho_1'\tau_N),$$

$$f = \tau_N\tau_1(\rho_N'\rho_1-\rho_1'\rho_N+\tau_N\rho_1-\tau_1\rho_N)+(\tau_1'\tau_N-\tau_N'\tau_1)\rho_N\rho_1,$$

and $\quad \tau_1' = d\tau_1/d\omega,\quad \rho_1' = d\rho_1/d\omega,\quad \tau_N' = d\tau_N/d\omega,$

$\quad \rho_N' = d\rho_N/d\omega.$

Hence, if a minimum P-condition number exists, it occurs at a value of $\bar{\omega}$ such that:

$$a\bar{\omega}^4+b\bar{\omega}^3+c\bar{\omega}^2+d\bar{\omega}+ f = 0, \qquad (3.23)$$

and then the expression,

$$P_{\bar{\omega}} = \frac{\tau_1(\rho_N(1-\bar{\omega})+\bar{\omega}\tau_N+\bar{\omega}^2k_N}{\tau_N(\rho_1(1-\bar{\omega})+\bar{\omega}\tau_1+\bar{\omega}^2k_1)} .$$

takes its minimum value at $\bar{\omega}$ and the eigenvalues of B_ω are bounded by:

$$\frac{\tau_N}{\rho_N(1-\bar{\omega})+\bar{\omega}_N^2+\bar{\omega}^2 k_N)} \leqslant \lambda_i(\omega) \leqslant \frac{\tau_1}{\rho_1(1-\bar{\omega})+\bar{\omega}\tau_1+\bar{\omega}^2 k_1} . \tag{3.24}$$

Assuming that τ_1, τ_N, k_1 and k_N to be independent of ω yields a simpler quadratic expression for ω which has been investigated earlier in Evans (1968). Benson (1968) has investigated this more general case and numerical evidence suggests that only for the symmetric case can one justify making this assumption.

The main difficulty in developing a general theory is that although the matrix $(I-\omega L)^{-1}A(I-\omega L^T)^{-1}$ is symmetric and hence possesses real eigenvalues, its eigenvectors cannot be made coincident to those of the Jacobi matrix by a simple transformation. However, we can build up a fairly rigorous and simple analysis by means of eigenvalue bounds which can be verified by numerical experiments to substantiate our claims.

A simplified but approximate analysis to derive a formulae which determines the minimum preconditioning factor can be obtained by assuming that A and LU are commutative. Then, by definition we have

$$(I-\omega L)^{-1}A(I-\omega U)^{-1}\underline{v} = \lambda\underline{v} , \tag{3.25}$$

where λ and $\underline{v}$ are the eigenvalues and eigenvectors of B_ω respectively. Thus, after some analysis, we have the result,

$$A\underline{u} = \lambda(I-\omega(L+U)+\omega^2 LU)\underline{u} , \tag{3.26}$$

where $\quad \underline{u} = (I-\omega U)^{-1}\underline{v}.$

Let the eigenvalues of A and LU be μ_i, ν_i, $i=1,2,\ldots,N$ respectively then we have

$$\lambda_i = \frac{\mu_i}{1-\omega(1-\mu_i)+\omega^2\nu_i} , \quad i=1,2,\ldots,N. \tag{3.27}$$

The optimal preconditioning factor $\bar{\omega}$ will be given when the expression

$$P_\omega = \frac{\lambda_1}{\lambda_N} = \frac{\mu_1[1-\omega(1-\mu_N)+\omega^2 k_N]}{\mu_N[1-\omega(1-\mu_1)+\omega^2 k_1]} \tag{3.28}$$

is a minimum.

However, simple upper and lower bounds can also be established in the following manner.

Now, λ_1 is minimised at $\omega=1$ when the denominator of (3.27) is maximised at which the value is

$$\lambda_1 = \frac{\mu_1}{\mu_1 + k_1} . \qquad (3.29)$$

In addition, λ_N is similarly maximised when the denominator is minimised at $\omega=\bar{\omega}$, i.e.,

$$\lambda_N = \frac{\mu_N}{\bar{\omega}\mu_N + (1-\bar{\omega}+\bar{\omega}^2 k_N)} ,$$

when the expression in brackets in the denominator is zero, which is

$$1-\bar{\omega}+\bar{\omega}^2 k_N = 0 ,$$

at which the value of λ_N is

$$\lambda_N = 1/\bar{\omega}. \qquad (3.30)$$

Thus, the minimum P-condition number occurs in the range $1<\omega<\bar{\omega}$.

Finally, a more detailed theoretical discussion of upper and lower eigenvalue bounds of B_ω is given in Evans & Missirlis (1980).

4. SPARSE PRECONDITIONED ITERATIVE METHODS

We now apply the results of Section 3 to modify the basic iterative methods discussed in Section 1. It follows immediately that if the techniques used in Section 3 are shown to apply to the methods given by equations such as (1.5) and (1.6), then an improvement in convergence rate must follow.

We now proceed from equation (3.1) and develop similar iterative processes to the method of Simultaneous Displacement, Richardson's method and the second order methods by working throughout in the transformed vector $\underline{x}$.

Denoting the matrix $(I-\omega L)^{-1}A(I-\omega U)^{-1}$ as before by B_ω then the equations describing iterative processes similar to (1.4), (1.5) and (1.6) can be written immediately. They are,

$$\underline{x}^{(n+1)} = \underline{x}^{(n)} + (\underline{d}_\omega - B_\omega \underline{x}^{(n)}), \tag{4.1}$$

$$\underline{x}^{(n+1)} = \underline{x}^{(n)} + \alpha_\omega (\underline{d}_\omega - B_\omega \underline{x}^{(n)}), \tag{4.2}$$

$$\underline{x}^{(n+1)} = \underline{x}^{(n)} + \alpha_n (d_\omega - B_\omega \underline{x}^{(n)}). \tag{4.3}$$

The iterations proceed in the $\underline{x}$ variable until a specified degree of accuracy is achieved. The final solution is then obtained by one application of the formula

$$\underline{u} = (I-\omega U)^{-1} \underline{x} . \tag{4.4}$$

The iteration (4.1) converges when

$$|1-\Lambda_i(\omega)| < 1 \quad (i=1,2,\ldots,p^2). \tag{4.5}$$

This gives the range $0<\omega<\omega_f$ for which the iteration method is convergent. For equation (4.2) and the optimum $\bar{\omega}$ specified in Section 3, the iteration will converge for a range of values of α_ω given by

$$0<\alpha_{\bar{\omega}}<(2/\bar{b}). \tag{4.6}$$

It will further possess an optimum value of α given by

$$\alpha_{\bar{\omega}} = \frac{2}{\bar{a}+\bar{b}} , \tag{4.7}$$

and a convergence rate approximately equal to

$$R_{\bar{\omega}} \simeq (2\bar{a}/\bar{b}). \tag{4.8}$$

Similarly for equation (4.3) the optimum values of α_i are given when the zeros of $(1-\alpha_i x)$ coincide with the zeros of the polynomial.

$$T_{n+1}\left(\frac{\bar{b}+\bar{a}-2\bar{z}}{\bar{b}-\bar{a}}\right).$$

This gives for the parameters α_i, the equation

$$\alpha_i = \frac{2}{[(\bar{a}+\bar{b})-(\bar{b}-\bar{a})\cos\frac{(2i-1)\pi}{2m}]} ,\ (i=1,2,\ldots,m), \tag{4.9}$$

where m is the number of Richardson parameters chosen to achieve the desired accuracy. This method will then have a rate of convergence approximately equal to

$$2\sqrt{\bar{a}/\bar{b}} . \tag{4.10}$$

The second order preconditioned Richardson method can similarly be written as

$$\underline{x}^{(n+1)} = \underline{x}^{(n)}+\alpha_\omega(\underline{d}_\omega-B_\omega\underline{x}^{(n)})+\beta_\omega(\underline{x}^{(n)}-\underline{x}^{(n-1)}), \qquad (4.11)$$

where

$$\alpha_\omega = \left(\frac{2}{\sqrt{a}+\sqrt{b}}\right)^2 \quad \text{and} \quad \beta_\omega = \left(\frac{\sqrt{a}-\sqrt{b}}{\sqrt{a}+\sqrt{b}}\right)^2 . \qquad (4.12)$$

The Chebyshev acceleration of this method also has the form

$$\underline{x}^{(n+1)} = \underline{x}^{(n)}+\alpha_n(\underline{d}_\omega-B_\omega\underline{x}^{(n)})+\beta_n(\underline{x}^{(n)}-\underline{x}^{(n-1)}), \qquad (4.13)$$

where

$$\bar{z} = \frac{(\bar{b}+\bar{a})}{(\bar{b}-\bar{a})} , \qquad (4.14)$$

and α_n and β_n are given by an equation similar to equation (1.23) (but with a and b replaced by $\bar{a}$ and $\bar{b}$). The rate of convergence of this method is given by (1.21) where

$$P(\bar{\omega}) = \frac{\bar{b}}{\bar{a}} . \qquad (4.15)$$

Finally, when α_ω is given by (4.7), the error operator of (4.2) possesses equal and opposite eigenvalues of magnitude $\bar{z}$ as given by (4.14). Since the eigenvalues are all real, this method can then be further accelerated by Chebyshev polynomials.

The result $\underline{x}^{(n+1)*}$ from equation (4.2) is now used in the formula

$$\underline{x}^{(n+1)} = \underline{x}^{(n)}+\alpha_n(\underline{x}^{(n+1)*}-\underline{x}^{(n)})+\beta_n(\underline{x}^{(n)}-\underline{x}^{(n-1)}), \qquad (4.16$$

or in standard form

$$\underline{x}^{(n+1)} = \underline{x}^{(n)}+\alpha_n(\underline{d}_\omega-B_\omega\underline{x}^{(n)})+\beta_n(\underline{x}^{(n)}-\underline{x}^{(n-1)}), \qquad (4.17$$

where the coefficients α_n and β_n are determined from the relations

$$\alpha_n = \frac{4}{(\bar{b}-\bar{a})} \; \frac{T_n(\bar{z})}{T_{n+1}(\bar{z})} \quad \text{and} \quad \beta_n = \frac{T_{n-1}(\bar{z})}{T_{n+1}(\bar{z})} . \qquad (4.18)$$

for n=2,3,..., and initially we choose $\alpha=1$. The equations (4.2) and (4.18) constitute the preconditioned form of the Chebyshev semi-iterative method described earlier in equatior (1.25).

5. EXPERIMENTAL RESULTS

In this section we shall present some experimental considerations to justify the procedures proposed in Section 3. Why do we expect the P-condition number to be reduced and minimized for some values of $\omega > 0$?

We now assume that the linear system (1.1) has been derived from simple finite difference approximations to an elliptic partial differential equation based on a rectangular mesh of grid lines, the model examples being considered are the solution of the boundary value problems:-

I.
$$\frac{\partial^2 u}{\partial x^2} + \frac{\partial^2 u}{\partial y^2} = 0 , \tag{5.1a}$$

over the unit square with the boundary conditions:

$$u(0,y) = u(1,y) = 0, \; u(x,1) = 0, \; u(x,0) = 1, \tag{5.1b}$$

and

II.
$$\frac{\partial^4 u}{\partial x^4} + 2\frac{\partial^4 u}{\partial x^2 \partial y^2} + \frac{\partial^4 u}{\partial y^4} = 0, \tag{5.2}$$

over a similar region with the values of u and $\frac{\partial^2 u}{\partial n^2}$ specified along the boundaries where n denotes the direction of the outward normal.

We choose a square grid of p rows and p columns parallel to the boundaries of the domain as our mesh with spacings $\Delta x = \Delta y = h$, and with the aid of central difference operators the finite difference equation for (5.1) at the point (i,j) on the mesh of (p+1)(p+1) points is

$$-\ell_{i,j}u_{i-1,j} - r_{i,j}u_{i+1,j} - t_{i,j}u_{i,j+1} - b_{i,j}u_{i,j-1} + u_{i,j} = d_{i,j}, \tag{5.3}$$

for i=1(1)p and j=1(1)p, with $\ell_{i,j} = b_{i,j} = r_{i,j} = t_{i,j} = 1/4$.

For a columnwise ordering of the grid points the coefficient matrix A for the finite difference equations has been shown to be positive definite and symmetric with the following tridiagonal partitioned form of order p with submatrix elements of order p (Varga, 1962), i.e.,

$$A = \begin{bmatrix} D_1 & -U_1 & & & & \\ -L_1 & D_2 & -U_2 & & 0 & \\ & \ddots & \ddots & \ddots & & \\ & & & & & \\ & 0 & & -L_{p-2} & D_{p-1} & -U_{p-1} \\ & & & & -L_{p-1} & D_p \end{bmatrix} , \qquad (5.4)$$

where

$$D_i = \begin{bmatrix} 1 & -t_{i,1} & & & & \\ -b_{i,2} & 1 & -t_{i,2} & & 0 & \\ & \ddots & \ddots & \ddots & & \\ & & & & & \\ & 0 & & -b_{i,p-1} & 1 & -t_{i,p-1} \\ & & & & -b_{i,p} & 1 \end{bmatrix} ,$$

$$L_{i-1} = \begin{bmatrix} \ell_{i,1} & & & \\ & \ell_{i,2} & & 0 \\ & & \ddots & \\ & 0 & & \ell_{i,p} \end{bmatrix} \text{ and } U_i = \begin{bmatrix} r_{i,1} & & & \\ & r_{i,2} & & 0 \\ & & \ddots & \\ & 0 & & r_{i,p} \end{bmatrix} .$$

Such matrices are derived from equations said to have σ_2-ordering. (Young, 1954).

A similar matrix system can be derived for problem II using a 13 point finite difference equation.

In order to assess the effectiveness of the preconditioning theory based on the 'splitting' principle, problems I and II were solved for a standard mesh size of $h^{-1}=20$ over the specified domains.

The structure of the coefficient matrix for each problem is shown in Figs. 1(a) and (b) and both are positive definite, sparse and ill conditioned, thus providing valuable test cases for the preconditioning theory.

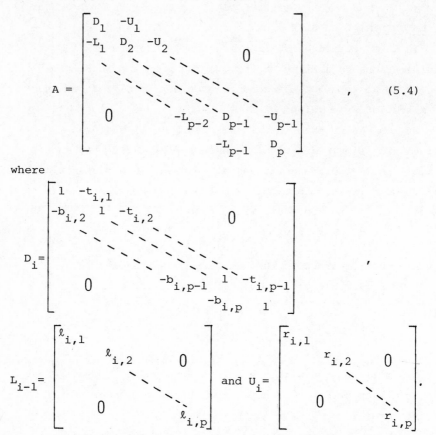

I. 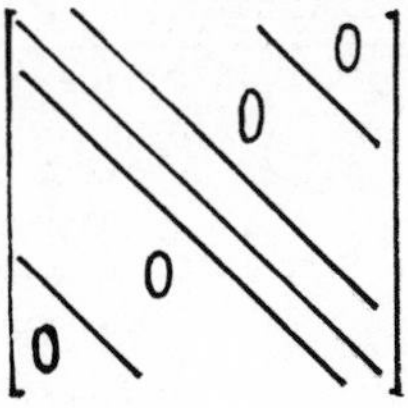 II. 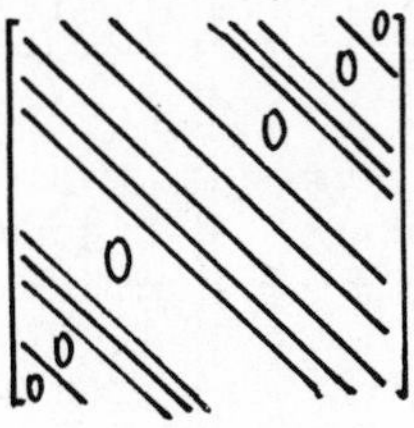

FIGURE 1

The performance of the preconditioned iterative methods (4.2), (4.11) and (4.12) on the chosen model problems are shown in the accompanying tables and diagrams. Table 5.1 gives the largest and smallest eigenvalues and P-condition numbers for the Laplace and Biharmonic operator with mesh size $h^{-1}=20$ which were obtained by the Power and inverse Wielandt methods for selected values of ω in the range [0,2]. Clearly, a minimum value for the P-condition number is achieved and this is portrayed for a smaller mesh size in Fig. 2 for the Biharmonic operator. Similar but less spectacular results were obtained for the Laplace operator and Fig.3 shows the P-condition numbers plotted for various values of the preconditioning parameter ω for the three chosen mesh sizes. The validity of equation (3.23) for the optimum preconditioning parameter $\bar{\omega}$ and minimum P-condition number for B_ω can be confirmed experimentally by determining the quantities ρ, $\underline{x}$ and k. However this is a quartic equation involving ω non-linear. More refined methods of determining $\bar{\omega}$ are given in Missirlis & Evans (1981). In each case, it can be seen that for $\omega=0$, the P-condition number is very large indeed and that for a certain value of the parameter ω, a minimum value of the P-condition number occurs which is roughly of the same order of magnitude as the $(m-1)/m^{th}$ power of the original P-condition number for an mth order differential operator. (Andersson, 1976).

ω Preconditioning Parameter	Laplace Equation λ_1 Maximum Eigenvalue	λ_N Minimum Eigenvalue	$P=\lambda_1/\lambda_N$ P-Condition Number
0	1.98769	0.01231	161.4695
0.2	1.64557	0.01531	107.4833
0.4	1.38504	0.01922	72.0624
0.6	1.18214	0.02491	47.4564
0.8	1.04154	0.03358	31.0167
1.0	1.00000	0.04766	20.9819
1.2	1.04163	0.07252	14.3633
1.4	1.19047	0.12225	9.7380
1.6	1.56190	0.23903	6.5343
1.8	2.77552	0.51243	5.4164
1.9	5.05802	0.54142	9.3421

continued...........

Biharmonic Equation λ_1 Maximum Eigenvalue	λ_N Minimum Eigenvalue	$P=\lambda_1/\lambda_N$ P-Condition Number
3.15216	0.001645	1916.2067
2.13342	0.001739	1226.8085
1.53845	0.001826	842.5246
1.19034	0.001859	640.3120
1.04163	0.002004	519.7754
1.00000	0.002409	415.1100
1.04163	0.003385	307.7194
1.19046	0.005385	221.0696
1.56249	0.010519	148.5398
2.77342	0.026595	104.2835
5.05815	0.035298	143.2985

TABLE 5.1: Tabulation of the maximum and minimum eigenvalues and P-condition numbers for the coefficient matrix B_ω versus the preconditioning parameter ω for the Laplace and Biharmonic operator on a 20×20 rectangular grid.

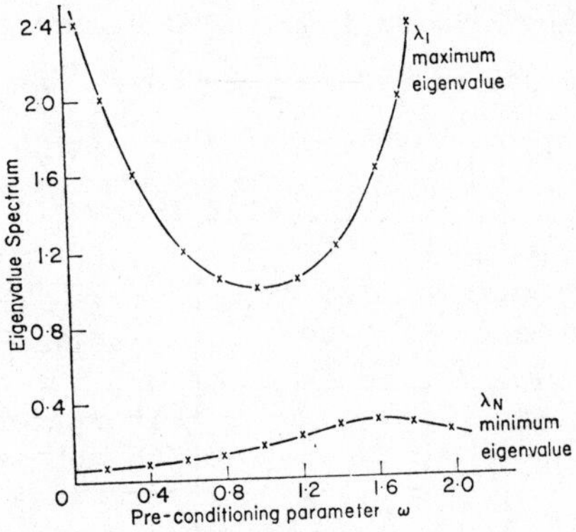

FIGURE 2

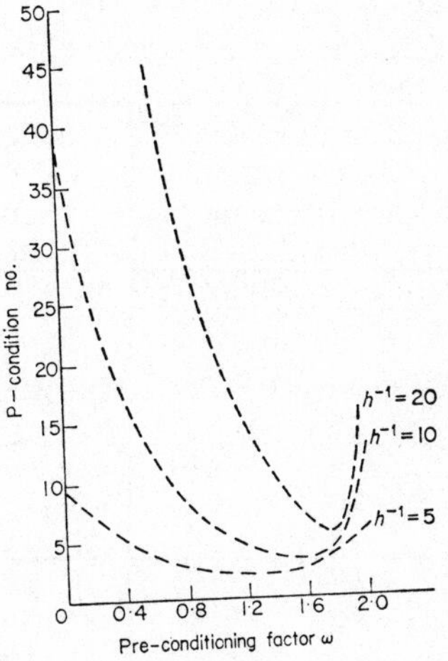

FIGURE 3

Further Table 5.2 compares the number of iterations
required with and without preconditioning. (Fig.4). Since
the results with optimal preconditioning are vastly superior
in iteration count, it is necessary to investigate the
feasibility of the proposed computational algorithms and
whether the extra arithmetic work involved does not offset
any increase in convergence rate which may have been achieved.

Problem	Grid Size h^{-1}	Precond-itioning Parameter ω	P condition Number P
Laplace operator	5	O	9.47
	10	O	39.86
	20	O	161.47
	5	1.3	1.65
	10	1.6	2.81
	20	1.8	5.42
Biharmonic operator	7	O	48.49
	10	O	286.44
	20	O	1916.21
	7	1.4	4.33
	10	1.6	13.39
	20	1.8	104.28

TABLE 5.2: Table
giving the P
condition numbers
and the number of
iterations required
to achieve an
accuracy of 5×10^{-6}
for the Precond-
itioned iterative
methods for values
of the pre-cond-
itioning parameter
$\omega = O$ (basic methods)
and $\omega = \bar{\omega}$, (optimal
preconditioning).

continued.......

Preconditioned Simult.Displ. Method	2nd Order Pre-Conditioned Richardson Method	2nd-Order Pre-Conditioned Chebyshev Method
Number of Iterations		
35	8	15
120	32	30
480	65	59
8	8	7
11	10	8
21	13	14
142	17	23
290	40	41
1432	124	120
18	8	8
54	15	14
174	40	36

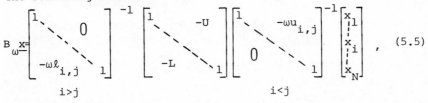

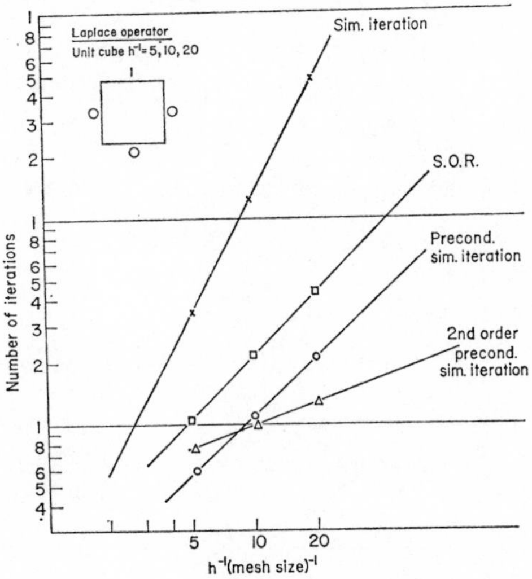

FIGURE 4

Computational considerations concerning the determination of the preconditioned matrix B_ω or the vector $B_\omega \underline{x}$ figure prominently into the reasons for restricting the choice of matrices P and Q to be inverse triangular in form. It can be shown that these forms permit simple forward and back substition processes which are easily facilitated on a computer in single or double precision arithmetic. Furthermore, these operations can also be shown to preserve any quality of sparsity in the original matrix A whether it is stored in its usual compact form as a square array in the computer memory or in generated form on a network of grid points. If A is stored in a compact form, then the determination of the vector $B_\omega \underline{x}$ is obtained in the following manner:

$$B_\omega \underline{x} = \begin{bmatrix} 1 & & 0 \\ & \ddots & \\ -\omega \ell_{i,j} & & 1 \end{bmatrix}^{-1}_{i>j} \begin{bmatrix} 1 & & -U \\ & \ddots & \\ -L & & 1 \end{bmatrix} \begin{bmatrix} 1 & & -\omega u_{i,j} \\ & \ddots & \\ 0 & & 1 \end{bmatrix}^{-1}_{i<j} \begin{bmatrix} x_1 \\ \vdots \\ x_i \\ \vdots \\ x_N \end{bmatrix} , \quad (5.5)$$

$\underline{y} = (I-\omega U)^{-1}\underline{x}$

Back substitution process

$\underline{v} = A\underline{y}$

Matrix-vector multiplication process

$\underline{x} = (I-\omega L)^{-1}\underline{v}$

Forward substitution process

The simple algorithms for carrying out these operations are: the back substitution process,

$$\begin{aligned} \underline{y} &= (I-\omega U)^{-1}\underline{x}\ , \\ y_N &= x_N\ , \\ y_j &= x_j + \omega \sum_{i=j+1}^{N} u_{j,i} y_i, \text{ for } j=N-1,N-2,\ldots,1, \end{aligned} \tag{5.6}$$

matrix vector multiplication,

$$\begin{aligned} \underline{v} &= A\underline{y}\ , \\ v_i &= \sum_{j=i}^{N} a_{i,j} y_j, \text{ for } i=1,2,\ldots,N, \end{aligned} \tag{5.7}$$

and forward substitution process,

$$\begin{aligned} \underline{x} &= (I-\omega L)^{-1}\underline{v}, \\ x_1 &= y_1\ , \\ x_j &= v_j + \omega \sum_{i=1}^{j-1} \ell_{j,i} x_i, \text{ for } j=2,3,\ldots,N. \end{aligned} \tag{5.8}$$

Alternatively, if A is a generated rather than a stored matrix then it has been shown previously that the vector $B_\omega \underline{x}$ can be obtained from the application of three simple computational molecules applied on the given grid of points which for the chosen problems are shown in Fig.5. The details of these operations have been given elsewhere (Evans, 1973a). To conclude, the application of preconditioning can lead to a two-fold increase in arithmetical work which must be offset against the greatly increased convergence rates obtained. However, as the experimental results show for large ill-conditioned systems, an order of magnitude improvement can be obtained.

I.

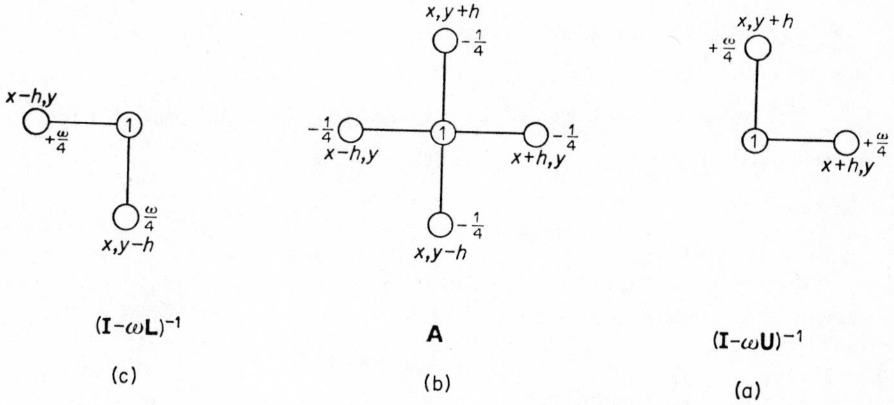
x−h,y
+ω/4
1
ω/4
x,y−h
(I-ωL)^-1
(c)
x,y+h
−1/4
−1/4
x−h,y
1
x+h,y
−1/4
x,y−h
A
(b)
x,y+h
+ω/4
1
+ω/4
x+h,y
(I-ωU)^-1
(a)

II.

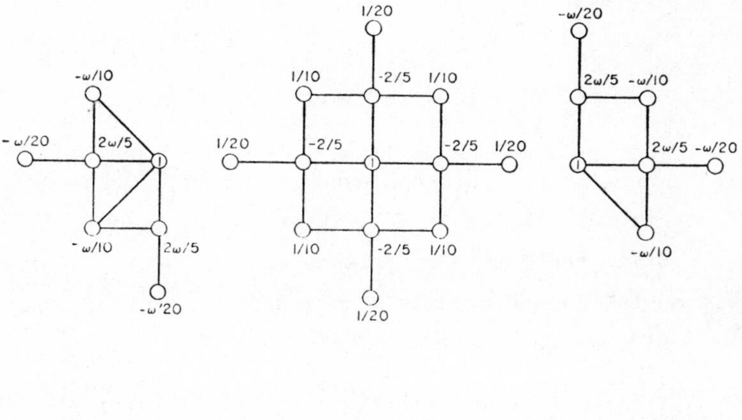
-ω/10
- ω/20
2ω/5
1
- ω/10
2ω/5
-ω/20
1/20
1/10
-2/5
1/10
1/20
-2/5
1
-2/5
1/20
1/10
-2/5
1/10
1/20
-ω/20
2ω/5
-ω/10
1
2ω/5
-ω/20
-ω/10

FIGURE 5

6. PRECONDITIONING USING AN INCOMPLETE ELIMINATION PROCESS

The preconditioning strategy developed in this section is based on factorisation techniques where sparse lower and upper triangular matrices L_s^{-1} and U_s (the approximate triangular factors of the coefficient matrix A) are determined in a simple manner. These can then be used in the manner indicated in Section 2. (Evans, 1974).

We consider the quasi-linear diffusion equation of the form,

$$u_t = (Ku_x)_x + (Ku_y)_y, \quad K=K(x,y), \tag{6.1}$$

in the region

$$\bar{R} = R\times[0<t\leq T],$$

where $\quad \bar{R} = \{(x,y) \; ; \; 0<x,y<1\},$

with the initial condition,

$$u(x,y,0) = f(x,y), \quad (x,y)\in R, \tag{6.2}$$

and the boundary condition,

$$u(x,y,t) = g(x,y,t), \quad (x,y,t)\in\partial R\times[0\leq t\leq T], \tag{6.3}$$

where ∂R is the boundary of R.

The region R is now covered by a rectilinear net with mesh spacings h_x, h_y and k in the x,y and t directions and mesh points (x_i, y_j, t_r) where,

$$x_i = ih_x, \; i=0,1,\ldots,m, \quad y_j = jh_y, \; j=0,1,2,\ldots,p,$$

$$t_r = rk, \quad r=0,1,\ldots,T/k.$$

Average central difference approximations to the partial derivatives in (6.1) are now used to derive a Crank-Nicolson type finite difference formula in two space dimensions. Thus, we can write the finite difference discretisation of equation (6.1) on the chosen grid as a series of five point linear finite difference equations of the form,

$$v_{i,j}u_{i,j-1,r+1} + a_{i,j}u_{i-1,j,r+1} + b_{i,j}u_{i,j,r+1} + c_{i,j}u_{i+1,j,r+1} + e_{i,j}u_{i,j+1,r+1} = d_{i,j,r}, \text{ for } 1\leq i\leq m-1 \text{ and } 1\leq j\leq p-1, \tag{6.4}$$

where $\quad v_{i,j} = -kK_{j-\frac{1}{2}}/2h_y^2, \; e_{i,j} = -kK_{j+\frac{1}{2}}/2h_y^2,$

$$a_{i,j} = -kK_{i-\frac{1}{2}}/2h_x^2; \quad c_{i,j} = -kK_{i+\frac{1}{2}}/2h_x^2;$$

$$b_{i,j} = 1+k[(K_{j-\frac{1}{2}}+K_{j+\frac{1}{2}})/2h_y^2+(K_{i-\frac{1}{2}}+K_{i+\frac{1}{2}})/2h_x^2],$$

and

$$d_{i,j,r} = u_{i,j,r}+k[K_{i+\frac{1}{2}}(u_{i+1,j,r}-u_{i,j,r})-K_{i-\frac{1}{2}}(u_{i,j,r}-u_{i-1,j,r})]/2h_x^2+k[K_{j+\frac{1}{2}}(u_{i,j+1,r}-u_{i,j,r})-K_{j-\frac{1}{2}}(u_{i,j,r}-u_{i,j-1,r})]/2h_y^2.$$

A pictorial representation of the computational molecule so derived is given below.

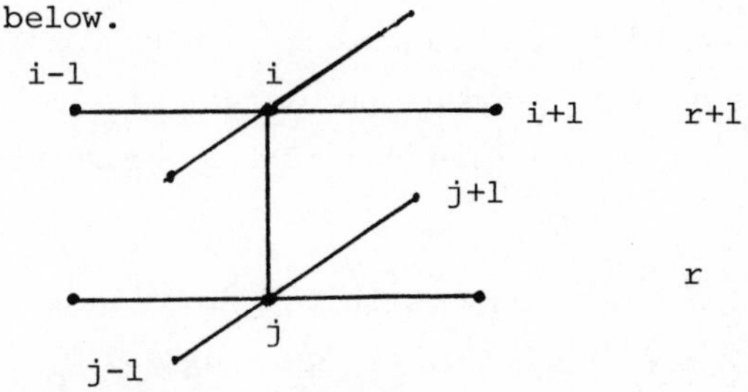

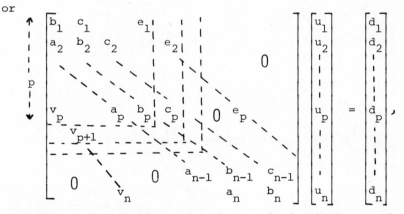

When we group the above system of finite difference equations into matrix form, we obtain a quindiagonal matrix of order n (n=mp) and of semi-bandwidth p of the form,

$$\underline{Au} = \underline{d}, \tag{6.5}$$

or

$$p\updownarrow\begin{bmatrix} b_1 & c_1 & & e_1 & & & & \\ a_2 & b_2 & c_2 & & e_2 & & 0 & \\ \ddots & \ddots & \ddots & & \ddots & & & \\ v_p & & a_p & b_p & c_p & 0 & e_p & \\ & v_{p+1} & & \ddots & \ddots & \ddots & & \ddots \\ 0 & \ddots & 0 & & a_{n-1} & b_{n-1} & c_{n-1} & \\ & & v_n & & & a_n & b_n & \end{bmatrix} \begin{bmatrix} u_1 \\ u_2 \\ \vdots \\ u_p \\ \vdots \\ u_n \end{bmatrix} = \begin{bmatrix} d_1 \\ d_2 \\ \vdots \\ d_p \\ \vdots \\ d_n \end{bmatrix},$$

which has to be solved to give the required solution to the problem at each time step.

A parallel investigation into the current solution processes involving the diffusion equation with one space dimension reveals the the fact that extensive use is made of the tridiagonal matrix algorithm (Varga, 1962) to solve the linear system involved. However, in the case of the quindiagonal matrix of semibandwidth p no such compact algorithmic solution is known to exist. However, we can derive one which is admirably suitable for our purposes by using the algorithm derived in Evans (1972) for successive peripheral overrelaxation in an echelon process as indicated by the partitioned dotted lines in the matrix (6.6).

Thus, the algorithm for the solution of a quindiagonal linear system of semi-bandwidth p which can be seen to be a natural extension to the tridiagonal algorithm is given below. We compute the following quantities recursively:

for i=1,

$$g_1 = c_1/b_1,\quad h_{1,p} = e_1/b_1,\quad f_1 = d_1/b_1, \tag{6.6}$$

$$G_{p,1} = v_p,\quad D_{p,1} = b_p,\quad g_p = c_p \text{ and } F_{p,1} = d_p,$$

for i=2,3,...,p-1,

$$w_i = b_i - a_i g_{i-1};\quad g_i = c_i/w_i;\quad f_i = (d_i - a_i f_{i-1})/w_i,$$

$$h_{i,p+k-2} = -a_i h_{i-1,p+k-2}/w_i, \text{ for } k=2,3,\ldots,i,$$

$$h_{i,p+i-1} = e_i/w_i,$$

$$G_{p,i} = -g_{i-1}G_{p,i-1};\quad D_{p,i} = D_{p,i-1} - G_{p,i-1}h_{i-1,p},$$

$$F_{p,i} = F_{p,i-1} - G_{p,i-1}f_{i-1};\quad g_p = g_p - G_{p,i-1}h_{i-1,p+1},$$

$$h_{p,p+k} = h_{p,p+k} - G_{p,i-1}h_{i-1,p+k}, \text{ for } k=2,3,\ldots,i;$$

for i=p

$$D_{p,p} = [D_{p,p-1} - (G_{p,p-1}+a_p)(g_{p-1}+h_{p-1,p})],$$

$$f_p = [F_{p,p-1} - (G_{p,p-1}+a_p)f_{p-1}]/D_{p,p},$$

$$g_p = [g_{p,p-1}+a_p)h_{p-1,p+1}]/D_{p,p},$$

$$h_{p,p+k-1} = [h_{p,p+k-1} - (G_{p,p+k-1}+a_p)h_{p-1,p+k-1}]/D_{p,p},$$

for k=3,4,...,p-1,

$$h_{p,2p-1} = e_p/D_{p,p},$$

for i=p+1,p+2,...,n

for j=i-p+1,

$$G_{i,j} = v_i,\ D_{i,j} = b_i,\ F_{i,j} = d_i \text{ and } G_i = c_i ,$$

for j=i-p+2,i-p+3,...,i-1,

$$G_{i,j} = G_{i,j}-[g_{j-1}+h_{j-1,j}]G_{i,j-1},\ a_i = a_i-G_{i,j-1}h_{j-1,i+1},$$

$$F_{i,j} = F_{i,j-1}-G_{i,j-1}f_{j-1},\ g_i = g_i-G_{i,j-1}h_{j-1,i+1},$$

$$D_{i,j} = D_{i,j-1}-G_{i,j-1}h_{j-1,i},\ G_{i,k} = G_{i,k}-G_{i,j-1}h_{j-1,k} \quad \text{for } k=j+1,\ldots,i-2,$$

$$h_{i,i+k} = h_{i,i+k}-G_{i,j-1}h_{j-1,i+k}, \text{ for } k=2,\ldots,p-1,$$

for j=i,

$$D_{j,j} = [D_{j,j-1}-(G_{j,j-1}+a_j)(g_{j-1}+h_{j-1,j})],$$

$$f_j = [F_{j,j-1}-(G_{j,j-1}+a_j)f_{j-1}]/D_{j,j} ,$$

$$g_j = [g_j-(G_{j,j-1}+a_j)h_{j-1,j+1}]/D_{j,j},\ h_{j,j+p-1} = e_j/D_{j,j},$$

$$h_{j,j+k-1} = [h_{j,j+k-1}-(G_{j,j-1}+a_j)h_{j-1,j+k-1}]/D_{j,j}, \quad \text{for } k=3,4,\ldots,(p-2).$$

The vector components u_i (i=1,2,...,n) of the solution of (6.5) are given recursively by a back substitution process expressed simply as

$$u_n = f_n , \tag{6.7}$$

for i=n-1,n-2,...,1,

$$u_i = f_i-gu_{i\,i+1}- \sum_{r=i+1}^{i+p} h_{i,r}u_r .$$

The above recursive sequence of equations is simply the Gaussian elimination process without interchanges(since they are not required for diagonally dominant matrix) expressed in algorithmic form. It is well known that such an elimination process transforms the original equations into another set of equations for which the transformed matrix is upper triangular. Further, it can be shown that the elimination process is valid provided that none of the leading minors of A, i.e. $b_i-a_ig_{i-1}$ etc. are zero. Since the matrix A derived from the finite difference discretisation of a self-adjoint partial differential equation is symmetric, and positive definite, pivoting techniques are unnecessary, thus preserving the band structure of the original matrix throughout the reduction process. Then, since the elimination procedure is effectively equivalent to multiplying the original matrix A by a lower triangular matrix L, to produce an upper triangular matrix U, we have

$$LA = U \,, \tag{6.8}$$

and

$$LA\underline{u} = U\underline{u} = L\underline{d} = \underline{f}, \tag{6.9}$$

are the equations from which the results are obtained by back substitution. In the notation used in the algorithm, we have the following layouts for the L and U matrices,

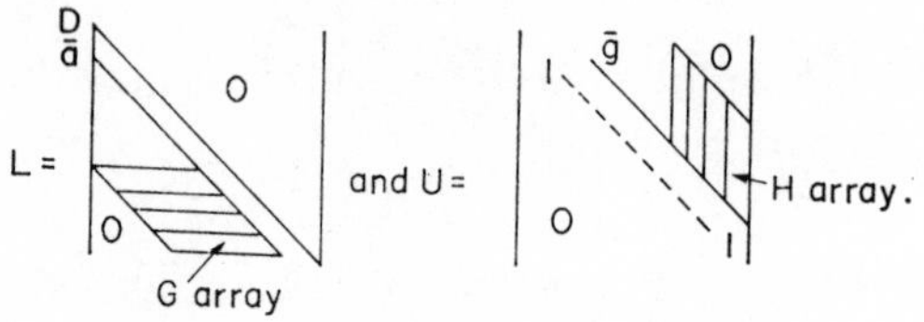

FIGURE 6

The algorithm (6.6)-(6.7) has a compact form which could be of universal use when coded and inserted in a computer program library. However, there are certain unfavourable features which we now enumerate and then attempt to eliminate.

Firstly, it consumes too much storage capacity for although the G, F and D arrays and $\bar{a}$ vector spaces can be overwritten, the $\underline{g}$ vector, $\underline{f}$ vector and H array spaces have to be strictly preserved. Secondly, the sparseness of the original matrix A has been irrevocably lost in the transformation to upper triangular form U and this shows itself in the inconvenience of having to store the H array in matrix form as well as the time consuming summation product term in the back substitution process. The amount of work involved is also prohibitively high, i.e.,

$O(p^2 n)$ multiplications for $n>p$.

This algorithm was applied to many of the standard partial differential equations involving two space dimensions of Mathematical Physics and the contents of the H and G arrays scrutinised, for matrices of order 50 and with bandwidths of 10,15,20,30 and 40 respectively. In particular, the error norms of the approximate solutions u_r, i.e., $[\Sigma(u-u_r)^2]^{\frac{1}{2}}$ obtained by including only r terms in the bandwidth is expressed graphically by Fig.7. These results clearly show that the error norm drops 50% in value between retaining two and three terms in the bandwidth. Clearly, an optimum value of r must exist. In addition, owing to the leveling off of each of the curves, there seems little point in retaining more than about r=4 fill-in terms of the bandwidth. In particular, for every case it was noticed that after the four outermost elements in each column of the H array, the magnitude of the remaining terms decreased rapidly as the matrix diagonal was approached. Thus, a large percentage of the fill-in terms are less than two per cent of the original extreme value occurring in the same position in the initial matrix A. Hence, it is a reasonable approach to re-design the algorithm (6.6)-(6.7) so that only the four outermost terms are retained in the G and H arrays, the remaining terms being just not computed at all. Consequently, a near or approximate factorisation of the matrix is achieved in which both sparseness is retained and storage requirements reduced to a minimum, for the H array can then be stored compactly in four vector spaces. Thus, we see an approximate sparse elimination process can be carried out in a similar way as before. In particular we have,

$$L_s A \simeq U_s , \qquad (6.11)$$

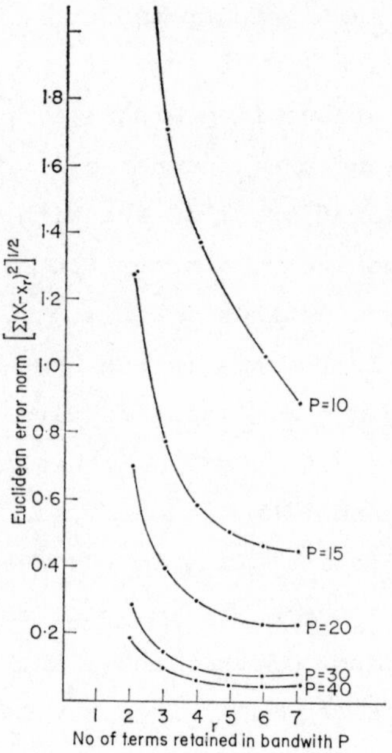

FIGURE 7

and the layout of the L_s and U_s matrices are as follows:

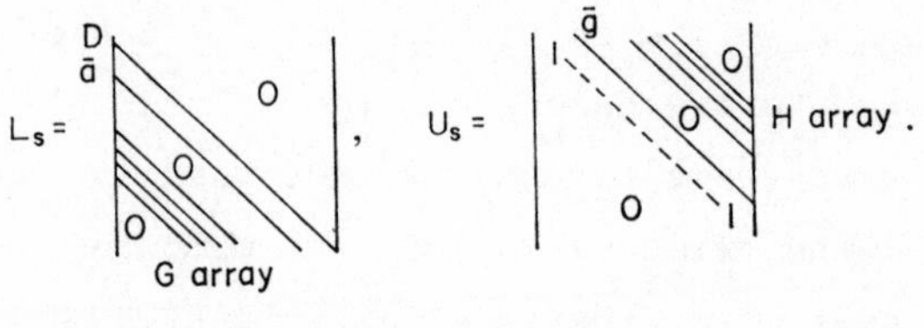

FIGURE 8

The approximate elimination algorithm (henceforth called the sparse matrix elimination scheme) so obtained for retaining r outermost off-diagonal entries can still be expressed in the same compact form as before. We again calculate the quantities recursively:

for i=1,

$$g_1 = c_1/b_1,\ h_{1,p} = e_1/b_1,\ \ f_1 = d_1/b_1\ ,$$

$$G_{p,1} = v_p,\ D_{p,1} = b_p,\ g_p = c_p \text{ and } F_{p,1} = d_p, \quad (6.13)$$

for i=2(1)p-1,

$$w_i = b_i - a_i g_{i-1},\ g_i = c_i/w_i,\ f_i = (d_i - a_i f_{i+1})/w_i,$$

$$h_{i,k} = -a_i h_{i-1,k}/w_i, \text{ for } k=i+p-r,\dots,i+p-2,$$

$$h_{i,p+i-1} = e_i/w_i,$$

$$G_{p,i} = -g_{i-1}G_{p,i-1},\ D_{p,i} = D_{p,i-1} - G_{p,i-1}h_{i-1,p},$$

$$F_{p,i} = F_{p,i-1} - G_{p,i-1}f_{i-1},\ g_p = g_p - G_{p,i-1}h_{i-1,p+1},$$

$$h_{p,p+k} = h_{p,p+k} - G_{p,i-1}h_{i-1,p+k}, \text{ for } k=p-r,\dots,i;$$

$$g_{p-1} = g_{p-1} + h_{p-1,p},$$

$$h_{p-1,p} = 0;$$

for i=p,

$$D_{p,p} = [D_{p,p-1} - (G_{p,p-1} + a_p)(g_{p-1} + h_{p-1,p})],$$

$$f_p = [F_{p,p-1} - (G_{p,p-1} + a_p) f_{p-1}]/D_{p,p},$$

$$g_p = [g_p - (G_{p,p-1} + a_p) h_{p-1,p+1}]/D_{p,p},$$

$$h_{p,p+k-1} = [h_{p,p+k-1} - (G_{p,p-1} + a_p) h_{p-1,p+k-1}]/D_{p,p}, \quad \text{for } k=p-r+1,\dots,p-1.$$

$$h_{p,2p-1} = e_p/D_{p,p};$$

for i=p+1(1)n,

for j=i-p+1,

$$G_{i,j} = v_i,\ D_{i,j} = b_i,\ F_{i,j} = d_i \text{ and } G_i = c_i;$$

for j=i-p+2(1)i-1,

$$G_{i,j} = G_{i,j} - (g_{j-1} + h_{j-1,j})G_{i,j-1},$$

$$a_i = a_i - G_{i,j-1}h_{j-1,i-1},$$

$$F_{i,j} = F_{i,j-1} - G_{i,j-1}f_{j-1},\ g_i = g_i - G_{i,j-1}h_{j-1,i+1},$$

$$D_{i,j} = D_{i,j-1} - G_{i,j-1}g_{j-1,i},$$

$$G_{i,k} = G_{i,k} - G_{i,j-1}h_{j-1,k}, \text{ for } k=j+1,j+2,\dots,i-2,$$

$$h_{i,i+k} = h_{i,i+k} - G_{i,j-1}h_{j-1,i+k}, \text{ for } k=p-r,\dots,p-1;$$

for j=i,

$$D_{j,j} = [D_{j,j-1} - (G_{j,j-1}+a_j)(g_{j-1}+h_{j-1,j})],$$

$$f_j = [F_{j,j-1} - (G_{j,j-1}+a_j)f_{j-1}]/D_{j,j} ,$$

$$g_j = [g_j - G_{j,j-1}+a_j)h_{j-1,j+1}]/D_{j,j} ,$$

$$h_{j,j+k-1} = [h_{j,j+k-1} - (G_{j,j-1}+a_j)h_{j-1,j+k-1}]/D_{j,j},$$

for k=p-r+1,...,p-1,

$$h_{j,j+p-1} = e_j/D_{j,j} .$$

The back substitution process is expressed simply as

$$u_n = f_n ,$$

for i=n-1(-1)n-p+r+1,

$$u_i = f_i - g_i u_{i+1} , \qquad (6.14)$$

for i=n-p+r(-1)n-p+1,

$$u_i = f_i - g_i u_{i+1} \sum_{j=i+p-r}^{n} h_{i,j} u_j ,$$

for i=n-p(-1)1,

$$u_i = f_i - g_i u_{i+1} \sum_{j=i+p-r}^{i+p-1} h_{i,j} u_j .$$

Thus, we have developed a sparse matrix elimination scheme in which the storage requirements and computational effort has been kept to a minimum. The amount of arithmetic work needed to implement this algorithm for $n>p$ is $O(n)$ multiplications and divisions, but if the coefficients of the matrix, i.e. a_i, b_i, c_i, e_i and v_i, $(i=1,2,\ldots,n)$ do not change then for subsequent applications the work reduces to $O(6n)$ multiplications.

7. SPARSE ELIMINATION PRECONDITIONED ITERATIVE METHODS

We now introduce iterative procedures based on incomplete elimination for solving numerically the usual five term sparse linear system,

$$A\underline{u} = \underline{d} ,$$

associated with (6.1). The iteration method (Preconditioned Simultaneous Displacement) we shall use has the following form,

$$L_s^{-1}U_s(\underline{u}^{(k+1)}-\underline{u}^{(k)}) = \alpha(\underline{d}-A\underline{u}^{(k)}) , \tag{7.1}$$

where L_s and U_s are the triangular matrices involved in the sparse elimination scheme discussed previously, α a predetermined acceleration parameter or sequence of parameters and the superscript k is an iteration index.

By making use of the well known concepts $\underline{r}^{(k)}=\underline{d}-A\underline{u}^{(k)}$, and $\delta\underline{u}^{(k+1)}=\underline{u}^{(k+1)}-\underline{u}^{(k)}$, a more convenient computational form to (7.1) can be expressed as

$$L_s^{-1}U_s\delta\underline{u}^{(k+1)} = \omega\underline{r}^{(k)} , \tag{7.2}$$

or

$$U_s\delta\underline{u}^{(k+1)} = \omega L_s\underline{r}^{(k)} . \tag{7.3}$$

This now corresponds closely to the algorithmic form given by (6.13)-(6.14). Thus, for a fixed choice of α, and an initial estimate to the solution $\underline{u}^{(0)}$ a sequence of approximate solutions $\underline{u}^{(1)},\underline{u}^{(2)},\underline{u}^{(3)},\ldots,\underline{u}^{(k)}$ can be obtained by using the algorithm (6.13)-(6.14).

In addition, 2nd order preconditioned methods of the form,

$$L_s^{-1}U_s(\underline{u}^{(k+1)}-\underline{u}^{(k)}) = \alpha(\underline{d}-A\underline{u}^{(k)})+\beta(\underline{u}^{(k)}-\underline{u}^{(k-1)}), \tag{7.4}$$

and

$$L_s^{-1}U_s(\underline{u}^{(k+1)}-\underline{u}^{(k)}) = \alpha_k(\underline{d}-A\underline{u}^{(k)})+\beta_k(\underline{u}^{(k)}-\underline{u}^{(k-1)}), \tag{7.5}$$

for the Richardson and Chebyshev methods can be formulated in a similar manner.

Initial results obtained from these methods using a fixed value of the acceleration parameter, α and a Chebyshev sequence α_k for the usual model problem are given in the accompanying Table 6.1.

The advantage to be gained in using an iterative method of solution (i.e. the sparse elimination scheme rather than a direct method, the full elimination scheme) for this type of problem is that the solution only differs slightly from plane to plane. Hence, the solution at the previous time step is a good initial approximation to commence the iteration scheme (7.1), which with the chosen acceleration parameters enables

the solution at each step to be obtained correct to an
accuracy of 5×10^{-6} in usually five iterations or less.

Method	h^{-1} mesh size	Basic method		
		Accel.factors α	β	No.of iterations
1st order simultaneous displacement method	10	1.0	0	120
	20	1.0	0	120
2nd order Richardson method	10	1.53	0.53	32
	20	1.73	0.73	65
2nd order Chebyshev method	10	Eqn.1.23		30
	20			59

continued......

Sparse Elimination Preconditioned Method		
Accel.factors α	β	No.of iterations
1.18	0	15
1.45	0	33
0.98	1.04	12
0.95	1.05	30
Eqn.1.23		12
		30

TABLE 6.1: Results for the Sparse Elimination algorithm
            when used to solve the discretised Laplace
            problem within the unit square with two mesh
            sizes giving sparse matrices of order 99 and
            361 respectively.

However, one must bear in mind the amount of work
required to seek the sparse L_s and U_s factors of a complicated
sparse matrix, because there are available from the matrix

itself, copies of the component L and U matrices which are, be definition, sparse, and which could be used in lieu, for similar purposes. These original L and U matrices are available with absolutely no work involved and can be obtained from the mesh itself. This approach is shown in Section 3 to lead to a powerful class of preconditioned iteration procedures based on the splitting principle which are applicable to the solution of sparse linear systems.

8. PRECONDITIONED VARATIONAL METHODS

Finally a class of iterative methods which can be re-appraised in the light of the preconditioning theory introduced in the previous paragraphs are those in which the solution of the linear system is obtained iteratively from vector corrections derived from the minimisation of a related quadratic function, i.e. the *gradient methods*. (Evans, 1973a).

Now, instead of choosing the initial linear system (1.1) we can choose its preconditioned form $B_\omega \underline{x} = \underline{d}_\omega$, as given in (3.3). Thus, for B_ω symmetric and positive definite, the variational methods for solving this system are derived from the fact that the quadratic functional,

$$F(\underline{x}) = -\tfrac{1}{2}\underline{x}B_\omega\underline{x} + \underline{x}\,\underline{d}_\omega = \text{constant}, \qquad (8.1)$$

defines a family of similar ellipsoids in Euclidean n-space, whose common centre is $B_\omega^{-1}\underline{d}_\omega$ the point at which F(x) takes its minimum value. Any arbitrary vector $\underline{x}^{(n)}$ will not, in general, be an exact solution of (3.3) but yield a residual $\underline{r}_\omega^{(n)}$, defined as,

$$\underline{r}_\omega^{(n)} = \underline{d}_\omega - B_\omega\underline{x}^{(n)} = [\text{Grad } F(\underline{x})]_{x^{(n)}} \; . \qquad (8.2)$$

Hence, we can attempt to proceed to the solution $B_\omega^{-1}\underline{d}_\omega$, the centre point of the ellipsoids, by a sequence of vector displacements of the form,

$$\underline{x}^{(n+1)} = \underline{x}^{(n)} + \omega_n\underline{p}^{(n)} , \qquad (8.3)$$

where $\underline{p}^{(n)}$ is an arbitrary direction and ω_n is chosen systematically at each iteration so that $F(\underline{x}^{(n+1)})$ is minimised.

The choice of the direction vector $\underline{p}^{(n)}$ differentiates many methods (all convergent for any given $\underline{p}^{(n)}$) which are based on this theory. When $\underline{p}^{(n)}$ is chosen to be $-\underline{r}^{(n)}$, we seek the minimum along the normal to the surface at the point $\underline{x}^{(n)}$. This is in the direction of steepest change of the quadratic function $F(\underline{x}^{(n+1)})$=constant. Thus, the *preconditioned steepest descent method* can be represented by the following simple algorithm:

Given an arbitrary vector $\underline{x}^{(0)}$, we compute the quantities:

$$\underline{r}_\omega^{(n)} = \underline{d}_\omega - B_\omega \underline{x}^{(n)},$$

$$\sigma_n = -\underline{r}_\omega^{(n)} \cdot \underline{r}_\omega^{(n)} / \underline{r}_\omega^{(n)} \cdot B_\omega \underline{r}_\omega^{(n)}, \leftarrow \leftarrow \leftarrow \leftarrow \leftarrow \leftarrow \leftarrow \leftarrow \leftarrow \leftarrow \leftarrow \leftarrow\uparrow$$

$$\underline{x}^{(n+1)} = \underline{x}^{(n)} - \sigma_n \underline{r}^{(n)}, \qquad n=n+1$$

$$\underline{r}_\omega^{(n+1)} = \underline{r}_\omega^{(n)} + \sigma_n B_\omega \underline{r}_\omega^{(n)},$$

to be continued until $|\underline{x}^{(n+1)} - \underline{x}^{(n)}| < 5\times 10^{-6}$, otherwise, $\rightarrow \rightarrow$ (8.4)

The method requires the storage of 3 vectors $\underline{x}$, $\underline{r}_\omega$, $B_\omega \underline{r}_\omega$ for its operation and the computation involves one matrix vector product, two vector inner products and two scalar by vector products. Unfortunately, without any preconditioning i.e. $\omega=0$, the method converges very slowly, which limits its practical value for problem solving. In this case A is ill-conditioned and the ellipsoids are long and thin and the normals along the surface are nearly coplanar. However, for $\omega>1$, the preconditioned form of the steepest descent method becomes quite competitive as the results in Table 8.1 and Fig. 9 for the chosen model problems confirm. In the light of these results and the simplicity of the algorithm it would appear that the Preconditioned Steepest Descent Method is worthy of reappraisal.

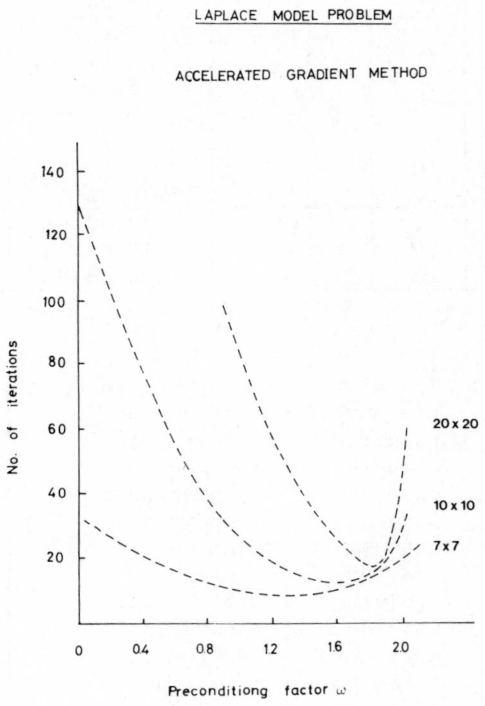
LAPLACE MODEL PROBLEM
ACCELERATED GRADIENT METHOD
No. of iterations
140
120
100
80
60
40
20
0
0.4
0.8
1.2
1.6
2.0
20 x 20
10 x 10
7 x 7
Preconditiong factor ω

FIGURE 9

Problem	Grid Size	Preconditioning parameter ω	P-condition number p	Preconditioned gradient method: No. of iterations	Preconditioned Conjugate Gradient Method: No. of iterations
Laplace	5	0	9.47	32	7
	10	0	39.86	142	19
	20	0	161.47	258	37
	5	1.3	1.65	8	6
	10	1.6	2.81	12	8
	20	1.8	5.42	17	11
Biharmonic	7	0	48.49	142	17
	10	0	286.44	290	40
	20	0	1916.21	>1300	124
	7	1.4	4.33	18	8
	10	1.6	13.39	54	15
	20	1.8	104.28	174	50

continued.......

TABLE 8.1: The P-condition number and the number of iterations required to achieve 5×10^{-6} accuracy for the preconditioned steepest descent method and preconditioned conjugate gradient method for a variety of problems involving the Laplace and Biharmonic operators on rectangular grids for values of the preconditioning parameter $\omega=0$ (basic methods) and $\omega=\omega_{opt}$ (optimal preconditioning.

A superior strategy for choosing the direction $\underline{p}^{(n)}$ is based on the knowledge that the centre of the ellipsoid lies in the plane conjugate to a given chord. Thus, by directing it towards the centre of the ellipsoids, we arrive at the *preconditioned conjugate gradient method* which is represented by the algorithm:

Given an arbitrary vector $x^{(0)}$, we compute the quantities:

$$\left.\begin{array}{l} \underline{r}_\omega^{(n)} = \underline{d}_\omega - B_\omega \underline{x}^{(n)} \\ \underline{d}_\omega^{(n)} = -\underline{r}_\omega^{(n)} \end{array}\right\}, \quad n=1,$$

$$\sigma_n = \underline{r}_\omega^{(n)} . \underline{d}_\omega^{(n)} / \underline{d}_\omega^{(n)} . B_\omega \underline{d}_\omega^{(n)},$$

$$\underline{x}^{(n+1)} = \underline{x}^{(n)} + \sigma_n \underline{d}_\omega^{(n)},$$

$$\underline{r}_\omega^{(n+1)} = \underline{r}_\omega^{(n)} - \sigma_n B_\omega \underline{d}_\omega^{(n)}, \qquad n=n+1$$

$$\tau_{n+1} = -\underline{r}_\omega^{(n+1)} . B_\omega \underline{d}_\omega^{(n)} / \underline{d}_\omega^{(n)} . B_\omega \underline{d}_\omega^{(n)},$$

$$\underline{d}_\omega^{(n+1)} = \underline{r}_\omega^{(n+1)} + \tau_{n+1} \underline{d}_\omega^{(n)},$$

to be continued until $|\underline{x}^{(n+1)} - \underline{x}^{(n)}| < 5\times 10^{-6}$, otherwise $\rightarrow$ (8.5)

BIHARMONIC MODEL PROBLEM

METHOD OF CONJUGATE GRADIENTS

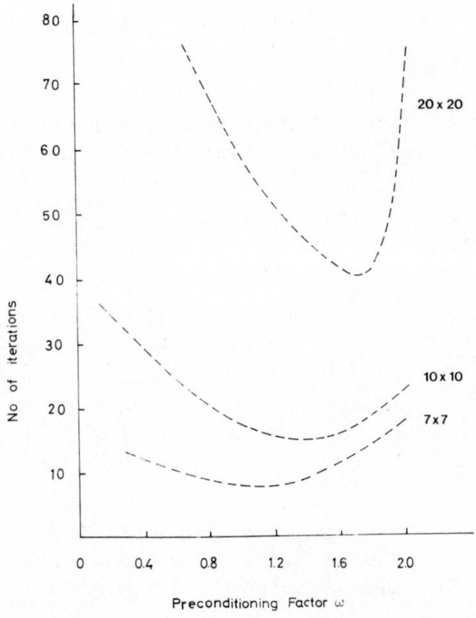

FIGURE 10

In both (8.4) and (8.5), the final solution $\underline{u}$ of the system (3.1) is given by the back substitution process $\underline{u}=(I-\omega L^T)^{-1}\underline{x}$.

The method requires the storage of 4 vectors $\underline{x}$, $\underline{r}_\omega$, $\underline{d}_\omega$, $B_\omega \underline{d}_\omega$ for its operation and involves one matrix vector product, two vector inner products and three scalar by vector products in computational effort. Reid (1971) has reported favourably on the conjugate gradient method as a powerful algorithm in its original form ($\omega=0$) and with the further application of preconditioning, an extremely attractive method of immense promise is revealed as the results for the Laplace and Biharmonic operators on rectangular networks show in Table 8.1 and Fig.10.

Finally, the alternative preconditioned form derived from the incomplete elimination process discussed in Section 6 can also be applied to the variational iterative schemes (8.4) and (8.5) where instead $B=(L_s^{-1}U_s)^{-1}A$. Similar results were obtained and these are compared in Evans et al (1980).

REFERENCES

1. Andersson, L., (1976) - "S.S.O.R. Preconditioning of Toeplitz Matrices", Ph.D. Thesis, Computer Sciences 76.02, Goteborg, Sweden.

2. Benson, A., (1968) - "The Numerical Solution of Partial Differential Equations by Finite Difference Methods", Ph.D. Thesis, University of Sheffield.

3. Dupont, T., Kendall, R. and Rachford, H.H. Jr., (1968) - "An Approximate Factorisation Procedure for Solving Self-Adjoint Elliptic Difference Equations", S.I.A.M., J.Numer. Anal. 5, 559-573.

4. Evans, D.J., (1963) - "The Extrapolated Modified Aitken Iteration Method for Solving Elliptic Difference Equations" Comp.J., 6, 193-201.

5. Evans, D.J., (1968) - "The Use of Preconditioning in Iterative Methods for Solving Linear Equations with Symmetric Positive Definite Matrices", J.I.M.A. 4, 295-314.

6. Evans, D.J. (1972) - "A New Iterative Procedure for the Solution of Sparse Systems of Linear Difference Equations", 'Sparse Matrices and their Applications', in Rose, D.J. and

Willoughby R.A. (eds.), 72-81, Plenum Press, New York.

7. Evans, D.J., (1973) - "The Analysis and Application of Sparse Matrix Algorithms in the Finite Element Method", 'The Mathematics of Finite Elements and Applications', Procs. edit. J.R. Whiteman, 427-447, Acad. Press, London, (presented at Mafelap Conference, April 1972).

8. Evans, D.J. (1973) - "Comparison of the Convergence Rates Of Iterative Methods for Solving Linear Equations with Preconditioning", Greek Math.Soc., Caratheodory Symp., 106-135.

9. Evans, D.J. (1974) - "Iterative Sparse Matrix Algorithms", in 'Software for Numerical Mathematics', ed. D.J. Evans, 49-83, Acad. Press.

10. Evans, D.J. and Lipitakis, E.A. (1979) - "A Sparse LU Factorization Procedure for the Solution of Parabolic Differential Equations", Numerical Methods in Thermal Problems, edit. R.W. Lewis & K. Morgan, 954-966, Pineridge Press, Swansea, U.K.

11. Evans, D.J. and Missirlis, N.M. (1980) - "The Preconditioned Simultaneous Displacement (PSD) Method for Elliptic Difference Equations", M.C.S. 22, 256-263.

12. Evans, D.J. and Lipitakis, E.A. (1980) - "A Normalised Implicit Conjugate Gradient Method for the Solution of Large Sparse Systems of Linear Equations", Comp.Meth. in Appl.Mech.Eng. 23, 1-20.

13. Evans, D.J. and Lipitakis, E.A. (1982) - "Implicit Semi-Direct Methods Based on Root-Free Sparse Factorisation Procedures", BIT 1982, (in press).

14. Evans, D.J., Lipitakis, E.A. and Missirlis, N.M., (1980) - "On Sparse and Compact Preconditioned Conjugate Gradient Methods for Partial Differential Equations", Int.Jour. Comp.Math. 9, 55-80.

15. Frankel, S.P. (1950) - "Convergence Rates of Iterative Treatments of Partial Differential Equations", M.T.A.C. 4, 65-75.

16. Forsythe, G. and Wasow, W.R. (1960) - "Finite Difference Methods for Partial Differential Equations", Wiley, N.Y.

17. Golub, G.H. and Varga, R.S. (1961) - "Chebyshev Semi-Iterative Methods, Successive Overrelaxation Iterative Methods and Second Order Richardson Iterative Methods", Num.Math. Pts I and II 3, 147-168.

18. Gustafsson, I., (1978) - "A Class of First Order Factorization Methods", BIT, 18, 142-156.

19. Hestenes, M.R. and Stiefel, E., (1952) - "Methods of Conjugate Gradients for Solving Linear Systems", NBS J. of Res. 49, 409-436.

20. Kershaw, D.S. (1978) - "The Incomplete Cholesky-Conjugate Gradient Method for the Iterative Solution of Systems of Linear Equations", Jour.Comp.Phys. 26, 43-65.

21. Meijerink, J.A. and Van Der Vorst, H.A., (1977) - "An Iterative Solution Method of Linear Systems of Which the Coefficient Matrix is a Symmetric M-Matrix", Math.Comp. 31, 148-162.

22. Missirlis, N.M. and Evans, D.J. (1981) - "On the Convergence of Some Generalised Preconditioned Iterative Methods", S.I.N.U.M. 18, 591-596.

23. Missirlis, N.M. and Evans, D.J., (1981) - "On the Dynamic Acceleration of the Preconditioned Simultaneous Displacement (PSD) Method", Int.J.Comp.Math., 10, 153-176.

24. Reid, J.K., (1971) - "On the Method of Conjugate Gradients for the Solution of Large Sparse Systems of Linear Equations", in Reid (ed.) 'Large Sparse Sets of Linear Equations', Acad. Press, London.

25. Sheldon, J. (1955) - "On the Numerical Solution of Elliptic Difference Equations", M.T.A.C. 9, 101-112.

26. Stiefel, E.L. (1958) - "Kernel Polynomials in Linear Algebra and Their Numerical Application", 1-22, NBS AMS 49

27. Stone, H.L., (1968) - "Iterative Solution of Implicit Approximations of Multi-dimensional Partial Differential Equations", SIAM J.Numer.Anal. 5, 530-558.

28. Varga, R.S. (1960) - "Factorisation and Normalised Iterative Methods", 'Boundary Problems in Differential Equations', (R.E. Langer), 121-142, Univ. Wisconsin Press, Madison, Wisconsin.

29. Varga, R.S. (1962) - "Matrix Iterative Analysis", Prentice Hall, Englewood Cliffs, New Jersey.

30. Young, D.M. (1954) - "Iterative Methods for Solving Partial Difference Equations of Elliptic Type", Trans. Amer.Math.Soc. 76, 92-111.

31. Young, D.M. (1971) - "Iterative Solution of Large Linear Systems", Academic Press, New York and London.

THE EXTRAPOLATION TECHNIQUE AS A PRECONDITIONING STRATEGY

A. HADJIDIMOS
Department of Mathematics, University of Ioannina, Ioannina, Greece.

Abstract: The discretisation of a Partial Differential Equation usually leads to a large sparse linear system for the solution of which iterative methods are most suitable. To speed up the convergence of an iterative method the extrapolation technique gives in many cases excellent results. This paper deals with the stationary extrapolation applied to first and second order methods and refers very briefly to the nonstationary one.

1. FIRST ORDER STATIONARY EXTRAPOLATION

1.1. Introduction and Preliminaries

Consider the linear system

$$Au = b, \tag{1.1}$$

where $A\in\mathbb{C}^{n,n}$ is a known nonsingular matrix, with entries $a_{jk}\in\mathbb{C}$ $|j,k=1(1)n$, and $u,b\in\mathbb{C}^n$ vectors with u unknown and b known. To solve (1.1) numerically we form a splitting of $A=P-Q$, with $P,Q\in\mathbb{C}^{n,n}$ and P easily inverted. Then the first order stationary iterative scheme, used for the solution of (1.1),

$$u^{(m+1)} = Tu^{(m)}+c \quad |m = 0,1,2,\ldots, \tag{1.2}$$

is constructed, where $T=P^{-1}Q$, $c=P^{-1}b$ and, in general, $u^{(0)}\in\mathbb{C}^n$ arbitrary. By using the splitting $A=P'-Q'$, with $P'=\frac{1}{\omega}P$, $Q'=\frac{1}{\omega}((1-\omega)P+\omega Q)$, where $\omega\in\mathbb{R}-\{0\}$, the following scheme

$$u^{(m+1)} = T_\omega u^{(m)}+\omega c \quad |m = 0,1,2,\ldots, \quad T_\omega\equiv(1-\omega)I+\omega T, \tag{1.3}$$

is yielded, with I the unit matrix. Scheme (1.3) is called the stationary extrapolated scheme of (1.2) and ω the extrapolation parameter. The very first extrapolation scheme was proposed by Richardson [62], corresponded to the splitting $P=I$, $Q=I-A$ and was nonstationary with respect to (wrt) ω.

The problem of determining an optimum $\omega(\hat{\omega})$ so that (1.3) converges and the spectral radius $\rho(T_\omega)$ of its iteration matrix becomes as small as possible has been tackled by many researchers. If $\lambda_j \mid j=1(1)n$ are the eigenvalues of T(those of T_ω are $1-\omega+\omega\lambda_j$)the problem posed possesses a solution iff $Re\lambda_j<1$(or $Re\lambda_j>1$)for all j (see e.g. [48]).The optimisation problem where at least one λ_j is complex had not been solved until recently [41],although $\hat{\omega}$ was known in some cases as for example when all λ_j i) are real (see e.g. [48]),ii) are purely imaginary [17],iii) have the same real part [76] and also in some others(see e.g. [76],[17],[43],[77] and [44]).The optimum solution obtained in [41] was given by means of an algorithm.The latter is also given in [13] together with an Algol 60 procedure and worked out examples.Here is the algorithm in question.

Step 1: Examine if $Re\lambda_j<1$(or $Re\lambda_j>1$) $\mid j=1(1)n$.If neither happens no convergent scheme can be obtained by extrapolation.If $Re\lambda_j>1 \mid j=1(1)n$ call λ_j the numbers $2-Re\lambda_j+iIm\lambda_j \mid j=1(1)n$.Step 2:Order the numbers $\mu_j=Re\lambda_j+i|Im\lambda_j| \mid j=1(1)n$ so that $Re\mu_j \mid j=1(1)n$ are decreasing.If two or more μ_j have the same real part keep only the one with the largest imaginary part.Call $\mu_j \mid j=1(1)m\leq n$ again the remaining numbers.If $m\leq 2$ go to Step 4.Step 3:Keep those μ_j which with the projections of μ_1 and μ_m on the real axis constitute the vertices of a convex polygon H containing all μ_j.Call $\mu_j \mid j=1(1)\ell\leq m$ again the vertices of H,which were among the μ_j of Step 2.Step 4: Find the one or the two μ_j(points)for which $Im\mu_j/(1-Re\mu_j) \mid j=1(1)\ell$ is a maximum.In the case of the one point (P) find the centre K(c,0) and the radius R of the circle with K the intersection of the perpendicular to QP at P with the real axis (Q is the point (1,0)) and R=(KP).If the circle at hand is the capturing one for all μ_j then go to Step 6 else go to Step 5.In the case of the two points (P_1,P_2) find the centre K(c,0) and the radius R of the circle which passes through these and then work as in the one point case. Step 5:Find all circles,with centres on the real axis,passing through any two points $P_j \mid j=1(1)\ell$,representing the numbers $\mu_j|j=$

$1(1)\ell$.For each circle examine if it is the capturing one and when this is found (let K(c,0) be its centre and R its radius) go to Step 6.Step 6:Find the optimum values $\hat{\omega}=(-1)^k/(1-c)=(-1)^k\rho(T_{\hat{\omega}})/R$, where k=1 if the transformation in Step 1 took place,otherwise k=0.

By applying the previous algorithm well-known results can be obtained.Thus if $\lambda_j \in \mathbb{R}\,|\,j=1(1)n,\lambda_m=\min\lambda_j,\lambda_M=\max\lambda_j$ and $\lambda_M<1$(or $\lambda_m>1$) we obtain

$$\hat{\omega} = 2/(2-\lambda_M-\lambda_m), \quad \rho(T_{\hat{\omega}}) = (\lambda_M-\lambda_m)/|2-\lambda_M-\lambda_m| \tag{1.4}$$

(see also [48]).If $\mathrm{Re}\lambda_j=\alpha\neq 1 \;|\, j=1(1)n$ we have

$$\hat{\omega}=(1-\alpha)/(1-2\alpha+\rho^2(T)),\rho(T_{\hat{\omega}})=((\rho^2(T)-\alpha^2)/(1-2\alpha+\rho^2(T)))^{\frac{1}{2}} \tag{1.5}$$

(see also [44]).Finally if all λ_j are purely imaginary (as in the case T is skew-Hermitian),we have from (1.5) and for $\alpha=0$ (see also [17])

$$\hat{\omega} = 1/(1+\rho^2(T)), \quad \rho(T_{\hat{\omega}}) = \rho(T)/(1+\rho^2(T))^{\frac{1}{2}}.$$

1.2. Extrapolation and Basic Iterative Methods

Let $A=D-C_L-C_U$,where $D=\mathrm{diag}(A),-C_L,-C_U$ the strictly lower and upper triangular parts of A and $\det(D)\neq 0$.If we put $L=D^{-1}C_L,U=D^{-1}C_U$ then the well-known methods of Jacobi(J),Gauss-Seidel (GS) and Successive Overrelaxation(SOR)correspond to the splittings

$$\begin{aligned} &\text{a. } P = D, \quad Q = D(L+U)\\ &\text{b. } P = D(I-L), \quad Q = DU\\ &\text{c. } P = \frac{1}{\omega}D(I-\omega L), \quad Q = \frac{1}{\omega}D((1-\omega)I+\omega U), \end{aligned} \tag{1.6}$$

where $\omega \in \mathbb{R}-\{0\}$ is the overrelaxation parameter.For $\omega=1$ the SOR method becomes the GS one so that it suffices to study the J and SOR methods.Hadjidimos [39] introduced the Accelerated Overrelaxation(AOR)method corresponding to the splitting

$$P = \frac{1}{\omega}D(I-rL), \quad Q = \frac{1}{\omega}D((1-\omega)I+(\omega-r)L+\omega U), \tag{1.7}$$

with $r\in\mathbb{R}$ the acceleration parameter.Because of (1.7) the AOR scheme is

$$u^{(m+1)} = L_{\omega,r} u^{(m)} + \omega(I-rL)^{-1} D^{-1} b \quad | m=0,1,2,\ldots \tag{1.8}$$

with iteration matrix

$$L_{\omega,r} = (I-rL)^{-1}((1-\omega)I+(\omega-r)L+\omega U). \tag{1.9}$$

In view of (1.6),(1.7) and (1.9),(1.8) gives,for various pairs (ω,r),the J,GS and SOR methods as well as their Extrapolated(E) counterparts.Thus for $(\omega,r)=(1,0),(1,1),(\omega,\omega),(\omega,0),(\omega,1)$ and $(\omega,r\neq 0)$ we obtain the J,GS,SOR,EJ,EGS and ESOR which is an extrapolated SOR with overrelaxation parameter r and extrapolation one $\frac{\omega}{r}$.The latter method not as an extrapolated one,was first studied by Sisler [65-67].Niethammer [56] studied the same method and found regious of convergence with $A\equiv I-L-U\equiv I-B$,B weakly 2-cyclic consistently ordered,$\rho(B)<1$ and $\omega r>0$ for i) A Hermitian positive definite and ii) B skew-Hermitian.Under these assumptions he obtained

$$\begin{aligned} &\text{i) } \hat{\omega} = \hat{r} = 2/(1+(1-\rho^2(B))^{\frac{1}{2}}), \quad \rho(L_{\hat{\omega},\hat{r}}) = \hat{r}-1 \\ &\text{ii) } \hat{\omega} = \hat{r} = 2/(1+(1+\rho^2(B))^{\frac{1}{2}}), \quad \rho(L_{\hat{\omega},\hat{r}}) = 1-\hat{r}\,. \end{aligned} \tag{1.10}$$

Avdelas and Hadjidimos [11] studied completely and quite independently case (i) and obtained in some cases better results than Niethammer's.More specifically if $\underline{\mu}=\min|\mu_j|\leqq\max|\mu_j|=\bar{\mu}$,where μ_j $|j=1(1)n$ the eigenvalues of B,the optimum parameters for the AOR method are (with $\hat{r}$ the same as in (1.10i))

a. If $(1-\bar{\mu}^2)^{\frac{1}{2}} \leq 1-\underline{\mu}^2$ then

$$\hat{\omega} = \hat{r}, \quad \rho(L_{\hat{\omega},\hat{r}}) = \hat{r}-1$$

b. If $1-\underline{\mu}^2 < (1-\bar{\mu}^2)^{\frac{1}{2}}$ then

$$\omega = \frac{1-\underline{\mu}^2+(1-\bar{\mu}^2)^{\frac{1}{2}}}{(1-\underline{\mu}^2)(1+(1-\bar{\mu}^2)^{\frac{1}{2}})}, \quad \rho(L_{\hat{\omega},\hat{r}}) = \frac{\underline{\mu}(\bar{\mu}^2-\underline{\mu}^2)^{\frac{1}{2}}}{(1-\underline{\mu}^2)^{\frac{1}{2}}(1+(1-\bar{\mu}^2)^{\frac{1}{2}})}$$

c. If $0<\underline{\mu}=\bar{\mu}=\mu$ then

$$\hat{\omega} = \pm 1/(1-\mu^2)^{\frac{1}{2}}, \quad \rho(L_{\hat{\omega},\hat{r}}) = 0.$$

Thus, only in case (a) the results of [11] are identical with the ones in [56]. In all other cases they are better. It should be added that for A irreducibly diagonally dominant, L,M, weakly 2-cyclic consistently ordered or Hermitian positive definite matrix convergence theorems for the AOR method are given in [39],[12],[76], and for A strictly diagonally dominant by Martins in [54].

The theory of the SOR method was extended by Varga [69] for A Hermitian with $a_{jj}>0 \mid j=1(1)n$ and D any Hermitian positive definite matrix who considered the scheme

$$u^{(m+1)} = \tilde{L}_{\omega,\omega} u^{(m)} + \omega(D-\omega E)^{-1} b \quad | m = 0,1,2,..., \tag{1.11}$$

$$\tilde{L}_{\omega,\omega} = (D-\omega E)^{-1}((1-\omega)D+\omega E^H), \tag{1.12}$$

with $E=\frac{1}{2}(D-A+S)$, S any skew-Hermitian and E^H the complex conjugate transpose of E. For the extended SOR method (1.11)-(1.12), which reduces to the classical one for $D\equiv diag(A)$ and $S\equiv 2C_L-D+A$, a theorem analogous to Ostrowski's [59] was stated and proved and also $\hat{\omega}$ and $\rho(\tilde{L}_{\hat{\omega},\hat{\omega}})$ were found. Based on the previous theory Hadjidimos and Yeyios [45] considered the ESOR(AOR) scheme corresponding to (1.11)-(1.12)

$$u^{(m+1)} = \tilde{L}_{\omega,r} u^{(m)} + \omega(D-rE)^{-1} b \quad | m = 0,1,2,..., \tag{1.13}$$

$$\tilde{L}_{\omega,r} = (D-rE)^{-1}((1-\omega)D+(\omega-r)E+\omega E^H). \tag{1.14}$$

For the above scheme the Householder-John theorem (see Ortega and Plemmons [57]) was applied and the result was the theorem below.

Theorem 1: Let $A=D-E-E^H$ be Hermitian with $a_{jj}>0 \mid j=1(1)n$, where D is Hermitian positive definite and $\det(D-rE)\neq 0$ for any permissible pair (ω,r) from the Table depending on the position of μ_m and μ_M wrt zero (μ_m and μ_M are the minimum and the maximum eigenvalues of $B=I-D^{-1}A=E+E^H$). Then for the parameters ω,r from the appropriate domains of the Table the AOR method (1.13)-(1.14) converges iff A is positive definite.

Regarding Theorem 1 some remarks should be made: i) For $r=\omega$ it reduces to the theorem by Varga [69] and refers to the interval (0,2). ii) The first part of Theorem 1 is of theoretical value unless

TABLE. The possible domains of the parameters ω and r.

Case	Relative position of μ_m and μ_M wrt zero	ω-domain	r-domain
1	$0<\mu_m \leq \mu_M$	$(-\infty,0)$	$(\omega+(2-\omega)/\mu_m,+\infty)$
		$(0,2)$	$(-\infty,\omega+(2-\omega)/\mu_M)$
		2	$(-\infty,2)$
		$(2,+\infty)$	$(-\infty,\omega+(2-\omega)/\mu_m)$
2	$0=\mu_m<\mu_M$	$(0,2)$	$(-\infty,\omega+(2-\omega)/\mu_M)$
3	$0=\mu_m=\mu_M$	$(0,2)$	$(-\infty,+\infty)$
4	$\mu_m<0<\mu_M$	$(0,2)$	$(\omega+(2-\omega)/\mu_m,\omega+(2-\omega)/\mu_M)$
5	$\mu_m<\mu_M=0$	$(0,2)$	$(\omega+(2-\omega)/\mu_m,+\infty)$
6	$\mu_m \leq \mu_M<0$	$(-\infty,0)$	$(-\infty,\omega+(2-\omega)/\mu_M)$
		$(0,2)$	$(\omega+(2-\omega)/\mu_m,+\infty)$
		2	$(2,+\infty)$
		$(2,+\infty)$	$(\omega+(2-\omega)/\mu_M,+\infty)$

knowledge of the position of μ_m and μ_M wrt zero is known in advance as for example when D=diag(A),D is a block diagonal part of A and D is positive definite.Then tr(B)=0 so that the ω- and r-domains are those of Cases 3 and 4 of the Table and iii) Members of our Numerical Analysis research team have been completing the determination of $\hat{\omega},\hat{r}$ in all cases.

To close this section we refer briefly to some generalised results of the M-matrix theory regarding the basic schemes.Varga [70-71] and others (see Plemmons [61]) proved equivalence theorems concerning the convergence of J,EJ and SOR schemes.To present one of them some definitions are needed:i) For a given $A\in\mathbb{C}^{n,n}$ the set $\Omega(A)$ of all $B\in\mathbb{C}^{n,n}$,for which $|b_{jk}|=|a_{jk}|$ $|j,k=1(1)n$,is the equimodular set of matrices associated with A.ii) For a given $A\in\mathbb{C}^{n,n}$ the matrix $\mathcal{M}(A)$,with $m_{jj}=|a_{jj}|,m_{jk}=-|a_{jk}||j,k=1(1)n,j\neq k$,is its comparison matrix and iii) A given $A \in \mathbb{C}^{n,n}$ is a nonsingular H-matrix iff $\mathfrak{M}(A)$ is a nonsingular M-matrix (this is due to Ostrowski [58]).

Theorem 2: Let $A \in \mathbb{C}^{n,n}$ with $a_{jj} \neq 0 \mid j=1(1)n$. The following three statements are equivalent: i) A is a nonsingular H-matrix. ii) For any $B \in \Omega(A)$ and any $\omega \in (0,2/(1+\rho(L_{1,0}(B)))$, $\rho(L_{\omega,0}(B)) \leqq \omega\rho(L_{1,0}(B)) + |1-\omega| < 1$. iii) For any $B \in \Omega(A)$ and any $\omega \in (0,2/(1+\rho(|L_{1,0}(B)|)))$, $\rho(L_{\omega,\omega}(B)) \leqq \omega\rho(|L_{1,0}(B)|) + |1-\omega| < 1$. (The notation $L_{\omega,r}(B)$ was used to stress that the iteration matrix is based on B and not on A). Without loss of generality it can be considered that Theorem 2 applies to the system $(I-T)u=c$, where T is the Jacobi matrix $L_{1,0}(B)$, with $\text{diag}(T)=0$. However, we may consider a more general T with $\text{diag}(T) \neq 0$. Based on this generalised Jacobi method Kulisch [50] (see also Marek [53]) obtained some results concerning the generalised GS and SOR schemes, where diag(T) was taken together with the strictly upper triangular part of T. By using the previous ideas Hadjidimos [40] generalised Theorem 2 so that to include: i) the ESOR method and ii) for all methods involved their generalised ones. For these the equimodular set of matrices associated with T, instead of that of A, is used and a generalisation of the comparison matrix is adopted as follows: Given a matrix $T' = I-T \in \mathbb{C}^{n,n}$ its generalised comparison matrix $\mathfrak{M}(T')$ is defined by $m_{jj} = 1-|t_{jj}|$, $m_{jk} = -|t_{jk}| \mid j,k=1(1)n, j \neq k$. (This reduces to the definition (ii) iff $t_{jj} \in [0,1) \mid j=1(1)n$). Let now that $S=L+D+U \in \Omega(T)$, where $D=\text{diag}(S)$ and L and U strictly lower and upper triangular matrices with T the matrix of (1.2). If $S'=I-S$ the generalised AOR matrix corresponding to a scheme of type (1.2) with iteration matrix S will be the following

$$L_{\omega,r}(S') = (I-rL)^{-1}((1-\omega)I + (\omega-r)L + \omega(D+U)).$$

A generalisation of Theorem 2, given below, is stated and proved in [40].

Theorem 3: The following four statements are equivalent: i) $T' = I-T$ is a nonsingular H-matrix. ii) For any $S \in \Omega(T)$ and any $\omega \in (0,2/(1+\rho(L_{1,0}(S'))))$, $\rho(L_{\omega,0}(S')) \leqq \omega\rho(L_{1,0}(S')) + |1-\omega| < 1$. iii) For any $S\in\Omega(T)$ and any $r\in(0,2/(1+\rho(|L_{1,0}(S')|)))$, $\rho(L_{r,r}(S'))\leq r\rho(|L_{1,0}(S')|)+|1-r|<1$. iv) For any $S\in\Omega(T)$, any $r\in(0,2/(1+\rho(L_{1,0}(S'))))$ and any $\omega\in(0,2r/(1+\rho(L_{r,r}(S'))))$, $\rho(L_{\omega,r}(S'))\leq \frac{\omega}{r}\rho(L_{r,r}(S'))+|1-\frac{\omega}{r}|<1$.

1.3. Extrapolation and Alternating Direction Implicit(ADI)Methods

Peaceman and Rachford [60] introduced the ADI methods for the numerical solution of second order parabolic and especially elliptic equations. An introduction to them can be found in Varga [68], Young [79], Wachspress [74] and Mitchell [55]. The early works by Douglas and Rachford [20], Douglas [18], Samarskii and Andreyev [63-64], Wachspress [72-73] and Conte and Dames [14] gave a tremendous interest in ADI methods so that many researchers have been tried to improve or generalise them. In order to outline the basic idea behind these methods and their connection with the extrapolation technique we work out a very simple problem.

Consider Laplace equation in the open unit square R (with vertices (0,0),(1,0),(1,1) and (0,1))

$$(\partial^2/\partial x_1^2+\partial^2/\partial x_2^2)u(x_1,x_2) = 0, \quad (x_1,x_2)\in R, \tag{1.15a}$$

where $u\equiv u(x_1,x_2)$ is given on the boundary ∂R of R

$$u(x_1,x_2) = g(x_1,x_2), \quad (x_1,x_2)\in\partial R \tag{1.15b}$$

and the derivatives $\partial^4 u/\partial x_1^4$, $\partial^4 u/\partial x_2^4$ exist and are continous on $\bar{R}=R\cup\partial R$. For the numerical solution of problem (1.15a)-(1.15b) a uniform grid of mesh size $h=1/N$(N the number of mesh subdivisions) is imposed on $\bar{R}$ and (1.15a) is approximated at each internal node by a 5-point difference formula. More specifically we have

$$\frac{1}{h^2}(\delta_{x_1}^2+\delta_{x_2}^2)u_{i_1 i_2} = 0 \quad |i_1,i_2 = 1(1)N-1, \tag{1.16}$$

where δ_{x_j} $|j=1,2$ is the central difference operator in x_j-direction and $u_{i_1 i_2}=u(i_1 h,i_2 h)=u(x_1,x_2)$ represents now the approximate solution of (1.15a)-(1.15b) at the nodal point (i_1,i_2). It can be shown that (1.16) approximates (1.15a) with a local truncation error $O(h^2)$ (see [68]). Provided that the equations (1.16) are taken in their natural ordering the totality of them yields a matrix equation problem of type (1.1). In this problem A is of order $n=(N-1)^2$ and of a block tridiagonal form with diagonal blocks $-J$, $U+2J, -J$. The unit matrix J is of order N-1, U is an $(N-1)\times(N-1)$ tridiagonal matrix with diagonals -1,2,-1, u an n-dimensional vector with components $u_{i_1i_2}$ and b a vector coming from the boundary conditions. By introducing the tensor product notation $\otimes$ (see Halmos [46]), used in connection with the ADI methods for the first time in [51], A can be written as

$$A = A_1 + A_2 \ , \ A_1 = J\otimes U, \ A_2 = U\otimes J \ . \tag{1.17}$$

It is readily seen that A_1 and A_2 are Hermitian and commute, therefore they possess a common orthonormal basis of eigenvectors (see [68]). The eigenvalues of A_1 and A_2 are those of U given by the expressions

$$\lambda_k = 4\sin^2(k\pi/(2N)) \,|\, k = 1(1)N-1 \tag{1.18}$$

(see [68]), with each eigenvalue repeated N-1 times. To solve the system (1.1) we consider the splitting of A, with $P=\frac{1}{r}(rI+A_1)(rI+A_2)$, $Q=\frac{1}{r}(r^2I+A_1A_2)$, where $r>0$ is an acceleration (iteration) parameter to be determined. The scheme corresponding to (1.2) has $T=(rI+A_2)^{-1}(rI+A_1)^{-1}(r^2I+A_1A_2)$, so that its extrapolated one is

$$\begin{aligned} u^{(m+1)} = {} & (I-\omega r(rI+A_2)^{-1}(rI+A_1)^{-1}(A_1+A_2))u^{(m)} + \\ & \omega r(rI+A_2)^{-1}(rI+A_1)^{-1}b \quad |m = 0,1,2,\ldots \ . \end{aligned} \tag{1.19}$$

Scheme (1.19) can be written as a two-step one

$$\begin{aligned} (rI+A_1)u^{(m+\frac{1}{2})} &= ((rI+A_1)(rI+A_2)-\omega r(A_1+A_2))u^{(m)}+\omega rb \\ (rI+A_2)u^{(m+1)} &= u^{(m+\frac{1}{2})} \quad |m = 0,1,2,\ldots, \end{aligned} \tag{1.20}$$

with $u^{(m+\frac{1}{2})}$ an auxiliary vector. Here some points have to be made: i) Scheme (1.20) is what is called Extrapolated (E)ADI scheme with extrapolation parameter ω (in fact the product ωr could be used as an extrapolation parameter). ii) (1.20) is written in the form suggested by D'Yakonov [21]. iii) (1.20) is stationary wrt r while in all the ADI schemes proposed in the papers quoted a set of acceleration parameters was used in a cyclic way.iv) In all those ADI schemes there was no extrapolation parameter but an a priori fixed number in its place. In Peaceman-Rachford scheme $\omega=2$ and in Douglas Rachford's $\omega=1$ and v) The first time ω was given values depending on the spectra of A_1 and A_2 was by Guittet [31]. Almost simultaneously Hadjidimos [32], [34] used ω as an extrapolation parameter. To come back to scheme (1.20) we have by using (1.17)

$$\begin{aligned} (J\otimes(rJ+U))u^{(m+\frac{1}{2})} &= ((J\otimes(rJ+U))(rJ+U)\otimes J) - \\ &\qquad \omega r(J\otimes U+U\otimes J))u^{(m)}+\omega rb \\ ((rJ+U)\otimes J)u^{(m+1)} &= u^{(m+\frac{1}{2})} \quad \Big| m=0,1,2,\ldots . \end{aligned} \tag{1.21}$$

The forms of the LHS's of (1.21) suggest that at each step of an iteration N-1 tridiagonal systems have to be solved for which efficient techniques exist as the one described in Faddeeva [24] and the one by Cuthill and Varga [15]. The iteration matrix of the procedure (1.19) or (1.20) is

$$T_\omega = I-\omega F,\quad F=r(A_1+A_2)(rI+A_1)^{-1}(rI+A_2)^{-1} . \tag{1.22}$$

Because of the forms of A_1 and A_2 both T_ω and F are Hermitian and commute with each one of A_1 and A_2. If a_1 and a_2 are the eigenvalues of A_1 and A_2 corresponding to the same eigenvector the eigenvalues f of the positive definite F and ρ of T_ω corresponding to that eigenvector will be

$$\begin{aligned} f &\equiv f(r,a_1,a_2) \equiv r(a_1+a_2)/((r+a_1)(r+a_2)), \\ \rho &\equiv \rho(\omega,r,a_1,a_2) \equiv 1-\omega f. \end{aligned} \tag{1.23}$$

Let f_m and f_M be the minimum and the maximum values of f. In view of (1.22) it is $T_\omega=(1-\omega)I+\omega(I-F)$. Thus because of (1.4) the best ω is given by $\omega=2/(2-(1-f_m)-(1-f_M))=2/(f_M+f_m)$ and the best $\rho(T_\omega)=(f_M-f_m)/(f_M+f_m)$. In view of these results $\rho(T_\omega)$ is minimised wrt r iff the ratio f_m/f_M is maximised [32] or equivalently iff the P-condition number of F, f_M/f_m, is minimised [22]. f_m and f_M are found in terms of r from the first of (1.23) taking into account that the bounds for a_1 and a_2, because of (1.18), are $\mu\equiv 4\sin^2(\frac{\pi}{2N})$ and $\nu\equiv 4\cos^2(\frac{\pi}{2N})$. Thus after some analysis it is obtained that

$$\hat{r} = \sqrt{\mu\nu} = 2\sin(\pi/N),$$

$$\hat{f}_m = \frac{2\sqrt{\mu\nu}}{(\sqrt{\mu}+\sqrt{\nu})^2} = \frac{\sin(\pi/N)}{1+\sin(\pi/N)}, \quad \hat{f}_M = \frac{\mu+\nu}{(\sqrt{\mu}+\sqrt{\nu})^2} = \frac{1}{1+\sin(\pi/N)} \qquad (1.24)$$

$$\hat{\omega} = 2, \quad \rho(T_{\hat{\omega}}) = \frac{(\sqrt{\mu}-\sqrt{\nu})^2}{(\sqrt{\mu}+\sqrt{\nu})^2} = \frac{1-\sin(\pi/N)}{1+\sin(\pi/N)} .$$

Regarding the results (1.24) and the analysis so far some remarks are made:i) In (1.24) $\hat{\omega} = 2$ shows that the optimum EADI scheme coincides with the (stationary) Peaceman-Rachford ADI one.ii) The optimum parameters (1.24) for the ADI scheme (1.20) with an a priori $\omega=2$ can be found in [68].iii) Guittet [31] gave $\hat{\omega}$ and $\hat{r}$ in the case of the p-dimensional unit cube,when the spectra of all A_j were the same.For p=2 he obtained (1.24).iv) The optimum parameters in the case of a rectangle (different spectra for A_1 and A_2) were obtained in [42].v) The present author [36-37] found also the otimum parameters for the Biharmonic equation in the unit square and cube R with the function u and the outward second normal derivative to the boundary ∂R of R being given on it and the same spectral for the A_j involved and vi) The same author [38] proved the equivalence between the EADI and the Alternating Direction Preconditioning (ADP) schemes of Gane and Evans [27] and gave optimum parameters for the aforementioned 2-dimensional Biharmonic problem with different spectra for A_1 and A_2.

An interesting problem concerning the EADI (and their equivalent ADP) methods in the 2-dimensional case when the elliptic equation is of a more general self-adjoint type,R is a rectangle,the boundary conditions on ∂R are those of a first or a second or even a third boundary value problem and the grid imposed is uniform with different mesh sizes in x_j-directions was considered and solved in [9].The basic idea,although not new with ADI problems,was to use two different acceleration parameters r_1 and r_2 in each iteration. More precisely the corresponding to (1.19) scheme is

$$u^{(m+1)} = (I-\omega r_1 r_2 (r_2 I+A_2)^{-1}(r_1 I+A_1)^{-1}(A_1+A_2))u^{(m)} + \omega r_1 r_2 (r_2 I+A_2)^{-1}(r_1 I+A_1)^{-1} b \quad |m = 0,1,2,\ldots$$

or equivalently

$$(r_1I+A_1)u^{(m+\frac{1}{2})} = ((r_1I+A_1)(r_2I+A_2)-\omega r_1r_2(A_1+A_2))u^{(m)}+\omega r_1r_2b \tag{1.25}$$
$$(r_2I+A_2)u^{(m+1)} = u^{(m+\frac{1}{2})} \quad |m=0,1,2,\ldots,$$

where ω (or ωr_1 or ωr_2 or even ωr_1r_2) can be considered as the extrapolation parameter. If $[\mu_j,\nu_j]$ is the domain for the eigenvalues a_j of A_j $|j=1,2$ and $|\mu_1|+|\mu_2|>0$, a similar but more complicated analysis to the one of the problem studied gave the following results

$$\hat{r}_1 = K/(\mu_1\nu_1-\mu_2\nu_2+L), \quad \hat{r}_2 = K/(\mu_2\nu_2-\mu_1\nu_1+L),$$
$$\hat{\omega} = 1/\hat{r}_1+1/\hat{r}_2 = 2L/K, \quad \rho(T_{\hat{\omega}}) = (1-M)/(1+M),$$
$$K = \mu_1\nu_1(\mu_2+\nu_2)+\mu_2\nu_2(\mu_1+\nu_1), \tag{1.26}$$
$$L = ((\mu_1+\nu_1)(\mu_1+\nu_2)(\mu_2+\nu_1)(\nu_1+\nu_2))^{\frac{1}{2}},$$
$$M = ((\mu_1+\mu_2)(\nu_1+\nu_2)/((\mu_1+\nu_2)(\mu_2+\nu_1)))^{\frac{1}{2}}.$$

The following remarks should be made: i) Because of (1.26) the optimum scheme (1.25) coincides in form with the one considered by Wachspress [73], who did not deal with the stationary case. Young [79], who tackles the stationary case of Wachspress, does not give the above explicit expressions. ii) For $\mu_j=\mu$ and $\nu_j=\nu$ $|j=1,2$ it is obtained that $\hat{r}_1=\hat{r}_2=\sqrt{\mu\nu}$, namely the results (1.24), where the product $\hat{\omega}\hat{r}=2\sqrt{\mu\nu}$ of (1.20) must be interpreted as the product $\hat{\omega}\hat{r}_1\hat{r}_2=2\sqrt{\mu\nu}$ of (1.25). iii) If $[\mu_1,\nu_1]\cap[\mu_2,\nu_2]\neq\emptyset$ and it is put $r_1=r_2$ then $\hat{r}_1=\hat{r}_2=((\mu_1\nu_1(\nu_2-\mu_2)+\mu_2\nu_2(\nu_1-\mu_1))/(\nu_2-\mu_2+\nu_1-\mu_1))^{\frac{1}{2}}$, which was obtained as a special case in [42] and iv) If $\mu_1=\mu_2=0$ the results (1.26) do not hold. However, "good" values for the various parameters can be found when $r_1=r_2=r$. For this let $\underline{\mu}_j$ be the positive lower bound for the positive eigenvalues of A_j $|j=1,2$, which can be found as is described in [1-2]. Let also that $\mu=\min\underline{\mu}_j$ and $\nu=\max\nu_j$. Since for the largest eigenvalue (=1) of the iteration matrix the corresponding eigenvector belongs to the nullspace of A_1+A_2 (see [19], [49], [35]) the pair $(a_1,a_2)=(0,0)$ does not have to be considered. So by letting $a_1 \in [0,\nu]$ and $a_2 \in [\mu,\nu]$ the formula of (ii) can be applied. Thus $r=\nu(\mu/(2\nu-\mu))^{\frac{1}{2}}$. The "good" ω can be obtained by the corresponding formula in [42].

Finally we mention that the biparametric problem for the Biharmonic equation considered earlier was solved in [8], where the rather unexpected result $\hat{r}_1=\hat{r}_2$ was obtained showing that the optimum biparametric scheme coincides with the optimum monoparametric one of [38].

Before we close this section we stress that the theory and the analysis so far is based on the commutativity of the A_j involved. In the noncommutative case, although the theory developed does not hold, the convergence rates do not seem to be very far from those of the commutative one (Widlund [75]).

2. SECOND ORDER STATIONARY EXTRAPOLATION

We study now the second order stationary extrapolated scheme

$$u^{(m+1)} = \omega T u^{(m)} + (1-\omega) u^{(m-1)} + \omega c \quad | m = 0,1,2,..., \tag{2.1}$$

with $u^{(-1)}, u^{(0)} \in \mathbb{C}^n$ arbitrary, which is yielded from the first order one (1.3) in an obvious way. By assuming that $|\mathrm{Re}\lambda_j|<1 \ | j=1(1)n$ de Pillis [16] found a predetermined scheme of type (2.1), which converges faster than the original (1.2). Based on his analysis Avdelas, Galanis and Hadjidimos [7] obtained a monoparametric class of schemes of de Pillis' type and succeeded in selecting the optimum one. For the latter $\hat{\omega}=2/(1+(1-\hat{M}_r^2+\hat{M}_i^2)^{\frac{1}{2}})$, where $\hat{M}_r$ and $\hat{M}_i$ are the semi-axes on the real and imaginary axes of the optimum capturing the spectrum $\sigma(T)$ ellipse centered at the origin and symmetric to the axes. The optimum ellipse is obtained after a very elaborate analysis [7] based on results by Young and Eidson [80] (see also [79]). Here we outline the way of obtaining the optimum capturing ellipse for $\sigma(T)$ which is also the optimum capturing one for the smallest convex polygon H, symmetrix wrt the axes and containing $\sigma(T)$. If H has one vertex $P(\alpha,\beta)$ in the first quadrant (one point problem) the semi-axes of the optimum capturing ellipse are the following

$$\hat{M}_r = (2\hat{\mu}/(1+\hat{\mu}^2))^{1/3}\alpha^{2/3}\ ,\ \hat{M}_i = (2\hat{\mu}/(1-\hat{\mu}^2))^{1/3}\beta^{2/3}\ , \qquad (2.2)$$

where $\hat{\mu}$ is the unique real root of

$$((1+\mu^2)/(2\mu))^{2/3}\alpha^{2/3}+((1-\mu^2)/(2\mu))^{2/3}\beta^{2/3} = 1 \qquad (2.3)$$

in (0,1).If H has two vertices $P_1(\alpha_1,\beta_1)$,$P_2(\alpha_2,\beta_2)$($\alpha_1<\alpha_2$ and $\beta_1>\beta_2$) in the first quadrant (two point problem) $\hat{\mu}$ will be given as follows [7]

a. If $\hat{M}_{r_2} < \min\{1,\tilde{M}_r\}$ then $\hat{\mu} = \hat{\mu}_2$

b. If $\tilde{M}_r < \hat{M}_{r_1}$ then $\hat{\mu} = \hat{\mu}_1$

c. If $\hat{M}_{r_1} \leq \tilde{M}_r \leq \hat{M}_{r_2}$ then $\hat{\mu} = \tilde{\mu}$,

where $\tilde{\mu}=(\tilde{M}_r+\tilde{M}_i)/(1+(1-\tilde{M}_r^2+\tilde{M}_i^2)^{\frac{1}{2}})\in(0,1)$, with

$$\tilde{M}_r = ((\alpha_2^2\beta_1^2-\alpha_1^2\beta_2^2)/(\beta_1^2-\beta_2^2))^{\frac{1}{2}},\ \tilde{M}_i = ((\alpha_2^2\beta_1^2-\alpha_1^2\beta_2^2)/(\alpha_2^2-\alpha_1^2))^{\frac{1}{2}}$$

and $\hat{M}_{r_j}$,$\hat{M}_{i_j}$ the expressions (2.2) with $\hat{\mu}_j\in(0,1)$ the root of (2.3) where (α,β) is replaced by (α_j,β_j) $|j=1,2$.If H has three or more vertices in the first quadrant the determination of $\hat{\mu}$ reduces to that of the one or two point problem ([80] and [79]).It should be added that $\hat{\mu}$ plays for (2.1) the rôle of $\rho(T)$ for (1.2) so that the smaller $\hat{\mu}$ is the faster the asymptotic convergence of (2.1) is.Also (2.1) can be compared with (1.2) by comparing the numbers $\hat{\mu}$ and $\rho(T)$.From the expression for $\hat{\omega}$ it is obvious that for $\hat{M}_r = \hat{M}_i$ (2.1) reduces to (1.2).

A special interest presents the more general problem for which $Re\lambda_j<1$(or $Re\lambda_j>1$) $|j=1(1)n$.In such a case we can apply to (1.2) a double extrapolation.Namely one of first order (1.3) with $\omega=\omega_1$,so that $|Re(1-\omega_1+\omega_1\lambda_j)|<1$ giving $\omega_1\in(0,2/(1-\min Re\lambda_j))$ if $Re\lambda_j<1$ or $\omega_1\in(2/(1-\max Re\lambda_j),0)$ if $Re\lambda_j>1$ $|j=1(1)n$,and one of second order (2.1) with T_{ω_1} in the place of T.The new scheme is

$$u^{(m+1)} = ((\omega-\omega_2)I+\omega_2 T)u^{(m)}+(1-\omega)u^{(m-1)}+\omega_2 c\ |m=0,1,2,\ldots, \qquad (2.4)$$

where $\omega_2=\omega\omega_1$.Our problem is that of finding $\hat{\omega}$ and $\hat{\omega}_1$ so that (2.4) converges asymptotically as fast as possible.This is achieved if we let ω_1 to vary in its domain.Then for each ω_1 we find $\hat{\mu}(\omega_1)$corresponding to the optimum capturing ellipse and select that for which $\hat{\mu}=\min_{\omega_1}\hat{\mu}(\omega_1)$.This problem has been solved by members of our research team and the optimum results worked out so far beat in many cases those of the nonstationary second order scheme by Manteuffel [52].Here we present two simple cases:i) Let $\lambda_j \in \mathbb{R}$ and $\lambda_j<1$(or $\lambda_j>1$) $|j=1(1)n$.Then $\hat{\omega}=2/(1+(1-\hat{M}_r^2)^{\frac{1}{2}})$,$\hat{\omega}_1=2/(2-\lambda_M-\lambda_m)$,$\hat{M}_r=(\lambda_M-\lambda_m)/|2-\lambda_M-\lambda_m|$ and $\lambda_M=\max\lambda_j$,$\lambda_m=\min\lambda_j$.The optimum scheme,which has $\hat{\mu}=\hat{M}_r/(1+(1-\hat{M}_r^2)^{\frac{1}{2}})$,was found in [16] and also in [7] (not in the sense of the optimum scheme (2.4)) and coincides asymptotically with the nonstationary second order scheme by Golub and Varga [28]. ii) Let $Re\lambda_j=\alpha\neq 1$.Then $\hat{\omega}=2/(1+(1-\hat{M}_i^2)^{\frac{1}{2}})$,$\hat{\omega}_1=1/(1-\alpha)$,$\hat{M}_i=(\rho^2(T)-\alpha)^{\frac{1}{2}}/(1-\alpha)$ and $\hat{\mu}=\hat{M}_i/(1+(1-\hat{M}_i^2)^{\frac{1}{2}})$.The corresponding scheme was found in [16] and [7] (not as an optimum one) and coincides asymptotically with the nonstationary scheme by Manteuffel [52] in the case of the one point.

In view of (i) above it can be proved that the stationary second order EADI schemes introduced in [23],studied more systematically in [6] (see also [5] and [3]) and suitably formulated so as to cover the stationary p-parametric(or monoparametric) EADI ones [10] are already the optimum in the sence of this section.Thus the optimum iteration parameters and the optimum $\omega(=\hat{\omega}_1)$ of the stationary first order EADI schemes taken together with $\hat{\omega}(=1+\hat{\ell}$ in [6]), from the formulae in (i),(where λ_M and λ_m must be interpreted as $1-\hat{f}_m$ and $1-\hat{f}_M$ (see §1.3)) produce the optimum second order ones, which coincide with the schemes of [6] as explained before.

3. FIRST AND SECOND ORDER NONSTATIONARY EXTRAPOLATION

In this last section we refer very briefly to the first and second order nonstationary extrapolated methods.

First of all Flanders and Shortley [26] introduced the use of Chebyshev polynomials in iterative methods.It was Young [78],however,who solved by means of them the minimisation problem of the spectral radius of the iteration matrix of a first order nonstationary extrapolated scheme corresponding to the splitting P=I,Q=I-A, with A Hermitian positive definite, after a predetermined number of ℓ iterations. The solution of the same problem with A having positive eigenvalues remains the same [79]. It is noted that the larger ℓ is the more sensitive to round-off errors this problem becomes, something which does not happen with second order nonstationary extrapolated schemes.

The idea of using Chebyshev polynomials with ADI methods is due to Gourlay [29-30]. Based on [30] the present author [33] showed in a special case that the set of ω's found in [78] optimises the nonstationary (wrt ω) EADI scheme for the same value of the iteration parameter r which optimises the stationary EADI one. Use of the latter was made by Fambo [25] and Iordanidis [47]. The problem of formulation and optimisation of a p-parametric (or monoparametric) nonstationary (wrt ω) EADI scheme was studied in [10]. The result was the same as in [33], namely that the set of ω's of [78] applied with the optimum iteration parameters of the stationary EADI scheme optimises the nonstationary one as well.

Golub and Varga [28] used Chebyshev polynomials to optimise the convergence rates of a second order nonstationary extrapolated scheme of type (2.1), with T Hermitian and $\rho(T)<1$, and gave an infinite sequence of ω's (see [68]). In [79] the solution to the previous problem is given under the assumption that the eigenvalues λ_j of T satisfy $\lambda_j \in \mathbb{R}, \lambda_j<1$ (or $\lambda_j>1$) $|j=1(1)n$. Manteuffel in [52] proposes a second order nonstationary method for the general case when A is real nonsymmetric and has eigenvalues with positive real parts

Finally Avdelas, Evans and Hadjidimos [4] used the idea by Gourlay [29-30] in connection with p-parametric (or monoparametric) stationary EADI schemes and discovered that the sequence of ω's of [28] used as external extrapolation parameters optimises the nonstationary EADI scheme provided that the optimum iteration parameters together with the optimum extrapolation parameter used as an internal one are those which optimise the corresponding stationary EADI scheme.

Acknowledgement

The author expresses his sincere thanks to his friends and colleagues Drs G.Avdelas,S.Galanis and A.Yeyios for helpful suggestions and constructive critisism on a first version of this paper.

References

1. AVDELAS, G. - Contribution to the Numerical Solution of the First Three Boundary Value Problems for the Equations of Laplace,Poisson and Helmholtz, Ph. D. Thesis (Greek), University of Ioannina, Ioannina, Greece, 1975.
2. AVDELAS, G. - Eigenvalue Spectrum of Some Special Matrices. J. Math. Anal. Appl., Vol. 68, pp. 1-11, 1979.
3. AVDELAS, G. - Stationary p-parametric E.A.D.I. Iterative Schemes, D. Sc. Thesis (Greek), University of Ioannina, Ioannina, Greece, 1979.
4. AVDELAS. G., EVANS, D.J. and HADJIDIMOS, A. - Second Order Chebyshev Semi-iteration in Connection with p-parametric E.A. D.I. Schemes, J.Comp.Appl.Maths., Vol. 7, pp.261-266, 1981.
5. AVDELAS, G. and GALANIS, S. - Optimum Three-level Stationary Iterative Schemes, TR 35, Department of Mathematics, University of Ioannina, Ioannina, Greece, 1980.
6. AVDELAS, G., GALANIS, S. and HADJIDIMOS, A. - Three-level Extrapolated Alternating Direction Implicit(E.A.D.I.)Methods for the Numerical Solution of Two-dimensional Second Order Elliptic Problems. J. Comp. Appl. Maths., Vol. 5, pp. 269-275, 1979.
7. AVDELAS, G., GALANIS, S. and HADJIDIMOS, A. - On the Optimization of a Class of Second Order Iterative Schemes, TR 66, Department of Mathematics, University of Ioannina, Ioannina, Greece, 1981.
8. AVDELAS, G. and HADJIDIMOS, A. - Optimum Biparametric E.A.D.I. and A.D.P. Schemes for the Numerical Solution of the Biharmonic Equation, TR 131, School of Physics and Mathematics, University of Ioannina, Ioannina, Greece, 1977.
9. AVDELAS, G. and HADJIDIMOS, A. - Optimum Biparametric E.A.D.I. and A.D.P. Schemes for the Numerical Solution of 2-dimensional Elliptic Problems. Rev. Roum. Math. Pures et Appl., Vol. XXIV, pp. 999-1012, 1979.
10. AVDELAS, G. and HADJIDIMOS, A. - On the General Problem of Formulation and Optimization of a p-parametric Extrapolated Alternating Direction Implicit Scheme. Comp. & Maths. with Appls., Vol. 5, pp. 51-57, 1979.
11. AVDELAS, G. and HADJIDIMOS, A. - Optimum Accelerated Overrelaxation in a Special Case. Math. Comp. Vol. 36, pp.183-187, 1981.
12. AVDELAS, G., HADJIDIMOS, A. and YEYIOS, A. - Some Theoretical and Computational Results Concerning the Accelerated Overrelaxation(AOR)Theory. Anal. Numér. et Théor. Approx., Vol 9, pp. 5-10, 1980.

13. AVDELAS, G., HADJIDIMOS, A. and YEYIOS, A. - An Algorithm for the Numerical Determination of the Optimum Extrapolation Parameter of a First Order Scheme, TR 55, Department of Mathematics, University of Ioannina, Ioannina, Greece, 1981.
14. CONTE, S.T. and DAMES, R.T. - An Alternating Direction Method for Solving the Biharmonic Equation. M.T.A.C., Vol.12, pp. 198-205, 1958.
15. CUTHILL, E.H. and VARGA, R.S. - A Method of Normalized Block Iteration. J. Assoc. Comp. Mach., Vol. 6, pp. 236-244, 1959.
16. De PILLIS, J. - How to Embrace your Spectrum for Faster Iterative Results. Linear Algebra Appl. Vol.34, pp.125-143, 1980.
17. De PILLIS, J. and NEUMANN, M. - Iterative Methods with k-part Splittings, IMA J. Numer.Anal., Vol. 1, pp.65-79, 1981.
18. DOUGLAS, J. - Alternating Direction Methods for Three Space Variables. Numer. Math., Vol. 4, pp. 41-63, 1962.
19. DOUGLAS, J. and PEARCY, C. - On Convergence of Alternating Direction Procedures in the Presence of Singular Operators. Numer. Math., Vol. 5, pp. 175-184, 1963.
20. DOUGLAS, J. and RACHFORD, H.H. - On the Numerical Solution of Heat Conduction Problems in Two and Three Space Variables. Trans. Amer. Math. Soc., Vol. 82, pp. 421-439, 1956.
21. D' YAKONOV, Y.G. - On the Application of Disintegrating Difference Operators. Z. Vycisl. Mat. i Mat. Fiz., Vol. 3, pp. 385-388, 1963.
22. EVANS, D.J. - The Use of Pre-conditioning in Iterative Methods for Solving Linear Equations with Symmetric Positive Definite Matrices. J. Inst. Maths. Applics., Vol. 4, pp. 295-314, 1968.
23. EVANS, D.J. and AVDELAS, G. - On Extrapolated Alternating Direction Implicit(E.A.D.I.)Methods of Second Order. BIT, Vol. 17, pp. 379-385, 1977.
24. FADDEEVA, V.N. - Computational Methods of Linear Algebra, Transl. Benster, C.D., Dover Publications Inc., New York, 1959.
25. FAMBO, F.F. - Acceleration of Alternating Direction Implicit Methods for the Numerical Solution of Elliptic Partial Differential Equations, Ph. D. Thesis, The New University of Ulster, Coleraine, Northern Ireland, 1971.
26. FLANDERS, D.A. and SHORTLEY, G. - Numerical Determination of Fundamental Models. J. Appl. Phys., Vol. 21, pp. 1326-1332, 1950.
27. GANE, C.R. and EVANS, D.J. - Alternating Direction Preconditioning Techniques, Computer Studies TR 18, Department of Computer Studies, University of Technology, Loughborough, England, 1974.
28. GOLUB, G.H. and VARGA, R.S. - Chebyshev Semi-Iterative Methods, Successive Overrelaxation Iterative Methods, and Second-Order Richardson Iterative Methods. Numer. Math., Part I and II, Vol. 3, pp. 147-168, 1961.
29. GOURLAY, A.R. - The Acceleration of the Peaceman Rachford Me-

thod by Chebyshev Polynomials. Comp. J., Vol. 10, pp. 378-382, 1968.

30. GOURLAY, A.R. - On Chebyshev Acceleration Procedures for ADI Methods. J. Inst. Maths. Applics., Vol. 6, pp. 1-11, 1970.

31. GUITTET, J. - Une Nouvelle Méthode de Directions Alternées à q Variables. J. Math. Anal. Appl., Vol. 17, pp. 199-213, 1967.

32. HADJIDIMOS, A. - Extrapolated Alternating Direction Implicit Methods for the Numerical Solution of Elliptic Partial Differential Equations, Ph. D. Thesis, University of Liverpool, England, 1968.

33. HADJIDIMOS, A. - A Note on the Determination of the Optimum Acceleration Parameter when Using Chebyshev Acceleration Procedures for the Stationary Extrapolated ADI Methods, (unpublished), 1968.

34. HADJIDIMOS, A. - Extrapolated Alternating Direction Implicit Iterative Methods. BIT, Vol 10, pp. 465-475, 1970.

35. HADJIDIMOS, A. - Optimum Extrapolated Alternating Direction Implicit Schemes in the Presence of Singular Matrices. J. Inst. Maths. Applics., Vol. 7, pp. 361-366, 1971.

36. HADJIDIMOS, A. - The Numerical Solution of a Model Problem Biharmonic Equation by Using Extrapolated Alternating Direction Implicit Methods. Numer. Math., Vol. 17, pp. 301-317, 1971.

37. HADJIDIMOS, A. - Extrapolated Alternating Direction Implicit Methods for Solving the Biharmonic Equation in Three Space Variables. Bul. Soc. Math. Grèce, Vol. 12, pp. 138-153, 1971.

38. HADJIDIMOS, A. - On Comparing Optimum Alternating Direction Preconditioning and Extrapolated Alternating Direction Implicit Schemes. J. Math. Anal. Appl., Vol. 59, pp. 573-586, 1977.

39. HADJIDIMOS, A. - Accelerated Overrelaxation Method. Math. Comp., Vol. 32, pp. 149-157, 1978.

40. HADJIDIMOS, A. - Some Basic Results on M-matrices in Connection with the Accelerated Overrelaxation(AOR) Method. Computing, Vol. 24, pp. 259-268, 1980.

41. HADJIDIMOS, A. - The Optimal Solution of the Extrapolation Problem of a First Order Scheme, TR 49, Department of Mathematics, University of Ioannina, Ioannina, Greece, 1980.

42. HADJIDIMOS, A. and IORDANIDIS, K. - Solving Laplace's Equation in a Rectangle by Alternating Direction Implicit Methods. J. Math. Anal. Appl., Vol. 48, pp. 353-367, 1974.

43. HADJIDIMOS, A. and YEYIOS, A. - The Principle of Extrapolation in Connection with the Accelerated Overrelaxation(AOR) Method. Linear Algebra Appl., Vol. 30, pp. 115-128, 1980.

44. HADJIDIMOS, A. and YEYIOS, A. - How to improve on the Convergence Rates of a First Order Scheme. Intern. J. Computer Maths (to appear).

45. HADJIDIMOS, A. and YEYIOS, A. - On Some Extensions of the Accelerated Overrelaxation(AOR) Theory. Internat. J. Math. & Math. Sci. (to appear).

46. HALMOS, P.R. - Finite Dimensional Vector Spaces, D. Van Nostrand Co., Princeton, 1958.

47. IORDANIDIS, K. - Numerical Solution of the Robin Problem by Extrapolated A.D.I. Methods. J. Inst. Maths. Applics., Vol. 12, pp. 91-96, 1973.

48. ISAACSON, E. and KELLER, H.B. - Analysis of Numerical Methods, John Wiley and Sons, New York, 1966.

49. KELLOGG, R.B. and SPANIER, J. - On Optimal Alternating Direction Parameters for Singular Matrices. Math. Comp., Vol. 19, pp. 448-452, 1965.

50. KULISCH, U. Über Reguläre Zerlegungen von Matrizen und Einige Anwendungen. Numer. Math., Vol. 11, pp. 444-449, 1968.

51. LYNCH, R.E., RICE, J.R. and THOMAS, D.H. - Tensor Product Analysis of Alternating Direction Implicit Methods. S.I.A.M., Vol. 13, pp. 995-1006, 1965.

52. MANTEUFFEL, T.A. - The Tchebychev Iteration for Nonsymmetric Linear Systems. Numer. Math., Vol. 28, pp. 307-327, 1977.

53. MAREK, I. - Frobenius Theory of Positive Operators: Comparison Theorems and Applications. SIAM J. Appl. Math., Vol.19, pp. 607-628, 1970.

54. MARTINS, M.M. - On an Accelerated Overrelaxation Iterative Method for Linear Systems with Strictly Diagonally Dominant Matrix. Math. Comp., Vol. 35, pp. 1269-1273, 1980.

55. MITCHELL, A.R. - Computational Methods in Partial Differential Equations, John Wiley & Sons Ltd, London, 1969.

56. NIETHAMMER, W. - On Different Splittings and the Associated Iterative Methods. SIAM J. Numer. Anal., Vol. 16, pp. 186-200, 1979.

57. ORTEGA, J.M. and PLEMMONS, R.J. - Extensions of the Ostrowski-Reich Theorem for SOR Iterations. Linear Algebra Appl., Vol. 28, pp. 177-191, 1979.

58. OSTROWSKI, A.M. - Über die Determinanten mit Überwiegender Hauptdiagonale. Comment. Math. Helv., Vol. 10, pp. 69-96, 1937.

59. OSTROWSKI, A.M. - On the Linear Iteration Procedures for Symmetric Matrices. Rend. Mat. et Appl., Vol. 13, pp. 140-163, 1954.

60. PEACEMAN, D.W. and RACHFORD, H.H. - The Numerical Solution of Parabolic and Elliptic Differential Equations. S.I.A.M., Vol.3, pp. 28-41, 1955.

61. PLEMMONS, R.J. - M-matrix Characterizations. I-Nonsingular M-matrices. Linear Algebra Appl., Vol. 18, pp. 175-188, 1977.

62. RICHARDSON, L.F. - The Approximate Arithmetical Solution by Finite Differences of Physical Problems Involving Equations with an Application to the Stresses in a Masonry Dam. Philos. Trans. Roy. Soc. London, Ser. A, Vol. 210, pp. 307-357, 1910.

63. SAMARSKII, A.A. and ANDREYEV, V.N. - A Difference Scheme of Higher Accuracy for an Equation of Elliptic Type in Several Space Variables. Z. Vycicl. Mat. i Mat. Fiz., Vol. 3, pp. 1006-1013, 1963.

64. SAMARSKII, A.A. and ANDREYEV, V.N. - Alternating Direction Iterational Schemes for the Numerical Solution of the Dirichlet Problem. Z. Vycisl. Mat. i Mat. Fiz., Vol. 4, pp. 1025-

1036, 1964.
65. SISLER, M. - Über ein Zweiparametriges Iterationsverfahrens. Ibid., Vol. 18, pp. 325-332, 1973.
66. SISLER, M. - Über die Optimierung eines Zweiparametrigen Iterationsverfahrens. Ibid., Vol. 20, pp. 126-142, 1975.
67. SISLER, M. Bemerkungen zur Optimierung eines Zweiparametrigen Iterationsverfahrens. Ibid., Vol. 21, pp. 213-220, 1976.
68. VARGA, R.S. - Matrix Iterative Analysis, Prentice-Hall, Englewood Cliffs, New Jersey, 1962.
69. VARGA, R.S. - Extensions of the Successive Overrelaxation Theory with Applications to Finite Element Approximations, Topics in Numerical Analysis, Ed. Miller, J.H.H., Academic Press, 1973.
70. VARGA, R.S. - On Recurring Theorems on Diagonal Dominance. Linear Algebra Appl., Vol. 13, pp. 1-9, 1976.
71. VARGA, R.S. - M-matrix Theory and Recent Results in Numerical Linear Algebra, Sparse Matrix Computations, Ed. Bunch, J.R. and Rose, D.J., Academic Press, 1976.
72. WACHSPRESS, E.L. - Optimum Alternating Direction Implicit Iteration Parameters for a Model Problem. S.I.A.M., Vol. 10, pp. 339-350, 1962.
73. WACHSPRESS, E.L. - Extended Application of Alternating Direction Implicit Iteration Model Problem. S.I.A.M., Vol. 11, pp. 994-1016, 1963.
74. WACHSPRESS, E.L. - Iterative Solution of Elliptic Systems and Applications to the Neutron Diffusion Equations of Reactor Physics, Prentice-Hall, Englewood Cliffs, New Jersey, 1966.
75. WIDLUND, O.B. - On the Rate of Convergence of an Alternating Direction Implicit Method in a Non-commutative Case. Math. Comp., Vol. 20, pp. 500-515, 1966.
76. YEYIOS, A. - On the Accelerated Overrelaxation (AOR) Method for Solving Large Linear Systems, Ph. D. Thesis, (Greek), University of Ioannina, Ioannina, Greece, 1979.
77. YEYIOS, A. - On the Accelerating Procedure of Extrapolation, Internat. J. Math. & Math. Sci. (in press).
78. YOUNG, D.M. - On Richardson's Method for Solving Linear Systems with Positive Definite Matrices. J. Math. Phys., Vol. 32, pp. 243-255, 1954.
79. YOUNG, D.M. - Iterative Solution of Large Linear Systems, Academic Press, New York, 1971.
80. YOUNG, D.M. and EIDSON, H.E. - On the Determination of the Optimum Relaxation Factor for the SOR Method when the Eigenvalues of the Jacobi Method are Complex, Report CNA-1, Center for Numerical Analysis, University of Texas, Austin, Texas, U.S.A., 1970.

FURTHER PRECONDITIONING STRATEGIES FOR THE NUMERICAL SOLUTION OF CERTAIN SPARSE LINEAR SYSTEMS

D.J. Evans & A. Benson

Department of Computer Studies, University of Technology, Loughborough, Leicestershire, U.K.

ABSTRACT

In this paper, an alternative preconditioning strategy is presented for use in the solution of large sparse linear systems possessing a periodic structure. An analysis of the optimum choice for the preconditioning parameter is presented and this is confirmed by numerical experiments for a simple example.

1. INTRODUCTION

In the solution of the system of linear algebraic equations,

$$A\underline{x} = \underline{d} \, , \tag{1.1}$$

by stationary iterative methods, the rate of convergence can be shown to depend inversely on the P-condition number of the coefficient matrix. In an expository paper, Evans (1968) showed that it was possible to minimise P by introducing a new acceleration parameter and solving instead, a system of preconditioned equations closely related to the original system. The strategy was given the name 'pre-conditioning' and has since been the subject of many research papers.

The purpose of this paper is to show that for special forms of the coefficient matrix A and appropriate choices of the preconditioning matrices, an alternative approach to the preconditioning strategy can be derived and the optimum value of the acceleration parameter obtained.

2. GENERAL THEORY OF PRECONDITIONING

If in the given system of equations (1.1), the matrix A satisfies the splitting,

$$A = I-L-U \text{ , where } U = L^T \, , \tag{2.1}$$

then by premultiplying (2.1) by $(I-\omega L)^{-1}$ and rewriting the system

as,
$$(I-\omega L)^{-1}A(I-\omega U)^{-1}(I-\omega U)\underline{x} = (I-\omega L)^{-1}\underline{d}\ , \qquad (2.2)$$

i.e.
$$B_\omega \underline{y} = \underline{d}_\omega\ , \qquad (2.3)$$

where,
$$B_\omega = (I-\omega L)^{-1}A(I-\omega U)^{-1}\ , \qquad (2.4)$$

$$\underline{y} = (I-\omega U)\underline{x} \text{ and } \underline{d}_\omega = (I-\omega L)^{-1}\underline{d}\ , \qquad (2.5)$$

we obtain an alternative or preconditioned system (2.3) to solve for the unknown vector x when $\omega>0$.

There are many iterative methods available for the solution of equation (2.3), though here only those which are particularly suitable will be considered bearing in mind that the matrix, B_ω given by equation (2.4) is of large order, symmetric, sparse and usually generated from the coefficient matrix A which is itself derived from discrete approximations to elliptic partial differential equations. Thus, of all the first order iterative methods, those which satisfy these constraints are of Simultaneous Displacement type, i.e.,

$$\underline{y}^{(n+1)} = \underline{y}^{(n)} + (\underline{d}_\omega - B_\omega \underline{y}^{(n)})\ , \qquad (2.6)$$
$$= \underline{y}^{(n)} + \underline{r}_\omega^{(n)}\ ,$$

which demonstrates that the change in each component of the transformed variable $\underline{y}$ is equal to the corresponding component of the preconditioned residual vector $\underline{r}_\omega = \underline{d}_\omega - B_\omega \underline{y}$. It can be easily seen from equation (2.6) that when the preconditioned parameter $\omega=0$, this method becomes the familiar Jacobi method.

Two forms of the above equation (2.6) in which a constant factor α or a different choice α_n for each iteration is multiplied by each component of the residual vector, and then added to each component of the present iterate $\underline{y}^{(n)}$ give,

$$\underline{y}^{(n+1)} = \underline{y}^{(n)} + \alpha \underline{r}_\omega^{(n)} = (I-\alpha B_\omega)\underline{y}^{(n)} + \alpha \underline{d}_\omega\ , \qquad (2.7)$$

and
$$\underline{y}^{(n+1)} = \underline{y}^{(n)} + \alpha_n \underline{r}_\omega^{(n)} = (I-\alpha_n B_\omega)\underline{y}^{(n)} + \alpha_n \underline{d}_\omega\ , \qquad (2.8)$$

which are preconditioned forms of the Simultaneous Displacement and Richardson's methods respectively.

If we assume that the spectrum of real eigenvalues λ_i of B_ω (which can be proved to be symmetric) is bounded by the values a_ω and b_ω such that,

$$0<a_\omega \leq \lambda_i \leq b_\omega < \infty \ , \quad (i=1,2,\ldots,N), \qquad (2.9)$$

then the criterion for convergence makes it necessary for

$$|1-\alpha\lambda_i| < 1, \qquad (2.10)$$

which gives the permissible range for values of α to be,

$$0<\alpha<2/b_\omega \ . \qquad (2.11)$$

The fastest convergence rate is obtained by choosing α so that the spectral radius of $(I-\alpha B_\omega)$ is minimized. Clearly the best choice of α is the one for which

$$1-\alpha a_\omega = -(1-\alpha b_\omega) \ ,$$

i.e.,

$$\alpha = 2/(a_\omega+b_\omega) \ . \qquad (2.12)$$

With this choice of α, the convergence factor or spectral radius $\rho(B_\omega)$, is, for both i=1 and N,

$$|1-\alpha\lambda_i| \leq \frac{b_\omega - a_\omega}{b_\omega + a_\omega} \ , \qquad (2.13)$$

Thus, if λ_i, i=1,2,...,N denote the eigenvalues of B_ω, then the optimum preconditioning parameter is that value for which,

$$P_\omega = \frac{\max|\lambda_i|}{\min|\lambda_i|} = \frac{b_\omega}{a_\omega} \ , \qquad (2.14)$$

is minimised.

We further define the asymptotic rate of convergence R_∞ by the formula,

$$R_\infty = -\log(\gamma) \ , \qquad (2.15)$$

where γ is the spectral radius of the error operator $(I-\alpha B_\omega)$ and is given by (2.13). For the method of Simultaneous Displacement, we obtain the rate of convergence from (2.13) and (2.14) given by,

$$R_\infty \simeq 2/P_\omega \ , \qquad (2.16)$$

which is clearly maximised when P_ω is minimised.

The basic iterative methods described above are usually arranged to be performed from a generated matrix rather than with the original sparse matrix stored in the computer memory. This is a relatively simple matter for the well known differential operators such as those derived from Laplace and Biharmonic equations. However, for the more general case, the non-zero matrix coefficients arising from partial differential equations with non-constant coefficients are computed initially at the onset of the solution process and pre-stored in a compact way so that the same storage efficiency can be achieved. In view of this, it is therefore important that the application of the preconditioning strategy and indeed the choice of preconditioning matrices can be achieved computationally on a generated rather than a stored matrix. Thus the most obvious choice for the preconditioning matrices is $(I-\omega L)$ and $(I-\omega U)$ as these are constructed from splittings derived from the original coefficient matrix (hence retaining sparsity) and are based on triangular matrices which are known to be easily inverted using forward and back substitution processes. (Evans, 1968).

3. AN ALTERNATIVE PRECONDITIONING STRATEGY

Consider now the solution of a simple periodic boundary value problem, e.g., the equation given by,

$$-\frac{d^2y}{dx^2} + \gamma y = \beta(x) \;, \quad \gamma > 0 \tag{3.1}$$

in the range $[0,1]$ with the boundary conditions,

$$y(0) = y(1) \;,$$
$$y'(0) = y'(1) \;,$$

which is particularly suitable for our purposes.

A further non-trivial example is given by the elliptic fourth boundary value problem (Benson & Evans, 1979).

By means of Taylor's theorem we can express d^2y/dx^2 in terms of a three-point central difference approximation on a grid of evenly spaced points plus a truncation error represented by:

$$-\left.\frac{d^2y}{dx^2}\right|_{x=x_i} = [2y(x_i)-y(x_i+h)-y(x_i-h)]/h^2 + \frac{h^2}{12}\left.\frac{d^4y}{dx^4}\right|_{x=x_i+\theta_i h} \;, \quad i=1,2,\ldots,N, \tag{3.2}$$

where the internal grid points are given by $x_i \equiv ih$, $1 \leq i \leq N$ and $h=1/N$, $|\theta_i|<1$, and we have assumed that the solution $y(x)$ is such that d^4y/dx^4 exists and is bounded $(<M_4)$ in the range $0 \leq x \leq 1$. A more accurate finite difference formula is known but (3.2) is sufficient for the purposes of our argument. In this manner, the

differential equation (3.1) can be written in the matrix form,

$$H\underline{y}(x) = \underline{d} + \underline{\tau} \quad , \tag{3.3}$$

where H is the (N×N) real matrix represented by,

$$H = \frac{1}{h^2}\begin{bmatrix} 2+\gamma h^2 & -1 & & & & -1 \\ -1 & \cdot & \cdot & & 0 & \\ & \cdot & \cdot & \cdot & & \\ & & \cdot & \cdot & \cdot & \\ & 0 & & \cdot & \cdot & \cdot \\ & & & \cdot & \cdot & -1 \\ -1 & & & & -1 & 2+\gamma h^2 \end{bmatrix} ,$$

and $\underline{y}(x)$, $\underline{d}$ and $\underline{\tau}$ are vectors given by,

$$\underline{y}(x) = \begin{bmatrix} y(x_1) \\ y(x_2) \\ \cdot \\ \cdot \\ \cdot \\ y(x_N) \end{bmatrix} , \quad \underline{d} = \begin{bmatrix} \beta(x_1) \\ \beta(x_2) \\ \cdot \\ \cdot \\ \cdot \\ \beta(x_N) \end{bmatrix} , \quad \underline{\tau} = \frac{-h^2}{12}\begin{bmatrix} y''(x_1+\theta_1 h) \\ y''(x_2+\theta_2 h) \\ \cdot \\ \cdot \\ \cdot \\ y''(x_N+\theta_N h) \end{bmatrix} .$$

In our solution process, we neglect the vector $\underline{\tau}$ and solve the system

$$H\underline{y} = \underline{d} \ , \tag{3.4a}$$

where $\underline{y}=(y_1,y_2,\ldots,y_N)^T$ and y_i is the discrete approximation to the solution $y(x_i)$ of the continuous problem given by (3.1). Further, equation (3.4a) can be simplified to the standard form,

$$(D-E-E^T)\underline{y} = \underline{d} \ , \tag{3.4b}$$

where D, E and E^T are the diagonal, lower and upper N-cyclic component matrices of H. Rewriting (3.4b) in the form,

$$D^{-\frac{1}{2}}(D-E-E^T)D^{-\frac{1}{2}}(D^{\frac{1}{2}}\underline{y}) = D^{-\frac{1}{2}}\underline{d} \ , \tag{3.5a}$$

assures us of symmetry in the final form, which is

$$A\underline{y}^* = (I-G-G^T)\underline{y}^* = (I-J)\underline{y}^* = \underline{g} \ , \tag{3.5b}$$

where J is an (N×N) real symmetric matrix $D^{-\frac{1}{2}}(E+E^T)D^{-\frac{1}{2}}$, $\underline{y}^*$ the intermediate transformation vector solution $D^{\frac{1}{2}}\underline{y}$ and $\underline{g}$ the (N×1) vector $D^{-\frac{1}{2}}\underline{d}$ given by,

$$J = \begin{bmatrix} 0 & a & & & & a \\ a & 0 & a & & & \\ & a & \cdot & \cdot & & \\ & & \cdot & \cdot & \cdot & \\ & & & \cdot & \cdot & \cdot \\ & & & & \cdot & \cdot & a \\ a & & & & a & 0 \end{bmatrix}, \quad \underline{y}^* = \frac{(2+\gamma h^2)^{\frac{1}{2}}}{h}\underline{y}, \quad \underline{g} = \frac{h}{(2+\gamma h^2)^{\frac{1}{2}}}\underline{d},$$

and

$$a = 1/(2+\gamma h^2) < 0.5. \tag{3.6}$$

Finally, since J possesses this simple circulant form, it is easy to show that for each eigenvalue μ_j given by,

$$\mu_j = \frac{2\cos(2j\pi h)}{(2+\gamma h^2)}, \quad 0 \leq j \leq N-1 \tag{3.7}$$

the corresponding eigenvector $\underline{x}^{(j)}$ with k^{th} component $x_k^{(j)}$ is defined by,

$$x_k^{(j)} = \exp(2\pi i j(k-1)h), \quad 1 \leq k \leq N, \quad i = \sqrt{-1} \tag{3.8}$$

From (3.7) it is readily seen that the spectral radius, $\rho(J)$, of the matrix J given by the expression,

$$\rho(J) \equiv \max_{0 \leq j \leq N-1} |\mu_j| \quad ,$$

satisfies the relationship,

$$\rho(J) < 1.$$

Thus, it follows from (3.5b) that the matrix A is positive definite and from (3.6), J is a non-negative matrix from which it follows immediately that all powers of J and thus A^{-1} can have only non-negative real entries. These facts form the fundamentals of the analysis to prove that the discrete solution $\underline{y}$ is close to the continuous solution $\underline{y}(x)$ if the mesh spacing h is sufficiently small, (Varga, 1962).

Now instead of considering A as possessing the splitting (2.1) we consider the alternative splitting,

$$A = I - G - G^T \quad , \tag{3.9}$$

where

$$G = \begin{bmatrix} 0 & & & & a \\ a & 0 & & & \\ & \cdot & \cdot & & \\ & & \cdot & \cdot & \\ & & & \cdot & \cdot \\ & & & a & 0 \end{bmatrix} \quad , \tag{3.10}$$

so that

$$GG^T = a^2 I \ . \tag{3.11}$$

Analogous to (2.3),(2.4),(2.5) we now have to solve the preconditioned system

$$B_1 \underline{z} = \underline{d}_1 \ ,$$

where $B_1 = (I-\omega G)^{-1} A (I-\omega G^T)^{-1}$, $\underline{z} = (I-\omega G^T)\underline{y}^*$ and $\underline{d}_1 = (I-\omega G)^{-1}\underline{g}$ (3.12)

by the iterative methods (2.7-2.8). Again, we note that B_1 is similar to the matrix T_1, where

$$T_1 = (I-\omega G^T)^{-1}(I-\omega G)^{-1} A \ .$$

i.e.,

$$T_1 = [(I-\omega G)(I-\omega G^T)]^{-1} A \ ,$$

$$= \left\{ \begin{bmatrix} 1 & & & & -\omega a \\ -\omega a & 1 & & & \\ & -\omega a & 1 & 0 & \\ & & \ddots & \ddots & \\ & 0 & & \ddots & \\ & & & -\omega a & 1 \end{bmatrix} \begin{bmatrix} 1 & -\omega a & & & \\ & 1 & -\omega a & 0 & \\ & & \ddots & \ddots & \\ & 0 & & \ddots & -\omega a \\ -\omega a & & & & 1 \end{bmatrix} \right\}^{-1} A$$

$$= \frac{1}{(1+\omega^2 a^2)} \begin{bmatrix} 1 & -c & & & & -c \\ -c & 1 & -c & & 0 & \\ & \cdot & \cdot & \cdot & & \\ & & \cdot & \cdot & \cdot & \\ & & & \cdot & \cdot & \cdot \\ & 0 & & -c & 1 & -c \\ -c & & & & -c & 1 \end{bmatrix}^{-1} A, \quad \text{where } c = \frac{\omega a}{1+\omega^2 a^2} \tag{3.13}$$

$$= \frac{1}{(1+\omega^2 a^2)} C^{-1} A.$$

From (2.14), (2.15), (2.16) we need to find the value of ω at which the P-condition number of B_1 (and hence T_1) is minimised though as we shall see, such a task now does not present a major problem with the new splitting (3.9) of the matrix A.

It is evident from (3.13) that if c=a, then C=A and

$$T_1 = \frac{1}{1+\omega^2 a^2} I , \tag{3.14}$$

that is T_1 (and consequently B_1 since they are similar) will have all its eigenvalues equal to

$$\lambda = \frac{1}{1+\omega^2 a^2} , \tag{3.15}$$

and so the P condition number would be unity (the smallest possible value). This would be the case, therefore, if c=a, i.e.,

$$\frac{\omega a}{1+\omega^2 a^2} = a,$$

so we have the result,

$$a^2\omega^2 - \omega + 1 = 0 . \tag{3.16}$$

The optimum value of ω, from (3.16) would consequently satisfy the relationship

$$\bar{\omega} = \frac{2}{1+\sqrt{1-4a^2}} , \tag{3.17}$$

and for this value, $P(\bar{\omega}) = 1,$ (3.18)

and from (3.15),(3.16),

$$\bar{\lambda} = \frac{1}{\bar{\omega}} = 0.5(1+\sqrt{1-4a^2}) , \tag{3.19}$$

which from equation (2.15) gives the rate of convergence.

In the practical application of the generalised preconditioning strategy we shall be faced with the computation of the iterative method (2.8), i.e.,

$$\underline{y}^{(n+1)} = (I-\alpha B_1)\underline{y}^{(n)} + \alpha \underline{d}_1 , \tag{3.20}$$

where B_1 and $\underline{d}_1$ are defined as in (3.12).

In the numerical application of the iterative procedure (3.20) it can be seen that it will involve the solution of linear systems of the form,

$$(I-\omega G)\underline{u} = \underline{v} \text{ and } (I-\omega G^T)\underline{u} = \underline{v} , \tag{3.21a,b}$$

where $\underline{u}$ and $\underline{v}$ are illustrative intermediate vectors in the solution process. Previously, these operations in the standard preconditioning strategy, (Evans, 1968), were triangular systems which utilised backward and forward substitition processes.

However, it can be shown that the system (3.21a), i.e.,

$$\begin{bmatrix} 1 & & & & & -\omega a \\ -\omega a & 1 & & & 0 & \\ & -\omega a & 1 & & & \\ & & \cdot & \cdot & & \\ & & & \cdot & \cdot & \\ & 0 & & & \cdot & \cdot \\ & & & & -\omega a & 1 \end{bmatrix} \begin{bmatrix} u_1 \\ u_2 \\ u_3 \\ \cdot \\ \cdot \\ \cdot \\ \cdot \\ u_N \end{bmatrix} = \begin{bmatrix} v_1 \\ v_2 \\ v_3 \\ \cdot \\ \cdot \\ \cdot \\ \cdot \\ v_N \end{bmatrix},$$

can also be obtained by stable elimination processes (Evans, 1971) which permit a simple algorithmic form, viz.

a) determine u_1 from the expression

$$u_1 = \{v_1+(\omega a)^{N-1}v_2+\ldots+(\omega a)^3 v_{N-2}+(\omega a)^2 v_{N-1}+\omega a v_N\}/[1-(\omega a)^N]. \tag{3.22a}$$

b) compute u_i from the recurrence relation

$$u_i = v_i+(\omega a)u_{i-1}, \text{ for } i=2,3,\ldots,N. \tag{3.22b}$$

Similar expressions of the form given above can be obtained for the solution of system (3.21b), i.e.,

$$u_N = [\omega a v_1+(\omega a)^2 v_2+\ldots(\omega a)^{N-1}v_{N-1}+v_N]/[1-(\omega a)^N], \tag{3.22c}$$

with each other component being given by,

$$u_i = v_i+\omega a u_{i+1}, \text{ for } i=N-1,N-2,\ldots,2,1. \tag{3.22d}$$

Both these recurrence relations and the polynomial expressions (3.22a) and (3.22c) are determined by a nesting procedure involving $\omega a<1$ ($a<\frac{1}{2}$, $\omega<2$) as a multiplier which ensures the stability of the processes against growth of rounding errors.

4. NUMERICAL RESULTS

In order to test the validity of the above hypothesis, a series of numerical experiments was carried out by taking different values of a and N and evaluating the eigenvalues of B_1, for a range of values of ω. In this case, the eigenvalues of T_1 and hence B_1 are given by,

$$\lambda_i = \frac{1+2a\cos\frac{2\pi i}{N}}{\omega^2 a^2+1+2\omega a\cos\frac{2\pi i}{N}} \quad , \ i=0,1,\ldots,N-1. \qquad (4.1)$$

The results given in Figure 1 show the maximum and minimum eigenvalue modulus plotted against ω for two representative values of a, i.e.,0.2 and 0.4 whilst in Figure 2, the values of the P condition numbers for the same values of a are shown for values of ω in the range $[0,2]$. In both cases the results are confirmed by the analysis culminating with equations (3.17)-(3.19).

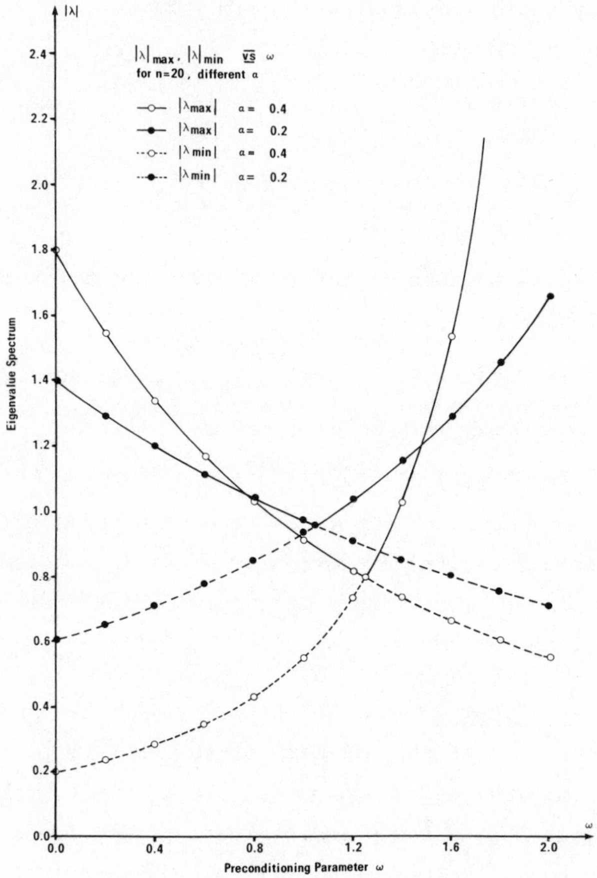

Figure 1

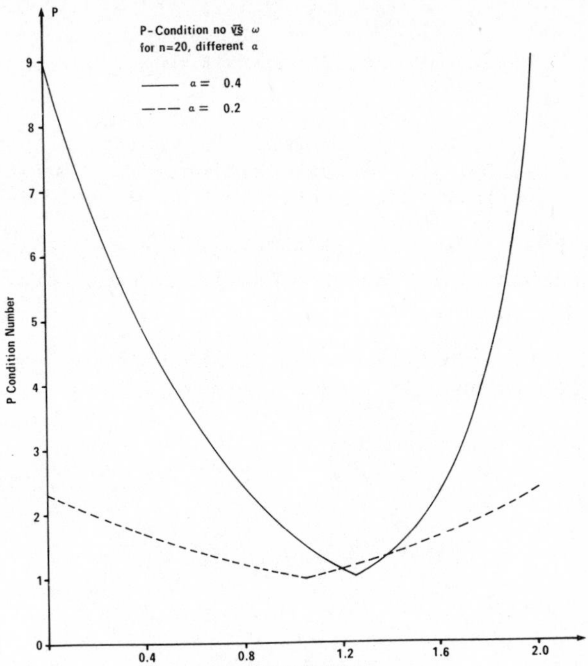

Figure 2

5. CONCLUSION

For certain periodic matrices A, defined and split according to (3.9), (3.10), an alternative preconditioning strategy has been shown to exist. In this case, there are certain important advantages over the standard preconditioning ideas of Section 2 namely,

(i) the new splitting gives superior results as the P-condition number of the preconditioned system can be reduced to the smallest value possible i.e. unity and

(ii) the optimum value of ω, at which (i) occurs, is obtained from a linear quadratic equation with simple coefficients.

REFERENCES

1. Benson, A. and Evans, D.J.- "Iterative Methods of Solution for the Elliptic Fourth Boundary Value Problems", M.C.S. 21, pp.282-288, 1979.

2. Evans, D.J. - "The Use of Preconditioning in Iterative Methods for Solving Linear Equations with Symmetric Positive Definite Matrices", J.I.M.A. 4, pp.295-314, 1968.

3. Evans, D.J. - "The Numerical Solution of the Fourth Boundary Value Problem for Parabolic Partial Differential Equations", J.I.M.A. 7, pp.61-73, 1971.

4. Varga, R.S. - "Matrix Iterative Analysis", Prentice Hall Inc., Englewood Cliffs, N.J., 1962.

ALTERNATING DIRECTION PRECONDITIONING METHODS FOR PARTIAL DIFFERENTIAL EQUATIONS

D.J. Evans & C.R. Gane†

Department of Computer Studies, University of Technology, Loughborough, Leicestershire, U.K.

ABSTRACT

In this paper, the application of preconditioning to improve the convergence rates of iterative methods for solving the large sparse linear systems of finite difference equations arising from the discretisation of elliptic partial differential equations of second and of fourth order on a rectangular grid is extended to the alternating direction implicit (A.D.I.) methods. Theoretical and experimental results are presented for both the model and non-commutative problems.

1. INTRODUCTION

This chapter is concerned with iterative solutions of linear systems of equations given in matrix form by $A\underline{u}=\underline{b}$, where A and $\underline{b}$ are known. The two iterative methods which will form the basis of this investigation are the Simultaneous Displacements method and Richardson's second degree method. These two processes are discussed in detail in Forsythe and Wasow (1960), and will be briefly described below. They are respectively defined by

$$\underline{u}^{(k+1)} = \underline{u}^{(k)}+p(\underline{b}-A\underline{u}^{(k)}) \ , \tag{1.1}$$

and

$$\underline{u}^{(k+1)} = u^{(k)}+p(\underline{b}-A\underline{u}^{(k)}+q(\underline{u}^{(k)}-\underline{u}^{(k-1)}) \ , \tag{1.2}$$

where p and q are acceleration parameters. We learn from the above reference that with the optimum choice of parameter(s) given by

---

† *Central Electricity Research Laboratories, Leatherhead, Surrey.*

$$\bar{p} = \frac{2}{a+b} \quad , \tag{1.3}$$

and $$\bar{p} = \left[\frac{2}{\sqrt{a}+\sqrt{b}}\right] \quad , \quad \bar{q} = \left[\frac{\sqrt{a}-\sqrt{b}}{\sqrt{a}+\sqrt{b}}\right]^2 \quad \text{, respectively} \tag{1.4}$$

(where a,b are the positive lower, upper bounds for the eigenvalues of the positive definite matrix A), the rates of convergence are respectively inversely proportional to P(A) and $\sqrt{P(A)}$, where P(A)=b/a is the P condition number of A.

Consequently, if some form of "preconditioning" to the original system can be effected so as to transform it into a new system whose coefficient matrix has a P-condition number which can be minimised (<<P(A)), then the above two methods will have improved rates of convergence. This idea was conceived by Evans (1968, 1974).

In the remainder of this chapter, we consider forms of preconditioning closely related to each other. The first is in connection with the iterative solution of linear systems of equations arising from the discretisation of self-adjoint elliptic partial differential equations in 2 space dimensions. The second, is applied to the iterative solution of the linear system obtained from discretising the Biharmonic equation in 2 space dimensions. In each case, a comparison of convergence rates is given between these new techniques and other well known iterative methods for solving the same problems.

2. SELF-ADJOINT ELLIPTIC PARTIAL DIFFERENTIAL EQUATIONS

We consider the solution of

$$\frac{\partial}{\partial x}(P(x,y)\frac{\partial U}{\partial x}) + \frac{\partial}{\partial y}(Q(x,y)\frac{\partial U}{\partial y}) + \sigma(x,y)U(x,y) + f(x,y) = 0,$$
$$(x,y)\in R, \tag{2.1}$$

where R is the rectangular region

$$0 \leq x \leq \ell_1, \; 0 \leq y \leq \ell_2 \; .$$

For boundary conditions, we assume that

$$U(x,y) = g(x,y) \quad , \quad (x,y)\in B \; , \tag{2.2}$$

where $g(x,y)$ is a prescribed function on the boundary B of R. We also have the conditions that P,Q,σ,f are continuous in $\bar{R}$ and satisfy

$$P(x,y)\ ,\ Q(x,y)>0\ ;\ \sigma(x,y)\leq 0,\ \text{for all}\ (x,y)\in R. \qquad (2.3)$$

We superimpose a rectangular grid (of length Δx in the x-direction and Δy in the y-direction) on R, such that

$$(n+1)\Delta x = \ell_1,\ (m+1)\Delta y = \ell_2\ . \qquad (2.4)$$

We consider the numerical solution of the problem defined in (2.1) associated with the set of mesh points on the rectangular grid. Defining the quantities $Hu(x,y)$, $Vu(x,y)$ by

$$Hu(x,y) = -a(x,y)u(x+\Delta x,y)+2b(x,y)u(x,y)-c(x,y)u(x-\Delta x,y), \qquad (2.5a)$$

$$Vu(x,y) = -\alpha(x,y)u(x,y+\Delta y)+2\beta(x,y)u(x,y)-\gamma(x,y)u(x,y-\Delta y), \qquad (2.5b)$$

where

$$\left.\begin{aligned} a &= (\Delta y/\Delta x)P(x+h/2,y), & c &= (\Delta y/\Delta x)P(x-h/2,y), & 2b &= a+c \\ \alpha &= (\Delta x/\Delta y)Q(x,y+h/2), & \gamma &= (\Delta x/\Delta y)Q(x,y-h/2), & 2\beta &= \alpha+\beta \end{aligned}\right\}. \qquad (2.6)$$

We see that they are respectively central difference approximations to $\{-\Delta x\Delta y\ \frac{\partial}{\partial x}(P(x,y)\frac{\partial U}{\partial x})\}$ and $\{-\Delta x\Delta y\ \frac{\partial}{\partial y}(Q(x,y)\frac{\partial U}{\partial y})\}$. Ordering the unknown mesh points along successive rows (x-lines) of the grid, the finite difference equations approximating (2.1) at each of these mesh points may then be written as

$$A\underline{u} = (H+V+\Sigma)\underline{u} = (H_1+V_1)\underline{u} = \underline{b}\ . \qquad (2.7)$$

where H_1 and V_1 are defined by

$$H_1 = H+\Sigma/2,\quad V_1 = V+\Sigma/2\quad . \qquad (2.8)$$

The matrices A,H_1 and V_1 have the following properties:

a) $A=H+V+\Sigma$ is an irreducible Stieltjes matrix, i.e. A is a real, symmetric and positive definite, irreducible matrix with non-positive off-diagonal entries.

b) H and V are real, symmetric, diagonally dominant matrices with positive diagonal entries and non-positive off-diagonal entries.

c) Σ is a non-negative diagonal matrix.

In order that H_1 and V_1 commute the differential equation (2.1) must be of the form

$$E_2(x)F_1(y)U(x,y)-F_1(y)\frac{\partial}{\partial x}[E_1(x)\frac{\partial U}{\partial x}]-E_2(x)\frac{\partial}{\partial y}[F_2(y)\frac{\partial U}{\partial y}]=S(x,y). \tag{2.9}$$

Hence, as this property is essential to the theoretical analysis which follows, we assume from now on that the differential equation is of the latter form.

3. THE ALTERNATING DIRECTION PRECONDITIONING TECHNIQUE

Let us firstly rewrite the coefficient matrix A as

$$A = I+(H_1-I/2) + (V_1-I/2)$$
$$= I+\hat{H}+\hat{V}\ , \tag{3.1}$$

where $\hat{H} = H_1-I/2\ ,\quad \hat{V} = V_1-I/2\ .$ (3.2)

If (2.7) is premultiplied by $(I+\omega\hat{H})^{-1}$ then it is equivalent to

$$(I+\omega\hat{H})^{-1}A(I+\omega\hat{V})^{-1}(I+\omega\hat{V})\underline{u} = (I+\omega\hat{H})^{-1}\underline{b}\ , \tag{3.3a}$$

or

$$B_\omega\underline{y} = \underline{d}, \tag{3.3b}$$

where

$$\left.\begin{aligned} B_\omega &= (I+\omega\hat{H})^{-1}A(I+\omega\hat{V})^{-1}\ ;\ \underline{y} = (I+\omega\hat{V})\underline{u} \\ \underline{d} &= (I+\omega\hat{H})^{-1}\underline{b} \end{aligned}\right\}. \tag{3.4}$$

The objective now is identical to that described for Triangular Preconditioning (Evans, 1968). It is obvious that because H_1, V_1 are symmetric and commutative then the same applies to $\hat{H},\hat{V}$. Hence B_ω is symmetric also. Because $A,\hat{H}$ and $\hat{V}$ are pairwise commutative, and all symmetric, they have a common set of orthonormal eigenvectors. Thus, from (3.4) the eigenvalues λ, of B_ω can be expressed as

$$\lambda = \frac{\mu+\nu}{(1-\omega/2+\omega\mu)(1-\omega/2+\omega\nu)}\ , \tag{3.5}$$

where μ,ν are the real, positive eigenvalues of H_1, V_1 respectively. We see from (3.5) that if $0\leq\omega\leq 2$, then $\lambda>0$ for all λ. Hence with ω in this range we can immediately state that the matrix B_ω is positive definite.

4. MINIMISATION OF P-CONDITION NUMBER OF COEFFICIENT MATRIX

As mentioned previously, the object of preconditioning the original system is to obtain a new system, viz. (3.3), whose coefficient matrix varies with a parameter ω. We then minimise the P-condition number of B_ω, denoted by $P(B_\omega)$, with respect to ω, for some "optimum" value $0\leq\omega=\bar{\omega}\leq 2$.

In order to investigate the P-condition number of B_ω, we must first examine its smallest eigenvalue given by $\min\limits_{\substack{a\leq\mu\leq b\\ \alpha\leq\nu\leq\beta}} \lambda(\mu,\nu,\omega) = \lambda_{min}(\mu,\nu,\omega)$ and then its largest eigenvalue given by $\max\limits_{\substack{a\leq\mu\leq b\\ \alpha\leq\nu\leq\beta}} \lambda(\mu,\nu,\omega) = \lambda_{max}(\mu,\nu,\omega)$ from (3.5).

We have assumed here that the eigenvalues μ of H and ν of V lie in the ranges

$$0<a\leq\mu\leq b,\ 0<\alpha\leq\nu\leq\beta. \tag{4.1}$$

To achieve these ends, let us consider the following continuous function

$$f(x,y,\omega) = \frac{x+y}{(1-\omega/2+\omega x)(1-\omega/2+\omega y)},\quad \left.\begin{matrix}0<a\leq x\leq b\\ 0<\alpha\leq y\leq\beta\end{matrix}\right\}. \tag{4.2}$$

Let us first consider the case where μ and ν (and hence x and y) lie in the same range, so that

$$a = \alpha,\ b = \beta\ . \tag{4.3}$$

(i) Minimum Eigenvalue of B_ω for $0<\omega<2$.

Now $f(x,y,\omega)$ from (4.2) can be rewritten as

$$f(x,y,\omega) = \frac{1}{\omega(2-\omega)}\left[1 - \frac{(1-\omega/2-\omega x)(1-\omega/2-\omega y)}{(1-\omega/2+\omega x)(1-\omega/2+\omega y)}\right],\quad 0<\omega<2. \tag{4.4}$$

Letting $r = \left(\frac{1-\omega/2}{\omega}\right)$ and $g(x;r) = \frac{r-x}{r+x}$, then, (4.5)

from (4.4) we have

$$f(x,y,\omega) = \frac{1}{\omega(2-\omega)}[1-g(x;r).g(y;r)],\ 0<\omega<2. \tag{4.6}$$

Hence,

$$\min_{\alpha\leq x,y\leq\beta} f(x,y,\omega) = \frac{1}{\omega(2-\omega)}[1- \max_{\alpha\leq x,y\leq\beta}|g(x;r),g(y;r)|], 0<\omega<2. \tag{4.7}$$

Since x and y lie within the same range,

$$\max_{\alpha\leq x,y\leq\beta} |g(x;r).g(y;r)| = \{\max_{\alpha\leq x\leq\beta} |g(x;r)|\}^2,\ 0<\omega<2. \tag{4.8}$$

Using results given by Varga (1963, p.215) we have

$$\max_{\alpha\leq x\leq\beta}|g(x;r)| = \begin{cases}\dfrac{\beta-r}{\beta+r}, & 0<r\leq\sqrt{\alpha\beta},\\[2ex] \dfrac{r-\alpha}{r+\alpha}, & r\geq\sqrt{\alpha\beta}.\end{cases} \tag{4.9}$$

From equations (4.7)-(4.9) we obtain

$$\min_{\alpha\leq x,y\leq\beta} f(x,y,\omega) = \begin{cases} \dfrac{1}{\omega(2-\omega)}\left[1 - \left(\dfrac{\beta-r}{\beta+r}\right)^2\right], & 0<r\leq\sqrt{\alpha\beta}, \\ \dfrac{1}{\omega(2-\omega)}\left[1 - \left(\dfrac{r-\alpha}{r+\alpha}\right)^2\right], & r\geq\sqrt{\alpha\beta}. \end{cases} \tag{4.10}$$

Substituting (4.5) in (4.10) yields

$$\min_{\alpha\leq x,y\leq\beta} f(x,y,\omega) = \begin{cases} \dfrac{1}{\omega(2-\omega)}\left[1 - \dfrac{(1-\omega/2-\omega\alpha)^2}{(1-\omega/2+\omega\alpha)^2}\right], & 0<\omega\leq\dfrac{2}{1+2\sqrt{\alpha\beta}}, \\ \dfrac{1}{\omega(2-\omega)}\left[1 - \dfrac{(1-\omega/2-\omega\beta)^2}{(1-\omega/2+\omega\beta)^2}\right], & \dfrac{2}{1+2\sqrt{\alpha\beta}}\leq\omega<2. \end{cases} \tag{4.11}$$

$$= \begin{cases} \dfrac{2\alpha}{(1-\omega/2+\omega\alpha)^2}, & 0<\omega\leq\bar{\omega}, \\ \dfrac{2\beta}{(1-\omega/2+\omega\beta)^2}, & \bar{\omega}\leq\omega<2, \end{cases} \tag{4.12}$$

where,

$$\bar{\omega} = \frac{2}{1+2\sqrt{\alpha\beta}} \quad . \tag{4.13}$$

Thus, returning to equation (3.5), we have that $\min_{\alpha\leq\mu,\nu\leq\beta}\lambda(\mu,\nu,\omega)$ is given by (4.12).

(ii) Maximum Eigenvalue of B_ω for $0<\omega<2$.

From (4.6) we have

$$\max_{\alpha\leq x,y\leq\beta} f(x,y,\omega) = \frac{1}{\omega(2-\omega)}\,[1 + M], \tag{4.14}$$

where
$$M = \max_{\alpha\leq x,y\leq\beta}\{-g(x;r).g(y;r)\} . \tag{4.15}$$

(a) For $r\geq\beta$, $g(x;r)$, $g(y;r)\geq 0$, for all $\alpha\leq x,y\leq\beta$. Hence, for r in this range

$$M = -\{\min_{\alpha\leq x\leq\beta} g(x;r)\}\{\min_{\alpha\leq y\leq\beta} g(y;r)\}$$
$$= -\{\min_{\alpha\leq x\leq\beta} g(x;r)\}^2 . \tag{4.16}$$

Clearly, for $r\geq\beta$

$$\min_{\alpha\leq x\leq\beta} g(x;r) = g(\beta;r) , \tag{4.17}$$

since the derivative of $g(x;r)$ with respect to x is negative for all $x\geq 0$. Hence, for $r\geq\beta$ or equivalently $0<\omega\leq\frac{2}{1+2\beta}$ (from (4.5)),

$$\max_{\alpha\leq\mu,\nu\leq\beta} \lambda(\mu,\nu,\omega) = \max_{\alpha\leq x,y\leq\beta} f(x,y,\omega) = f(\beta,\beta,\omega) . \tag{4.18}$$

(b) For $\alpha \leq r \leq \beta$, $g(x;r)$ and $g(y;r)$ will take on positive and negative values for $0<\alpha\leq x,y\leq\beta$. Hence

$$M = \{-\min_{\alpha\leq x\leq\beta} g(x;r)\}\{\max_{\alpha\leq y\leq\beta} g(y;r)\} = \{\max_{\alpha\leq x\leq\beta} g(x;r)\}\{-\min_{\alpha\leq y\leq\beta} g(y;r)\}. \tag{4.19}$$

Clearly for $\alpha\leq r\leq\beta$,

$$\min_{\alpha\leq x\leq\beta} g(x;r) = g(\beta;r); \quad \max_{\alpha\leq y\leq\beta} g(y;r) = g(\alpha;r) . \tag{4.20}$$

Since x and y are in the same range, their roles can be reversed in (4.20). Thus, for $\alpha\leq r\leq\beta$, or equivalently $\frac{2}{1+2\beta}\leq\omega\leq\frac{2}{1+2\alpha}$, (from (4.5)),

$$\max_{\alpha\leq\mu,\nu\leq\beta} \lambda(\mu,\nu,\omega) = \max_{\alpha\leq x,y\leq\beta} f(x,y,\omega) = f(\beta,\alpha,\omega) = f(\alpha,\beta,\omega). \tag{4.21}$$

(c) Lastly, for $0<r\leq\alpha$, both $g(x;r)$ and $g(y;r)$ will be non-positive for all $\alpha\leq x,y\leq\beta$. Hence

$$\begin{aligned} M &= \{-\max_{\alpha\leq x\leq\beta} g(x;r)\}\{\max_{\alpha\leq y\leq\beta} g(y;r)\} \\ &= -\{\max_{\alpha\leq x\leq\beta} g(x;r)\}^2 . \end{aligned} \tag{4.22}$$

Now for $0<r\leq\alpha$,

$$\max_{\alpha\leq x\leq\beta} g(x;r) = g(\alpha;r) . \tag{4.23}$$

Therefore, for $0<r\leq\alpha$ or equivalently $\frac{2}{1+2\alpha}\leq\omega<2$,

$$\max_{\alpha\leq\mu,\nu\leq\beta} \lambda(\mu,\nu,\omega) = \max_{\alpha\leq x,y\leq\beta} f(x,y,\omega) = f(\alpha,\alpha,\omega) . \tag{4.24}$$

Collecting together equations (4.18), (4.21) and (4.24) and using (3.5) yields

$$\max_{\alpha\leq\mu,\nu\leq\beta} \lambda(\mu,\nu,\omega) = \max_{\alpha\leq x,y\leq\beta} f(x,y,\omega) = \begin{cases} \lambda(\beta,\beta,\omega), & 0<\omega\leq\frac{2}{1+2\beta}, \\ \lambda(\alpha,\beta,\omega), & \frac{2}{1+2\beta}\leq\omega\leq\frac{2}{1+2\alpha}, \\ \lambda(\alpha,\alpha,\omega), & \frac{2}{1+2\alpha}\leq\omega<2. \end{cases} \tag{4.25}$$

(iii) The P-Condition Number of B_ω for $0<\omega<2$.

It now follows from equations (4.12) and (4.25) that we must examine $P(B_\omega)$ in four successive intervals namely, $0<\omega\leq 2/(1+2\beta)$, $2/(1+2\beta)\leq\omega\leq 2/(1+2\sqrt{\alpha\beta})$, $2/(1+2\sqrt{\alpha\beta})\leq\omega\leq 2/(1+2\alpha)$ and $2/(1+2\alpha)\leq\omega<2$. (N.B. $0<\alpha<\sqrt{\alpha\beta}<\beta$ since $\alpha<\beta$). Since B_ω $(0\leq\omega\leq 2)$ is positive definite, then $P(B_\omega)>0$ for $0\leq\omega\leq 2$.

(a) When $0<\omega\leq 2/(1+2\beta)$, we have from (4.12) and (4.25)

$$P(B_\omega) = \frac{\beta(1-\omega/2+\omega\alpha)^2}{\alpha(1-\omega/2+\omega\beta)^2} \quad . \tag{4.26}$$

Therefore, in this range,

$$\frac{dP}{d\omega} = \frac{2\beta}{\alpha}(\alpha-\beta)\frac{(1-\omega/2+\omega\alpha)}{(1-\omega/2+\omega\beta)^3} < 0, \text{ since } \beta>\alpha. \tag{4.27}$$

Hence, for $0<\omega\leq 2/(1+2\beta)$,

$$\min_\omega P(B_\omega) = P(B_{\omega=2/(1+2\beta)}) \ . \tag{4.28}$$

(b) When $2/(1+2\beta)\leq\omega\leq 2/(1+2\sqrt{\alpha\beta})$, equations (4.12) and (4.25) yield

$$P(B_\omega) = \frac{(\alpha+\beta).(1-\omega/2+\omega\alpha)}{2\alpha(1-\omega/2+\omega\beta)} \ . \tag{4.29}$$

Thus, in this range

$$\frac{dP}{d\omega} = \frac{(\alpha^2-\beta^2)}{2\alpha(1-\omega/2+\omega\beta)^2} < 0, \text{ since } \alpha<\beta. \tag{4.30}$$

Hence, for $2/(1+2\beta)\leq\omega\leq 2/(1+2\sqrt{\alpha\beta})$,

$$\min_\omega P(B_\omega) = P(B_{\omega=2/(1+2\sqrt{\alpha\beta})}) \ . \tag{4.31}$$

(c) When $2/(1+2\sqrt{\alpha\beta})\leq\omega\leq 2/(1+2\alpha)$ we obtain from (4.12) and (4.25),

$$P(B_\omega) = \frac{(\alpha+\beta).(1-\omega/2+\omega\beta)}{2\beta(1-\omega/2+\omega\alpha)} \quad . \tag{4.32}$$

So in this inverval,

$$\frac{dP}{d\omega} = \frac{-(\alpha^2-\beta^2)}{2\beta(1-\omega/2+\omega\alpha)^2} > 0, \text{ since } \alpha<\beta \quad . \tag{4.33}$$

Therefore, when $2/(1+2\sqrt{\alpha\beta})\leq\omega\leq 2/(1+2\alpha)$,

$$\min_\omega P(B_\omega) = P(B_{\omega=2/(1+2\sqrt{\alpha\beta})}) \ . \tag{4.34}$$

(d) Lastly, when $2/(1+2\alpha)\leq\omega<2$, we have from (4.12) and (4.25),

$$P(B_\omega) = \frac{\alpha(1-\omega/2+\omega\beta)^2}{\beta(1-\omega/2+\omega\alpha)^2} \ .$$

In this interval

$$\frac{dP}{d\omega} = \frac{2\alpha(\beta-\alpha)(1-\omega/2+\omega\beta)}{\beta(1-\omega/2+\omega\alpha)^3} > 0. \text{ since } \beta>\alpha. \tag{4.35}$$

Hence, for $2/(1+2\alpha)\leq\omega<2$,

$$\min_\omega P(B_\omega) = P(B_{\omega=2/(1+2\alpha)}) \ . \tag{4.36}$$

It is now apparent from equations (4.28), (4.31), (4.34) and (4.36) that

$$\min_{0<\omega<2} P(B_\omega) = P(B_{\bar{\omega}}) = \frac{\alpha+\beta}{2\sqrt{\alpha\beta}} , \tag{4.37}$$

where $\bar{\omega}=2/(1+2\sqrt{\alpha\beta})$, and $P(B_{\bar{\omega}})$ is given by (4.29) or (4.32). We shall refer to $\bar{\omega}$ as the optimum preconditioning parameter.

(iv) Case When Eigenvalue Ranges of H_1 and V_1 are Different

Up to this point, we have assumed that $0<a\leq\mu\leq b$ and $0<\alpha\leq\nu\leq\beta$ are such that $a=\alpha$ and $b=\beta$. This will not always occur in practice, e.g. when solving Laplace's equation on a rectangle with adjacent sides of different length, using a square mesh.

As far as the minimum eigenvalue of B_ω is concerned an approximate analysis can be achieved by noticing that $0<\alpha'\leq\mu,\nu\leq\beta'$, where $\alpha'=\min(a,\alpha)$ and $\beta'=\max(b,\beta)$. Then, $\min_{\substack{a\leq\mu\leq b\\ \alpha\leq\nu\leq\beta}} \lambda(\mu,\nu,\omega) \geq \min_{\alpha'\leq\mu,\nu\leq\beta'} \lambda(\mu,\nu,\omega)$ and the value of the function on the right can be determined from 4(i) for $0<\omega<2$.

Next, we have

$$\max_{\substack{a\leq\mu\leq b\\ \alpha\leq\nu\leq\beta}} \lambda(\mu,\nu,\omega) \leq \max_{\alpha'\leq\mu,\nu\leq\beta'} \lambda(\mu,\nu,\omega) ,$$

and the function on the right for $0<\omega<2$ can be evaluated from 4(ii). Hence from subsections 4(i), 4(ii), we obtain

$$\min_{0<\omega<2} P(B_\omega) \leq \min_{0<\omega<2} \left\{ \frac{\max_{\alpha'\leq\mu,\nu\leq\beta'} \lambda(\mu,\nu,\omega)}{\min_{\alpha'\leq\mu,\nu\leq\beta'} \lambda(\mu,\nu,\omega)} \right\} = P(B_{\bar{\omega}'}),$$

where $\bar{\omega}'=2/(1+2\sqrt{\alpha'\beta'})$.

5. APPLICATION OF THE PRECONDITIONING TECHNIQUE TO THE BASIC ITERATIVE METHODS

We will now apply the results of the previous section to modify the two original basic iterative methods given by equations (1.1) and (1.2), i.e. the Simultaneous Displacement and Richardson's second degree methods, respectively, It follows immediately that if the techniques used in Section 4 are shown to apply to the above two methods, then an improvement in convergence rate must follow.

We now proceed from equation (3.3) and develop similar iterative processes to the above two methods by working throughout in the transformed variable y. From equation (3.3) then, equations describing iterative processes similar to (1.1) and (1.2) are

$$\underline{y}^{(k+1)} = \underline{y}^{(k)} + \bar{\alpha}(\underline{d}-B_{\bar{\omega}}\underline{y}^{(k)}) , \tag{5.1}$$

and $$\underline{y}^{(k+1)} = \underline{y}^{(k)} + \bar{\alpha}(\underline{d}-B_{\bar{\omega}}\underline{y}^{(k)}) + \bar{\beta}(\underline{y}^{(k)}-\underline{y}^{(k-1)}) , \tag{5.2}$$

respectively. The iterations proceed in the y variable until a specified degree of accuracy is achieved. The final solution is then obtained by one application of the formula,

$$\underline{u} = (I+\bar{\omega}\hat{V})^{-1}\underline{y} . \tag{5.3}$$

The optimum value of $\alpha=\bar{\alpha}$ for method (5.1) is given from (1.3) by

$$\bar{\alpha} = \frac{2}{\bar{a}+\bar{b}} , \tag{5.4}$$

where $0<\bar{a}\leqslant\lambda_i(\bar{\omega})\leqslant\bar{b}$, and the $\lambda_i(\bar{\omega})$ are the positive real eigenvalues of $B_{\bar{\omega}}$. With this choice of α, the spectral radius of the iteration matrix $(I-\bar{\alpha}B_{\bar{\omega}})$ is

$$\max_i |1-\bar{\alpha}\lambda_i(\bar{\omega})| \leqslant \frac{\bar{b}-\bar{a}}{\bar{b}+\bar{a}} = \frac{P(B_{\bar{\omega}})-1}{P(B_{\bar{\omega}})+1} . \tag{5.5}$$

The values $\bar{a} = \min_{\alpha\leqslant\mu,\nu\leqslant\beta} \lambda(\mu,\nu,\bar{\omega})$ and $\bar{b} = \max_{\alpha\leqslant\mu,\nu\leqslant\beta} \lambda(\mu,\nu,\bar{\omega})$ are given by equations (4.12) and (4.21) respectively. From (5.5) the rate of convergence is

$$R(I-\alpha B_{\bar{\omega}}) = -\log\left\{\frac{P(B_{\bar{\omega}})-1}{P(B_{\bar{\omega}})+1}\right\} . \tag{5.6}$$

Similarly, for equation (5.2) (preconditioned Richardson's second degree method) we have from (1.4),

$$\bar{\alpha} = \left[\frac{2}{\sqrt{\bar{a}}+\sqrt{\bar{b}}}\right]^2 \quad\text{and}\quad \bar{\beta} = \left[\frac{\sqrt{\bar{a}}-\sqrt{\bar{b}}}{\sqrt{\bar{a}}+\sqrt{\bar{b}}}\right]^2 , \tag{5.7}$$

and the spectral radius of the iteration matrix is

$$\sqrt{\bar{\beta}} , \tag{5.8}$$

so that the rate of convergence satisfies,

$$-\log(\sqrt{\beta}) = -\log\left\{\frac{\sqrt{P(B_{\bar{\omega}})}-1}{\sqrt{P(B_{\bar{\omega}})}+1}\right\} . \tag{5.9}$$

There are other methods to which we could apply this new preconditioning technique such as the Chebyshev acceleration of (5.2) which with $\bar{\alpha}_k, \bar{\beta}_k$ replacing $\bar{\alpha}, \bar{\beta}$ respectively, is defined by that equation. This is a non-stationary method, since the acceleration parameters $\bar{\alpha}_k, \bar{\beta}_k$ vary at each iteration. Chebyshev polynomials are used to optimise $\bar{\alpha}_k, \bar{\beta}_k$ with the result that we again have an asymptotic rate of convergence equal to $2/\sqrt{P(B_{\bar{\omega}})}$. However, in what is to follow, we restrict our attention to methods (5.1) and (5.2).

The Computational Procedure

We will consider the computational procedure for method (5.1) only, since it is analogous for method (5.2). Let us assume that we have ordered our grid points horizontally along successive x-lines of the grid. We also assume that we have first obtained our optimum parameters $\bar{\omega}, \bar{\alpha}$ from (4.13), (5.4) respectively. Initially, from (3.4) we have

$$(I+\omega\hat{H})\underline{d} = \underline{b} \ , \quad (\underline{b} \text{ known}) \ , \tag{5.10}$$

since we have ordered the mesh points horizontally, we see from (2.5a) that $\hat{H}=H_1-I/2=(H+\Sigma/2)-I/2$ is tridiagonal. Thus, since $(I+\omega\hat{H})$ is positive definite $(0<\omega<2)$ we can solve (5.10) directly for $\underline{d}$ using the Gaussian elimination (Thomas) algorithm. Next, from (5.1) and (3.4) we see that at each iteration we must evaluate

$$(I+\omega\hat{H})^{-1}A(I+\omega\hat{V})^{-1}\underline{y}^{(k)} \ , \ k\geq 0 \ . \tag{5.11}$$

Let $\underline{x}^{(k)}=(I+\omega\hat{V})^{-1}\underline{y}^{(k)}$, i.e. $(I+\omega\hat{V})\underline{x}^{(k)}=\underline{y}^{(k)}$. Because of the present ordering of the unknown mesh points, we see from (2.5b) that $\hat{V}=V_1-I/2=(V+\Sigma/2)-I/2$ is a block tridiagonal matrix, where each block element is a diagonal matrix. By re-ordering the elements of the vector $\underline{y}^{(k)}$ and thus $\underline{x}^{(k)}$ also, to correspond with a vertical ordering of the grid points along successive y-lines, a new system is derived, namely,

$$(I+\omega\hat{V}_T)\underline{x}_T^{(k)} = \underline{y}_T^{(k)} \quad . \tag{5.12}$$

From (2.5b) we see that $\hat{V}_T$ is a block diagonal matrix, each block being a tridiagonal matrix, whose size corresponds to the number of unknown mesh points on each y-line. Since for $0<\omega<2$, $(I+\omega\hat{V}_T)$ is positive definite (and diagonally dominant), we can solve (5.12) directly for $\underline{x}_T^{(k)}$ using the Thomas algorithm. Next we must solve

$$\underline{z}^{(k)} = A\underline{x}^{(k)}, \; k\geq 0 \; . \tag{5.13}$$

The vector $\underline{z}^{(k)}$ can be generated, by remembering that the matrix A of (2.7) was formed from applying the five point difference approximation of (2.1) at each unknown grid point with a horizontal ordering. Thus (5.13) can be evaluated by applying the same difference analogue at each mesh point in the same order with $\underline{x}^{(k)}$ (not $\underline{x}_T^{(k)}$) replacing $\underline{u}$ (see (2.7)). Finally, we must solve

$$\underline{q}^{(k)} = (I+\omega\hat{H})^{-1}\underline{z}^{(k)}, \text{ i.e. } (I+\omega\hat{H})\underline{q}^{(k)} = \underline{z}^{(k)}, \; k\geq 0. \tag{5.14}$$

System (5.14) is solved in exactly the same way as system (5.10), to obtain the vector $\underline{q}^{(k)}$ which is the value(s) of the expression in (5.11).

Method (5.1) can now be written as

$$\underline{y}^{(k+1)} = \underline{y}^{(k)} + \bar{\alpha}(\underline{d}-\underline{q}^{(k)}) \; , \; k\geq 0 \; , \tag{5.15}$$

which is easily solved for $\underline{y}^{(k+1)}$, $k\geq 0$ since each vector on the right hand side is known. This process continues until convergence (to the desired accuracy) is achieved. The original solution vector $\underline{u}$ is then found from (5.3) in the same way as system (5.12) was derived and then solved for. We now see the reason for calling this an "Alternating Direction Preconditioning (A.D.P.) technique".

6. THE MODEL PROBLEM: COMPARATIVE RATES OF CONVERGENCE

The model problem we are considering is the Laplace equation in the unit square with Dirichlet boundary conditions. For this problem we have $\Sigma=0$ in (2.7). Also, dividing each difference equation by 4, and denoting the new coefficient matrix again by A, we have

$$A = H+V \quad (=H_1+V_1) \ , \tag{6.1}$$

where

$$H=(H_1)=\begin{bmatrix} B & & \\ & B & 0 \\ & & \ddots & \\ 0 & & & B \end{bmatrix}_{(n^2\times n^2)} , \quad B=\begin{bmatrix} 1/2 & -1/4 & & 0 \\ -1/4 & \ddots & \ddots & \\ & \ddots & \ddots & -1/4 \\ 0 & & -1/4 & 1/2 \end{bmatrix}_{(n\times n)} , \tag{6.2}$$

and

$$V=(V_1)=\begin{bmatrix} I/2 & -I/4 & & 0 \\ -I/4 & I/2 & \ddots & \\ & \ddots & \ddots & -I/4 \\ 0 & & -I/4 & I/2 \end{bmatrix}_{(n^2\times n^2)} , \quad \text{I is the unit matrix of order n.} \tag{6.3}$$

Here we have assumed a square mesh size of $h=1/(n+1)$ so that there are n^2 internal mesh points. If again μ,ν denote the eigen-values of $H(=H_1),V(=V_1)$, respectively, then,

$$\mu_i = \sin^2\{i\pi/(2(n+1))\}, \ \nu_j = \sin^2\{j\pi/(2(n+1))\}, \ 1\leq i,j\leq n. \tag{6.4}$$

Hence, if $a\leq\mu_i\leq b, \alpha\leq\nu_j\leq\beta, 1\leq i,j\leq n$, then

$$a = \alpha = \sin^2\left(\frac{\pi}{2(n+1)}\right), \ b = \beta = \cos^2\left(\frac{\pi}{2(n+1)}\right) \ . \tag{6.5}$$

Thus, for the model problem, equation (4.13) yields,

$$\bar{\omega} = \frac{2}{1+\sin(\pi/(n+1))} = \frac{2}{1+\sin(\pi h)} \ . \tag{6.6}$$

Substituting the values of α,β from (6.3) in equation (4.37) yields

$$\min_{0<\omega<2} P(B_\omega) = P(B_{\bar{\omega}}) = \frac{1}{\sin(\pi/(n+1))} = \frac{1}{\sin(\pi h)} \ . \tag{6.7}$$

Using the optimum acceleration parameter $\bar{\alpha}$ defined by (5.4), ($\bar{a}$ and $\bar{b}$ defined by (4.12) and (4.21) for $\omega=\bar{\omega}$) from (5.5) the Alternating Direction Preconditioned (A.D.P.) Simultaneous Displacement method has a spectral radius given by

$$\max_{1\leq i,j\leq n} \left|1-\bar{\alpha}\lambda_{i,j}(\bar{\omega})\right| = \frac{1-\sin(\pi h)}{1+\sin(\pi h)} \ . \tag{6.8}$$

Thus, we have that

$$\min_{\omega,\alpha} \rho(\text{A.D.P. Simultaneous Displacements})$$

$$= \min_{\omega} \rho(\text{point S.O.R.}) = \min_{r>0} \ (\text{one parameter P.R. (A.D.I.)}) . \quad (\text{Varga,1963}) \qquad (6.9)$$

Hence, as optimised one parameter iterative methods, the Peaceman-Rachford method, the point S.O.R. process and the A.D.P. Simultaneous Displacements method have identical asymptotic rates of convergence for all $h>0$ for the model problem. In fact, in each case from (6.8) we have,

$$R_\infty \approx \frac{2\pi}{n+1} = 2\pi h \ , \text{ as } h\to 0 \ . \qquad (6.10)$$

We also notice in this context, that the optimum preconditioning parameter $\bar{\omega}$ given by (6.6) is identical to the optimum over-relaxation parameter for the point S.O.R. method (Young, 1954).

In view of equation (5.9), improved rates of convergence can be obtained by using the Alternating Direction Preconditioned second degree Richardson method defined by (5.2). Thus, using the optimum preconditioning factor $\bar{\omega}$ from (6.6) and the optimum parameters $\bar{\alpha}$ and $\bar{\beta}$ in (5.7) we have from (5.9)

$$R_\infty \ (\text{A.D.P. second degree Richardson method})$$
$$= 2\sqrt{\sin(\pi h)} \approx 2\sqrt{\pi h} = 2\sqrt{\frac{\pi}{n+1}} \quad \text{as } n\to\infty . \qquad (6.11)$$

We see (Varga, 1963) that this quantity is identical to the asymptotic average rate of convergence of the 2-parameter P.R. (A.D.I.) iterative method (where we are using the $M=2^k$ $(k\geq 0)$ type parameters derived by Wachpress (1962)). Further, for larger n, increased asymptotic average rates of convergence can be obtained by using $M>>2$ parameters.

Hence for the model problem our two new preconditioned methods (5.1) and (5.2) would appear to offer similar gains in of convergence as existing methods such as the $M=2^k$-parameter P.R. (A.D.I.) iterative procedure.

7. NUMERICAL RESULTS

The two problems we solved were taken from Young and Ehrlich (1960). Both involved the approximate solution of

Laplace's equation. The A.D.P. Simultaneous Displacement and A.D.P. second degree Richardson methods given by (5.1) and (5.2) were compared with the point S.O.R. method and the Peaceman Rachford procedure using a fixed number M(=1,2 and 3) of parameters (the results for the last two methods being taken from the previously mentioned reference). By "fixed" number of parameters, we mean that M is independent of h, while the values of the parameters themselves vary with h. Varga (1962), p.229 shows that with M>1, this technique converges faster than the point S.O.R. method with optimum ω, when h^{-1} is large (i.e. for the model problem). It should be noted that these type of parameters will tend to give slower convergence than the Wachpress ($M=2^k$) parameters, since the former were obtained using a much cruder analysis than the latter (e.g. see Varga (1962), pp.226-229).

The first region chosen was the unit square. The boundary conditions were,

$$U(x,1) = 1, \; 0\leqslant x\leqslant 1 \;, \; U(0,y) = U(1,y) = U(x,0) = 0, \quad 0\leqslant x,y\leqslant 1. \tag{7.1}$$

For the former two methods mentioned above h^{-1} took the values 5(5)30, whereas the latter two had h^{-1}=5,10,20 and 40. The convergence criteria were

$$\max_{1\leqslant i,j\leqslant n} |y_{i,j}^{(k+1)}-y_{i,j}^{(k)}| < 10^{-6} \;. \tag{7.2a}$$

for the former two methods, and

$$\max_{1\leqslant i,j\leqslant n} |u_{i,j}^{(k+1)}-u_{i,j}^{(k)}| < 10^{-6}, \quad k\geqslant 0, \tag{7.2b}$$

for the latter two. The respective iterative processes started with $y_{i,j}^{(0)},u_{i,j}^{(0)}=0$, for all $1\leqslant i,j\leqslant n$. The results are displayed in Figure 1, the logarithm of N, the number of iterations, being plotted against $\log(h^{-1})$. They bear out the theoretical comparison made between the rates of convergence of the optimised one parameter A.D.P. Simultaneous Displacements, point S.O.R. and P.R. (A.D.I.) methods. The Figure also shows the expected increase in rate of convergence over the previous three methods just mentioned, of the A.D.P. second degree Richardson method.

The second region chosen was the unit square with a 0.4×0.4 square removed from the centre. For the first two methods mentioned at the beginning of the section, h^{-1}=10,20,30,40 and for the second pair, h^{-1}=10,20,40. The boundary conditions were

$$U(x,0) = U(1,y) = U(x,1) = U(0,y) \;, \; 0 \leq x,y \leq 1,$$
$$U(x,0.3) = U(0.7,y) = U(x,0.7) = U(0.3,y) = 1; 0.3 \leq x,y \leq 0.7. \tag{7.3}$$

The results, using the same axes as before, are given in Figure 2. For this problem, the matrices H and V and therefore $\hat{H}$ and $\hat{V}$ do <u>not</u> commute, hence the coefficient matrix B_ω of the preconditioned system (3.3) is unsymmetric, and thus could possess complex eigenvalues. A theoretical proof of convergence and theoretical treatment of optimum parameters and rates of convergence has not yet been devised for the "non-commutative case". In this respect, we have inherent difficulties analogous to those of the "non-commutative case" of the P.R. (A.D.I.) iterative method.

The above problem is intended to serve as an example of the viability of our preconditioned iterative methods in practice, when "model problem" conditions do not prevail. Emulating Young and Ehrlich (1960), it was decided in <u>all</u> methods, to use the respective "optimum" parameters obtained for the first problem i.e. on the unit square. Whilst, from Figure 2, the rates of convergence are seen to be worse for the P.R. (A.D.I.) method (M=1,2,3) as compared to those given for the unit square (see Figure 1), the two preconditioned methods give almost identical results on the two regions in question.

Lastly, we compared the P-condition numbers of the A.D.P. and Triangular Preconditioning techniques for the model problem. The latter procedure has been outlined in Evans (1968). The results are tabulated in Table 1, the T.P. results being taken from Evans (1968).

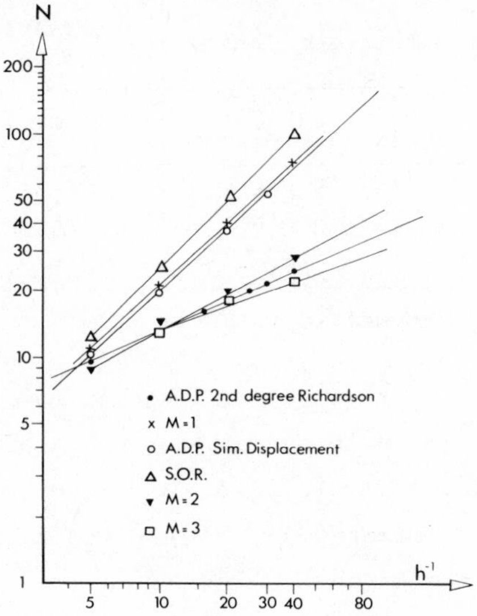

Figure 1

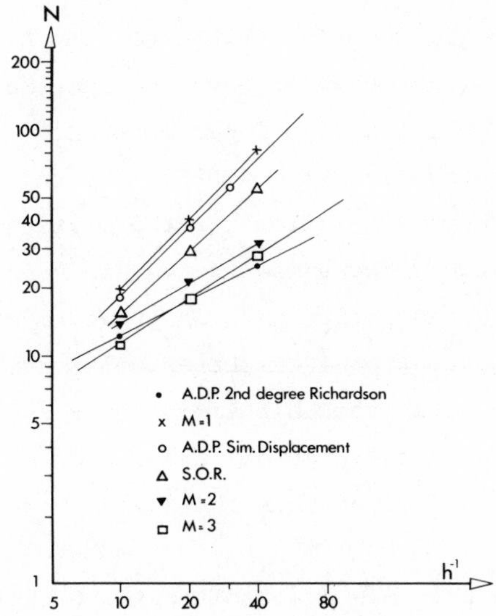

Figure 2

h^{-1}	Triangular Preconditioning		Alternating Direction Preconditioning	
	$\bar{\omega}$	$P(B_{\bar{\omega}})$	$\bar{\omega}$	$P(B_{\bar{\omega}})$
5	1.3	1.655	1.26	1.702
10	1.6	2.816	1.53	3.250
20	1.75	5.274	1.73	6.405

TABLE 1: P-condition numbers

We see that for $h^{-1}=5,10,20$, Triangular Preconditioning (T.P.) have smaller P-condition numbers than A.D.P. Thus, in view of equations (5.6) and (5.9) we can expect faster rates of convergence from the former technique applied to the two iterative methods discussed in section 5.

8. CONCLUSIONS

For the model problem, we have shown that the A.D.P. Simultaneous Displacement method has identical rates of convergence with the optimised point S.O.R. and one parameter P.R. (A.D.I.) iterative procedures. Likewise, the A.D.P. second order Richardson process was shown to have the same rate of convergence (for the model problem) as the 2 parameter P.R. (A.D.I.) using $M=2^k$ type parameters (i.e. k=1). Since for the model problem, the amount of work per iteration for the latter two methods can be shown to be roughly the same, the equality in rates of convergence of the respective methods above, must very nearly apply to total computational effort as well. Hence methods (5.1) and (5.2) provide no appreciable improvement over existing procedures (for the model problem) and indeed as $h \to 0$, could be considerably less efficient than the $M(>>2)$-parameter P.R. (A.D.I.) process.

An advantage this new preconditioning technique has is that we have been able to develop an exact theoretical treatment of rates of convergence and optimum parameters, for the commutative case. Only for one case has this been possible for Triangular Preconditioning. However, provided the coefficient matrix A is positive definite, then for T.P., B_ω is positive definite also, and our two preconditioned methods (5.1) and (5.2) (using T.P.) will therefore converge, by (5.5) and (5.8). For A.D.P. this result has only been proved when $\hat{H}$ and $\hat{V}$ commute. If they do not, the matrix B_ω (for A.D.P.) is not symmetric, and therefore could possess complex eigenvalues. Also in the non-commutative case, a theoretical analysis of the eigenvalues of B_ω seems improbable. Nevertheless the problem relating to Figure 2 is a "non-commutative one", but the rates of convergence using the parameters from the unit square (for the model problem) are almost identical to those convergence rates for the model problem Little or nothing is known however, about the behaviour of these two A.D.P. iterative methods on more general regions in the (x-y) plane.

9. THE BIHARMONIC EQUATION

We consider the solution of

$$\nabla^4 U(x,y) = \frac{\partial^4 U}{\partial x^4} + 2\frac{\partial^4 U}{\partial x^2 \partial y^2} + \frac{\partial^4 U}{\partial y^4} = f(x,y), \quad (x,y) \in R, \qquad (9.1a)$$

where R is the rectangular region defined by (2.2). With f(x,y)=0 for example, the biharmonic equation (9.1a), together with appropriate boundary conditions, describes the small deflections in a thin homogeneous plate under various forces exerted on the boundary of the plate (rectangle).

The particular boundary conditions we shall be considering in connection with the approximate solution of equation (9.1a), are given by,

$$\left.\begin{aligned} U(x,y) &= e(x,y) \\ \frac{\partial^2 U}{\partial n^2}(x,y) &= g(x,y) \end{aligned}\right\} \text{ for all } (x,y) \in B, \qquad (9.1b)$$

where B is the boundary of R, and $\partial/\partial n$ is the normal derivative to B.

Using Taylor series, we obtain the following finite

difference analogue to (9.1a) at the point $(x=i\Delta x,\ y=j\Delta y)$

$$\{\frac{(\Delta y)^2}{(\Delta x)^2}[u_{i+2,j}-4u_{i+1,j}+6u_{i,j}-4u_{i-1,j}+u_{i-2,j}]+\frac{(\Delta x)^2}{(\Delta y)^2}[u_{i,j+2}-4u_{i,j+1}+6u_{i,j}-4u_{i,j-1}+u_{i,j-2}]+2\{u_{i-1,j-1}+u_{i+1,j-1}+u_{i-1,j+1}+u_{i+1,j+1}-2(u_{i-1,j}+u_{i+1,j}+u_{i,j-1}+u_{i,j+1})+4u_{i,j}\}=(\Delta x\Delta y)^2 f_{i,j}\ . \qquad (9.2)$$

Similar finite difference approximations with Δx and Δy replacing h, respectively, are used to approximate $\partial U/\partial n$ on the boundary. Then, superimposing a rectangular grid (as defined by (2.4)) on the region R, and ordering the mesh points horizontally, yields the system of equations

$$A\underline{u} = H\underline{u} + V\underline{u} + 2M\underline{u} = \underline{b}\ , \qquad (9.3)$$

where the matrices H,V,M correspond to the curly brackets in (9.3) in an analogous manner to the derivative of equations (2.5a) and (2.5b). More precisely, if

$$H_L = \begin{bmatrix} 2 & -1 & & & 0 \\ -1 & \ddots & \ddots & & \\ & \ddots & \ddots & \ddots & \\ & & \ddots & \ddots & -1 \\ 0 & & & -1 & 2 \end{bmatrix}_{(n\times n)}\ , \qquad (9.4)$$

then $H = (\frac{\Delta y}{\Delta x})^2$ diagonal $\{H_L^2\}$ in block $(n\times n)$ form. (9.5)

Also if,

$$V_L = \begin{bmatrix} 2I & -I & & & 0 \\ -I & \ddots & \ddots & & \\ & \ddots & \ddots & \ddots & \\ & & \ddots & \ddots & -I \\ 0 & & & -I & 2I \end{bmatrix}_{(nm\times nm)} \qquad (9.6)$$

(I=unit matrix of order n), then

$$V = (\Delta x/\Delta y)^2\ V_L^2\ , \qquad (9.7)$$

and

$$M = V_L\ .\ \text{diagonal}\ \{H_L\}. \qquad (9.8)$$

It is easily verified that the matrices H,V and M are

symmetric and commute, and so have a common set of orthonormal eigenvectors. In particular from equations (9.3)-(9.8) we see that

$$A = \left[\left(\frac{\Delta y}{\Delta x}\right) \text{diagonal}\{H_L\} + \left(\frac{\Delta x}{\Delta y}\right) V_L \right]^2 . \tag{9.9}$$

It thus follows from (9.9) that A is symmetric and positive definite, so system (9.3) has a unique solution.

10. THE ALTERNATING DIRECTION PRECONDITIONING (A.D.P.) TECHNIQUE

Let us re-write the coefficient matrix A as

$$A = I + \hat{H} + \hat{V} + 2M ,$$

where $\quad \hat{H} = H - I/2, \quad \hat{V} = V - I/2 .$ (10.1)

If (9.3) is pre-multiplied by $(I+\omega\hat{H})^{-1}$ then it is equivalent to

$$(I+\omega\hat{H})^{-1} A (I+\omega\hat{V})^{-1} (I+\omega\hat{V})\underline{u} = (I+\omega\hat{H})^{-1}\underline{b} . \tag{10.2a}$$

or

$$B_\omega \underline{y} = \underline{d} , \tag{10.2b}$$

where $B_\omega \underline{y}$ and $\underline{d}$ (with $\hat{H},\hat{V}$ specified by (10.1) are defined by (3.4). The objectives now are exactly the same as before, i.e. to minimise $P(B_\omega)$ with respect to $0 \lesssim \omega \lesssim 2$. Since H,V and M commute it follows from (10.1), that $\hat{H},\hat{V}$ and A are pairwise commutative. Thus if μ,ν denote the real positive eigenvalues of $(\Delta y/\Delta x)$ diagonal$\{H_L\}$and $(\Delta x/\Delta y)V_L$ respectively, with

$$0 < a \lesssim \mu \lesssim b, \quad 0 < \alpha \lesssim \nu \lesssim b, \tag{10.3}$$

we have from equations (9.5)-(10.2),

$$\lambda = \frac{(\mu+\nu)^2}{(1-\omega/2+\omega\mu^2)(1-\omega/2+\omega\nu^2)} , \tag{10.4}$$

where λ is an eigenvalue of B_ω. We see that if $0 \lesssim \omega \lesssim 2$, then $\lambda > 0$ for all λ, then B_ω is positive definite. Also, for the problem in question, the eigenvalues μ and ν are given by

$$\begin{aligned} \mu_i &= \frac{\Delta y}{\Delta x} \cdot 4\sin^2 \frac{i\pi}{2(n+1)} , \text{ for } i=1,2,\ldots,n , \\ \nu_j &= \frac{\Delta x}{\Delta y} \cdot 4\sin \frac{j\pi}{2(m+1)} , \text{ for } j=1,2,\ldots,m . \end{aligned} \tag{10.5}$$

So from (10.3) we have

$$a = \frac{\Delta y}{\Delta x} 4\sin^2 \frac{\pi}{2(n+1)}\ , \quad b = \frac{\Delta y}{\Delta x} 4\cos^2 \frac{\pi}{2(n+1)}\ ,$$
$$\alpha = \frac{\Delta x}{\Delta y} 4\sin^2 \frac{\pi}{2(m+1)}\ , \quad \beta = \frac{\Delta x}{\Delta y} 4\cos^2 \frac{\pi}{2(m+1)}\ . \tag{10.6}$$

To simplify the argument, let us firstly assume that the region of solution is the unit square, so that $\Delta x=\Delta y=h$ and $m=n$. It now follows for this region that $a=\alpha$ and $b=\beta$ from (10.6).

11. MINIMISATION OF P-CONDITION NUMBER OF COEFFICIENT MATRIX

Again we examine the maximum and minimum eigenvalues of $\lambda(B_\omega)$ given by (10.4) with respect to μ_i and ν_j $(1\leqslant i,j\leqslant n)$ and $\min_{0\leqslant\omega\leqslant 2} P(B_\omega)$, we consider the continuous function

$$f(x,y,\omega) = \frac{(x+y)^2}{(1-\omega/2+\omega x^2)(1-\omega/2+\omega y^2)}$$
$$= g(x,y,\omega)+h(x,y,\omega)\,,\quad 0<\alpha\leqslant x,y\leqslant\beta\ , \tag{11.1}$$

where
$$g(x,y,\omega) = \frac{x^2+y^2}{(1-\omega/2+\omega x^2)(1-\omega/2+\omega y^2)}\ ,$$

and
$$h(x,y,\omega) = \frac{2xy}{(1-\omega/2+\omega x^2)(1-\omega/2+\omega y^2)}\ . \tag{11.2}$$

Since $0\leqslant\omega\leqslant 2$ and $0<\alpha\leqslant x,y\leqslant\beta$,

$$g(x,y,\omega),\ h(x,y,\omega) > 0\ . \tag{11.3}$$

(i) Minimum Eigenvalue of B_ω for $0<\omega<2$.

From (11.1) and (11.3) we have

$$\min_{\alpha\leqslant x,y\leqslant\beta} f(x,y,\omega) = \min_{\alpha\leqslant x,y\leqslant\beta} (g(x,y,\omega)+h(x,y,\omega))$$

$$\geqslant \min_{\alpha\leqslant x,y\leqslant\beta} g(x,y,\omega) + \min_{\alpha\leqslant x,y\leqslant\beta} h(x,y,\omega)\,,\quad 0<\omega<2. \tag{11.4}$$

From (4.11) and (4.2) we have

$$\min_{\alpha\leqslant x,y\leqslant\beta} g(x,y,\omega) = \begin{cases} g(\alpha,\alpha,\omega)\ , & 0<\omega\leqslant\bar{\omega} \\ g(\beta,\beta,\omega)\ , & \bar{\omega}\leqslant\omega<2, \end{cases} \tag{11.5}$$

where, in this case $\bar{\omega}=2/(1+2\alpha\beta)$ since $0<\alpha^2\leqslant x^2,y^2\leqslant\beta^2$.

Letting
$$p(x) = \frac{\sqrt{2}x}{(1-\omega/2+\omega x^2)}\ , \tag{11.6}$$

we have from (11.2)

$$h(x,y,\omega) = p(x)p(y) . \tag{11.7}$$

Next, we obtain

$$\frac{dp}{dx} = \sqrt{2}\,\frac{\omega((1-\omega/2)/\omega-x^2)}{(1-\omega/2+\omega x^2)^2} ,$$

$$\begin{cases} \leq 0 \text{ when } x^2 \geq \dfrac{(1-\omega/2)}{\omega} , \\ \geq 0 \text{ when } x^2 \leq \dfrac{(1-\omega/2)}{\omega} . \end{cases} \tag{11.8}$$

Hence if $\alpha^2 \geq (1-\omega/2)/\omega$, i.e. $2/(1+2\alpha^2) \leq \omega < 2$, then from (11.8), $dp/dx \leq 0$. Since $p(x) > 0$ for all $0 < \alpha \leq x \leq \beta$, $0 < \omega < 2$, we have

$$\left.\begin{array}{l} \min\limits_{\alpha\leq x\leq\beta} p(x) = p(\beta) \\ \text{and} \\ \max\limits_{\alpha\leq x\leq\beta} p(x) = p(\alpha) \end{array}\right\} , \quad \frac{2}{1+2\alpha^2} \leq \omega < 2. \tag{11.9}$$

Likewise, if $\beta^2 \leq (1-\omega/2)/\omega$, i.e. $0 < \omega \leq 2/(1+2\beta^2)$, then from (11.8) $dp/dx \geq 0$. Therefore

$$\left.\begin{array}{l} \min\limits_{\alpha\leq x\leq\beta} p(x) = p(\alpha) \\ \text{and} \\ \max\limits_{\alpha\leq x\leq\beta} p(x) = p(\beta) \end{array}\right\} , \quad 0 < \omega \leq \frac{2}{1+2\beta^2} . \tag{11.10}$$

Consider $2/(1+2\beta^2) \leq \omega \leq 2/(1+2\alpha^2)$. In this range,

$$\frac{dp}{dx} = 0 \text{ when } \bar{x} = \sqrt{\frac{1-\omega/2}{\omega}} . \tag{11.11}$$

It is easily verified that this value of x gives a maximum i.e.

$$\max_{\alpha\leq x\leq\beta} p(\bar{x}) = p\left(\sqrt{\frac{1-\omega/2}{\omega}}\right) , \tag{11.12}$$

when $$\frac{2}{1+2\beta^2} \leq \omega \leq \frac{2}{1+2\alpha^2} .$$

Hence, for ω in this range,

$$\min_{0<\alpha\leq x\leq\beta} p(x) = p(\alpha) \text{ or } p(\beta) .$$

It is then also easily shown that

$$\min_{\alpha\leq x\leq\beta} p(x) = \begin{cases} p(\alpha) , & \dfrac{2}{1+2\beta^2} \leq \omega \leq \bar{\omega} \\ p(\beta) , & \bar{\omega} \leq \omega < 2, \end{cases} \tag{11.13}$$

where $\bar{\omega} = 2/(1+2\alpha\beta)$. Hence from equations (11.7), (11.9), (11.10) and (11.13) we obtain

$$\min_{\alpha\leq x,y\leq\beta} h(x,y,\omega) = \begin{cases} h(\alpha,\alpha,\omega), & 0<\omega\leq\bar{\omega} \\ h(\beta,\beta,\omega), & \bar{\omega}\leq\omega<2. \end{cases} \tag{11.14}$$

Combining this result with (11.5) we see that equality holds in (11.4) i.e.

$$\min_{\alpha\leq x,y\leq\beta} f(x,y,\omega) = \begin{cases} f(\alpha,\alpha,\omega), & 0<\omega\leq\bar{\omega} \\ f(\beta,\beta,\omega), & \bar{\omega}\leq\omega<2 . \end{cases} \tag{11.15}$$

(ii) Maximum Eigenvalue of B_ω for $0<\omega<2$.

From (11.1) and (11.3) we have

$$\max_{\alpha\leq x,y\leq\beta} f(x,y,\omega) \leq \max_{\alpha\leq x,y\leq\beta} g(x,y,\omega) + \max_{\alpha\leq x,y\leq\beta} h(x,y,\omega) ,\quad 0<\omega<2. \tag{11.16}$$

By (4.25) and (4.2)

$$\max_{\alpha\leq x,y\leq\beta} g(x,y,\omega) = \begin{cases} g(\beta,\beta,\omega), & 0<\omega\leq 2/(1+2\beta^2) , \\ g(\alpha,\beta,\omega), & 2/(1+2\beta^2)\leq\omega\leq 2/(1+2\alpha^2) , \\ g(\alpha,\alpha,\omega), & 2/(1+2\alpha^2)\leq\omega<2. \end{cases} \tag{11.17}$$

Next, from equations (11.6)-(11.12) we have

$$\left\{\max_{\alpha\leq x\leq\beta} p(x)\right\}^2 = \max_{\alpha\leq x,y\leq\beta} h(x,y,\omega) = \begin{cases} h(\beta,\beta,\omega), & 0<\omega\leq 2/(1+2\beta^2) , \\ h(\bar{x},\bar{x},\omega), & 2/(1+2\beta^2)\leq\omega\leq 2/(1+2\alpha^2), \\ h(\alpha,\alpha,\omega), & 2/(1+2\alpha^2)\leq\omega<2, \end{cases} \tag{11.18}$$

where $\bar{x}=\sqrt{(1-\omega/2)/\omega}$. Combining equations (11.16)-(11.18) yields

$$\max_{\alpha\leq x,y\leq\beta} f(x,y,\omega) \begin{cases} = f(\beta,\beta,\omega), & 0<\omega\leq\frac{2}{1+2\beta^2} , \\ \leq g(\alpha,\beta,\omega) + h(\bar{x},\bar{x},\omega), & \frac{2}{1+2\beta^2}\leq\omega\leq\frac{2}{1+2\alpha^2}, \\ = f(\alpha,\alpha,\omega), & \frac{2}{1+2\alpha^2}\leq\omega<2. \end{cases} \tag{11.19}$$

(iii) The P-Condition Number of B_ω $(0<\omega<2)$.

From (10.4) we see that

$$\min_{\alpha\leq\mu,\nu\leq\beta} \lambda(\mu,\nu,\omega) = \min_{\alpha\leq x,y\leq\beta} f(x,y,\omega), \text{ given by (11.15)} \tag{11.20}$$

Next, $\max_{\alpha\le\mu,\nu\le\beta} \lambda(\mu,\nu,\omega)$ is given approximately by (11.19).

Combining equations (11.15) and (11.19), we obtain

$$P(B_\omega) = \frac{\max \lambda(\mu,\nu,\omega)}{\min \lambda(\mu,\nu,\omega)}, \quad 0<\alpha\le\mu,\nu\le\beta$$

$$\begin{cases} = \dfrac{f(\beta,\beta,\omega)}{f(\alpha,\alpha,\omega)}, & 0<\omega\le\dfrac{2}{1+2\beta^2}, \\ \le \dfrac{g(\alpha,\beta,\omega)+h(\bar{x},\bar{x},\omega)}{f(\alpha,\alpha,\omega)}, & \dfrac{2}{1+2\beta^2}\le\omega\le\bar{\omega}, \\ \le \dfrac{g(\alpha,\beta,\omega)+h(\bar{x},\bar{x},\omega)}{f(\beta,\beta,\omega)}, & \bar{\omega}\le\omega\le\dfrac{2}{1+2\alpha^2}, \\ = \dfrac{f(\alpha,\alpha,\omega)}{f(\beta,\beta,\omega)}, & \dfrac{2}{1+2\alpha^2}\le\omega<2. \end{cases} \tag{11.21}$$

($\bar{\omega}=2/(1+2\alpha\beta)$). Using exactly the same approach as that used in Section 4, we arrive at

$$\min_{0\le\omega\le 2} P(B_\omega) \le \frac{g(\alpha,\beta,\bar{\omega})+h(\bar{x},\bar{x},\omega)}{f(\alpha,\alpha,\bar{\omega})} . \tag{11.22}$$

We also have

$$\bar{x}(\bar{\omega}) = \sqrt{\frac{1-\bar{\omega}/2}{\bar{\omega}}} = \sqrt{\alpha\beta} . \tag{11.23}$$

Substituting for x and ω in equations (11.1)-(11.2), (11.22) becomes

$$\min_{0\le\omega\le 2} P(B_\omega) \le \frac{(3\alpha^2+3\beta^2+2\alpha\beta)}{8\alpha\beta} . \tag{11.24}$$

In fact, from (11.1) it is easily verified that

$$\max_{\alpha\le x,y\le\beta} f(x,y,\bar{\omega}) = f(\bar{x},\bar{x},\omega) . \tag{11.25}$$

This yields from (11.21) and (11.15)

$$\min_{0\le\omega\le 2} P(B_\omega) \le \frac{f(\bar{x},\bar{x},\omega)}{f(\alpha,\alpha,\bar{\omega})} = \frac{(\alpha+\beta)^2}{4\alpha\beta} , \tag{11.26}$$

a smaller upper bound than in (11.24) provided $\alpha<\beta$.

Considering specifically the case when R is the unit square, from (10.6)

$$\alpha=4\sin^2(\pi h/2), \ \beta=4\cos^2(\pi h/2), \ h=1/(n+1) . \tag{11.27}$$

Therefore,

$$\bar{\omega} = \frac{2}{1+2\alpha\beta} = \frac{2}{1+8\sin^2(\pi h)} , \tag{11.28}$$

and from (11.26)

$$\min_{0<\omega<2} P(B_\omega) \leq \frac{(\alpha+\beta)^2}{4\alpha\beta} = \frac{1}{\sin^2(\pi h)} . \tag{11.29}$$

When $\omega=0$, we have from (10.4)

$$P(A) = P(B_{\omega=0}) = \cot^4(\pi h/2) >> 1/(\sin^2 \pi h) . \tag{11.30}$$

If the eigenvalues $\lambda_{i,j}$ $(i,j=1,2,\ldots,n)$ of $B_{\bar{\omega}}$ are such that

$$0<a\leq\lambda_{i,j}\leq b , \tag{11.31}$$

then from (11.15) and (11.25) we have

$$\left.\begin{aligned} \bar{a} &= \min_{1\leq i,j\leq n} \lambda_{i,j}(\bar{\omega}) = f(\alpha,\alpha,\bar{\omega}) = f(\beta,\beta,\bar{\omega}) , \\ \bar{b} &= \max_{1\leq i,j\leq n} \lambda_{i,j}(\bar{\omega}) \leq \max_{\alpha\leq x,y\leq\beta} f(x,y,\omega) = f(\bar{x},\bar{x},\bar{\omega}) \end{aligned}\right\} \tag{11.32}$$

where from (11.23), $\bar{x}(\bar{\omega})=\sqrt{\alpha\beta}=2\sin(\pi h)$.

(iv) Case When Eigenvalue Ranges of μ,ν are Different

The approximate analysis for this case follows along similar lines to Section 4. If μ,ν are such that $a\leq\mu\leq b$, $\alpha\leq\nu\leq\beta$ and $\lambda(\mu,\nu,\omega)$ is an eigenvalue of B_ω (defined by (10.2)) then

$$\min_{0<\omega<2} P(B_\omega) \leq \min_{0<\omega<2} \left\{ \frac{\max_{\alpha'\leq\mu,\nu\leq\beta'} \lambda(\mu,\nu,\omega)}{\min_{\alpha'\leq\mu,\nu\leq\beta'} \lambda(\mu,\nu,\omega)} \right\} \leq P(B_{\bar{\omega}'}) .$$

Here we have $\alpha'=\min(a,\alpha)$, $\beta'=\max(b,\beta)$ and $\bar{\omega}'=2/(1+2\alpha'\beta')$.

Such a case with $a\neq\alpha$ and $b\neq\beta$ occurs for the solution of problem (9.1) on the rectangle with $\Delta x = \Delta y$ and $m\neq n$: see equations (10.6).

12. APPLICATION OF THE A.D.P. TECHNIQUE TO TWO BASIC ITERATIVE METHODS

After deriving the optimum (or "near" optimum) preconditioning parameter $\omega=\bar{\omega}$ and the spectral bounds $\bar{a},\bar{b}$, for the positive real eigenvalues $\lambda_{i,j}$ $(1\leq i,j\leq n)$ of $B_{\bar{\omega}}$, we can again use the Simultaneous Displacement method and Richardson's second degree method to solve system (10.2).

The Computational Procedure

This proceeds in an analogous manner to the procedure outlined in Section 5. Since the two are so similar, we will not repeat it in detail, but only point out the differences. We first assume that for this case, $\bar{\omega},\bar{\alpha},A,B_\omega,\hat{H}$ and $\hat{V}$ are defined by (11.28) (5.4),(9.9),(10.2) and (10.1) respectively. The only difference between the two cases is that for the present one, we now have quindiagonal systems of equations to solve instead of tridiagonal ones (this is apparent from the definitions of $\hat{H}$ and $\hat{V}$ in (10.1) and (9.4)-(9.7).

Let us assume then that we have ordered the unknown grid points along successive x-lines of the grid. The matrix $(I+\omega\hat{H})$ is then symmetric, positive definite and quindiagonal, so that we can use a Gaussian elimination procedure to solve system (5.10) given thus,

$$A\underline{u} = \underline{k}\ ,\ \text{where}$$

$$A = \begin{bmatrix} c_1 & d_1 & e_1 & & & & \\ b_2 & c_2 & d_2 & e_2 & & 0 & \\ a_3 & b_3 & c_3 & d_3 & e_3 & & \\ & \ddots & \ddots & \ddots & \ddots & \ddots & \\ & & \ddots & \ddots & \ddots & \ddots & e_{n-2} \\ & 0 & & \ddots & \ddots & \ddots & d_{n-1} \\ & & & & a_n & b_n & c_n \end{bmatrix} .$$

The system may be solved by the following (Gaussian) elimination scheme: Let

$$\omega_1 = c_1\ ,\ \beta_1 = d_1/\omega_1,\ \ \beta_0 = 0\ ,\ \beta_n = 0\ ,$$

$$\gamma_1 = e_1/\omega_1,\ \gamma_0 = 0,\ \gamma_n = \gamma_{n-1} = 0\ ,$$

and define recursively,

$$\delta_j = b_j - a_j\beta_{j-2}\ ,$$

$$\omega_j = c_j - a_j\gamma_{j-2} - \gamma_j\beta_{j-1}\ ,\ \text{for } 2\leqslant j\leqslant n\ ,$$

$$\beta_j = (d_j - \gamma_j\gamma_{j-1})/\omega_j$$

$$\gamma_j = e_j/\omega_j \; .$$

Next form

$$h_0 = 0 \; , \; h_1 = k_1/\omega_1, \; h_j = (k_j - a_j h_{j-2} - \delta_j h_{j-1})/\omega_j, \; 2 \leq j \leq n.$$

The values of $\underline{u}$ are then obtained recursively from the formula,

$$u_n = h_n, \; u_j = h_j - \beta_j u_{j+1} - \gamma_j u_{j+2} \; , \; j=n-1, n-2, \ldots, 1.$$

This algorithm was first presented by Conte and Dames (1958). It requires not more than 8 multiplications, 3 divisions and 8 additions per unknown u_i, and not more than 4 multiplications, 1 division and 4 additions per unknown for subsequent applications if the matrix A remains unchanged.

Then to evaluate the first part of (5.11), i.e. $\underline{x}^{(k)} = (I+\omega\hat{V})^{-1}\underline{y}^{(k)}$, we re-order both vectors to correspond to a column (or y-line) ordering of the mesh points. A new system, $(I+\omega\hat{V}_Q)\underline{x}_Q^{(k)} = y_Q^{(k)}$ results, the coefficient matrix being quindiagonal and positive definite (this follows particularly from (9.2)), so we can again solve this system by an algorithmic procedure. System (5.13) is evaluated in an analogous manner as before, except we now have a 13-point difference formula (from (9.2)) instead of a 5-point one. Expression (5.11) is finally evaluated by solving (5.14) in the manner already illustrated. We now solve for $\underline{y}^{(k+1)}$ from equation (5.15) the process continuing until convergence (to the desired accuracy) in $\underline{y}$ is obtained. The original solution vector $\underline{u}$ is then obtained from $(I+\omega\hat{V})\underline{u}=\underline{y}$ in the way described above.

13. RATES OF CONVERGENCE ON THE SQUARE

The rate of convergence of the present A.D.P. Simultaneous Displacement method for solving problem (9.1) is given by (5.6). Substituting $P(B_{\bar{\omega}}) = 1/\sin^2(\pi h)$ from (11.29) in the former equation yields

$$R_\infty(I-\tilde{\alpha}B_{\bar{\omega}}) \sim 2\pi^2 h^2 \; , \text{ as } h \to 0 \; . \qquad (13.1)$$

Substituting the same value of $P(B_{\bar{\omega}})$ in equation (5.9) gives the asymptotic rate of convergence of the A.D.P. second degree Richardson iterative process for approximating the solution of problem (9.1) namely,

$$R_\infty \sim 2\pi h \ , \text{ as } h \to 0 \ . \qquad (13.2)$$

We see from (6.10) that this is identical to the asymptotic rate of convergence of the "tridiagonal" A.D.P. Simultaneous Displacements procedure for solving the model problem of section 6.

The Douglas-Rachford (A.D.I.) Iterative Method

The method used by Conte and Dames (1958) to approximate the solution of problem (9.1) is the A.D.I. scheme which corresponds to the Douglas-Rachford method for second order equations. Applied to the iterative solution of system (9.3) this is

$$(r_{k+1}I+H)\underline{u}^{(k+\frac{1}{2})} = (r_{k+1}I-2M-V)\underline{u}^{(k)} \ , \ k \geq 0 \ , \qquad (13.3a)$$

and

$$(r_{k+1}I+V)\underline{u}^{(k+1)} = V\underline{u}^{(k)} + r_{k+1}\underline{u}^{(k+\frac{1}{2})} \ , \qquad (13.3b)$$

where r_{k+1} is a positive parameter chosen to accelerate convergence. The recommended set of iteration parameters given by the above two methods are

$$r_{k+1} = 16x^k , \ k=1,2,\ldots,M \ , \qquad (13.4)$$

for $0<x<1$. The number M of parameters for a given value of the mesh size h (N.B. their region of problem solution was the unit square) satisfies

$$M \geq 1 + \frac{4\log(\sin(\pi h/2))}{\log x} \ . \qquad (13.5)$$

They come to the conclusion, that for the above scheme (13.3) to give the best results, $0<x<0.2$.

We notice from (13.3) and (9.4)-(9.7) that the coefficient matrices on the right of equations (13.3a) and (13.3b) are both direct sums of positive definite symmetric quindiagonal matrices after suitable permutations of their rows and columns, so that we can solve both systems directly using the compact algorithm for quindiagonal systems.

14. NUMERICAL RESULTS

The problem solved was the biharmonic equation

$$\nabla^4 U(x,y) = 0, \quad (x,y) \in R,$$

where the region R was the unit square. The boundary conditions were

$$U(x,y) = \frac{\partial^2 U}{\partial n^2}(x,y) = 0, \quad \text{for all } (x,y) \in B,$$

where B is the boundary of the square, and $\partial/\partial n$ is the normal derivative to B. The number of iterations which were required to make

$$\max_{1 \leq i,j \leq n} |y_{i,j}^{(k+1)} - y_{i,j}^{(k)}| < 10^{-6}$$

for the two above preconditioned iterative methods, and

$$\max_{1 \leq i,j \leq n} |u_{i,j}^{M(k+1)} - u_{i,j}^{(Mk)}| < 10^{-6},$$

(M=number of acceleration parameters) for the D.R. (A.D.I.) process, starting with $y_{i,j}^{(0)}=1$, $u_{i,j}^{(0)}=1$, $1 \leq i,j \leq n$, are given in Figure 3. The logarithm of N, the number of iterations are plotted against $\log(h^{-1})$.

For the A.D.P. Simultaneous Displacement and A.D.P. second degree Richardson methods the optimum preconditioning parameter $\bar{\omega}$ used, was that specified in (11.28), while the optimum acceleration parameters used are those defined by (5.4) and (5.7) respectively, using the values of $\bar{a}, \bar{b}$ from (11.32).

Regarding the Douglas-Rachford A.D.I. scheme (or D.R. (A.D.I.) for short), we at first choose x=0.2 in (13.4). The number of parameters then to be used cyclically, for each value of h, was taken to be the smallest integer satisfying (13.5). The results from Figure 3 show that the A.D.P. Simultaneous Displacements method converges (as expected) much slower than the other two. We also notice that for this value of x, the A.D.P. second order Richardson scheme converges in less iterations than D.R. (A.D.I.) for approximately $h^{-1} \leq 12$, and in more for $h^{-1} > 12$, (i.e. the "cross over point" is at $h^{-1}=12$).

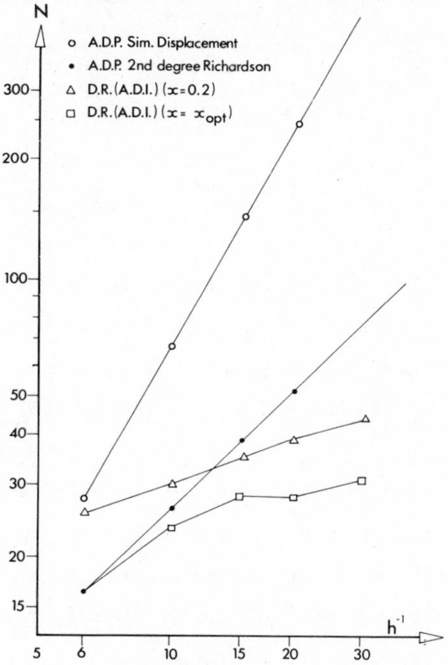

Figure 3

In view of the fact that Conte and Dames (1958) recommended $x<0.2$, it was decided to try values of $x=0.1(0.01)0.2$ in (13.4) and (13.5) for the same values of h^{-1} as before, to find the 'best' values. These were found to be x_{opt} (say)=0.15,0.16,0.17, 0.18 and 0.19 for h^{-1}=6,10,15,20 and 30, respectively. The "cross over point" now for the two afore-mentioned methods is seen to be about $h^{-1}=6$, although there is no significant difference in the number of iterations until $h^{-1}>>10$.

We now look at the comparative amounts of work per iteration of the above two iterative methods. On the square, the D.R. (A.D.I.) method (13.a),(13.b) requires the implicit solution of 2n quindiagonal systems of equations. The coefficient matrices have the form,

$$A = \begin{bmatrix} c_1 & d_1 & e_1 & & & & \\ b_2 & c_2 & d_2 & e_2 & & 0 & \\ a_3 & b_3 & c_3 & d_3 & e_3 & & \\ & \ddots & \ddots & \ddots & \ddots & \ddots & \\ & & & & & & e_{n-2} \\ & 0 & & & & & d_{n-1} \\ & & & & a_n & b_n & c_n \end{bmatrix}$$

and for this particular problem

$$d_i = b_{i+1} = -4,\ i=1,2,\dots,n-1,\ e_i = a_{i+2} = 1,\ i=1,2,\dots,m-2. \tag{14.1}$$

Making use of relations (14.1), for one complete iteration (i.e. a double sweep of the grid) the D.R. (A.D.I.) method requires approximately $13n^2$ multiplications, $6n^2$ divisions and $30n^2$ additions (see Conte and Dames (1958), p.204).

The A.D.P. second order Richardson method is defined by equation (5.2) where for this problem $B_{\bar{\omega}}$, in particular, is defined by (10.2). From the latter equation we can write

$$\begin{aligned} B_{\omega}\underline{y}^{(k)} &\equiv [I(1-\omega/2)+H]^{-1}A[I(1-\omega/2)+\omega V]^{-1}\underline{y}^{(k)} \\ &\equiv \omega^{-2}[rI+H]^{-1}A[rI+V]^{-1}\underline{y}^{(k)}, \end{aligned} \tag{14.2}$$

where $r=(1-\omega/2)/\omega$. Adopting the same procedure as that described in Section 12, we can save n^2 multiplications per iteration by expressing $B_{\omega}\underline{y}^{(k)}$ as in (14.2) and making use of the last relation in (14.1). Method (5.2) then requires approximately $12n^2$ multiplications, $2n^2$ divisions and $24n^2$ additions, per iteration. In order to compare this amount of work with that given for D.R. (A.D.I.) we ignore the additions in both and regard a multiplication and a division as being about the same, in terms of computing time. Thus, the respective amounts of work are now approximately $19n^2$ multiplications and $14n^2$ multiplications. In other words D.R. (A.D.I.) requires about $1\frac{1}{3}$ times the work per iteration of the

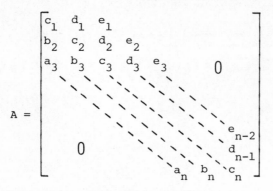

A.D.P. second degree Richardson method.

Thus, returning our attention to Figure 3, it would appear that with x(fixed)=0.2 in (13.4), the former process is more efficient (in terms of total arithmetic operations) than the latter for approximately $h^{-1}>20$. For $h^{-1}\leq 20$ (approximately) the roles are reversed. In the case when the "optimum" value of x (say x_{opt}) is used in (13.4), as mentioned above, the "cross over point" in terms of overall efficiency will be at about $h^{-1}=12$.

REFERENCES

1. Birkhoff, G., Varga, R.S. and Young D.M. - "Alternating Direction Implicit Methods", Vol.3, Advances in Computers, ed. F. Alt., Academic Press Inc., 189-273 (1962).

2. Conte S.A. and Dames R.T. - "An Alternating Direction Method for Solving the Biharmonic Equation", Math.Tables Aids.Comp., 12 198-205, (1958).

3. Evans, D.J. - "The Use of Preconditioning in Iterative Methods for Solving Linear Equations with Symmetric, Positive Definite Matrices", J.Inst.Maths.Applics., 4, 295-315, (1968).

4. Evans, D.J. - "Iterative Sparse Matrix Algorithms", 'Software for Numer.Maths.' (edit. D.J. Evans), Academic Press, 49-83 (1974).

5. Evans, D.J. - "Alternating Direction Implicit Preconditioning Methods for Self-Adjoint Elliptic Differential Equations', Comp. & Math. with Applics. 7, 151-158, (1981).

6. Forsythe G.E. and Wasow W.R. - "Finite Difference Methods for Partial Differential Equations", Wiley, New York, London, (1960).

7. Thomas L.H. - "Elliptic Problems in Linear Difference Equations Over a Network", Watson Scientific Computing Laboratory, Columbia University, New York, (1949).

8. Varga R.S. - "Matrix Iterative Analysis", Prentice-Hall Inc., Englewood Cliffs, N.J., (1962).

9. Wachspress E.L. - "Optimum Alternating-Direction Implicit Iteration Parameters for a Model Problem", J.Soc.Indust. Appl.Math., 10, 339-350, (1962).

10. Wachspress E.L. - "Iterative Solution of Elliptic Systems", Prentice-Hall, N.J. (1966).

11. Wachspress E.L. and Habetler G.J. - "An Alternating Direction Implicit Iteration Technique", J.Soc.Indust. Appl.Math., 8, 403-424, (1960).

12. Young D.M. - "Iterative Solution of Large Linear Systems", Academic Press, New York, (1971).

13. Young D.M. and Ehrlich L. - "Some Numerical Studies of Iterative Methods for Solving Elliptic Difference Equations", Boundary Problems in Differential Equations, University of Wisconsin Press, Madison. Edit. R.E. Langer, 143-162, (1960).

14. Young D.M. and Frank T.G. - "A Survey of Computer Methods for Solving Elliptic and Parabolic Partial Differential Equations", ICC Bulletin, 2, 11, (1963).

PRECONDITIONED ITERATIVE METHODS FOR THE NUMERICAL SOLUTION OF ELLIPTIC PARTIAL DIFFERENTIAL EQUATIONS

D.J. Evans and N.M. Missirlis*
Department of Computer Studies, University of Technology, Loughborough, Leicestershire, U.K.

ABSTRACT

This chapter further develops the preconditioning concept previously introduced by the authors [4], [8], in the Preconditioned Simultaneous Displacement (PSD) iterative method and the Alternating Direction Preconditioning (ADP) method for the numerical solution of the large, sparse, linear systems which arise from the discretisation by finite difference/element methods of boundary value problems involving elliptic partial differential equations. The convergence properties of the methods are analysed under certain assumptions on the coefficient matrix and optimum values of the involved preconditioning parameters are determined in terms of bounds on the eigenvalues of certain matrices. The application of various accelerated techniques are also considered and experimental results on the model Dirichlet problem are shown to agree with the theoretical predictions. Finally, the development of a dynamic algorithm for improving the estimates of the preconditioning parameters is presented which are then used to accelerate the PSD method by employing semi-iterative techniques.

1. INTRODUCTION

In this chapter, we are concerned with the solution of large linear systems arising from the discretisation by finite difference methods of boundary value problems involving elliptic partial differential and biharmonic equations.

---

*Unit for Applied Mathematics, University of Athens, Athens 621, Greece.

Let us therefore consider systems of the form

$$Au = b \ , \tag{1.1}$$

where A is a given (N×N) real sparse matrix, b is the (N×1) right hand side vector and u is the unknown vector to be determined. We assume throughout our analysis that the matrix A is symmetric and positive definite. For the solution of such systems, iterative methods are widely used [10],[20],[21],[23]. In particular, we consider the concept of preconditioning first introduced by Evans [4] and later on developed in [5],[6], [7], in a more general way than before which allow us to show that all the known linear first degree methods are based on this technique [16]. We pre-multiply (1.1) by

$$R^{-1}Au = R^{-1}b \ , \tag{1.2}$$

where R is a non-singular matrix which satisfies the following properties:

(i) The spectral condition number* of the matrix $R^{-1}A$ becomes smaller than the spectral condition number of A.

(ii) For any vectors s and t it is "computationally convenient" to solve the system Rs=t,i.e., R is easily solvable.

We are now able to construct an iterative scheme with respect to the preconditioned system (1.2) as follows:

$$u^{(n+1)} = u^{(n)} + \tau R^{-1}(b-Au^{(n)}) \ , \tag{1.3}$$

where τ is a real parameter with $\tau \neq 0$. The type of iterative scheme (1.3) represents all the known linear first degree iterative methods for certain values of τ and different forms of the matrix R. This can be observed if we assume that A has the form I-L-U or H+V, where L,U are strictly lower and upper triangular matrices, respectively, as defined in (1.7) and H,V are defined as in (12.10). Then, from (1.3) we can verify Table 1.1 where the $\bar{a}$'s and $\bar{b}$'s are the minimum and maximum eigenvalues $\lambda(R^{-1}A)$ of the corresponding preconditioned

---

The spectral condition number of a given matrix A is given by $k(A)=||A||.||A^{-1}||$.

τ	R	Iterative Method
1	I	Jacobi (J) [20],[23]
$2/(\bar{a}+\bar{b})$	I	Simultaneous Displacement (SD) [10]
1	(I-L)	Gauss-Seidel (GS) [20],[23]
ω	$(I-\omega L)$	Successive Overrelaxation (SOR [20], [23]
$\omega(2-\omega)$	$(I-\omega L)(I-\omega U)$	Symmetric SOR (SSOR) [23]
1	"	Preconditioned Jacobi(PJ) [4],[16]
$2/(\bar{a}+\bar{b})$	"	Preconditioned Simultaneous Displacement (PSD) [4],[16]
r	(I+rH)(I+rV)	Douglas-Rachford ADI (DR-ADI) [12]
2r	(I+rH)(I+rV)	Peaceman-Rachford ADI(PR-ADI) [21]
$2/(\bar{a}+\bar{b})$	(I+rH)(I+rV)	Alternating Direction Preconditioning (ADP) [7], (or EADI [14])

TABLE 1.1

matrix $R^{-1}A$, respectively, i.e.,

$$\bar{a} \leq \lambda(R^{-1}A) \leq \bar{b} \ . \tag{1.4}$$

From the above, it is concluded that all known convergent iterative methods can be interpreted as improving the "condition" of the given original system (1.1) by using different types of conditioning matrices R in (1.2).

If the matrix A can be suitably split into the sum of two matrices, then from Table 1.1 we see that the effectiveness of the conditioning matrix R* in its most general form (e.g. see the form of R for the SSOR, PSD, ADP, PR-ADI methods) depends upon the value of the involved parameters and the product of these two matrices.

Therefore, if A has the splitting

$$A = D-C_L-C_U \ , \tag{1.5}$$

where D is the diagonal matrix with the same diagonal elements as A and C_L, C_U are strictly lower and upper triangular matrices, respectively, then a suitable form of R for further investigation is

$$R = (D-\omega C_L)D^{-1}(D-\omega C_U) \ . \tag{1.6}$$

If R=A, then we obtain the solution of (1.1) immediately.

Evidently, for the different values of the parameters τ and ω we can easily construct Table 1.1 where

$$L = D^{-1}C_L \quad \text{and} \quad U = D^{-1}C_U , \qquad (1.7)$$

to incorporate many of the well known iterative methods.

Alternatively, we can consider the matrix R in the Alternating Direction Implicit methods to have the form,

$$R = (I+rH)(I+rV) = I+rA+r^2HV , \qquad (1.8)$$

where $A = H+V$.

In our attempt to produce a more effective conditioning matrix R we introduce and examine in Section 1 the Preconditioned Simultaneous Displacement Method and in Section 2 the Modified Alternating Direction Preconditioning (MADP) method for the numerical solution of self-adjoint second order elliptic and biharmonic equations.

SECTION 1

2. THE PRECONDITIONED SIMULTANEOUS DISPLACEMENT METHOD (PSD)

We now consider the iterative method which is produced by (1.3) when R is given by (1.6), i.e. we consider the iterative scheme defined by

$$u^{(n+1)} = u^{(n)} + \tau(I-\omega U)^{-1}(I-\omega L)^{-1}D^{-1}(b-Au^{(n)}) , \qquad (2.1)$$

where $L = D^{-1}C_L$, $\qquad U = D^{-1}C_U$, (2.2)

and $\tau(\neq 0),\omega$ are real parameters. The iterative process (2.1) can be regarded as belonging to a generalised class referred to as the Preconditioned Simultaneous Displacement method (PSD method). At this point we can state that if we let τ and ω obtain their optimum values τ_o and ω_o, respectively, then the PSD method will possess smaller spectral radius, than any other iterative scheme which employs the same conditioning matrix R. Therefore, there may be a possibility of PSD to produce a better rate of convergence than SSOR. However any comparison must take into acccount the computational work involved per iteration and in [16] it is shown that one PSD iteration can be computed as two half iterations such that a reduction scheme similar to Neithammer's

[19] can be effectively applied.

Next we present another form for the computation of one PSD iteration in terms of intermediate vector approximation as follows:

$$u^{(n+\frac{1}{2})} = (1-\tau)u^{(n)} + \omega L u^{(n+\frac{1}{2})} + (\tau-\omega)Lu^{(n)} + \tau(Uu^{(n)}+c) ,$$
$$u^{(n+1)} = u^{(n+\frac{1}{2})} + \omega U u^{(n+1)} - \omega U u^{(n)}, \text{where } c = D^{-1}b. \qquad (2.3)$$

By expressing one PSD iteration as in (2.3) we can exploit the fact that when ω is fixed it is not necessary to recompute the vector $Uu^{(n)}$ in the second half iteration. Thus, if sufficient computer storage is available to accommodate an extra N-vector, the computational work involved for the PSD method after n iterations is reduced such that to be approximately the same as with the SOR method [16].

An alternative form of PSD is given by

$$u^{(n+1)} = \mathcal{D}_{\tau,\omega} u^{(n)} + \delta_{\tau,\omega} , \qquad (2.4)$$

where $$\mathcal{D}_{\tau,\omega} = I - \tau \mathcal{B}_{\omega} = I - \tau(I-\omega U)^{-1}(I-\omega L)^{-1}D^{-1}A , \qquad (2.5)$$

and $$\delta_{\tau,\omega} = \tau(I-\omega U)^{-1}(I-\omega L)^{-1}D^{-1}b. \qquad (2.6)$$

Since the conditioning matrix of PSD, given by (2.5) is non-singular the method is completely consistent with (1.1) for $\tau \neq 0$ and A being non-singular.

3. CONVERGENCE ANALYSIS FOR PSD

Now a study of the convergence properties of the PSD method is given under the assumption that A is real and symmetric with positive diagonal elements, (Missirlis [16]).

Lemma 3.1: Let A be a real symmetric matrix with positive diagonal elements, then $\mathcal{B}_{\omega}$ defined by (2.5) has real eigenvalues. Moreover, $\mathcal{B}_{\omega}$ has positive eigenvalues if and only if A is positive definite.

Proof: Evidently, $\mathcal{B}_{\omega}$ is similar to the matrix

$$\overline{\mathcal{B}}_{\omega} = D^{-\frac{1}{2}}(D-\omega C_U)\mathcal{B}_{\omega}(D-\omega C_U)^{-1}D^{\frac{1}{2}}$$
$$= [D^{\frac{1}{2}}(D-\omega C_L)]^{-1} A [D^{\frac{1}{2}}(D-\omega C_L)^{-1}]^T , \qquad (3.1)$$

which implies that $\overline{\mathcal{B}}_{\omega}$ has real eigenvalues and hence the same is true for B_ω.

Moreover, if we assume that A is positive definite, then

$$\bar{B}_\omega = [D^{\frac{1}{2}}(D-\omega C_L)^{-1}A^{\frac{1}{2}}][D^{\frac{1}{2}}(D-\omega C_L)^{-1}A^{\frac{1}{2}}]^T \quad , \qquad (3.2)$$

and since $D^{\frac{1}{2}}(D-\omega C_L)A^{\frac{1}{2}}$ is non-singular, it follows (see Theorem 2-2.6 in [23]), that $\bar{B}_\omega$ is positive definite. Since B_ω is similar to $\bar{B}_\omega$, it follows that its eigenvalues are also positive. Alternatively, if we assume that B_ω is positive definite, then the same is true for $\bar{B}_\omega$ and from (3.1) we can easily see that A must be positive definite, thus completing the proof of the Lemma.

In the following, we attempt to determine the range for the parameters τ and ω such that PSD to converge. From (2.5) and the proof of Lemma 3.1, it follows that $\lambda>0$ is an eigenvalue of B_ω, if and only if A is positive definite and

$$d = 1-\tau\lambda \quad , \qquad (3.3)$$

is an eigenvalue of $D_{\tau,\omega}$. On the other hand $S(D_{\tau,\omega})$, the spectral radius of $D_{\tau,\omega}$, is less than unity if and only if $|d|<1$ or equivalently if

$$0<\tau<2/\Lambda(B_\omega) \quad , \qquad (3.4)$$

where $\Lambda(B_\omega)$ denotes the maximum eigenvalue of B_ω. Also if we happen to know an eigenvector v associated with the maximum eigenvalue of B_ω, then we would be able to determine $\Lambda(B_\omega)$ exactly from the expression

$$\Lambda(B_\omega) = \max_{v\neq 0} \frac{(v,B_\omega v)}{(v,v)} \quad , \qquad (3.5)$$

but this is an unrealistic expectation, thus we will attempt to determine an upper bound of $\Lambda(B_\omega)$ using (3.5) and (2.5).

<u>Theorem 3.2</u>: Let A be a real symmetric positive definite matrix, then for any ω in the range $0<\omega<2$ we have

$$\Lambda(B_\omega) \leq \frac{1}{\omega(2-\omega)} \quad . \qquad (3.6)$$

<u>Proof</u>: Let us assume that λ is an eigenvalue of B_ω and $v\neq 0$ is an associated eigenvector, then $B_\omega v=\lambda v$ which by taking inner products of both sides and substituting B_ω by (2.5) yields the following expression for λ

$$\lambda = \frac{(v,Av)}{(v,D(I-\omega L)(I-\omega U)v} \quad . \qquad (3.7)$$

However,

$$D(I-\omega L)(I-\omega U) = \omega(2-\omega)A + D[(1-\omega)I + \omega L][(1-\omega)I+\omega U], \tag{3.8}$$

which when substituted in (3.7) gives

$$\lambda = \frac{(v,Av)}{(v,\omega(2-\omega)Av)+(v,D[(1-\omega)I+\omega L][(1-\omega)I+\omega U]v)} \; . \tag{3.9}$$

It is readily seen that the second term in the denominator of (3.9) is always non-negative thus (3.9) yields $\lambda \leq (v,Av)/(v,\omega(2-\omega)Av)$ and the proof of the theorem is complete since if A is positive definite, then by Lemma 3.1, $\lambda>0$.

Next, the following theorem gives sufficient and necessary conditions for the PSD method to converge.

Theorem 3.3: Let A be a real, symmetric and positive definite matrix, then the PSD method converges if

$$0<\omega<2 \quad \text{and} \quad 0<\tau<2\omega(2-\omega)\leq 2 \; . \tag{3.10}$$

Proof; By combining the estimated bound on $\Lambda(B_\omega)$, given by (3.6) with the inequalities (3.4), it is readily seen that (3.10) are satisfied, thus completing the proof of the theorem.

Theorem 3.4: Let A be a real, symmetric matrix with positive diagonal elements. If the PSD method converges then, τ and ω satisfy the relationships (3.10) and A is positive definite.

Proof: If A is not positive definite, then there exists $v\neq 0$ and $\alpha\leq 0$, such that $Av=\alpha v$. From (3.1) and if we let $w=D^{-\frac{1}{2}}(D-\omega C_U)v$ then we have,

$$(w,\bar{B}_\omega w)/(w,w) = \alpha(v,v)/(w,w) \; . \tag{3.11}$$

By (2.5) we also have that $(w,\bar{D}_{\tau,\omega}w)/(w,w)=1-\tau\alpha(v,v)/(w,w)$ which implies that $S(\bar{D}_{\tau,\omega})\geq 1$ provided $0<\tau$ where $\bar{D}_{\tau,\omega}=D^{-\frac{1}{2}}(D-\omega C_U)D_{\tau,\omega}(D-\omega C_U)^{-1}D$. However, we have seen that $S(D_{\tau,\omega})<1$ when (3.4) is satisfied. This contradiction proves that A is positive definite and the proof of Theorem 3.4 is complete.

The problem of determining "good" estimates τ_1,ω_1 for τ and ω, respectively such that $S(D_{\tau,\omega})$ to attain the value $S(D_{\tau_1,\omega_1})$ which is near to the actual minimum value $S(D_{\tau_0,\omega_0})$ is of paramount importance. The development of the analysis will be under the condition that A is a real, symmetric and positive definite matrix.

Here we recall [3] that we will attempt to determine a "good" choice of ω_1 of ω by minimising $P(B_\omega)$, the bound of the P-condition number of B_ω given by

$$P(B_\omega) = \Lambda(B_\omega)/\lambda(B_\omega) \ , \qquad (3.12)$$

where $\Lambda(B_\omega),\lambda(B_\omega)$ denote the maximum and minimum eigenvalue bounds of B_ω, respectively. On the other hand, τ will be determined by the expression

$$\tau = \tau_1 = 2/(\Lambda(B_{\omega_1}) + \lambda(B_{\omega_1})) \ , \qquad (3.13)$$

in order for $S(D_{\tau,\omega})$ to attain its minimum value.

By Theorem 3.2 we have an upper bound for $\Lambda(B_\omega)$ and the following theorem determines a lower bound for $\lambda(B_\omega)$.

Theorem 3.5: Let A be a real symmetric and positive definite matrix and $\bar{\beta}$, M and m be numbers such that [26]

$$-2\sqrt{\bar{\beta}} \leq m \leq m(B), \quad M(B) \leq M \leq \min(1, 2\sqrt{\bar{\beta}}) \text{ and } S(LU) \leq \bar{\beta}, \qquad (3.14)$$

where M(B), m(B), denote the maximum and minimum eigenvalues of B, respectively.

Then, for any ω in the range $0<\omega<2$, a lower bound on $\lambda(B_\omega)$ is given by

$$\lambda(B_\omega) \geq \begin{cases} (1-M)/(1-\omega M+\omega^2\bar{\beta}) = \phi_1(\omega), & \text{if } \bar{\beta}\geq\frac{1}{4} \text{ or if } \bar{\beta}<\frac{1}{4} \text{ and } \omega\leq\omega^*, \\ (1-m)/(1-\omega m+\omega^2\bar{\beta}) = \phi_2(\omega), & \text{if } \bar{\beta}<\frac{1}{4} \text{ and } \omega>\omega^*, \end{cases} \qquad (3.15)$$

where for $\bar{\beta}<\frac{1}{4}$ we define ω^* by

$$\omega^* = 2/(1+(1-4\bar{\beta})^{\frac{1}{2}}) \ . \qquad (3.16)$$

Proof: By expanding the numerator and denominator in (3.7) we obtain $((v,Dv)\neq 0)$,

$$\lambda = \frac{1-\hat{\alpha}(v)}{1-\omega\hat{\alpha}(v)+\omega^2\hat{\beta}(v)} \equiv \lambda(\omega,\hat{\alpha},\hat{\beta}) \ , \qquad (3.17)$$

where

$$\hat{\alpha}(v) = \frac{(v,DBv)}{(v,Dv)} \quad \text{and} \quad \hat{\beta}(v) = \frac{(v,DLUv)}{(v,Dv)} \ . \qquad (3.18)$$

It can be readily seen that [26]

$$m(B) \leq \hat{\alpha}(v) \leq M(B) \text{ and } 0 \leq \hat{\beta}(v) \leq S(LU) \ , \qquad (3.19)$$

thus our problem is reduced to the following

$$\lambda(B_\omega) \geq \min_{\hat{\alpha},\hat{\beta}} \lambda(\omega,\hat{\alpha},\hat{\beta}) \ . \qquad (3.20)$$

From (3.17) and for fixed $\omega,\hat{\alpha}$ we have that $\text{sign}(\partial\lambda/\partial\hat{\beta})<0$ which by taking into consideration the inequalities (3.14) and

(3.19) implies that $\lambda(B_\omega)\geq\min_{\hat{\alpha}}\lambda(\omega,\hat{\alpha},\bar{\beta})$. Further, we have $\text{sign}(\partial\lambda/\partial\hat{\alpha})=\text{sign}(-\omega^2\bar{\beta}+\omega-1)$ and we can easily study the behaviour of $-\omega^2\bar{\beta}+\omega-1$. This is shown in Table 3.1 where we can see the validity of (3.15) and the proof of the theorem is now complete.

$\bar{\beta}$-Domain	ω-Domain	$\omega^2\bar{\beta}-\omega+1$	$\lambda(B_\omega)$ Bound
$\bar{\beta}\geq\frac{1}{4}$	$0<\omega<2$	≥ 0	$\phi_1(\omega)=\lambda(\omega,M,\bar{\beta})$
$0\leq\bar{\beta}<\frac{1}{4}$	$0<\omega<\omega^*$	>0	$\phi_1(\omega)=\lambda(\omega,M,\bar{\beta})$
	$\omega=\omega^*$	$=0$	$\phi_1(\omega)=\lambda(\omega,\hat{\alpha},\bar{\beta})$
	$\omega^*<\omega<2$	<0	$\phi_2(\omega)=\lambda(\omega,m,\bar{\beta})$

TABLE 3.1

By combining Theorems 3.2 and 3.5 we can form an upper bound for $P(B_\omega)$ which can be studied as a function of ω.

Theorem 3.6: Let A be a real, symmetric and positive definite matrix and $\bar{\beta}$,M and m satisfying (3.14). Then, an upper bound on $P(B_\omega)$ is given by

$$P(B_\omega)\leq\begin{cases}\dfrac{1-\omega M+\omega^2\bar{\beta}}{\omega(2-\omega)(1-M)}=\theta_1(\omega)\ , & \text{if } \bar{\beta}\geq\frac{1}{4} \text{ or if } \bar{\beta}<\frac{1}{4} \text{ and } \omega\leq\omega^*,\\ \dfrac{1-\omega m+\omega^2\bar{\beta}}{\omega(2-\omega)(1-m)}=\theta_2(\omega)\ , & \text{if } \bar{\beta}<\frac{1}{4} \text{ and } \omega>\omega^*,\end{cases} \tag{3.21}$$

where for $\bar{\beta}<\frac{1}{4}$, ω^* is defined by (3.16). Moreover, the bound on $P(B_\omega)$ is minimised if we let $\omega=\omega_1$ where

$$\omega_1=\begin{cases}2/(1+(1-2M+4\bar{\beta})^{\frac{1}{2}})=\omega_M, & \text{if } M\leq 4\bar{\beta},\\ 2/(1+(1-4\bar{\beta})^{\frac{1}{2}})=\omega^* & , \text{if } M\geq 4\bar{\beta},\end{cases} \tag{3.22}$$

and its corresponding value is given by

$$P(B_{\omega_1})\leq\begin{cases}[1+(1-2M+4\bar{\beta})^{\frac{1}{2}}/(1-M)]/2=[(2-M\omega_M)/(\omega_M-M\omega_M)]/2\ , & \text{if } M\leq 4\bar{\beta},\\ [1+1/(1-4\bar{\beta})^{\frac{1}{2}}]/2=1/(2-\omega^*)\ , & \text{if } M\geq 4\bar{\beta}.\end{cases} \tag{3.23}$$

On the other hand, if we let

$$\tau = \tau_1 = 2\omega_1(2-\omega_1)/(1+1/P(B_{\omega_1})) \; , \tag{3.24}$$

then $S(\mathcal{D}_{\tau,\omega})$ attains its minimum value and is given by

$$S(\mathcal{D}_{\tau_1,\omega_1}) \leq \begin{cases} [2(1-\omega_M)+M\omega_M]/[2(1+\omega_M)-3M\omega_M], & \text{if } M\leq 4\bar{\beta} \; , \\ 1/[2/(\omega^*-1)-1] & , \text{if } M\geq 4\bar{\beta} \; . \end{cases} \tag{3.25}$$

Proof: Evidently, (3.21) is derived from (3.12) and Theorems 3.2 and 3.5. By letting $p(\omega,\hat{\alpha},\hat{\beta})=1/[\omega(2-\omega)\lambda(\omega,\hat{\alpha},\hat{\beta})]$, then from (3.17) we have

$$p(\omega,\hat{\alpha},\hat{\beta}) = \frac{1-\omega\hat{\alpha}+\omega^2\hat{\beta}}{\omega(2-\omega)(1-\hat{\alpha})} \; , \tag{3.26}$$

hence,

$$p(\omega,M,\bar{\beta}) = \theta_1(\omega) \quad \text{and} \quad p(\omega,m,\bar{\beta}) = \theta_2(\omega) \; . \tag{3.27}$$

From (3.26) it follows that $\text{sign}(\partial p(\omega,\hat{\alpha},\hat{\beta})/\partial\omega)=\text{sign}(\omega^2(2\hat{\beta}-\hat{\alpha})+2\omega-2)$ and therefore, the extreme points of $\theta_1(\omega)$ and $\theta_2(\omega)$ for $\omega\in(0,2)$ are $\omega_M=2/(1+(1-2M+4\bar{\beta})^{\frac{1}{2}})$ and $\omega_m=2/(1+(1-2m+4\bar{\beta})^{\frac{1}{2}})$, respectively. Furthermore, the behaviour of $\theta_1(\omega)$ and $\theta_2(\omega)$ as functions of ω are summarised in Tables 3.2 and 3.3 where we have let $\gamma(\omega,\hat{\alpha},\hat{\beta})=\omega^2(2\hat{\beta}-\hat{\alpha})+2\omega-2$. It can be easily verified that $\omega_m\leq\omega_M$ and $\omega_m\leq\omega^*$. Since $\text{sign}(\partial p(\omega,\hat{\alpha},\hat{\beta})/\partial\hat{\alpha})=\text{sign}(\omega^2\bar{\beta}-\omega+1)$ we can establish Table 3.4 showing the relation of $\theta_1(\omega)$ and $\theta_2(\omega)$ for the different ranges of $\bar{\beta}$ and ω.

ω-Domain	$\gamma(\omega,M,\bar{\beta})$	$\theta_1(\omega)$
$0<\omega<\omega_M$	>0	Decreasing
$\omega=\omega_M$	$=0$	Stationary
$\omega_M<\omega<2$	<0	Increasing

TABLE 3.2

ω-Domain	$\gamma(\omega,m,\bar{\beta})$	$\theta_2(\omega)$
$0<\omega<\omega_m$	>0	Decreasing
$\omega=\omega_m$	$=0$	Stationary
$\omega_m<\omega<2$	<0	Increasing

TABLE 3.3

It is observed that if $\bar{\beta}\geq 1/4$, then the bound on $P(B_\omega)$ is represented by $\theta_1(\omega)$, for all $\omega\in(0,2)$, thus the best choice of ω (see Table 3.2) is $\omega=\omega_M$. If $\bar{\beta}<1/4$, then we consider two cases: Case I, $\omega_m\leq\omega^*\leq\omega_M$ and Case II, $\omega_m\leq\omega_M\leq\omega^*$. Clearly, $\omega=\omega^*$ and $\omega=\omega_M$ is the best value of ω for Case I and Case II, respectively. Next, we note that if $\bar{\beta}\geq 1/4$ and $M<1$, then these conditions imply $M\leq 4\bar{\beta}$.

In Case II, we have $\bar{\beta}<1/4$ and $\omega_M \leq \omega^*$ which imply $M \leq 4\bar{\beta}$ as well. Finally, in Case I the conditions $\bar{\beta}<1/4$ and $\omega_M \geq \omega^*$ imply $M \geq 4\bar{\beta}$. Consequently, the best value of ω, namely ω_1, is given by (3.22).

$\bar{\beta}$-Domain	ω-Domain	$\omega^2\bar{\beta}-\omega+1$	Relation	$P(B_\omega)$ Bound
$\bar{\beta} \geq 1/4$	$0<\omega<2$	≥ 0	$\theta_1 \geq \theta_2$	θ_1
$0 \leq \bar{\beta} \leq 1/4$	$0<\omega \leq \omega^*$	≥ 0	$\theta_1 \geq \theta_2$	θ_1
	$\omega^* \leq \omega<2$	≤ 0	$\theta_1 \leq \theta_2$	θ_2

TABLE 3.4

By substituting directly ω_M, ω^* into the first and second part in (3.21), respectively we obtain the minimised bound on $P(B_\omega)$ given by (3.23). The determination of the value τ_1 for τ can be achieved by combining (3.13),(3.12) and it is easily seen that τ_1 is given by (3.24). For this value of τ, we can find the minimum value of the bound for $S(D_{\tau,\omega})$, given by (3.25), from the formula [5]

$$S(D_{\tau_1,\omega_1}) = \frac{P(B_{\omega_1})-1}{P(B_{\omega_1})+1} , \tag{3.28}$$

thus completing the proof of the theorem.

By considering the additional cases where $\bar{\beta} \geq 1/4$ and $\bar{\beta} \leq 1/4$ (3.23) yields the following modified bound on $P(B_{\omega_1})$

$$P(B_{\omega_1}) \leq \begin{cases} [1+1/(1-M)^{\frac{1}{2}}]/2 & , \text{ if } \bar{\beta} \leq M/4 , \\ [1+(2/(1-M))^{\frac{1}{2}}]/2 & , \text{ if } M/4 \leq \bar{\beta} \leq 1/4, \\ [1+\gamma^{-1}(2/(1-M))^{\frac{1}{2}}]/2 & , \text{ if } 1/4 \leq \bar{\beta}, \end{cases} \tag{3.29}$$

where

$$\gamma = [1+2(\bar{\beta}-1/4)/(1-M)]^{-\frac{1}{2}} . \tag{3.30}$$

From the above theorem we see that good estimates for ω and τ to be used in the PSD method are given by ω_1 and τ_1, respectively in terms of an upper bound M for S(B) and an upper bound $\bar{\beta}$ for S(LU).

4. COMPARISON OF RECIPROCAL RATES OF CONVERGENCE OF PSD,SOR AND SSOR

We now compare the reciprocal rates of convergence* of SOR and SSOR with PSD in order to obtain an idea of the effectiveness of the latter method. The comparison is carried out under the general condition that A is a Stieltjes matrix as well as when A is a real, symmetric, positive definite and consistently ordered matrix. In the general case (when A is a Stieltjes matrix) it can be proved [24] that asymptotically as $S(B)\to 1^-$,

$$RR(B) \lesssim 2RR(B_{\bar{\omega}}) \text{ and } RR(\mathcal{L}_{\omega_b}) \lesssim \sqrt{RR(B_{\bar{\omega}})} , \quad (4.1)$$

where $B_{\omega}, \mathcal{L}_{\omega_b}$ denote the iteration matrices of the Jacobi Overrelaxation (JOR) and SOR methods, respectively. Moreover, in the case where A has Property A we find (see Theorem 6.8 p.51 in [16])

$$RR(B_{\bar{\omega}}) = RR(B) \text{ and } RR(\mathcal{L}_{\omega_b}) \sim \frac{1}{2\sqrt{2}} \sqrt{RR(B_{\bar{\omega}})} . \quad (4.2)$$

By combining the first parts of (4.1) and (4.2) successively with (3.29) we can establish Table 4.1. A simple examination of Table 4.1 and the table given by (4.55) in [26] reveals the following relationship between the asymptotic bounds on $RR(\mathcal{D}_{\tau_1,\omega_1})$ and the reciprocal rate of convergence for the SSOR method, denoted by $RR(\mathcal{G}_{\omega_1})$**

$$RR(\mathcal{D}_{\tau_1,\omega_1}) \sim RR(\mathcal{G}_{\omega_1})/2 . \quad (4.3)$$

Range of $\bar{\beta}$	Asymptotic Bounds on $RR(\mathcal{D}_{\tau_1,\omega_1})/\sqrt{RR(B_{\bar{\omega}})}$	
	General Case	Property A
$\bar{\beta}\leq\frac{M}{4}$	$\frac{1}{2\sqrt{2}}$	$\frac{1}{4}$
$\frac{M}{4}<\bar{\beta}\leq\frac{1}{4}$	$\frac{1}{2}$	$\frac{1}{2\sqrt{2}}$
$\frac{1}{4}<\bar{\beta}$	$\frac{1}{2}\gamma^{-1}$	$\frac{1}{2\sqrt{2}}\gamma^{-1}$

TABLE 4.1

---

* *For any convergent iterative method of the form* $u^{(n+1)}=Gu^{(n)}+k$ *the quantity* $R(G)=-\log S(G)$ *is the rate of convergence whereas the quantity* $RR(G)=-/R(G)$ *is the reciprocal rate of convergence.*

** $\mathcal{G}_{\omega_1}$ *denotes the iteration matrix of SSOR.*

From relationship (4.3) it follows that the number of iterations of the PSD method is asymptotically half the number of iterations of SSOR for both methods to achieve the same level of accuracy.

If we proceed to compare the asymptotic bounds on $RR(L_{\omega_b})$ and $RR(D_{\tau_1,\omega_1})$ we establish Table 4.2.

Range of $\bar{\beta}$	Asymptotic Bounds on $RR(D_{\tau_1,\omega_1})/RR(L_{\omega_b})$	
	General Case	Property A
$\bar{\beta}\leq\frac{M}{4}$	$\frac{1}{2\sqrt{2}}$	$\frac{1}{\sqrt{2}}$
$\frac{M}{4}\leq\bar{\beta}\leq\frac{1}{4}$	$\frac{1}{2}$	1
$\frac{1}{4}<\bar{\beta}$	$\frac{1}{2}\gamma^{-1}$	γ^{-1}

TABLE 4.2

From Table 4.2 we see that: i) In the general case and for $\bar{\beta}\leq 1/4$ we have that the SOR method converges approximately at least half as fast as the PSD method and ii) In the case of a consistently ordered matrix and for $\bar{\beta}\leq M/4$, the PSD method yields an improvement of approximately 40% in the range of convergence over SOR whereas for $M/4\leq\bar{\beta}\leq 1/4$ both methods yield approximately the same rate of convergence. As it is observed, from both Tables 4.1 and 4.2, for $\bar{\beta}>1/4$, the asymptotic bounds on $RR(D_{\tau_1,\omega_1})$ depend strongly upon γ^{-1} and are reliable if this quantity is not very large [26].

In order to test the theoretical results obtained we considered the numerical solution of the Laplace equation, where the boundary values were taken to be zero on all sides of the unit square. The starting vector $u^{(0)}$ used was the vector with all its components equal to unity. For purposes of comparison, we considered the application of the SOR, SSOR and PSD methods with optimum parameters to the system of equations derived from the discretization of the above equation. In each case the natural ordering of the points was used and the iterative procedure was terminated when the inequality $||u^{(n)}||_\infty\leq 10^{-6}$ was satisfied. The number of iterations for the different mesh sizes of the numerical experiments for the optimum $(\omega_o, \tau_o, \omega_b)$ cases are presented in Figure 4.1. For the determination of the rate of convergence of the SSOR, SOR and PSD methods with optimum parameters, we plot the logarithm of the number of iterations versus log h^{-1}.

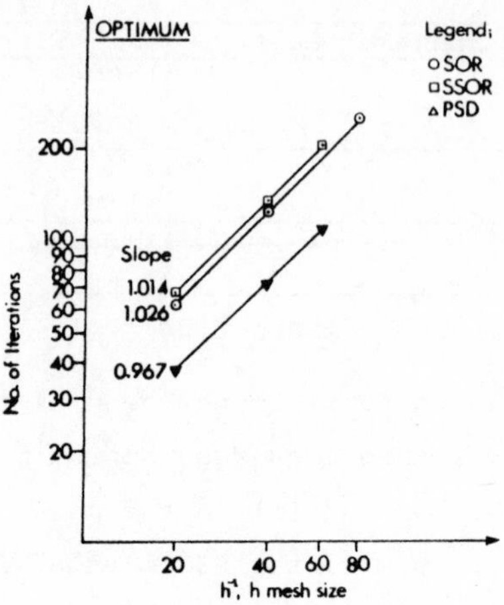

FIGURE 4.1

The slope α indicates $0(h^{\alpha})$
Rate of Conveyance

From Figure 4.1, it seems clear that for the model problem the number of iterations required for convergence with the PSD method behaves approximately as h^{-1}. This is also the case for the SOR and SSOR methods as expected. Also we see that when τ, ω are optimum, then as the mesh size h decreases, the number of iterations of PSD tends to become half the number of iterations of SSOR. Further details of the numerical results and their comparison with the theory are given in [9].

In the next section, we will proceed to examine the effectiveness of the PSD method accelerated by a variety of techniques.

5. THE PSD SEMI-ITERATIVE METHOD (PSD-SI METHOD)

Let us consider the completely consistent linear stationary iterative method defined by

$$u^{(n+1)} = Gu^{(n)}+k \ , \qquad (5.1)$$

where I-G is non-singular and $k=(I-G)A^{-1}b$. We assume that the eigenvalues μ of G are real and lie in the interval

$$\alpha \leq \mu \leq \beta < 1 \ , \qquad (5.2)$$

where α,β are real numbers. It can be shown [23] that we can improve the rate of convergence, often by an order of magnitude, by using the optimum semi-iterative method based on (5.1). This method is defined by

$$u^{(n+1)} = \rho_{n+1}[\bar{\rho}(Gu^{(n)}+k)+(1-\bar{\rho})u^{(n)}]+(1-\rho_{n+1})u^{(n-1)}, \qquad (5.3)$$

where

$$\left.\begin{aligned} &\bar{\rho} = 2/[2-(\alpha+\beta)], \quad \rho_1 = 1 \ , \quad \rho_2 = (1-\sigma^2/2)^{-1} \ , \\ &\rho_{n+1} = (1-\sigma^2\rho_n/4)^{-1} \ , \ n=2,3,4,\ldots \text{ and } \sigma=(\beta-\alpha)/[2-(\beta+\alpha)] \end{aligned}\right\}. \qquad (5.4)$$

For the PSD method we have that if A is positive definite, all the eigenvalues λ of B_ω are positive and there exist real positive numbers $\lambda(B_\omega),\Lambda(B_\omega)$ such that

$$0<\lambda(B_\omega)\leq\lambda\leq\Lambda(B_\omega) \ . \qquad (5.5)$$

Therefore, all the eigenvalues ν of $D_{\tau,\omega}$ are real and if $\tau>0$, lie in the range

$$\alpha = 1-\tau\Lambda(B_\omega)\leq\nu\leq 1-\tau\lambda(B_\omega) = \beta<1. \qquad (5.6)$$

By letting $G=D_{\tau,\omega}$ in (5.3), the formula for the optimum semi-iterative method based on PSD, denoted by PSD-SI, is given by

$$u^{(n+1)} = u^{(n-1)}+\rho_{n+1}(u^{(n)}-u^{(n-1)})+\rho_{n+1}\bar{\rho}(I-\omega U)^{-1}(I-\omega L)^{-1}D^{-1}(b-Au^{(n)}) \ , \qquad (5.7)$$

where $\bar{\rho}$ and the parameter sequence $\rho_1,\rho_2,\rho_3,\ldots$ can be determined by (5.4), (5.6) and are given by

$$\left.\begin{aligned} &\bar{\rho} = 2/(\Lambda(B_\omega)+\lambda(B_\omega)) \ , \ \rho_1 = 1, \ \rho_2 = (1-\sigma^2/2)^{-1}, \\ &\rho_{n+1} = (1-\sigma^2\rho_n/4)^{-1}, \ n=2,3,4,\ldots \text{ and } \sigma = (P(B_\omega)-1)/(P(B_\omega)+1), \end{aligned}\right\} \qquad (5.8)$$

where $P(B_\omega)$ denotes the P-condition number of B_ω, i.e.

$$P(B_\omega) = \Lambda(B_\omega)/\lambda(B_\omega) \ . \qquad (5.9)$$

Evidently, (5.7) is independent of τ. However by (5.8) we see that $\bar{\rho}=\tau_0$, thus a more compact form of the PSD-SI method is given by the following scheme

$$u^{(n+1)} = (1-\rho_{n+1})u^{(n-1)} + \rho_{n+1}(D_{\tau_0,\omega}u^{(n)} + \delta_{\tau_0,\omega}) . \quad (5.10)$$

Moreover, we have $\sigma=S(D_{\tau_0,\omega})<1$. As can be seen from (5.10) the PSD-SI method is a linear non-stationary method of second degree and the improvement in the convergence rate comes at the expense of requiring storage for one additional vector. An alternative form of the PSD-SI method in terms of the corresponding iteration matrix is the following

$$u^{(n)} = P_n(D_{\tau,\omega})u^{(0)} + k_n , \quad (5.11)$$

where

$$P_n(D_{\tau,\omega}) = T_n\left(\frac{2D_{\tau,\omega}-(\beta+\alpha)I}{\beta-\alpha}\right)/T_n\left(\frac{2-(\beta+\alpha)}{\beta-\alpha}\right) , \quad (5.12)$$

$k_n=(I-P_n(D_{\tau,\omega}))A^{-1}b$ and $T_n(x)$ is the classical Chebyshev polynomial of degree n [23].

By expressing $D_{\tau,\omega}$ in terms of B_ω and taking into account (5.6), (5.12) yields

$$P_n(D_{\tau,\omega}) = T_n\left(\frac{\Lambda(B_\omega)+\lambda(B_\omega)-2B_\omega}{\Lambda(B_\omega)-\lambda(B_\omega)}\right)\bigg/T_n\left(\frac{\Lambda(B_\omega)+\lambda(B_\omega)}{\Lambda(B_\omega)-\lambda(B_\omega)}\right) . \quad (5.13)$$

Next, we determine the virtual spectral radius $\bar{S}(P_n(D_{\tau,\omega}))$ from the expression

$$\bar{S}(P_n(D_{\tau,\omega})) = (T_n(1/\sigma))^{-1} = 2r^{n/2}/(1+r^n) , \quad (5.14)$$

where $$r^{\frac{1}{2}} = \sigma/(1+(1-\sigma^2)^{\frac{1}{2}}) = (\sqrt{P(B_\omega)}-1)/(\sqrt{P(B_\omega)}+1) . \quad (5.15)$$

In addition, for $P(B_\omega)>>1$, we have $r\sim 1-4/\sqrt{P(B_\omega)}$, thus the asymptotic average rate of convergence for the PSD-SI method is given by

$$R_\infty(P_n(D_{\tau,\omega})) = -\frac{1}{2}\log r \sim 2/\sqrt{P(B_\omega)} . \quad (5.16)$$

Since by (3.28) the rate of convergence of the PSD method $R(D_{\tau,\omega})$ behaves asymptotically as

$$R(D_{\tau_0,\omega}) \sim 2P(B_\omega) , \quad (5.17)$$

the following relationship holds between the reciprocal rates of convergence

$$RR_\infty(P_n(\mathcal{D}_{\tau,\omega})) \sim \sqrt{RR(\mathcal{D}_{\tau_0,\omega})/2} \quad . \tag{5.18}$$

Therefore, the use of the PSD-SI method results in an order of magnitude improvement in the reciprocal rate of convergence as compared to the repeated use of PSD.

Moreover from (5.18), we see that the improvement in the rate of convergence of the PSD-SI method compares favourably with the frequently used SOR method as it has approximately the same rate of convergence with PSD [9]. This can be seen more explicitly by a simple comparison between the asymptotic bounds on $RR_\infty(P_n(\mathcal{D}_{\tau_1,\omega_1}))$ with the best possible bound on $RR(\mathcal{L}_{\omega_b})$. These comparisons are shown in Table 5.2, from which we see that for $0\leq\bar{\beta}\leq 1/4$ we have substantial improvements of the rate of convergence of PSD-SI over SOR, while for the case where $\bar{\beta}\geq 1/4$ the results again depend strongly upon the quantity γ.

Range of $\bar{\beta}$	Asymptotic Bounds on $RR_\infty(P_n(\mathcal{D}_{\tau_1,\omega_1}))/(RR(B_{\bar{\omega}}))^{1/4}$	
	General Case	Property A
$\bar{\beta}\leq M/4$	$2^{-5/4}$	$2^{-3/2}$
$M/4\leq\bar{\beta}\leq 1/4$	2^{-1}	$2^{-5/4}$
$1/4<\bar{\beta}$	$2^{-1}\gamma^{-1/2}$	$2^{-5/4}\gamma^{-1/2}$

TABLE 5.1

A comparison between the PSD-SI and JOR methods. These comparisons are obtained from (5.18) using Table 4.1.

Range of $\bar{\beta}$	Asymptotic Bounds on $RR_\infty(P_n(\mathcal{D}_{\tau_1,\omega_1}))/\sqrt{RR(\mathcal{L}_{\omega_b})}$	
	General Case	Property A
$\bar{\beta}\leq M/4$	$2^{-5/4}$	$2^{-3/4}$
$M/4\leq\bar{\beta}\leq 1/4$	2^{-1}	$2^{-1/2}$
$1/4<\bar{\beta}$	$2^{-1}\gamma^{-1/2}$	$(2\gamma)^{-1/2}$

TABLE 5.2

A comparison between the PSD-SI and SOR methods. These comparisons are obtained from (5.18) using Table 4.2.

6. THE PSD SECOND DEGREE METHOD (PSD-SD METHOD)

It is known (see [11]) that in many cases one can replace the non-stationary second-degree method (5.3) which is equivalent to the optimum semi-iterative method, by a stationary second-degree method without a significant loss in the convergence rate. The method is similar to (5.3) and if we consider (5.1) as the basic method it is defined by

$$u^{(n+1)} = u^{(n)} + \xi_o(u^{(n)} - u^{(n-1)}) + \eta_o(Gu^{(n)} + k - u^{(n)}) , \quad (6.1)$$

where $\xi_o = \hat{\omega}_o - 1,\ \hat{\omega}_o = 2/[1+(1-\sigma^2)^{\frac{1}{2}}],\ \eta_o = 2\hat{\omega}_o/[2-(\alpha+\beta)],$ (6.2)

and σ is defined by (5.8). Let us now assume that A is positive definite and $G=D_{\tau,\omega}$, then from (5.8), (5.6) and (6.2) we have

$$\hat{\omega}_o = 1+[(\sqrt{P(B_\omega)}-1)/(\sqrt{P(B_\omega)}+1)]^2 \text{ and } \eta_o = \hat{\omega}_o\bar{\rho}/\tau . \quad (6.3)$$

Therefore, the formula for the optimum second-degree method based on PSD, denoted by PSD-SD, is given by

$$u^{(n+1)} = u^{(n)} + (\hat{\omega}_o - 1)(u^{(n)} - u^{(n-1)}) + \hat{\omega}_o\bar{\rho}(I-\omega U)^{-1}(I-\omega L)^{-1}D^{-1}(b-Au^{(n)}) . \quad (6.4)$$

Evidently, (6.4) is independent of τ. However, by (5.8) we see that $\bar{\rho}=\tau_o$, thus the PSD-SD method can be written in the alternative form

$$u^{(n+1)} = (1-\hat{\omega}_o)u^{(n-1)} + \hat{\omega}_o(D_{\tau_o,\omega}u^{(n)} + \delta_{\tau_o,\omega}) . \quad (6.5)$$

If $\hat{G}$ denotes the iteration matrix of (6.5), then we have [16]

$$S(\hat{G}) = (\hat{\omega}_o - 1)^{\frac{1}{2}} = (\sqrt{P(B_\omega)}-1)/(\sqrt{P(B_\omega)}+1). \quad (6.6)$$

From (6.6), we see that if $P(B_\omega)$ is sufficiently large, as is frequently the case then

$$R(\hat{G}) \sim 2/\sqrt{P(B_\omega)} . \quad (6.7)$$

From relationship (6.7), it follows that by using the PSD-SD method we can attain an order of magnitude improvement in the rate of convergence as compared with the PSD method.

For a direct comparison with the PSD-SI iterative scheme, we specify the first iterant $u^{(1)}$ to be determined by a PSD iteration. Thus, the PSD-SD method is defined by

$$u^{(1)} = D_{\tau_o,\omega} u^{(0)} + \delta_{\tau_o,\omega}, \quad u^{(n+1)} = (1-\hat{\omega}_o)u^{(n-1)} + \hat{\omega}_o(D_{\tau_o,\omega}u^{(n)} + +\delta_{\tau_o,\omega}), n=1,2,\ldots \quad (6.8)$$

In this case, the virtual spectral radius of the PSD-SI method can be proved to be given by (see [23] p.490-491),

$$\bar{S}(Q_n(D_{\tau,\omega})) = \frac{2\hat{r}^{n/2}}{1+\hat{r}^n}[1+(\frac{n-1}{2})(1-\hat{r})], \quad (6.9)$$

where

$$\hat{r} = \hat{\omega}_o - 1 \quad ,$$

and the polynomials $Q_n(D_{\tau,\omega})$ satisfy the recurrence relation

$$\left.\begin{aligned} &Q_o(D_{\tau,\omega}) = I, \quad Q_1(D_{\tau,\omega}) = D_{\tau,\omega} \quad , \\ &Q_{n+1}(D_{\tau,\omega}) = \hat{\omega}_o D_{\tau,\omega} Q_n(D_{\tau,\omega}) + (1-\hat{\omega}_o)Q_{n-1}(D_{\tau,\omega}) \end{aligned}\right\}. \quad (6.10)$$

By recalling (5.14), the virtual spectral radius of the PSD-SI method is given by

$$\bar{S}(P_n(D_{\tau,\omega})) = 2\hat{r}^{n/2}/(1+\hat{r}^{(n)}) \quad . \quad (6.11)$$

Thus, the PSD-SI method converges faster than the PSD-SD method since $\bar{S}(P_n(D_{\tau,\omega})) \lessapprox \bar{S}(Q_n(D_{\tau,\omega}))$.

7. THE PSD VARIABLE EXTRAPOLATION METHOD (PSD-VE METHOD)

Now, we construct a non-stationary iterative method based on PSD which necessitates the storage of just one vector iterant and attains approximately the same rate of convergence as the PSD-SI and the PSD-SD methods. The variable extrapolation method is a modified version of Richardson's method [22] and if we consider again (5.1) as the basic method, it is defined by

$$u^{(n+1)} = \theta_{n+1}(Gu^{(n)} + k) + (1-\theta_{n+1})u^{(n)} \quad , \quad (7.1)$$

where we choose different values of the θ_k and then we use them in the periodic cyclic order $\theta_1, \theta_2, \ldots, \theta_m, \theta_1, \theta_2, \ldots$. The values of the θ_k are given by the following expression

$$\theta_k = 2/[2-(\beta-\alpha)\cos\frac{(2k-1)\pi}{2m} - (\beta+\alpha)], \; k=1,2,\ldots,m. \quad (7.2)$$

By letting $G=D_{\tau,\omega}$ in (7.1), the optimum variable extrapolation method based on PSD, denoted by PSD-VE is defined by

$$u^{(n+1)} = \theta_{n+1}(\mathcal{D}_{\tau,\omega}u^{(n)}+\delta_{\tau,\omega})+(1-\theta_{n+1})u^{(n)},$$

which by (2.5) and (2.6) yields (7.3)

$$u^{(n+1)} = u^{(n)}+\hat{\theta}_{n+1}(I-\omega U)^{-1}(I-\omega L)^{-1}D^{-1}(b-Au^{(n)}), \quad (7.4)$$

where $\hat{\theta}_{n+1}=\theta_{n+1}\tau$ and the iteration parameters $\hat{\theta}_k$ are determined by (7.2),(5.6),(5.8), hence,

$$\hat{\theta}_k = \bar{\rho}/(1-\sigma\cos\frac{(2k-1)\pi}{2m}), \quad k=1,2,\ldots,m. \quad (7.5)$$

In order to determine the efficiency of PSD-VE we determine its virtual spectral radius which can be verified to be [19]

$$\bar{S}(P_{tm}(\mathcal{D}_{\tau,\omega})) = [2r^{m/2}/(1+r^m)]^t, \quad (7.6)$$

where r is defined by (5.15) and m,t are integers (t specifies the number of cycles). Although, the method seems to be quite promising, care must be taken with the values of m if one uses the aforementioned cyclic ordering of the iteration parameters θ_k since numerical instability may occur [22]. However, this difficulty can be overcome and as we will see later the PSD-VE method combined with a technique which reduces the computational work involved makes this the most desirable accelerated version of the PSD method.

8. THE PSD CONJUGATE GRADIENT METHOD (PSD-CG METHOD)

In this section, we will consider the application of the Conjugate Gradient method (CG method) [15],[25] as an alternative form of accelerating the convergence rate of PSD. Next, we present the CG method based on (5.1), where we assume that G has the form

$$G = I-(QQ^T)^{-1}A, \quad (8.1)$$

with Q being a non-singular matrix. In this case we define the CG method as the non-stationary second degree method [25]

$$u^{(n+1)} = \rho_{n+1}[\gamma_{n+1}(Gu^{(n)}+k)+(1-\gamma_{n+1})u^{(n)}]+(1-\rho_{n+1})u^{(n-1)},$$

where (8.2)

$$\left.\begin{aligned} \gamma_{n+1} &= (\hat{r}^{(n)},\hat{r}^{(n)})/(\hat{s}^{(n)},A\hat{s}^{(n)}), \quad \hat{r}^{(n)} = Q^{-1}r^{(n)}, \\ r^{(n)} &= b-Au^{(n)}, \\ \hat{s}^{(n)} &= Gu^{(n)}+k-u^{(n)}, \end{aligned}\right\}$$

and
$$\rho_{n+1} = \left[1 - \frac{\gamma_{n+1}}{\gamma_n} \cdot \frac{(\hat{r}^{(n)},\hat{r}^{(n)})}{(\hat{r}^{(n-1)},\hat{r}^{(n-1)})} \cdot \frac{1}{\rho_n}\right]^{-1} \quad n=1,2,\ldots \;. \qquad (8.3)$$

The advantages of the CG method over the previously considered accelerated techniques are:

i) It is not necessary to know the maximum and minimum eigenvalues of G and

ii) The CG method appears to be better than the SI method applied to (5.1) in the sense of minimising the $A^{\frac{1}{2}}$-norm of the error vector.

In fact, it can be proved that [23]

$$||u^{(n)}-\bar{u}||_{A^{\frac{1}{2}}} \leq ||\tilde{u}^{(n)}-\bar{u}||_{A^{\frac{1}{2}}} \;, \qquad (8.4)$$

where $\bar{u}$ is the exact solution of (1.1), $u^{(n)}, \tilde{u}^{(n)}$ are the approximate solutions obtained by the CG and SI methods with respect to (5.1), respectively.

Let us now assume, without loss of generality (see [23], p.112), that the matrix A in (1.1) has the splitting A=I-L-U, where $L=U^T$ and L,U are the strictly lower and upper triangular parts of A, respectively. We note that if we let

$$Q = \tau^{-\frac{1}{2}}(I-\omega L) \;, \qquad (8.5)$$

the iteration matrix of the PSD method has the form (8.1). By (8.3) and (8.5) we have

$$\hat{r}^{(n)} = \sqrt{\tau}(I-\omega L)^{-1}r^{(n)} \text{ and } \hat{s}^{(n)} = \tau(I-\omega U)^{-1}(I-\omega L)^{-1}r^{(n)} \;, \qquad (8.6)$$

since the PSD method is given by

$$u^{(n+1)} = u^{(n)}+\tau(I-\omega U)^{-1}(I-\omega L)^{-1}r^{(n)} \;. \qquad (8.7)$$

Therefore by (8.2) and (8.7) we readily see that the PSD-CG scheme is defined by

$$u^{(n+1)} = \rho_{n+1}(u^{(n)}+\hat{\gamma}_{n+1}(I-\omega U)^{-1}(I-\omega L)^{-1}r^{(n)})+(1-\rho_{n+1})u^{(n-1)}, \qquad (8.8)$$

where the iteration parameters $\hat{\gamma}_{n+1}=\gamma_{n+1}\tau$ and ρ_{n+1} are determined by (8.3) as follows,

$$\hat{\gamma}_{n+1} = (r^{(n)},\tilde{s}^{(n)})/(\tilde{s}^{(n)},A\tilde{s}^{(n)}),$$
$$\tilde{s}^{(n)} = (I-\omega U)^{-1}(I-\omega L)^{-1}r^{(n)} \;, \qquad (8.9)$$

and

$$\rho_{n+1} = \left[1 - \frac{\gamma_{n+1}}{\gamma_n} \cdot \frac{(r^{(n)}, r^{(n)})}{(r^{(n-1)}, r^{(n-1)})} \cdot \frac{1}{\rho_n}\right]^{-1} . \tag{8.10}$$

Evidently, by (8.4) it follows that with the PSD-CG method we will also have an order of magnitude improvement in the rate of convergence as compared to SOR.

9. COMPUTATIONAL RESULTS FOR THE MODEL PROBLEM

In order to obtain some information about the aforementioned accelerated iterative procedures in practice, we consider now the numerical solution of the Laplace equation in the unit square with zero boundary values. The starting vector $u^{(0)}$ used was the vector with all its components equal to unity. For purposes of comparison we considered the application of the PSD-SI, PSD-VE and PSD-CG to the system of equations derived from the discretisation of the aforementioned equation. The number of iterations for the different mesh sizes of the numerical experiments are shown in Figure 9.1. For all the cases, we used the optimum parameters τ_o, ω_o [9] whereas for the PSD-VE method the value

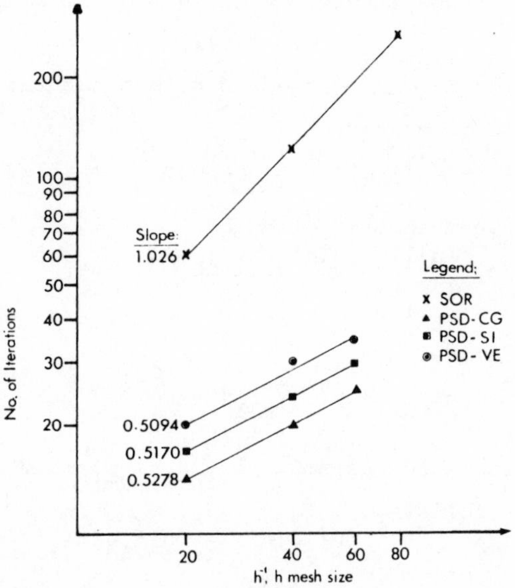

FIGURE 9.1

The slope α indicates $O(h^{\alpha})$ Rate of Conveyance of m_o was determined as the smallest integer such that (see [ ])

$$\left[-\frac{1}{m_o}\log\frac{2r^{m_o/2}}{1+r^{m_o}}\right]^{-1} \leq 1.25/(-\tfrac{1}{2}\log r) \ . \qquad (9.1)$$

This convergence criterion guarantees that the reciprocal rate of convergence of PSD-VE does not exceed 125% of the reciprocal rate of convergence of the PSD-SI method. The determination of the rate of convergence which has been attained by these methods is also shown in Figure 9.1 where we plot the number of iterations versus $\log(h^{-2})$. From Figure 9.1 we note that the number of iterations required for convergence using the PSD-SI, PSD-VE and PSD-CG methods behaves approximately as $h^{-\frac{1}{2}}$ which is a substantial improvement over the behaviour h^{-1} of SOR. However, if we take into account the computational work involved as the other criterion of comparison, then we have that (see Appendix A in [16]) the number of operations required per iteration using the PSD-SI method is approximately twice that required by the SOR method. Moreover, the PSD-CG method involves considerable computational effort which does not seem to offset the relatively small gain in the number of iterations compared with PSD-SI. On the other hand, we note that the form (1.3) of PSD enables us to apply a similar technique to Niethammer's [19] in order to reduce the computational effort so as to be competitive with the work involved in SOR. This work-saving technique cannot be effectively applied for the PSD-SD, PSD-SI and the PSD-CG methods since they are all second-degree methods [16]. Consequently the advantage of Niethammer's technique can be exploited only in the PSD-VE method since (7.4) can be written alternatively as (see also (2.3))

$$u^{(n+\frac{1}{2})} = \theta_{n+1}\left(\frac{\omega}{\theta_{n+1}}Lu^{(n+\frac{1}{2})} + \left(1-\frac{\omega}{\theta_{n+1}}\right)Lu^{(n)} + Uu^{(n)} + c\right) + \\ + (1-\theta_{n+1})u^{(n)} \ ,$$

and

$$u^{(n+1)} = u^{(n+\frac{1}{2})} + \omega(Uu^{(n+1)} - Uu^{(n)}) \ , \qquad (9.2)$$

where we see that if during the nth iteration the vector $Uu^{(n)}$ is stored, it can be used in the next iteration. Thus, the number of operations of the PSD-VE procedure is approximately the same as the number of operations of the PSD method which for n iterations are given by the expression $(22n+8)J^2$ as compared to $17nJ^2$ for the SOR method (see Table A.1, Appendix A in [16]), where h is the mesh size and $h=1/J$.

Further details of the application of these accelerated preconditioned iterative methods to a class of more generalised problems are given in [17].

10. THE ADAPTIVE DETERMINATION OF THE PSD PARAMETERS

From Section 3 we recall that the preconditioning parameter is optimum for that value of ω for which $P(B_\omega)$ is minimised. We have also shown that for the largest and smallest eigenvalue of B_ω we can let, respectively

$$\Lambda(B_\omega) = \frac{1}{\omega(2-\omega)} \quad , \quad \lambda(B_\omega) = \frac{1-\alpha}{1-\omega\alpha+\omega^2\beta} = \phi(\omega,v) \quad , \tag{10.1}$$

where

$$\alpha = \frac{(v,Bv)}{(v,v)} \quad , \quad \beta = \frac{(v,LUv)}{(v,v)} \quad , \tag{10.2}$$

and $B_\omega v=\lambda(B_\omega)v$. Therefore, the P-condition number of B_ω is given by the expression

$$P(B_\omega) \lesssim \frac{1-\omega\alpha+\omega^2\beta}{\omega(2-\omega)(1-\alpha)} \quad , \tag{10.3}$$

where α and β are given by (10.2). From (10.3) and (10.1) we see that finding the optimum parameters ω_o, $P(B_{\omega_o})$ depends upon the availability of an eigenvector corresponding to the smallest eigenvalue of B_ω. Evidently, if we happen to know this eigenvector, then we would be able to determine $\lambda(B_\omega)$ from the formula $\lambda(B_\omega)=(v,B_\omega v)/(v,v)$ and then compute $P(B_\omega)$ from the expression $P(B_\omega)=[\omega(2-\omega)\lambda(B_\omega)]^{-1}$.

It is therefore clear that the determination of v such that $B_\omega v=\lambda(B_\omega)v$ is satisfied, is essential for obtaining the optimum parameters. Thus, we are motivated by this observation to seek for a vector which is automatically calculated in the practical implementation of the PSD-SI method and can be made to approach an eigenvector of B_ω corresponding to $\lambda(B_\omega)$.

Next, our task will be to compute an approximation to the quantities α and β defined by (10.2) thus computing by (10.3) approximations (using the analysis of Section 3) to the optimum parameters ω_o and $P(B_{\omega_o})$.

Theorem 10.1: Without loss of generality, let $A=I-L-U$ be a positive definite matrix, then for any vector $v\neq 0$, the representation $\phi(\omega,v)$ given by (10.1) is a Rayleigh quotient with respect to the vector $w=(I-\omega U)v$ and the positive definite matrix $\bar{B}_\omega=(I-\omega U)B_\omega(I-\omega U)^{-1}$ that is $\phi(\omega,v)=\frac{(w,\bar{B}_\omega w)}{(w,w)}$. Furthermore,

$$\lambda(B_\omega)=\lambda(\bar{B}_\omega)\lesssim\phi(\omega,v)\ . \tag{10.4}$$

Proof: We first show that $\bar{B}_\omega$ is positive definite. We recall that

$$B_\omega=(I-\omega U)^{-1}(I-\omega L)^{-1}A\ . \tag{10.5}$$

Hence

$$\begin{aligned}\bar{B}_\omega&=(I-\omega U)B_\omega(I-\omega U)^{-1}=(I-\omega L)^{-1}A(I-\omega U)^{-1}\\&=(I-\omega L)^{-1}A\,[(I-\omega L)^{-1}]^T\ .\end{aligned} \tag{10.6}$$

Therefore, $\bar{B}_\omega$ is positive definite. From (10.5) we also obtain

$$A=(I-\omega L)\bar{B}_\omega(I-\omega U)\ . \tag{10.7}$$

If we take inner products of both sides of the last equation with respect to $v\neq 0$ we have

$$(v,Av)=(v,(I-\omega L)\bar{B}_\omega(I-\omega U)v)=(w,\bar{B}_\omega w)\ , \tag{10.8}$$

where $w=(I-\omega U)v$. Dividing by (w,w) both sides of (10.8) we have

$$\frac{(v,Av)}{(w,w)}=\frac{(w,\bar{B}_\omega w)}{(w,w)}\ . \tag{10.9}$$

Expanding the inner product (w,w) we have successively

$$\begin{aligned}(w,w)&=((I-\omega U)v,(I-\omega U)v)=(v,(I-\omega(L+U)+\omega^2LU)v),\\&=(v,v)-\omega(v,Bv)+\omega^2(v,LUv)\ .\end{aligned} \tag{10.10}$$

Hence, by using (10.10) the left hand side of (10.9) yields

$$\frac{(v,Av)}{(w,w)}=\frac{(v,v)-(v,Bv)}{(v,v)-\omega(v,Bv)+\omega^2(v,LUv)}=\frac{1-\alpha}{1-\omega\alpha+\omega^2\beta}=\phi(\omega,v)\ .$$

Thus (10.9) becomes

$$\phi(\omega,v)=\frac{(w,\bar{B}_\omega w)}{(w,w)}\ .$$

Finally, since $\bar{B}_\omega$ is similar to B_ω, we have (10.4) and the proof of the theorem is complete.

Corollary 10.2: Under the hypotheses of Theorem 10.1 any eigenvalue of B_ω can be represented by $\phi(\omega,v)$.

From the above analysis we see that the inequality (10.4) is satisfied for any non-zero vector v. On the other hand, we observe from (10.4) that the closer we approach an eigenvector corresponding to the smallest eigenvalue of B_ω, the better we will be able to determine $\lambda(B_\omega)$ from $\phi(\omega,v)$. It is evident now there is a strong need for finding this eigenvector. However, we have to devise another approach other than using the power method since the power iterations require extensive computational effort and do not contribute directly to the solution of the system (1.1). The answer to this problem was given in [3] for the general case. Here we have modified this approach to suit our purposes.

Next, we will first show that the pseudo-residual vector, as defined in [3] satisfies the relationship $\delta^{(n)}=P_n(B_\omega)\delta^{(0)}$ and secondly that $\delta^{(n)}$ approaches the vector v which satisfies the relationship $B_\omega v=\lambda v$.

Theorem 10.3: The pseudo-residual vector

$$\delta^{(n)}=D_{1,\omega}u^{(n)}+\delta_{1,\omega}-u^{(n)}, \tag{10.11}$$

where $u^{(n)}$ is the latest PSD-SI iterate, satisfies the relationship $\delta^{(n)}=P_n(B_\omega)\delta^{(0)}$, with $D_{1,\omega}=I-{}_\omega$ and $\delta_{1,\omega}=(I-\omega U)^{-1}(I-\omega L)^{-1}b$. (10.12)

Proof: From (10.11) we have

$$\delta^{(n)}=B_\omega(u^{(n)}-\bar{u}), \tag{10.13}$$

where $u^{(n)}$ is the latest PSD-SI iterate and $\bar{u}=(I-B_\omega)\bar{u}+\delta_{1,\omega}$ with $\bar{u}$ being the exact solution. Alternatively we can write the PSD-SI method in the form

$$u^{(n)}=P_n(B_\omega)u^{(0)}+k_n, \tag{10.14}$$

where

$$P_n(B_\omega)=\frac{T_n\left(\dfrac{\Lambda(B_\omega)+\lambda(B_\omega)-2B_\omega}{\Lambda(B_\omega)-\lambda(B_\omega)}\right)}{T_n\left(\dfrac{\Lambda(B_\omega)+\lambda(B_\omega)}{\Lambda(B_\omega)-\lambda(B_\omega)}\right)}, \tag{10.15}$$

$k_n=(I-P_n(B_\omega))A^{-1}b$ and $T_n(x)$ is the classical Chebyshev polynomial of degree n.

Moreover, since (10.14) is consistent, we have the relationship

$$\bar{u} = P_n(B_\omega)\bar{u} + k_n \ . \tag{10.16}$$

Subtracting (10.16) from (10.14) yields the result

$$u^{(n)} - \bar{u} = P_n(B_\omega)(u^{(0)} - \bar{u}) \ . \tag{10.17}$$

Further, by combining (10.13) and (10.17) we obtain the following expression for $\delta^{(n)}$

$$\delta^{(n)} = B_\omega P_n(B_\omega)(\bar{u} - u^{(0)}) \ . \tag{10.18}$$

Letting n=0 in (10.13) we obtain

$$u^{(0)} - \bar{u} = B_\omega^{-1}\delta^{(0)} \ , \tag{10.19}$$

which on substitution in (10.18) yields (10.12) and the proof of the theorem is complete. The next theorem will establish the fact that $\delta^{(n)}$ does converge to a multiple of the eigenvector corresponding to the smallest eigenvalue of B_ω.

Theorem 10.4: The pseudo-residual vector $\delta^{(n)}$ given by (10.11) approaches a multiple of the eigenvector associated with $\lambda(B_\omega)$ as $n\to\infty$.

Proof: Let v_k, k=1,2,...,N, form a complete set of eigenvectors of B_ω corresponding to the positive eigenvalues

$$\Lambda(B_\omega) = \lambda_1 > \lambda_2 \geq \ldots \geq \lambda_{N-1} > \lambda_N = \lambda(B_\omega) > 0. \tag{10.20}$$

Next, we express $\delta^{(0)}$ as a linear combination of v_k, thus

$$\delta^{(0)} = \sum_{k=1}^{N} d_k v_k \ , \ d_N \neq 0 \ , \tag{10.21}$$

and let

$$\lambda_E(B_\omega) > \lambda_N \ , \tag{10.22}$$

where $\lambda_E(B_\omega)$ is an estimate of $\lambda(B_\omega)$.

From Theorem 10.3 we have that the pseudo-residual vector defined by (10.11) satisfies the relationship (10.12) which by using (10.21) and (10.15) yields successively

$$\delta^{(n)} = P_n(B_\omega)\sum_{k=1}^{N} d_k v_k = \sum_{k=1}^{N} d_k P_n(\lambda_k) v_k = P_n(\lambda_N)\left\{ d_N v_N + \sum_{k=1}^{N-1} d_k \frac{T_n\left(1 - \dfrac{2(\lambda_k - \lambda_E(B_\omega))}{\lambda_1 - \lambda_E(B_\omega)}\right)}{T_n\left(1 - \dfrac{2(\lambda_N - \lambda_E(B_\omega))}{\lambda_1 - \lambda_E(B_\omega)}\right)} v_k \right\} .$$

Consequently,

$$\frac{\delta^{(n)}}{P_n(\lambda_N)} = d_N v_N + \sum_{\lambda_k \in [\lambda_E(B_\omega),\lambda_1]} d_k \frac{T_n\left(1-\frac{2(\lambda_k-\lambda_E(B_\omega))}{\lambda_1-\lambda_E(B_\omega)}\right)}{T_n\left(1-\frac{2(\lambda_N-\lambda_E(B_\omega))}{\lambda_1-\lambda_E(B_\omega)}\right)} v_k$$

$$+ \sum_{\substack{\lambda_k \notin [\lambda_E(B_\omega),\lambda_1] \\ \lambda_k \neq \lambda_N}} d_k \frac{T_n\left(1-\frac{2(\lambda_k-\lambda_E(B_\omega))}{\lambda_1-\lambda_E(B_\omega)}\right)}{T_n\left(1-\frac{2(\lambda_N-\lambda_E(B_\omega))}{\lambda_1-\lambda_E(B_\omega)}\right)} v_k . \tag{10.23}$$

Applying the classical Chebyshev Theorem (see e.g. [23]) to the terms of the first sum we see that as n increases these terms are decreasing at an optimal rate. Let us now concentrate on the terms of the second sum where we have that $\lambda_k \in [\lambda_N, \lambda_E(B_\omega)]$, i.e.,

$$\lambda_E(B_\omega) > \lambda_k > \lambda_N . \tag{10.24}$$

We note that by letting

$$x = 1 - \frac{2(\lambda_k-\lambda_E(B_\omega))}{\lambda_1-\lambda_E(B_\omega)} , \quad y = 1 - \frac{2(\lambda_N-\lambda_E(B_\omega))}{\lambda_1-\lambda_E(B_\omega)} ,$$

using (10.24) we can easily find the following relationship between x and y

$$y > x > 1 . \tag{10.25}$$

On the other hand, since $x>1$, $T_n(x)=\frac{1}{2}[e^{n\cosh^{-1}x}+e^{-n\cosh^{-1}x}]$.

Finally, by using (10.25) we obtain

$$\lim_{n\to\infty} \frac{T_n(x)}{T_n(y)} = \lim_{n\to\infty} e^{n(\cosh^{-1}x-\cosh^{-1}y)} = 0 , \tag{10.26}$$

which indicates that the terms of the second sum in (10.23) vanish also as $n\to\infty$ and therefore the proof of the theorem is complete.

The above theorem established the theoretical basis for the use of the PSD-SI method to solve Au=b and simultaneously compute an approximate eigenvector associated with $\lambda(B_\omega)$. This can be more explicitly seen if we express the pseudo-residual vector as

$$\delta^{(n)} = (I-\omega U)^{-1}(I-\omega L)^{-1}(b-Au^{(n)}) , \tag{10.27}$$

where $u^{(n)}$ is the latest PSD-SI iteration. But, the next PSD-SI iteration $u^{(n+1)}$ is given by

$$u^{(n+1)} = (1-\rho_{n+1})u^{(n-1)}+\rho_{n+1}(u^{(n)}+\bar{\rho}\delta^{(n)}), \tag{10.28}$$

which shows that the pseudo-residual vector is essentially obtained as a by-product of the application of the PSD-SI method. This last observation is the main advantage of determining the parameters ω and $P(B_\omega)$ adaptively since we obtain the fundamental eigenvector by exploiting the iteration used to improve the accuracy of the approximate solution of $Au=b$.

Furthermore, it remains to be shown that any approximation to an eigenvector of B_ω yields a corresponding eigenvalue approximation. This is derived from a theorem in [3] and is presented here without proof.

Theorem 10.5: If A and B are positive definite matrices, then the eigenvectors x_k and the corresponding eigenvalues λ_k of the generalised eigenvalue problem $Ax_k=\lambda_k Bx_k$ satisfy the following properties:

a) The eigenvalues λ_k are all positive i.e., $\lambda_k>0$ for $k=1,2,\ldots,N$.

b) The eigenvectors x_k of $B^{-1}A$ are orthogonal with respect to B, i.e. $(x_k,Bx_j)=0$ for $j\neq k$.

Finally, the next Lemma defines the approximate eigenvalue $\lambda(B_\omega)$.

Lemma 10.6: Let λ_k, $k=1,2,\ldots,N$, be the eigenvalues of B_ω and v_k the corresponding eigenvectors. If $\mu = \dfrac{(y,Ay)}{(y,(I-\omega L)(I-\omega U)y)}$ where y is approximately equal to v_N with errors ε_k in the direction v_k and $\varepsilon_N \gg \varepsilon_k$, $k\neq N$, then $\mu \simeq \lambda_N$ with error of order $\left(\dfrac{\varepsilon_k}{\varepsilon_N}\right)^2$.

Proof: From the hypotheses of the above lemma we have that

$$B_\omega v_j = \lambda_j v_j \quad , \tag{10.29}$$

also from Theorem 10.5 we have

$$(v_k,(I-\omega L)(I-\omega U)v_j) = \delta_{j,k} \quad , \tag{10.30}$$

where $\delta_{j,k}$ is the Kronecker delta. Next, we express y in terms of v_k, hence

$$y = \sum_{k=1}^{N} \varepsilon_k v_k \, ,$$

and

$$\mu = \frac{\left(\sum_{k=1}^{N}\varepsilon_k v_k,\ A\sum_{j=1}^{N}\varepsilon_j v_j\right)}{\left(\sum_{k=1}^{N}\varepsilon_k v_k,(I-\omega L)(I-\omega U)\sum_{j=1}^{N}\varepsilon_j v_j\right)}$$

$$= \frac{\left(\sum_{k=1}^{N}\varepsilon_k v_k,(I-\omega L)(I-\omega U)B_\omega\sum_{j=1}^{N}\varepsilon_j v_j\right)}{\left(\sum_{k=1}^{N}\varepsilon_k v_k,(I-\omega L)(I-\omega U)\sum_{j=1}^{N}\varepsilon_j v_j\right)} . \qquad (10.31)$$

Combining now (10.29),(10.20) and (10.31) we obtain the result

$$\mu = \frac{\sum_{k=1}^{N}\varepsilon_k^2\lambda_k}{\sum_{k=1}^{N}\varepsilon_k^2} = \frac{\lambda_N + \sum_{k=1}^{N-1}\varepsilon_k^2\lambda_k/\varepsilon_N^2}{1 + \sum_{k=1}^{N-1}\varepsilon_k^2/\varepsilon_N^2} .$$

which completes the proof of the lemma.

As a result of Theorem 10.4 and Lemma 10.6 we have that an approximation to $\lambda(B_\omega)$ can be calculated by the expression

$$\mu^{(n)} = \frac{(\delta^{(n)},A\delta^{(n)})}{(\delta^{(n)},(I-\omega L)(I-\omega U)\delta^{(n)})} . \qquad (10.32)$$

11. THE ADAPTIVE ALGORITHM

In this section, a precise definition of the algorithm which uses the PSD-SI method and simultaneously improves the parameters ω and $P(B_\omega)$ is given.

From Theorem 10.1 which gives an upper bound $\phi(\omega,v)$ for the smallest eigenvalue $\lambda(B_\omega)$ we conclude that we can determine a lower bound for the P-condition number $P(B_\omega)$ from the relationship

$$p(\omega,v) \leq P(B_\omega) , \qquad (11.1)$$

where

$$p(\omega,v) = \frac{1}{\omega(2-\omega)\phi(\omega,v)} . \qquad (11.2)$$

The lower bound $p(\omega,v)$ of $P(B_\omega)$ as defined in (11.2) indicates that it should be possible to approximate $P(B_\omega)$ by using the PSD-SI method. In order to obtain more information about the role of α and β given by (10.2) we examine the behaviour of $\phi(\omega,v)$ with respect to these quantities. We recall from (10.1) that,

$$\phi(\omega,v) = \frac{1-\omega\alpha+\omega^2\beta}{1-\alpha} = \omega + \frac{1-\omega+\omega^2\beta}{1-\alpha} \quad , \tag{11.3}$$

hence we can easily construct Table 11.1

β-Domain	ω-Domain	$\omega^2\beta-\omega+1$
β≤1/4	0<ω≤ω*	≥0
	ω=ω*	=0
	0<ω*≤ω	≤0
β>1/4	0<ω<2	>0

TABLE 11.1: Behaviour of $\omega^2\beta-\omega+1$ as a function of ω

where
$$\omega^* = \frac{2}{1+\sqrt{1-4\beta}} \quad . \tag{11.4}$$

From (11.2),(11.3) and a cursory examination of Table 11.1 reveals that for β≤1/4 we have i) if ω≤ω*, then p(ω,v) is maximised when α is maximised, ii) if ω=ω*, then $p(\omega,v)=\frac{1}{2-\omega^*}$ and iii) if ω≥ω*, then p(ω,v) is maximised when α is minimised. Finally, if β>1/4, then p(ω,v) is maximised when α is maximised for 0<ω<2. Furthermore, if we maximise φ(ω,v) then we have an approximation to $P(B_\omega)$ which can be minimised with respect to ω.

Next, let us assume that we have an approximation or an initial guess $v=v^{(1)}\neq 0$ to the eigenvector of $\lambda(B_\omega)$. If A is an L-matrix, the quantity α is maximised for v≥0 (i.e., all the components of v are non-negative). Thus, for β≤1/4 and ω≤ω* or β>1/4, where p(ω,v) is maximised if α is maximised, we may always let v have the form $v^{(1)}=(v_1^{(1)},v_2^{(1)},v_3^{(1)},\ldots,v_N^{(1)})^T$ where $v_i^{(1)}\geq 0$. This choice of v gives
$$\alpha_1 = \frac{(v^{(1)},Bv^{(1)})}{(v^{(1)},v^{(1)})} \geq 0 \quad , \tag{11.5}$$

if A is an L-matrix, [23]. On the other hand, it is required that the quantity α be minimised (if ω*≤ω and β≤1/4) which can be achieved if one chooses v to have the alternative form
$$v^{(2)} = (v_1^{(2)},v_2^{(2)},v_3^{(2)},\ldots,v_N^{(2)})^T,$$

where

$$v_k^{(2)} = \begin{cases} v_k^{(1)}, & \text{on even points,} \\ -v_k^{(1)}, & \text{on odd points.} \end{cases}$$

The above choice of v gives

$$\alpha_2 = \frac{(v^{(2)}, Bv^{(2)})}{(v^{(2)}, v^{(2)})} \lesssim 0 , \tag{11.6}$$

which tends to maximise $p(\omega,v)$ if $\omega^* \lesssim \omega$ and $\beta \lesssim 1/4$.

Finally, we see that if A is an L-matrix with Property A, then

$$\beta_1 = \frac{(v^{(1)}, LUv^{(1)})}{(v^{(1)}, v^{(1)})} = \frac{(v^{(2)}, LUv^{(2)})}{(v^{(2)}, v^{(2)})} = \beta_2 . \tag{11.7}$$

Consequently, a lower bound on $P(\mathcal{B}_\omega)$ is given by the following expression

$$P(\mathcal{B}_\omega) \geqslant \begin{cases} \dfrac{1-M}{1-\omega M+\omega^2\beta}, & \text{if } \omega \lesssim \omega^* \\ \dfrac{1-m}{1-\omega m+\omega^2\beta}, & \text{if } \omega \gtrsim \omega^* \end{cases} , \tag{11.8}$$

where

$$\left.\begin{aligned} M &= \alpha_1 \geqslant 0 , \\ m &= \alpha_2 \leqslant 0 , \\ \beta &= \beta_1 = \beta_2 . \end{aligned}\right\} \tag{11.9}$$

and

Thus, (see Theorem 3.6) we have that a good choice of preconditioning parameter ω in the sense of minimising the bound (11.8) is given by

$$\omega_1 = \begin{cases} \dfrac{2}{1+\sqrt{1-2M+4\beta}} = \omega_M , & \text{if } M \lesssim 4\beta, \\ \dfrac{2}{1+\sqrt{1-4\beta}} = \omega^* , & \text{if } M \gtrsim 4\beta, \end{cases} \tag{11.10}$$

whereas the corresponding value of $P(\mathcal{B}_\omega)$ is given by

$$P(\mathcal{B}_{\omega_1}) \gtrsim \begin{cases} \dfrac{1}{2}\left(1+\dfrac{\sqrt{1-2M+4\beta}}{1-M}\right) = \dfrac{1}{2}\left(\dfrac{2-M\omega_M}{(1-M)\omega_M}\right), & \text{if } M \lesssim 4\beta, \\ \dfrac{1+\sqrt{1-4\beta}}{2\sqrt{1-4\beta}} = \dfrac{1}{2-\omega^*} , & \text{if } M \gtrsim 4\beta. \end{cases} \tag{11.11}$$

Another approach which does not require the above analysis for finding a good estimate to the optimum value of the preconditioning parameter is the use of a direct search technique, such as the Fibonacci method. Following this approach we can determine an approximation to the optimum parameter ω by minimising

$$P_1(\omega) = \max\{p_1,p_2\} = \max_i\{p(\omega,v^{(i)})\} = \max_i\left\{\frac{1-\omega\alpha_i+\omega^2\beta_i}{\omega(2-\omega)(1-\alpha_i)}\right\}, i=1,2. \tag{11.12}$$

As soon as we determine a good estimate $\omega=\omega_1$, we can immediately obtain our first estimate of $P(B_\omega)$ by evaluating

$$P_E(B_{\omega_1}) = P_1(\omega_1) , \tag{11.13}$$

and then we can apply the PSD-SI method using ω_1 and $P_E(B_{\omega_1})$. As we have seen, at the same time we can determine $v^{(3)}$ to be another estimate for the eigenvector v and we proceed by forming $v^{(4)}=(v_1^{(4)},v_2^{(4)},\ldots,v_N^{(4)})^T$, where

$$v_k^{(4)} = \begin{cases} v_k^{(3)} & \text{, on even points,} \\ -v_k^{(3)} & \text{, on odd points.} \end{cases}$$

At this stage we determine a good estimate $\omega=\omega_2$ by minimising

$$p_2(\omega) = \max\{p_1,p_2,p_3,p_4\} = \max_i\{p_i\},\ i=1,2,3,4,$$

and computing the corresponding estimate of $P(B_{\omega_2})$ by

$$P_E(B_{\omega_2}) = P_2(\omega_2) . \tag{11.14}$$

It becomes clear after this analysis that if there are r available estimates of the eigenvector v, then we have 2r vectors $v^{(1)},v^{(2)},\ldots,v^{(2r)}$, where for i odd, $v^{(i)}$ is an eigenvector approximation whereas for i even, $v^{(i)}$ is given by

$$v^{(i)} = (v_i^{(i)},v_2^{(i)},\ldots,v_N^{(i)})^T ,$$

where

$$v_k^{(i)} = \begin{cases} v_k^{(i-1)} & \text{, on even points,} \\ -v_k^{(i-1)} & \text{, on odd points.} \end{cases}$$

The rth estimate ω_r is determined by minimising the quantity

$$P_r(\omega) = \max_i\ \{p_i\},\ i=1,2,\ldots,2r$$

and the corresponding estimate $P(B_{\omega_r})$ is given by

$$P_E(B_{\omega_r}) = P_r(\omega_r) . \tag{11.15}$$

In addition, we note that since we have

$$p(\omega,v) \leq P(B_\omega) \ , \tag{11.16}$$

it follows that the inequalities

$$P_E(B_{\omega_r}) \leq P(B_{\omega_r}) \leq P(B_{\omega_o}) \ , \tag{11.17}$$

are valid, where ω_o is the optimum preconditioning parameter.

We will now present an adaptive algorithm which uses the PSD-SI (or PJ-SI) method to solve the system Au=b and automatically improve the parameters $(\omega, P(B_\omega))$. The algorithm will use the PSD-SI iterative scheme and a sequence of parameters $(\omega_i, P(B_{\omega_i}))$ which converge to the optimum parameter set $(\omega_o, P(B_{\omega_o}))$. The theoretical basis of the algorithm has been developed in Theorems 10.3, 10.4 and Lemma 10.6.

Algorithm 11.1

1. Choose an initial approximation $u^{(0)}$ such that $||u^{(0)}-\bar{u}||_{A^{1/2}} \leq ||\bar{u}||_{A^{1/2}}$ and choose a convergence tolerance ζ.

   Also, let $v = (v_1,v_2,\ldots,v_N)^T = (1,1,\ldots,1)^T$,

   and $\hat{v} = (\hat{v}_1,\hat{v}_2,\ldots,\hat{v}_N)^T$,

   where
   $$\hat{v}_k = \begin{cases} v_k & \text{, on even points,} \\ -v_k & \text{, on odd points.} \end{cases}$$

   Set i=1.

2. For the latest two vectors $v, \hat{v}$ compute

$$\alpha_i = \frac{(v,Bv)}{(v,v)} \ , \ \alpha_{i+1} = \frac{(\hat{v},B\hat{v})}{(\hat{v},\hat{v})} \ , \ \beta_i = \frac{(v,LUv)}{(v,v)} \ , \ \beta_{i+1} = \frac{(\hat{v},LU\hat{v})}{(\hat{v},\hat{v})}, \tag{11.18}$$

   if they have not been previously computed.

3. Use a Fibonacci search technique to determine ω by minimising the function

$$P(\omega) = \max_i \left\{ \frac{1-\omega\alpha_i+\omega^2\beta_i}{\omega(2-\omega)(1-\alpha_i)} \right\}^* \ , \tag{11.19}$$

   for all available pairs (α_i, β_i). Moreover, compute the corresponding value $P_E(B_\omega)$ from the expression

$$P_E(B_\omega) = P(\omega) \ . \tag{11.20}$$

---

* *For the unimodality of the function* P(ω) *see Appendix D in* [16].

4. Choose n_q to be the least integer n which satisfies the inequality

$$\frac{1}{n}\log\frac{2\bar{r}^n}{1+\bar{r}^{2n}} \geq -0.9\log\bar{r}\ , \qquad (11.21)$$

where

$$\bar{r} = \frac{\sqrt{P_E(\mathcal{B}_\omega)}-1}{\sqrt{P_E(\mathcal{B}_\omega)}+1}\ . \qquad (11.22)$$

5. Iterate n_q times with the PSD-SI method using the latest parameters ω and $P_E(\mathcal{B}_\omega)$. After each iteration, check for convergence by computing the pseudo-residual vector

$$\delta^{(n)} = (I-\omega U)^{-1}(I-\omega L)^{-1}(b-Au^{(n)}),\ n\leq n_q\ , \qquad (11.23)$$

and testing by the stopping procedure whether or not

$$\frac{P_E(\mathcal{B}_\omega)\,||\delta^{(n)}||_{A^{\frac{1}{2}}}}{||u^{(n)}||_{A^{\frac{1}{2}}}} \leq \zeta\ . \qquad (11.24)$$

If (11.24) is satisfied terminate the algorithm, otherwise continue to the next step.

6. In this step we test whether we should update the parameters $\omega, P_E(\mathcal{B}_\omega)$ or not. From the previous step we have obtained the pseudo-residual vector $\delta^{(n_q)}$, thus we now compute,

$$\alpha = \frac{(\delta^{(n_q)}, B\delta^{(n_q)})}{(\delta^{(n_q)}, \delta^{(n_q)})}\ ,\ \beta = \frac{(\delta^{(n_q)}, LU\delta^{(n_q)})}{(\delta^{(n_q)}, \delta^{(n_q)})}\ , \qquad (11.25)$$

and

$$p = \frac{1-\omega\alpha+\omega^2\beta}{\omega(2-\omega)(1-\alpha)}\ . \qquad (11.26)$$

If the following inequality is satisfied

$$p \leq P_E(\mathcal{B}_\omega)\ , \qquad (11.27)$$

then goto step 5 and note that the next PSD-SI iteration can be computed from

$$u^{(n+1)} = (1-\rho_{n+1})u^{(n-1)}+\rho_{n+1}(u^{(n)}+\bar{\rho}\delta^{(n)}),\ (11.28)$$

where $\delta^{(n)}$ has already been computed in step 5. Otherwise, continue to the next step before altering the parameters.

7. In order not to waste the computational work for the determination of $\delta^{(n_q)}$ in (11.23), apply a PSD-SI iteration using

(11.28) with the old parameters ω and $P_E(\mathcal{B}_\omega)$. Furthermore, let

$$v = \delta^{(n_q)} = \left(\delta_1^{(n_q)}, \delta_2^{(n_q)}, \ldots, \delta_N^{(n_q)}\right)^T,$$

and $$\hat{v} = (\hat{v}_1, \hat{v}_2, \ldots, \hat{v}_N)^T,$$

where

$$\hat{v}_k = \begin{cases} \delta_k^{(n_q)}, & \text{on even points,} \\ -\delta_k^{(n_q)}, & \text{on odd points.} \end{cases}$$

Then, set i=i+2 and goto step 2 to compute new quantities for α and β in order to update ω and $P_E(\mathcal{B}_\omega)$. Evidently, in step 2, α_i and β_i have already been computed by (11.25) in step 6.

Further details of the performance of Algorithm 11.1 for a number of numerical experiments involving the generalised Dirichlet problem on the unit square are given in [18].

SECTION 2

12. THE PRECONDITIONED ALTERNATING DIRECTION IMPLICIT (PADI) METHODS FOR SELF-ADJOINT ELLIPTIC P.D.E.'s

Let us consider the solution of partial differential equations of the form,

$$\frac{1}{E_2(x)}\frac{\partial}{\partial x}\left(E_1(x)\frac{\partial U}{\partial x}\right) + \frac{1}{F_1(y)}\frac{\partial}{\partial y}\left(F_2(y)\frac{\partial U}{\partial y}\right) - kU = \frac{G(x,y)}{E_2(x)F_1(y)}, \tag{12.1}$$

in the rectangular region $$0 \leq x \leq L_x, \quad 0 \leq y \leq L_y, \tag{12.2}$$

where $U=U(x,y)$, $k \geq 0$ and the functions $E_1(x), E_2(x), F_1(y), F_2(y)$ are assumed positive and continuous in the region (12.2). For boundary conditions we assume

$$U(x,y) = g(x,y), \quad (x,y) \in \partial R, \tag{12.3}$$

where $g(x,y)$ is a prescribed function on the boundary ∂R of the rectangle (12.2). In order to use finite difference methods for the solution of the above problem we superimpose a uniform grid of mesh size h_x and h_y in the x- and y-directions, respectively such that,

$$N_a = L_a/h_a \quad , \quad a=x,y \ , \tag{12.4}$$

where N_a is an integer.

The five point difference analogue of (12.1),[1] is

$$H_o[u](x,y)+V_o[u](x,y) +E_o[u](x,y) = -h_x h_y \frac{G(x,y)}{E_2(x)F_1(y)} \ , \tag{12.5}$$

where

$$H_o[u](x,y) = \frac{h_y}{h_x}\left\{\left[\frac{E_1(x+\frac{1}{2}h_x)+E_1(x-\frac{1}{2}h_x)}{E_2(x)}\right]u(x,y) - \frac{E_1(x+\frac{1}{2}h_x)}{E_2(x)}u(x+h_x,y) - \frac{E_1(x-\frac{1}{2}h_x)}{E_2(x)}u(x-h_x,y)\right\}, \tag{12.6}$$

$$V_o[u](x,y) = \frac{h_x}{h_y}\left\{\left[\frac{F_2(y+\frac{1}{2}h_y)+F_2(y-\frac{1}{2}h_y)}{F_1(y)}\right]u(x,y) - \frac{F_2(y+\frac{1}{2}h_y)}{F_1(y)}u(x,y+h_y) - \frac{F_2(y-\frac{1}{2}h_y)}{F_1(y)}u(x,y-h_y)\right\}, \tag{12.7}$$

and

$$E_o[u](x,y) = h_x h_y k u(x,y) \ . \tag{12.8}$$

By using the natural ordering the difference equations (12.5) can be written in the matrix form

$$Au = (H_o+V_o+\Lambda)u = (H+V)u = b \ , \tag{12.9}$$

where H_o, V_o and Λ correspond to the operators $H_o[u], V_o[u]$ and $E_o[u]$ respectively and

$$H = H_o + \tfrac{1}{2}\Lambda \ , \quad V = V_o + \tfrac{1}{2}\Lambda \ . \tag{12.10}$$

Let us consider the case where the matrix R has the form

$$R = (I+r_1H)(I+r_2V) \ , \tag{12.11}$$

with r_1, r_2 real parameters, then by (1.3) we have the iterative scheme,

$$u^{(n+1)} = u^{(n)} + \tau(I+r_2V)^{-1}(I+r_1H)^{-1}(b-Au^{(n)}), \tag{12.12}$$

where the matrices H,V are symmetric, positive definite and commute [1]. The iterative process defined by (12.12) will be referred to as the Modified Alternating Direction Preconditioning method (MADP method) [16]. From (12.12) we also have

$$u^{(n+1)} = T_{\tau,r_1,r_2}\, u^{(n)} + k \ , \tag{12.13}$$

where $$T_{\tau,r_1,r_2} = I-\tau(I+r_2V)^{-1}(I+r_1H)^{-1}A, \tag{12.14}$$

and $$k = \tau(I+r_2V)^{-1}(I+r_1H)^{-1}b . \tag{12.15}$$

Since H and V each have strong diagonal dominance, $(I+r_1H)$, $(I+r_2V)$ are non-singular for all $r_1,r_2 \in (0,\infty)$ (e.g. see [23]p.41). Moreover, if $\tau \neq 0$ then by (12.14) $I-T_{\tau,r_1,r_2}$ is non-singular and therefore the MADP method is completely consistent ([23] Theorem 3-2.6). Alternatively, the MADP method can be written in a computable form as

$$(I+r_1H)u^{(n+\frac{1}{2})} = [I+(r_1-\tau)H]u^{(n)}+\tau(b-Vu^{(n)}) ,$$

and $$(I+r_2V)u^{(n+1)} = u^{(n+\frac{1}{2})}+r_2Vu^{(n)} , \tag{12.16}$$

where we observe that it is not necessary to recompute $Vu^{(n)}$ in the second half iteration, thus we can use Neithammer's scheme [19] to reduce the computational work for each MADP iteration by providing storage space for an extra vector iterant.

Let us assume that the real eigenvalues μ,ν of H,V lie in the ranges, $$0 < a \leq \mu \leq b, \quad 0 < \alpha \leq \nu \leq \beta \quad , \tag{12.17}$$

respectively, then the preconditioned matrix

$$B_{r_1,r_2} = (I+r_2V)^{-1}(I+r_1H)^{-1}A , \tag{12.18}$$

has real eigenvalues which are given by the expression

$$\lambda \equiv \lambda(\mu,\nu,r_1,r_2) = \frac{\mu+\nu}{(1+r_1\mu)(1+r_2\nu)} , \tag{12.19}$$

and are positive for all $r_1,r_2 \in (0,\infty)$. On the other hand, we have that the P-condition number of B_{r_1,r_2} is given by

$$P(B_{r_1,r_2}) = \frac{\lambda_M}{\lambda_m} , \tag{12.20}$$

$$\left.\begin{aligned} \text{where} \quad \lambda_M &= \max_{\mu,\nu} \{\lambda(\mu,\nu,r_1,r_2)\}, \\ \text{and} \quad \lambda_m &= \min_{\mu,\nu} \{\lambda(\mu,\nu,r_1,r_2)\} . \end{aligned}\right\} \tag{12.21}$$

In order to maximise the rate of convergence of the MADP method we shall seek to select r_1,r_2 to minimise $P(B_{r_1,r_2})$ and determine the optimum value for τ by the expression (see Section 3).

$$\tau = 2/(\lambda_M+\lambda_m) \ . \tag{12.22}$$

Finally, we observe that the MADP iterative scheme is convergent for all $r_1,r_2 \in (0,\infty)$ and τ lying in the interval [3],

$$0 < \tau < 2/\lambda_M \ . \tag{12.23}$$

We distinguish two cases in our analysis (i) the eigenvalue ranges of the matrices H and V are the same and (ii) the eigenvalue ranges are different.

13. CASE WHERE THE EIGENVALUE RANGES OF H AND V ARE THE SAME

In this case we prove the following theorem:

Theorem 13.1: Let H,V be the matrices defined by (12.10) with eigenvalues μ,ν, respectively such that

$$0 < a \lesssim \mu,\nu \lesssim b. \tag{13.1}$$

Then, the P-condition number of B_{r_1,r_2} is given in Tables 13.1 and 13.3 for the different ranges of the parameters $r_1,r_2 \in (0,\infty)$. Moreover, $P(B_{r_1,r_2})$ is minimised if we let

$$r_1 = r_2 = r' = 1/(ab)^{\frac{1}{2}} \ , \tag{13.2}$$

and its corresponding value is given by the expression

$$P(B_{r',r'}) = (a+b)r'/2 \ . \tag{13.3}$$

On the other hand, if we also let

$$\tau = \bar{\tau} = 2r' \ , \tag{13.4}$$

then the spectral radius of the iteration matrix T_{τ,r_1,r_2} attains its minimum value and is given by the expression

$$S(T_{\bar{\tau},r',r'}) = \left(\frac{b^{\frac{1}{2}}-a^{\frac{1}{2}}}{b^{\frac{1}{2}}+a^{\frac{1}{2}}}\right)^2 \ . \tag{13.5}$$

Proof: As it can be seen from (12.21) we have to examine the behaviour of λ given by (12.19) as a function of μ,ν. By taking partial derivatives of the function λ with respect to μ and ν we obtain the following results

$$\text{sign}(\frac{\partial\lambda}{\partial\mu}) = \text{sign}(1-\nu r_1) \ ,$$

and

$$\text{sign}(\frac{\partial\lambda}{\partial\nu}) = \text{sign}(1-\mu r_2) \ . \tag{13.6}$$

From (13.6) we can readily see that for fixed $r_1,r_2>0$ neither of the expressions $\frac{\partial\lambda}{\partial\mu}, \frac{\partial\lambda}{\partial\nu}$ changes sign as μ and ν vary in the interval (13.1). We therefore conclude that the possible extreme values of λ will occur at the points (a,a), (a,b), (b,a) and (b,b). (see Guittet [9], Lemma 1). On the other hand, the values of λ at these points are the following

$$A = \lambda(a,a,r_1,r_2), \quad B = \lambda(a,b,r_1,r_2),$$
$$D = \lambda(b,a,r_1,r_2) \text{ and } C = \lambda(b,b,r_1,r_2). \tag{13.7}$$

Hence from (13.7), (12.21) and for fixed r_1,r_2 we have that

$$\lambda_M = \max\{A,B,C,D\} \text{ and } \lambda_m = \min\{A,B,C,D\}. \tag{13.8}$$

However, the order of the quantities A,B,C and D is determined from the following relationships

$$\begin{aligned} \text{sign}(A-B) &= \text{sign}(r_2-1/a), \\ \text{sign}(B-C) &= \text{sign}(r_1-1/b), \\ \text{sign}(A-D) &= \text{sign}(r_1-1/a), \\ \text{sign}(D-C) &= \text{sign}(r_2-1/b), \end{aligned} \tag{13.9}$$

which suggest that we have to examine the relative positions of r_1 and r_2 with respect to the values 1/a and 1/b. We therefore have to distinguish nine cases which are presented in Table 13.1 together with the values λ_M, λ_m and $P(B_{r_1,r_2})$ for each case.

r_1-Domain	r_2-Domain	λ_M	λ_m	$P(B_{r_1,r_2})$
$0<r_1\leq 1/b$	$0<r_2\leq 1/b$	C	A	C/A
	$1/b\leq r_2\leq 1/a$	D	A	D/A
	$1/a\leq r_2<\infty$	D	B	D/B
$1/b\leq r_1\leq 1/a$	$0<r_2\leq 1/b$	B	A	B/A
	$1/b\leq r_2\leq 1/a$	max{B,D}	min{A,C}	max{B,D}/min{A,C}
	$1/a\leq r_2<\infty$	D	C	D/C
$1/a\leq r_1<\infty$	$0<r_2\leq 1/b$	B	D	B/D
	$1/b\leq r_2\leq 1/a$	B	C	B/C
	$1/a\leq r_2<\infty$	A	C	A/C

TABLE 13.1

In Table 13.1 we have determined $P(B_{r_1,r_2})$ thus we can now study its behaviour as a function of r_1 and r_2. By assuming that r_1

is kept fixed, then we obtain the results summarised in Table 13.2 for i=1, whereas if r_2 is fixed, we have the same results where now i=2.

r_i-Domain*	r_j-Domain**	$\text{sign}(\partial P(B_{r_1,r_2})/\partial r_i)$	$P(B_{r_1,r_2})$
$0<r_i\leq 1/b$	$0<r_j\leq 1/b$	sign (a-b)	Decreasing
	$1/b\leq r_j\leq 1/a$	0	Stationary
	$1/a\leq r_j<\infty$	sign (b-a)	Increasing
$1/b\leq r_i\leq 1/a$	$0<r_j\leq 1/b$	sign (a-b)	Decreasing
	$1/b\leq r_j\leq 1/a$	-	-
	$1/a\leq r_j<\infty$	sign (b-a)	Increasing
$1/a\leq r_i<\infty$	$0<r_j\leq 1/b$	sign (a-b)	Decreasing
	$1/b\leq r_j\leq 1/a$	0	Stationary
	$1/a\leq r_j<\infty$	sign (b-a)	Increasing

*i=1,2

**j={1,2}-{i}

TABLE 13.2

From Table 13.2 we conclude that $P(B_{r_1,r_2})$ attains its minimum value when r_1 and r_2 lie in the following range

$$1/b\leq r_1,r_2\leq 1/a. \tag{13.10}$$

But for this range $P(B_{r_1,r_2})$ is given by the expression (see Table 13.1)

$$P(B_{r_1,r_2}) = \frac{\max\{B,D\}}{\min\{A,C\}}. \tag{13.11}$$

On the other hand, the order of B,D and A,C, when r_1,r_2 lie in the range (13.10), is determined from the relationships

$$\text{sign}(B-D) = \text{sign}(r_1-r_2),$$

and

$$\text{sign}(A-C) = \text{sign}(r_1r_2ab-1). \tag{13.12}$$

Consequently, we have the following expressions for λ_M and λ_m.

$$\lambda_M = \begin{cases} D, & \text{if } r_1\leq r_2 \\ B, & \text{if } r_1\geq r_2 \end{cases}, \tag{13.13}$$

and

$$\lambda_m = \begin{cases} A, & \text{if } \frac{1}{b} \leq r_2 \leq \frac{1}{r_1 ab} \\ C, & \text{if } \frac{1}{r_1 ab} \leq r_2 \leq \frac{1}{a}. \end{cases} \tag{13.14}$$

Evidently, the quantity $\frac{1}{r_1 ab}$ belongs to the interval $[\frac{1}{b},\frac{1}{a}]$ for all $r_1 \in [\frac{1}{b},\frac{1}{a}]$. Moreover, we note that

$$\frac{1}{b} \leq r_1 \leq \frac{1}{r_1 ab} \leq \frac{1}{a}, \quad \text{if } r_1 \leq \frac{1}{\sqrt{ab}},$$

and

$$\frac{1}{b} \leq \frac{1}{r_1 ab} \leq r_1 \leq \frac{1}{a}, \quad \text{if } r_1 \geq \frac{1}{\sqrt{ab}}. \tag{13.15}$$

From the inequalities (13.15) it follows that we have to consider six cases which emerge for the different values of $r_2 \in [\frac{1}{b},\frac{1}{a}]$ by keeping r_1 fixed. The results which are obtained after the examination of these cases are summarised in Table 13.3.

r_1-Domain	r_2-Domain	λ_M	λ_m	$P(B_{r_1,r_2})$	$\text{sign}(\partial P(B_{r_1,r_2})/\partial r_2)$
$1/b \leq r_1 \leq 1/(ab)^{1/2}$	$1/b \leq r_2 \leq r_1$	B	A	B/A	sign (a-b)
	$r_1 \leq r_2 \leq 1/(r_1 ab)$	D	A	D/A	O
	$1/(r_1 ab) \leq r_2 \leq 1/a$	D	C	D/C	sign (b-a)
$1/(ab)^{1/2} \leq r_1 \leq 1/a$	$1/b \leq r_2 \leq 1/(r_1 ab)$	B	A	B/A	sign (a-b)
	$1/(r_1 ab) \leq r_2 \leq r_1$	B	C	B/C	O
	$r_1 \leq r_2 \leq 1/a$	D	C	D/C	sign (b-a)

TABLE 13.3

From Table 13.3 we conclude that for fixed r_1, $P(B_{r_1,r_2})$ attains its minimum value for r_2 such that

$$r_2 \in [\min(r_1, \frac{1}{r_1 ab}), \max(r_1, \frac{1}{r_1 ab})]. \tag{13.16}$$

On the other hand, by keeping r_2 fixed in the above interval we have that

$$\text{i) if } r_1 \in [\frac{1}{b}, \frac{1}{\sqrt{ab}}], \text{ then } \text{sign}(\frac{\partial P(B_{r_1,r_2})}{\partial r_1}) = \text{sign}(a-b),$$

$$\text{ii) if } r_1 \in [\frac{1}{\sqrt{ab}}, \frac{1}{a}], \text{ then } \text{sign}(\frac{\partial P(B_{r_1,r_2})}{\partial r_1}) = \text{sign}(b-a), \tag{13.17}$$

which indicate that $P(B_{r_1,r_2})$ is minimised when r_1 becomes equal to the quantity

$$r_1 = r' = \frac{1}{\sqrt{ab}} , \tag{13.18}$$

and from (13.16) we have

$$r_2 = r_1 , \tag{13.19}$$

hence (13.2) follows.

In addition, from (13.13),(13.14),(13.7) and (12.19) we obtain the following expressions for the smallest and largest eigenvalues of $B_{r',r'}$

$$\lambda_M = B = D = \frac{(a+b)\sqrt{ab}}{(\sqrt{a}+\sqrt{b})^2} ,$$

and (13.20)

$$\lambda_m = A = C = \frac{2ab}{(\sqrt{a}+\sqrt{b})^2} .$$

By combining (12.20) and (13.20) we easily obtain (13.3), whereas by (12.22) and (13.20) the optimum value of τ is given in (13.4). Finally, for this value of τ the spectral radius of $T_{\bar{\tau},r',r'}$ is given by the expression

$$S(T_{\bar{\tau},r',r'}) = \frac{P(B_{r',r'})-1}{P(B_{r',r'})+1} , \tag{13.21}$$

which gives (13.5) if combined with (13.3) and therefore the proof of the theorem is complete.

14. THE CASE WHERE THE EIGENVALUE RANGES OF H AND V MAY BE DIFFERENT

In this case we prove the following theorem:

Theorem 14.1: Let H,V be the matrices defined by (12.10) with eigenvalues μ,ν, respectively such that

$$0 < a \lesssim \mu \lesssim b \quad \text{and} \quad 0 < \alpha \lesssim \nu \lesssim \beta . \tag{14.1}$$

Then the P-condition number of B_{r_1,r_2} is minimised when the parameters r_1,r_2 take the values

$$r_1^* = \frac{1-\Sigma sc^{\frac{1}{2}}}{-t+\Sigma qc^{\frac{1}{2}}} , \quad r_2^* = \frac{1+\Sigma sc^{\frac{1}{2}}}{t+\Sigma qc^{\frac{1}{2}}} , \tag{14.2}$$

where

$$c = \frac{1}{1+\theta+[\theta(2+\theta)]^{\frac{1}{2}}} , \tag{14.3}$$

$$\theta = \frac{2(\beta-\alpha)(b-a)}{(a+\alpha)(b+\beta)} , \tag{14.4}$$

$$\Sigma s = \frac{(\beta-\alpha)-(b-a)}{(b+\beta)-(a+\alpha)c} , \tag{14.5}$$

$$\Sigma q = \frac{(b+\beta)+(b-\beta)\Sigma s}{2}, \tag{14.6}$$

$$t = \frac{(b-\beta)+(b+\beta)\Sigma s}{2}, \tag{14.7}$$

and its corresponding value is given by

$$P(B_{r_1^*,r_2^*}) = (c^{\frac{1}{2}}+c^{-\frac{1}{2}})/2 , \tag{14.8}$$

On the other hand, if we also let

$$\tau = \tau^* = 2/(\Sigma c^{\frac{1}{2}}) , \tag{14.9}$$

where Σ is any positive value, then the spectral radius of T_{τ,r_1,r_2} attains its minimum value which is given by the expression

$$S(T_{\tau^*,r_1^*,r_2^*}) = \left(\frac{1-c^{\frac{1}{2}}}{1+c^{\frac{1}{2}}}\right)^2 . \tag{14.10}$$

Proof: Under the hypothesis of the theorem we have that μ,ν lie in the different ranges given by (14.1). Since we have solved the problem for the case where μ,ν lie in the same range (section 13), we attempt to find a technique of transforming our present problem so that we return to the previous case of the "single range". This will prevent us repeating the laborious procedure (see proof of Theorem 13.1) of the more complex problem in the present case. The technique for achieving this is known and is due to Wachspress and Jordan [21],[23].

We commence our analysis by noting that the function defined by (12.19) can be written alternatively as

$$\lambda = \frac{\omega_1\omega_2}{\omega_1+\omega_2}\left[1 - \frac{(\mu-\omega_2)(\nu-\omega_1)}{(\mu+\omega_1)(\nu+\omega_2)}\right] , \tag{14.11}$$

where

$$\omega_1 = \frac{1}{r_1} \quad \text{and} \quad \omega_2 = \frac{1}{r_2} . \tag{14.12}$$

We note that because of the form of λ in (14.11) our interest is focused on the second term in the brackets. By adhering to the analysis of Wachspress and Jordan [21], we seek to introduce new variables $\hat{\mu}$ and $\hat{\nu}$ such that

$$\mu = \frac{t+q\hat{\mu}}{1+s\hat{\mu}} \quad , \qquad \nu = \frac{t'+q'\hat{\nu}}{1+s'\hat{\nu}} \quad , \tag{14.13}$$

so that for some $\hat{\omega}_1$ and $\hat{\omega}_2$ we have

$$\left(\frac{\mu-\omega_2}{\mu+\omega_1}\right)\left(\frac{\nu-\omega_1}{\nu+\omega_2}\right) = \left(\frac{\hat{\mu}-\hat{\omega}_2}{\hat{\mu}+\hat{\omega}_1}\right)\left(\frac{\hat{\nu}-\hat{\omega}_1}{\hat{\nu}+\hat{\omega}_2}\right) \quad , \tag{14.14}$$

and where $\hat{\mu}$ and $\hat{\nu}$ vary over the ranges

$$\sigma \lessgtr \hat{\mu} \lessgtr \Sigma \quad , \quad \sigma \lessgtr \hat{\nu} \lessgtr \Sigma \, . \tag{14.15}$$

It can be shown (see [23]p.511) that if (14.14) holds, then the relationships (14.13) become

$$\mu = \frac{t+q\hat{\mu}}{1+s\hat{\mu}} \quad , \qquad \nu = \frac{-t+q\hat{\nu}}{1-s\hat{\nu}} \quad . \tag{14.16}$$

In order to determine t,q,s,σ and Σ we require that $\mu=a$ corresponds to $\hat{\mu}=\sigma$, $\mu=b$ corresponds to $\hat{\mu}=\Sigma$, $\nu=\alpha$ corresponds to $\hat{\nu}=\sigma$ and $\hat{\nu}=\beta$ corresponds to $\hat{\nu}=\Sigma$ hence by (14.16) we have

$$\begin{aligned} a &= \frac{t+q\sigma}{1+s\sigma} \, , \qquad b = \frac{t+q\Sigma}{1+s\Sigma} \, , \\ \alpha &= \frac{-t+q\sigma}{1-s\sigma} \, , \qquad \beta = \frac{-t+q\Sigma}{1-s\Sigma} \, . \end{aligned} \tag{14.17}$$

From the above relationships and after some algebraic manipulation (see [23]p.512) we obtain

$$c+1/c = 2(1+\theta) \; , \tag{14.18}$$

where

$$c = \sigma/\Sigma , \tag{14.19}$$

and θ given by (14.4). The quantities $\Sigma s, \Sigma q$ and t are also determined and are given by (14.5),(14.6), and (14.7) respectively. Therefore, we can rewrite (14.16) to yield

$$\mu = \frac{t+(\Sigma q)(\hat{\mu}/\Sigma)}{1+(\Sigma s)(\hat{\mu}/\Sigma)} \, , \qquad \nu = \frac{-t+(\Sigma q)(\hat{\nu}/\Sigma)}{1-(\Sigma s)(\hat{\nu}/\Sigma)} \, , \tag{14.20}$$

whereas by combining (14.13) and (14.14) we find

$$\omega_1 = \frac{-t+(\Sigma q)(\hat{\omega}_1/\Sigma)}{1-(\Sigma s)(\hat{\omega}_1/\Sigma)} \quad \text{and} \quad \omega_2 = \frac{t+(\Sigma q)(\hat{\omega}_2/\Sigma)}{1+(\Sigma s)(\hat{\omega}_2/\Sigma)} \, . \tag{14.21}$$

At this stage we note that we have transformed our problem to be identical with the one discussed in the previous section, where now instead of μ,ν, we have the transformed variables $\hat{\mu},\hat{\nu}$

respectively possessing the same range given by (14.15). Evidently, from Theorem 13.1 and the relationships (14.12), (14.15) we see that the optimum parameters $\hat{\omega}_1$ and $\hat{\omega}_2$ for the transformed problems are

$$\hat{\omega}_1 = \hat{\omega}_2 = (\sigma\Sigma)^{\frac{1}{2}} , \tag{14.22}$$

thus from (14.22),(14.21),(14.19), we find that the optimum values of ω_1,ω_2 for the present problem are given by

$$\omega_1^* = \frac{-t+\Sigma qc^{\frac{1}{2}}}{1-\Sigma sc^{\frac{1}{2}}} , \quad \omega_2^* = \frac{t+\Sigma qc^{\frac{1}{2}}}{1+\Sigma sc^{\frac{1}{2}}} . \tag{14.23}$$

Finally, from (14.12) and (14.23) we can readily see that (14.2) follows. It is a trivial matter now, from the above analysis and using the relationships (13.3), (13.4), (13.5) to show the validity of (14.8),(14.9) and (14.10) respectively, thus the proof of the theorem is complete.

Evidently, we can choose any positive value for Σ, e.g. $\Sigma=1$, As we have shown (Theorem 13.1) in the case where A is given by

$$A = H + V , \tag{14.24}$$

where H,V are defined in (12.10), MADP method coincides with the Peaceman-Rachford ADI method ([1],[21],[23]) at the optimum stage when all the parameters are kept fixed during the iterations.

However, the advantage of the MADP method over the PR-ADI comes when more accurate finite difference equations are used to approximate (12.1) (e.g. the nine-point difference formula). In this case we have $\bar{\tau}\neq 2r'$ [16], which implies that (i) the MADP method does not coincide with the PR-ADI scheme and (ii) the parameter τ in the PR-ADI method does not take its optimum value as it is always forced to be $\tau=r_1+r_2$ (this is not the case for the MADP method). It is therefore expected that the MADP method has a slightly better rate of convergence than the PR-ADI scheme. It should be mentioned that similar results to Section 13 for the case $r_1=r_2=r$ and for the EADI method have been obtained by Guittet [12]. For the same case but with the eigenvalues of H and V lying in different ranges the optimum parameters have been found in [13]. Furthermore, some of the results obtained in this section.

have also been found in [23] but for the Peaceman-Rachford-ADI method, whereas in [7] the optimum parameters for the ADP method (i.e. $r_1=r_2$) were given, but for the case of different eigenvalue ranges, the problem was solved under the assumption that $0<\alpha'\leq\mu,\nu\leq\beta'$ with $\alpha'=\min(a,\alpha)$ and $\beta'=\max(b,\beta)$. The approach given in Section 13 and 14 solves the general problem in the stationary case and also serves as a standard technique of tackling similar problems. This is justified in the remainder of this paper where we determine the optimum values of the involved parameters for the solution of the biharmonic equation by using the corresponding MADP method.

15. FOURTH ORDER PARTIAL DIFFERENTIAL EQUATION - BIHARMONIC EQUATION

Let us consider the numerical solution of the biharmonic equation

$$\frac{\partial^4 U}{\partial x^4} + 2\frac{\partial^4 U}{\partial x^2 \partial y^2} + \frac{\partial^4 U}{\partial y^4} = f(x,y) , \tag{15.1}$$

in the rectanglar region defined by (12.2). In connection with the solution of (15.1) we also consider the following boundary conditions

$$\left.\begin{aligned} U(x,y) &= e(x,y) \\ \frac{\partial^2 U}{\partial n^2} &= g(x,y) \end{aligned}\right\} , \tag{15.2}$$

where $e(x,y)$ and $g(x,y)$ are prescribed functions on the boundary ∂R and $\frac{\partial}{\partial n}$ is the normal derivative to ∂R. By imposing a uniform grid of mesh sizes h_x and h_y in the x- and y-directions, respectively such that

$$N_a = L_a/h_a , \quad a=x,y , \tag{15.3}$$

where N_a is an integer, the application of the thirteen point finite difference analogue approximating (15.1) yields the difference equation

$$\hat{H}_0[u](x,y)+\hat{V}_0[u](x,y)+\hat{E}_0[u](x,y) = h_x^2 h_y^2 f(x,y), \tag{15.4}$$

where

$$\hat{H}_0[u](x,y) = \left(\frac{h_y}{h_x}\right)^2[u(x+2h_x,y)-4u(x+h_x,y)+6u(x,y) -4u(x-h_x,y)+u(x-2h_x,y)], \tag{15.5}$$

$$\hat{V}_O[u](x,y) = \left(\frac{h_x}{h_y}\right)^2 [u(x,y+2h_y)-4u(x,y+h_y)+6u(x,y)$$
$$-4u(x,y-h_y)+u(x,y-2h_y)],$$

and (15.6)

$$\hat{E}_O[u](x,y) = 2\{u(x-h_x,y-h_y)+u(x+h_x,y-h_y)+u(x-h_x,y+h_y)$$
$$+u(x+h_x,y+h_y)-2[u(x-h_x,y)+u(x+h_x,y)+u(x,y+h_y)$$
$$+u(x,y-h_y)]+4u(x,y)\}. \qquad (15.7)$$

On the other hand, if we approximate the second normal derivative such that (15.4) incorporates the boundary conditions, then by using the natural ordering of the grid points, we arrive at a system of equations which have the form,

$$Au = (\hat{H}_O + \hat{V}_O + \hat{\Lambda})u = b\ , \qquad (15.8)$$

where the matrices $\hat{H}_O, \hat{V}_O$ and $\hat{\Lambda}$ correspond to the operators $\hat{H}_O[u]$, $\hat{V}_O[u]$ and $\hat{E}_O[u]$, respectively. More precisely, using tensor products [14] we have that the coefficient matrix A has the form

$$A = (H + V)^2\ , \qquad (15.9)$$

where

$$H = I_x \otimes U_x,\quad V = U_y \otimes I_y\ , \qquad (15.10)$$

and

$$U_a = \frac{h_b}{h_a}\begin{bmatrix} 2 & -1 & & & & \\ -1 & 2 & -1 & & 0 & \\ & \ddots & \ddots & \ddots & & \\ & & \ddots & \ddots & \ddots & \\ & 0 & & -1 & 2 & -1 \\ & & & & -1 & 2 \end{bmatrix}\ , \qquad (15.11)$$

with $a=x,y$, $b\in\{x,y\}-\{a\}$ and I_a is the unit matrix of order N_b-1. From (15.10), (15.11) it can be verified that the matrices H,V are symmetric and commute, thus we can apply the MADP iterative scheme for the numerical solution of the matrix equation (15.8).

Further, the form of A can be given more explicitly as

$$A = H^2 + V^2 + 2HV\ , \qquad (15.12)$$

which implies that if we consider the conditioning matrix R to have the form,

$$R = (I+r_1H^2)(I+r_2V^2) = I+r_1H^2+r_2V^2+r_1r_2(HV)^2\ , \qquad (15.13)$$

then by (15.12) and (15.13) we see that R approximates A

reasonably well. Moreover, from (1.3) we have the iterative scheme

$$u^{(n+1)} = u^{(n)} + \tau(I+r_2V^2)^{-1}(I+r_1H^2)^{-1}(b-Au^{(n)}) , \quad (15.14)$$

which is the MADP method corresponding to matrix equation (15.8). In order to compute an MADP iteration we employ the two-level form (similar to (12.16))

$$(I+r_1H^2)u^{(n+\frac{1}{2})} = [I+(r_1-\tau)H^2]u^{(n)} - \tau[V^2u^{(n)}+2HVu^{(n)}-b],$$

and $$(I+r_2V^2)u^{(n+1)} = u^{(n+\frac{1}{2})} + r_2V^2u^{(n)} , \quad (15.15)$$

where again it is not necessary to recompute $V^2u^{(n)}$ in the second half iteration and therefore the Niethammer's scheme can be used to reduce the computational work for each MADP iteration by providing storage space for an extra vector iterant.

An alternative form of (15.14) is the following

$$u^{(n+1)} = \hat{T}_{\tau,r_1,r_2}u^{(n)} + \hat{k} , \quad (15.16)$$

where $$\hat{T}_{\tau,r_1,r_2} = I-\tau(I+r_2V^2)^{-1}(I+r_1H^2)^{-1}A , \quad (15.17)$$

and $$\hat{k} = \tau(I+r_2V^2)^{-1}(I+r_2H^2)^{-1}b . \quad (15.18)$$

From (15.17) we see that the preconditioned matrix is

$$\hat{B}_{r_1,r_2} = (I+r_2V^2)^{-1}(I+r_1H^2)^{-1}A , \quad (15.19)$$

and since H,V are symmetric and commute we have that B_{r_1,r_2} has real eigenvalues which are given by the expression

$$\lambda \equiv \lambda(\mu,\nu,r_1,r_2) = \frac{(\mu+\nu)^2}{(1+r_1\mu^2)(1+r_2\nu^2)} , \quad (15.20)$$

where μ,ν denote the eigenvalues of H and V respectively. If $r_1,r_2 \in (0,\infty)$, then from (15.20) we have that $\lambda>0$ which implies that $\hat{B}_{r_1,r_2}$ is a positive definite matrix. Here we should note that the eigenvalues of μ and ν are given by the expressions

$$\mu = (h_y/h_x)4\sin^2(i\pi/2N_x) , \text{ for } i=1,2,\ldots,N_x-1,$$

$$\nu = (h_x/h_y)4\sin^2(j\pi/2N_y) , \text{ for } j=1,2,\ldots,N_y-1, \quad (15.21)$$

and therefore they are bounded as follows:

$$0<a = (h_y/h_x)4\sin^2(\pi/2N_x)\leq\mu\leq(h_y/h_x)4\cos^2(\pi/2N_x) = b,$$
$$0<\alpha = (h_x/h_y)4\sin^2(\pi/2N_y)\leq\nu\leq(h_x/h_y)4\cos^2(\pi/2N_y) = \beta. \tag{15.22}$$

The optimum values for the parameters r_1, r_2 such that the rate of convergence of the iterative scheme (15.15) is maximised, are determined such that

$$P(\hat{B}_{r_1,r_2}) = \frac{\lambda_M}{\lambda_m}, \tag{15.23}$$

where $\lambda_M = \max_{\mu,\nu}\{\lambda(\mu,\nu,r_1,r_2)\}$ and $\lambda_m = \min_{\mu,\nu}\{\lambda(\mu,\nu,r_1,r_2)\}$, (15.24)
is minimised, whereas the optimum value of τ is again

$$\tau = 2/(\lambda_M+\lambda_m) . \tag{15.25}$$

We proceed to develop a similar analysis for the determination of the optimum parameters as in sections 13 and 14.

16. CASE WHERE THE EIGENVALUE RANGES OF H AND V ARE THE SAME

In this case we prove the following theorem:

Theorem 16.1: Let H,V be the matrices defined by (15.10) and (15.11) with eigenvalues μ,ν, respectively such that

$$0 < a \leq \mu,\nu \leq b , \tag{16.1}$$

then $P(\hat{B}_{r_1,r_2})$ is given in Tables 16.4 and 16.5 for the different ranges of the parameters $r_1,r_2\in(0,\infty)$.

Moreover, $P(\hat{B}_{r_1,r_2})$ is minimised if we let

$$r_1 = r_2 = r' = 1/(ab) , \tag{16.2}$$

and its corresponding value is given by the expression

$$P(\hat{B}_{r',r'}) = (a+b)^2 r'/4 . \tag{16.3}$$

On the other hand, if we also let,

$$\tau = \bar{\tau} = 2r'/(1+1/P(\hat{B}_{r',r'})) , \tag{16.4}$$

then the spectral radius $S(\hat{T}_{\tau,r_1,r_2})$ attains its minimum value which is given by the expression

$$S(\hat{T}_{\bar{\tau},r',r'}) = \frac{(b-a)^2}{(a+b)^2+4ab} . \tag{16.5}$$

Proof: We notice that (15.20) can be rewritten as

$$\lambda(\mu,\nu,r_1,r_2) = g(\mu,\nu,r_1,r_2)+h(\mu,\nu,r_1,r_2) , \qquad (16.6)$$

where
$$g \equiv g(\mu,\nu,r_1,r_2) = \frac{\mu^2+\nu^2}{(1+r_1\mu^2)(1+r_2\nu^2)} , \qquad (16.7)$$

and
$$h \equiv h(\mu,\nu,r_1,r_2) = \frac{2\mu\nu}{(1+r_1\mu^2)(1+r_2\nu^2)} . \qquad (16.8)$$

From (16.6) we have that

$$\lambda_M = \max_{\mu,\nu}\{g+h\} \leqslant \max_{\mu,\nu}\{g\} + \max_{\mu,\nu}\{h\} , \qquad (16.9)$$

and similarly,

$$\lambda_m = \min_{\mu,\nu}\{g+h\} \geqslant \min_{\mu,\nu}\{g\} + \min_{\mu,\nu}\{h\} . \qquad (16.10)$$

But from (16.7) and (12.19) we see that the behaviour of the function g can be summarised in Table 16.1 (which is similar to Table 13.1).

r_1-Domain	r_2-Domain	max{g}	min{g}
$0<r_1\leqslant 1/b^2$	$0<r_2\leqslant 1/b^2$	C'	A'
	$1/b^2\leqslant r_2\leqslant 1/a^2$	D'	A'
	$1/a^2\leqslant r_2<\infty$	D'	B'
$1/b^2\leqslant r_1\leqslant 1/a^2$	$0<r_2\leqslant 1/b^2$	B'	A'
	$1/b^2\leqslant r_2\leqslant 1/a^2$	max{B',D'}	min{A',C'}
	$1/a^2\leqslant r_2<\infty$	D'	C'
$1/a^2\leqslant r_1<\infty$	$0<r_2\leqslant 1/b^2$	B'	D'
	$1/b^2\leqslant r_2\leqslant 1/a^2$	B'	C'
	$1/a^2\leqslant r_2<\infty$	A'	C'

TABLE 16.1

where
$$A' = g(a,a,r_1,r_2), \quad B' = g(a,b,r_1,r_2) ,$$
$$C' = g(b,b,r_1,r_2), \quad D' = g(b,a,r_1,r_2) . \qquad (16.11)$$

We therefore continue with the study of the behaviour of the function $h(\mu,\nu,r_1,r_2)$. By taking partial derivatives of h with respect to μ and ν we obtain the following results

$$\text{sign}(\frac{\partial h}{\partial \mu}) = \text{sign}(1/\mu^2 - r_1) \;,$$

and
$$\text{sign}(\frac{\partial h}{\partial \nu}) = \text{sign}(1/\nu^2 - r_2) \;. \qquad (16.12)$$

From the above relationships we see that for fixed $r_1, r_2 > 0$ neither of the expressions $\frac{\partial h}{\partial \mu}$, $\frac{\partial h}{\partial \nu}$ changes sign as μ and ν vary in the interval (16.1). Consequently, the possible extreme values of h will occur at the points (a,a),(a,b),(b,a) and (b,b). On the other hand, if we let

$$A = h(a,a,r_1,r_2),\quad B = h(a,b,r_1,r_2)\;,$$
$$C = h(b,b,r_1,r_2) \text{ and } D = h(b,a,r_1,r_2)\;, \qquad (16.13)$$

then the order of the quantities A,B,C and D is determined by the following relationships

$$\text{sign}(A-B) = \text{sign}(D-C) = \text{sign}(r_2 - 1/(ab)) \;,$$

and
$$\text{sign}(A-D) = \text{sign}(B-C) = \text{sign}(r_1 - 1/(ab)) \;. \qquad (16.14)$$

In view of (16.14) we construct Table 16.2 which presents the maximum and minimum values of h with respect to μ,ν for the different values of r_1 and r_2 in the interval $(0,\infty)$.

r_1-Domain	r_2-Domain	max{h}	min{h}
$0<r_1\leq 1/(ab)$	$0<r_2\leq 1/(ab)$	C	A
	$1/(ab)\leq r_2<\infty$	D	B
$1/(ab)\leq r_1<\infty$	$0<r_2\leq 1/(ab)$	B	D
	$1/(ab)\leq r_2<\infty$	A	C

TABLE 16.2

In order to form the function λ (using the relationships (16.9) and (16.10)) we note from Tables 16.2,16.1 that we have to examine further the relative positions of r_1 and r_2 with respect to the value 1/(ab) in the study of the function $g(\mu,\nu,r_1,r_2)$. As a first step towards this direction we extend the case where $1/b^2\leq r_1\leq 1/a^2$ in Table 16.1 by constructing (in a similar manner to Table 13.3) Table 16.3. Taking also into consideration the position r_2 with respect to the value 1/(ab), then we extend Table 16.3 to yield Table 16.4, whereas the cases $0<r_1\leq 1/b^2$, $1/a^2\leq r_1<\infty$

in Table 16.1, yield Table 16.5.

r_1-Domain	r_2-Domain	max{g}	min{g}
$\frac{1}{b^2} \leq r_1 \leq \frac{1}{ab}$	$\frac{1}{b^2} \leq r_2 \leq r_1$	B'	A'
	$r_1 \leq r_2 \leq \frac{1}{r_1(ab)^2}$	D'	A'
	$\frac{1}{r_1(ab)^2} \leq r_2 \leq \frac{1}{a^2}$	D'	C'
$\frac{1}{ab} \leq r_1 \leq \frac{1}{a^2}$	$\frac{1}{b^2} \leq r_2 \leq \frac{1}{r_1(ab)^2}$	B'	A'
	$\frac{1}{r_1(ab)^2} \leq r_2 \leq r_1$	B'	C'
	$r_1 \leq r_2 \leq \frac{1}{a^2}$	D'	C'

TABLE 16.3

r_1-Domain	r_2-Domain	max{g}	max{h}	min{g}	min{h}	$P(\hat{B}_{r_1,r_2})$
$\frac{1}{b^2} \leq r_1 \leq \frac{1}{ab}$	$\frac{1}{b^2} \leq r_2 \leq r_1$	B'	C	A'	A	(B'+C)/(A'+A)
	$r_1 \leq r_2 \leq \frac{1}{ab}$	D'	C	A'	A	(D'+C)/(A'+A)
	$\frac{1}{ab} \leq r_2 \leq \frac{1}{r_1(ab)^2}$	D'	D	A'	B	(D'+D)/(A'+B)
	$\frac{1}{r_1(ab)^2} \leq r_2 \leq \frac{1}{a^2}$	D'	D	C'	B	(D'+D)/(C'+B)
$\frac{1}{ab} \leq r_1 \leq \frac{1}{a^2}$	$\frac{1}{b^2} \leq r_2 \leq \frac{1}{r_1(ab)^2}$	B'	B	A'	D	(B'+B)/(A'+D)
	$\frac{1}{r_1(ab)^2} \leq r_2 \leq \frac{1}{ab}$	B'	B	C'	D	(B'+B)/(C'+D)
	$\frac{1}{ab} \leq r_2 \leq r_1$	B'	A	C'	C	(B'+A)/(C'+C)
	$r_1 \leq r_2 \leq \frac{1}{a^2}$	D'	A	C'	C	(D'+A)/(C'+C)

TABLE 16.4

r_1-Domain	r_2-Domain	max{g}	max{h}	min{g}	min{h}	$P(\hat{B}_{r_1,r_2})$
$0<r_1\leq\frac{1}{b^2}$	$0<r_2\leq\frac{1}{b^2}$	C'	C	A'	A	(C'+C)/(A'+A)
	$\frac{1}{b^2}\leq r_2\leq\frac{1}{ab}$	D'	C	A'	A	(D'+C)/(A'+A)
	$\frac{1}{ab}\leq r_2\leq\frac{1}{a^2}$	D'	D	A'	B	(D'+D)/(A'+B)
	$\frac{1}{a^2}\leq r_2<\infty$	D'	D	B'	B	(D'+D)/(B'+B)
$\frac{1}{a^2}\leq r_1<\infty$	$0<r_2\leq\frac{1}{b^2}$	B'	B	D'	D	(B'+B)/(D'+D)
	$\frac{1}{b^2}\leq r_2\leq\frac{1}{ab}$	B'	B	C'	D	(B'+B)/(C'+D)
	$\frac{1}{ab}\leq r_2\leq\frac{1}{a^2}$	B'	A	C'	C	(B'+A)/(C'+C)
	$\frac{1}{a^2}\leq r_2<\infty$	A'	A	C'	C	(A'+A)/(C'+C)

TABLE 16.5

If one studies the behaviour of $\frac{\partial P(\hat{B}_{r_1,r_2})}{\partial r_2}$ (assuming r_1 is fixed for all the cases in Tables 16.4 and 16.5, then it can be easily verified that the minimum value of $P(\hat{B}_{r_1,r_2})$ is attained if

$$r_2 = \frac{1}{ab} \quad . \tag{16.15}$$

Because of the symmetry of our problem we can work similarly for determining the optimum value of r_1 and conclude also that this is identical with the value of r_2 given by (16.15), hence (16.2) follows. Evidently, from Table 16.4 we have that the smallest and largest eigenvalues of $\hat{B}_{r',r'}$, are given by the expression

$$\lambda_M = D'+C = D'+D = B'+B = B'+A = ab \ ,$$
$$\lambda_m = A'+A = A'+B = C'+D = C'+C = \frac{4a^2b^2}{(a+b)^2} \ . \tag{16.16}$$

Consequently, from (16.16) and (15.23) we see that $P(\hat{B}_{r',r'})$ is given by (16.3) while from (15.25) the optimum value $\bar{\bar{\tau}}$ of τ is given by (16.4). But for this value of τ, the spectral radius of

$\hat{T}_{\tilde{\tau},r',r'}$ is given by the formula

$$S(\hat{T}_{\tilde{\tau},r',r'}) = \frac{P(\hat{B}_{r',r'})-1}{P(\hat{B}_{r',r'})+1}, \tag{16.17}$$

which by (16.3) yields (16.5) and the proof of the theorem is complete.

17. THE CASE WHERE THE EIGENVALUE RANGES OF H AND V MAY BE DIFFERENT

In this case we prove the following theorem:

Theorem 17.1: Let H and V be the matrices defined by (15.10), (15.11) with eigenvalues μ,ν, respectively such that

$$0 < a \lesssim \mu \lesssim b \text{ and } 0 < \alpha \lesssim \nu \lesssim \beta. \tag{17.1}$$

Then, the P-condition number of $\hat{B}_{r_1,r_2}$ is minimised when the parameters r_1,r_2 take the values

$$r_1^* = \frac{1-\Sigma^2 sc^{\frac{1}{2}}}{-t+\Sigma^2 qc^{\frac{1}{2}}}, \quad r_2^* = \frac{1+\Sigma^2 sc^{\frac{1}{2}}}{t+\Sigma^2 qc^{\frac{1}{2}}}, \tag{17.2}$$

where

$$c = \frac{1}{1+\theta+[\theta(2+\theta)]^{\frac{1}{2}}}, \tag{17.3}$$

$$\theta = \frac{2(\beta^2-\alpha^2)(b^2-a^2)}{(a^2+\alpha^2)(b^2+\beta^2)}, \tag{17.4}$$

$$\Sigma^2 s = \frac{(\beta^2-\alpha^2)(b^2-a^2)}{(b^2+\beta^2)-(a^2+\alpha^2)c}, \tag{17.5}$$

$$\Sigma^2 q = \frac{(b^2+\beta^2)+(b^2-\beta^2)\Sigma^2 s}{2}, \tag{17.6}$$

$$t = \frac{(b^2-\beta^2)+(b^2+\beta^2)\Sigma^2 s}{2}, \tag{17.7}$$

and its corresponding value is given by

$$P(\hat{B}_{r_1^*,r_2^*}) = \frac{(1+c^{\frac{1}{2}})^2}{4c^{\frac{1}{2}}}. \tag{17.8}$$

On the other hand, if we also let

$$\tau = \tau^* = \frac{2}{\Sigma^2 c^{\frac{1}{2}}[1+1/P(\hat{B}_{r_1^*,r_2^*})]}, \tag{17.9}$$

then the spectral radius of $\hat{T}_{\tau,r_1,r_2}$ attains its minimum value which is given by the expression

$$S(\hat{T}_{\tau^*,r^*,r^*}) = \frac{(1-c^{\frac{1}{2}})^2}{(1+c^{\frac{1}{2}})^2+4c^{\frac{1}{2}}} \quad . \tag{17.10}$$

<u>Proof</u>: From the previous section it can be noticed that the value of the optimum parameters which minimise the P-condition number of the matrix $\hat{B}_{r_1,r_2}$ is identical with the one which minimises the ratio

$$G = \frac{\max\{g\}}{\min\{g\}} \quad . \tag{17.11}$$

Indeed, we observe that the function $g(\mu,\nu,r_1,r_2)$ is obtained from the function $\lambda(\mu,\nu,r_1,r_2)$ given by (12.19) with μ,ν being replaced by μ^2,ν^2, respectively. Consequently, from Theorem 13.1 we have that G is minimised if we let r_1,r_2 have the values given by (16.2), since $0<a^2\leq\mu^2,\nu<^2\leq b^2$. On the other hand, the behaviour of $P(\hat{B}_{r_1,r_2})$ is not affected by the bilinear transformation of the form (14.13), in the sense that it is the same between the original and the corresponding transformed intervals. Thus, if we transform our problem (using a similar analysis to Section 14 so that we return to the previous case of the "single range", then the optimum values of the corresponding transformed parameters r_1 and r_2 will still remain the same as the ones which minimise the transformed ratio G. In other words, our problem is identical with the one tackled in Section 14, the only difference being that instead of having μ,ν in (14.11) here we have μ^2,ν^2. Thus, by adhering again to the analysis of Wachspress and Jordan [21] we seek to introduce new variables $\hat{\mu}^2,\hat{\nu}^2$ such that

$$\mu^2 = \frac{t+q\hat{\mu}^2}{1+s\hat{\mu}^2} \; , \qquad \nu^2 = \frac{t'+q'\hat{\nu}^2}{1+s\hat{\nu}^2} \; , \tag{17.12}$$

so that for some $\hat{\omega}_1$ and $\hat{\omega}_2$ we have

$$\left(\frac{\mu^2-\omega_2}{\mu^2+\omega_1}\right)\left(\frac{\nu^2-\omega_1}{\nu^2+\omega_2}\right) = \left(\frac{\hat{\mu}^2-\hat{\omega}_2}{\hat{\mu}^2+\hat{\omega}_1}\right)\left(\frac{\hat{\nu}^2-\hat{\omega}_1}{\hat{\nu}^2+\hat{\omega}_2}\right) \; , \tag{17.13}$$

where $\hat{\mu}^2$ and $\hat{\nu}^2$ vary over the ranges

$$\sigma^2 \leq \hat{\mu}^2 \leq \Sigma^2 \; , \quad \sigma^2 \leq \hat{\nu}^2 \leq \Sigma^2 \quad . \tag{17.14}$$

By following the same analysis as in Section 14 we can easily show the validity of Theorem 17.1.

From Theorem 16.1 we see that as $P(\hat{B}_{r',r'})$ increases, then $\bar{\tau}$ tends to become equal to 2r'. In other words, for sufficiently small mesh size the MADP method for the numerical solution of the biharmonic equation tends to attain the same rate of convergence as the iterative scheme

$$u^{(n+1)} = u^{(n)}+2r'(I+r'V^2)^{-1}(I+r'H^2)^{-1}(b-Au^{(n)}). \tag{17.15}$$

However, it is believed that the significance of the above results will be further realised when we consider the more general problem of varying the involved parameters τ, r_1, r_2 [2]. It should be mentioned that some of the results of Section 16 (for the case where the eigenvalue ranges of H and V are the same) have been obtained in [7]. However, in the case where the eigenvalue ranges of H,V were different, the optimum parameters were found in [7] under the assumption that

$$0 < \alpha' \lesssim \mu, \mu \lesssim \beta' ,$$

where $\alpha'=\min(a,\alpha)$ and $\beta'=\max(b,\beta)$. Moreover in [14] optimum parameters for the EADI method are obtained for the same problem but $r_1=r_2=r$ and for the case where the eigenvalue ranges are different.

18. COMPARISON OF RATES OF CONVERGENCE FOR THE MODEL PROBLEM

If we consider the solution of the biharmonic equation in the unit square with $h_x=h_y=h$, then by (15.22) we have

$$a = \alpha = 4\sin^2(\frac{\pi h}{2}) \text{ and } b = \beta = 4\cos^2(\frac{\pi h}{2}) , \tag{18.1}$$

hence from Theorem 16.1 we obtain successively,

$$r' = \frac{1}{4\sin^2(\pi h)} , \tag{18.2}$$

$$P(\hat{B}_{r',r'}) = \frac{1}{\sin^2(\pi h)} , \tag{18.3}$$

and

$$\tau_0 = \frac{1}{2\sin^2(\pi h)[1+\sin^2(\pi h)]} . \tag{18.4}$$

Finally, the spectral radius is given by the expression

$$S(\hat{T}_{\tau_0,r',r'}) = \frac{1-\sin^2(\pi h)}{1+\sin^2(\pi h)} , \tag{18.5}$$

thus the rate of convergence of the iterative scheme (15.14) is

$$R(\hat{T}_{\tau_0,r',r'}) \sim 2\pi^2 h^2 , \tag{18.6}$$

for sufficiently small h.

If, however, the 25-point difference analogue is used to approximate the biharmonic equation, then the matrix A has the following splitting

$$A = (H+V-kHV)^2 , \tag{18.7}$$

where $k=1/6$. Evidently, for $k=0$ the iterative scheme (15.14) is fourth order correct in h, while for $k=1/6$ it is eighth order correct in h. Thus we can find that the eigenvalues of $\hat{B}_{r_1,r_2}$ are given by the expression

$$\lambda = \frac{(\mu+\nu-k\mu\nu)^2}{(1+r_1\mu^2)(1+r_2\nu^2)} , \tag{18.8}$$

and can be bound as follows

$$(1-kb/2)^2\phi \leq \lambda \leq (1-ka/2)^2\phi , \tag{18.9}$$

where

$$\phi = \frac{(\mu+\nu)^2}{(1+r_1\mu^2)(1+r_2\nu)^2} . \tag{18.10}$$

From the above we find again that if we let

$$r_1 = r_2 = r' = \frac{1}{ab} , \tag{18.11}$$

then $P_k(\hat{B}_{r_1,r_2})$ is minimised and its corresponding value is given by

$$P_k(\hat{B}_{r',r'}) = k''P(\hat{B}_{r',r'}) , \tag{18.12}$$

where

$$k'' = \left(\frac{1-ka/2}{1-kb/2}\right)^2 . \tag{18.13}$$

Moreover, from (16.16) and (18.9) we find that if we let

$$\tau_0 = \frac{(a+b)^2 r'^2}{2(1-kb/2)^2[1+(k'')^2 P(\hat{B}_{r',r'})]} , \tag{18.14}$$

the spectral radius is also minimised and given by the expression

$$S_k(\hat{T}_{\tau_O,r',r'}) = \frac{k''P(\hat{B}_{r',r'})-1}{k''P(\hat{B}_{r',r'})+1} , \tag{18.15}$$

therefore the rate of convergence is

$$R_k(\hat{T}_{\tau_O,r',r'}) \sim \frac{2}{k''P(\hat{B}_{r',r'})} . \tag{18.16}$$

Finally, if we consider the application of the MADP-SI method for the solution of the present problem, then the rate of convergence is

$$R_{k,\infty}(P_n(\hat{T}_{\tau_O,r',r'})) \sim 2/\sqrt{k''P(\hat{B}_{r',r'})} , \tag{18.17}$$

which for k=O, (18.3),(18.12) and (18.13) give the result

$$R_\infty(P_n(\hat{T}_{O,r',r'})) \sim 2\pi h , \tag{18.18}$$

for sufficiently small h.

In order to verify the theoretical results presented here we solved the biharmonic equation

$$\frac{\partial^4 U}{\partial x^4} + 2\frac{\partial^4 U}{\partial x^2 \partial y^2} + \frac{\partial^4 U}{\partial y^4} = 0 , \quad (x,y)\in R, \tag{18.19}$$

where the region R was the unit square. The boundary conditions were given as

$$\left.\begin{array}{l}
\left.\begin{array}{l} U(x,y) = \sin(\pi x)/e^{\pi} \\ \dfrac{\partial^2 U}{\partial y^2}(x,y) = \pi^2 \sin(\pi x)/e^{\pi} \end{array}\right\} 0\leqslant x\leqslant 1, \; y=0 , \\
\left.\begin{array}{l} U(x,y) = \sin(\pi x) \\ \dfrac{\partial^2 U}{\partial y^2}(x,y) = \pi^2 \sin(\pi x) \end{array}\right\} 0\leqslant x\leqslant 1, \; y=1, \\
\text{and} \\
\left.\begin{array}{l} U(x,y) = 0 \\ \dfrac{\partial^2 (x,y)}{\partial x^2} = 0 \end{array}\right\} 0\leqslant y<1, \; x=0,1.
\end{array}\right\} \tag{18.20}$$

By applying the 13-point difference analogue we approximated (18.1) and the produced system was solved with the MADP method as defined by (15.15) and the MADP-SI method defined by

$$u^{(n+1)} = (1-\rho_{n+1})u^{(n+1)} + \rho_{n+1}(\hat{T}_{\tau_O,r',r'}u^{(n)} + \hat{t}) , \tag{18.21}$$

where

$$\left.\begin{aligned}\rho_1 &= 1\,,\\ \rho_2 &= \left(1-\frac{\sigma^2}{2}\right)^{-1},\\ \rho_{n+1} &= 1-\left(\frac{\sigma^2\rho_n}{4}\right)^{-1},\quad n=2,3,\ldots\\ \text{and}\qquad \sigma &= \frac{P(\hat{B}_{r',r'})-1}{P(\hat{B}_{r',r'})+1}\,.\end{aligned}\right\}\qquad(18.22)$$

Again we choose as starting vector $u^{(0)}$ the vector with all its components equal to unity while for convergence the following criterion was required to be satisfied

$$\max|u^{(n+1)}-u^{(n)}|\leq 10^{-6}\,.$$

In Figure 18.1 we present the number of iterations required to solve the present problem with the iterative procedures mentioned above for the different mesh sizes shown. To determine the rates of convergence we plot the logarithm of the observed number of iterations versus h^{-1} for the MADP and MADP-SI methods.

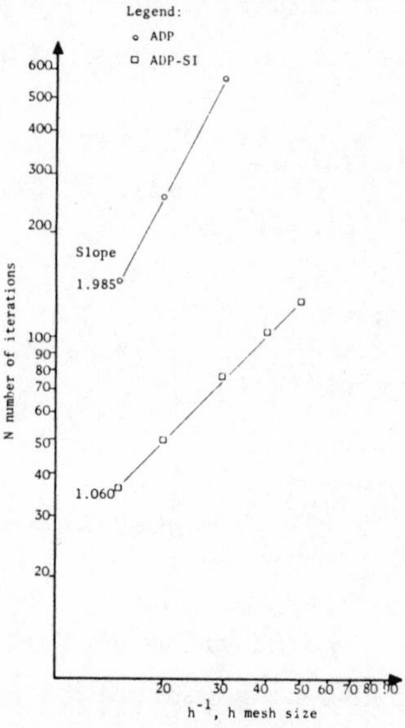

FIGURE 18.1

DETERMINATION OF RATE OF CONVERGENCE ATTAINED FOR THE BIHARMONIC EQUATION USING ADP AND ADP-SI METHODS

The results verify our expectations (see 18.6) and (18.18)) by showing that the number of iterations of the MADP method for the biharmonic equation (15.1) varies approximately like $O(h^{-2})$, whereas for the MADP-SI method like $O(h^{-1})$. Finally, a further improvement (perhaps by an order of magnitude) on the rate of convergence can be achieved by considering a sequence of parameters $\{r_i\}$ (see Conte and Dames [2]).

REFERENCES

1. Birkhoff, G., Varga, R.S. and Young, D.M. - "Alternating Direction Implicit Methods", Advances in Computers 3, pp.189-273, 1962.

2. Conte, S.D. and Dames, R.T. - "An Alternating Direction Method for Solving Linear Equations with Symmetric Positive Matrices", M.T.A.C. 12, pp.198-205, 1958.

3. Diamond, M.A. - "An Economical Algorithm for the Solution of Finite Difference Equations Independent of User-Supplied Parameters", Ph.D. Thesis, Dept. of Computer Science, Univ. of Illinois, 1971.

4. Evans, D.J. - "The Use of Preconditioning in Iterative Methods for Solving Linear Equations with Symmetric Positive Matrices", J.Inst.Math.Applics. 4, pp.295-314, 1968.

5. Evans, D.J. - "Comparison of the Convergence Rates of Iterative Methods for Solving Linear Equations with Preconditioning", Greek Mathematical Society, Carathéodory Symposium, pp.106-135, 1973.

6. Evans, D.J. - "Iterative Sparse Matrix Algorithms", in 'Software in Numerical Mathematics', D.J. Evans (Edit.), Academic Press, pp.49-83, 1974.

7. Evans, D.J. and Gane, C.R. - "Alternating Direction Preconditioning Techniques", Technical Report No.18, Dept. of Computer Studies, Loughborough University of Technology, U.K., 1974.

8. Evans, D.J. and Missirlis, N.M. - "The Modified Alternating Direction Preconditioning Method for the Numerical Solution of Elliptic Self-Adjoint Partial Differential and Biharmonic Equations", B.I.T., 19, pp.172-185, 1979.

9. Evans, D.J. and Missirlis, N.M. - "The Preconditioned Simultaneous Displacement Method (P.S.D.)", Math.Comp.Sim. 22, pp.256-263, 1980.

10. Forsythe, G.E. and Wasow, W.R. - "Finite Difference Methods for Partial Differential Equations", Wiley, New York, 1960.

11. Golub, G.H. and Varga, R.S. - "Chebyshev Semi-Iterative Methods, Successive Over-Relaxation Iterative Methods and Second Order Richardson Iterative Methods", Num.Math., Parts I and II, 3, pp.147-168, 1961.

12. Guittet, J. - "Une Nouvelle Methode de Directions Alternees á q Variables", J.Math.Anal.Appl. 17, pp.199-213, 1967.

13. Hadjidimos, A. and Iordanidis, K. - "Solving Laplace's Equation in a Rectangle by Alternating Direction Implicit Methods", J.Math.Appl. 48, pp.353-367, 1974.

14. Hadjidimos, A. - "On Comparing Optimum Alternating Direction Preconditioning and Extrapolated Direction Implicit Schemes", J.Math.Anal.Appl. 59, pp.573-586, 1975.

15. Hestenes, M.R. and Stiefel, E. - "Method of Conjugate Gradients for Solving Linear Systems", J.Res.Nat.Bur.Stand. 49, pp.409-436, 1952.

16. Missirlis, N.M. - "Preconditioned Iterative Methods for Solving Elliptic Partial Differential Equations", Ph.D. Thesis, Loughborough University of Technology, Loughborough, U.K., 1978.

17. Missirlis, N.M. and Evans, D.J. - "On the Acceleration of the Preconditioned Simultaneous Displacement Method", Math. Comp.Sim. 23, pp.191-198, 1981.

18. Missirlis, N.M. and Evans, D.J. - "On the Dynamic Acceleration of the Preconditioned Simultaneous Displacement Method", Int.Jour.Comp.Math. 10, pp.153-176, 1981.

19. Niethammer, W. - "Relaxation bei Komplexen Matrizen", Math. Zeitsch. 86, pp.34-40, 1964.

20. Varga, R.S. - "Matrix Iterative Analysis", Prentice-Hall, Englewood Cliffs, New Jersey, 1962.

21. Wachspress, E.L. - "Iterative Solution of Elliptic Schemes and Application to Neutron Diffusion Equations of Reactor Physics", Prentice-Hall, Englewood Cliffs, New Jersey, 1966.

22. Young, D.M. - "On Richardson's Method for Solving Linear Systems with Positive Definite Matrices, J.Math.Phys. XXXII pp.243-255, 1954.

23. Young, D.M. - "Iterative Solution of Large Linear Systems", Academic Press, 1971.

24. Young, D.M. - "On the Solution of Large Systems of Linear Algebraic Equations with Sparse, Positive Definite Matrices", in 'Numerical Solution of Systems of Non-Linear Algebraic Equations', (G.D. Byrne & C.A. Hall, edits.), Academic Press, N.Y. pp.101-156, 1974.

25. Young, D.M. - "Notes on the Conjugate Gradient Method", unpublished, 1975.

26. Young, D.M. - "On the Accelerated S.S.O.R. Method for Solving Large Linear Systems", Advances in Mathematics, 23, pp.215-271, 1977.

SSOR PRECONDITIONING OF TOEPLITZ MATRICES

LENNART ANDERSSON
Department of Mathematics
University of Luleå, Sweden

ABSTRACT

SSOR preconditioning of Toeplitz and block Toeplitz matrices is considered. It is shown that the spectral condition number is thus reduced by one order of magnitude for the case when the matrix corresponds to an elliptic difference operator of order 2m.

1. INTRODUCTION

Preconditioning may be used to obtain faster convergence for the solution of linear systems by iterative methods. Iterative methods are of special interest when the coefficient matrix is large and sparse. The rate of convergence generally depends on the condtion number of the matrix. The condition number may be very large when the system appears as the result of discretization of an elliptic partial differential equation. The purpose of this paper is to derive asymptotic bounds for such systems when using SSOR preconditioning.

We shall consider Toeplitz matrices only. This is a very special type of matrices, but they are important in this connection since discretization of a differential operator such as the Laplacian on a regular mesh by finite difference or finite element methods leads to matrices with similar spectral properties.

The present paper is a summary of [1].

2. PRELIMINARIES

2.1. SSOR preconditioning

The number of iterations for solving a linear system of equations with a positive definite matrix, A, depends on the spectral condition number of the matrix, k(A). By preconditioning it may be possible to reduce the condition number before applying some iterative method such as the conjugate gradient method.

Here we shall consider SSOR preconditioning [2,4]. The name originates from the Symmetric Successive Over-Relaxation method. We shall not penetrate the connections here.

2.2. Toeplitz matrices

Let $A = (\alpha_{ij})$ be an n*n symmetric Toeplitz matrix, i.e. $\alpha_{ij} = a_{j-i}$, where the numbers a thus fully determine the matrix. We shall consider matrices with a fixed bandwidth 2p+1 and let the order, n, vary. With the matrix A we associate a trignometric polynomial, cf the symbol of a difference operator,

$$A(\theta) = a_0 + 2 \sum_{j=1}^{p} a_j \cos j\theta$$

The extreme eigenvalues of the matrix A may be estimated by using the following theorem, cf [5].

Theorem 2.

Suppose that $A(\theta)$ attains its minimal value at $\theta=0$ where the first nonvanishing derivative is of order 2m. Let λ_1 be the smallest eigenvalue of the corresponding n*n matrix. Then there exist positive constants C_0 and C_1 such that

$$A(0) + C_0 n^{-2m} < \lambda_1 < A(0) + C_1 n^{-2m} \quad .$$

The point $\theta=0$ can be replaced in the theorem; the important fact is that $A(\theta)$ has a global minimum (or maximum for the largest eigenvalue) at the point of consideration.

The proof of this theorem uses Parseval's formula in order to get an equivalent problem of estimating an integral. Let $x \in R^n$ and consider the function

$$x(\theta) = \sum_{j=1}^{n} x_j \exp(i\theta j) \quad .$$

Then

$$(x,Ax) = (2\pi)^{-1} \int_{-\pi}^{\pi} A(\theta) |x(\theta)|^2 \, d\theta \quad .$$

3. SSOR PRECONDITIONING OF TOEPLITZ MATRICES

Throughout this section we shall suppose that there are constants C_0 and C_1 such that

$$C_0 \theta^{2m} \leq A(\theta) \leq C_1 \theta^{2m} \quad , \quad |\theta| \leq \pi \quad . \tag{1}$$

According to Theorem 2 this implies that the condition number of the corresponding Toeplitz matrix of order n is $O(n^{2m})$. We shall show that SSOR preconditiong will reduce the condition number by one power of n. We shall first prove two technical lemmas.

Lemma 1.

If $A(\theta)\geq 0$ with equality for $\theta=0$ then

$$\sum_{j=1}^{p} ja_j < 0 \quad .$$

Proof.

Let (in this proof only)

$$A(z) = 1 + 2 \sum_{j=1}^{p} a_j z^j$$

The assumptions of the Lemma may now be restated as

$$\text{Re } A(z) \geq 0 \quad , \quad |z| = 1 \quad , \text{ with equality for } z=1$$

The assertion will be

$$A'(1) < 0 \ .$$

By the maximum principle it follows that

$$\text{Re } A(z) \geq 0 \quad \text{for} \quad |z| \leq 1 \quad .$$

Hence the unit disc of the z-plane is mapped by A into the right half of the complex plane. It follows that

$$A'(1) \leq 0 \quad .$$

Now suppose that $A'(1)=0$ in order to get a contradiction. Consider a small sector with opening angle $3\pi/4$ inside the unit disc of the z-plane with its apex at $z=1$. This sector is consequently mapped by A into the right half of the complex plane and the apex is mapped onto the origin. However

$$A(z) = C(z-1)^j + O((z-1)^{j+1}) \quad , \quad z \to 1 \quad ,$$

where $C>0$ and $j>1$. Hence the opening angle of the mapped sector will not be smaller than $3\pi/2$, which contradicts the fact that the sector is mapped into the right half plane.

Lemma 2.

Let $T=(t_{ij})$ be an infinite $(i,j\in Z)$ Toeplitz matrix and $T^{(n)}$ the corresponding $n*n$ Toeplitz matrix. Then

$$(TT^*)^{(n)} - T^{(n)}T^{(n)*}$$

is positive semidefinite.

Proof.

Let $T^{(n)}$ be extended by zero elements and formally regarded as an infinite matrix. Define

$$M = T - T^{(n)} \quad .$$

Then

$$TT^* = T^{(n)}T^{(n)*} + MM^* + MT^{(n)*} + T^{(n)}M^*$$

We notice that the two last terms contribute nothing to the relevant submatrix, while MM^* is positive semidefinite and the proof is complete.

In order to estimate the condition number of A after preconditioning we have to estimate the quantities a and K of Theorem 1.

Without loss of generality we may assume that D=I, so that $A=I+L+L^*$ and

$$A(\theta) = 1+L(\theta)+L(-\theta) = 1 + 2\sum_{j=1}^{p} a_j \cos j\theta$$

It follows from Theorem 2 that

$$a < Cn^{2m} \quad .$$

Lemma 3.

Suppose that $A(\theta)$ satisfies (1). Then there is a positive constant C such that

$$K < Cn^{2m-2} \quad , \quad \text{for all } n.$$

Proof.

First we observe that the matrix $(I+2L)(I+2L^*)$ is not exactly a Toeplitz matrix. The Toeplitz structure is somewhat disturbed at the upper left corner of the matrix. Lemma 2 can now be used to justify the following result

$$\|(I+2L^*)x\|^2 \leq (2\pi)^{-1} \int |1+2L(-\theta)|^2 |x(\theta)|^2 \, d\theta \quad .$$

Now

$$\mathrm{Re}\ (1+2L(-\theta)) = A(\theta)$$

and

$$\mathrm{Im}\ (1+2L(-\theta)) = 2\ \mathrm{Im}\ L(-\theta) \quad .$$

By definition we obtain

$$\mathrm{Im}\ L(-\theta) = i\theta \sum ja_j \ .$$

(We may remark that by Lemma 1 the coefficient of 0 cannot vanish.) We can now conclude that there is a positive constant C such that

$$|1+2L(-\theta)|^2 \leq C\theta^2 \ .$$

The rest of the proof rely on the one-dimensional version of Lemma 4 of the next section.

4. PRECONDITIONING OF BLOCK TOEPLITZ MATRICES

4.1. Two-dimensional Toeplitz matrices

Let A be a two-dimensional Toeplitz matrix, i.e. a block matrix where each block is a Toeplitz matrix and the blocks themselves form a Toeplitz matrix. We write

$$A = (A_{j-i}) \quad , \quad i,j=1,\dots,n \quad ,$$

where

$$A_k = (a_{k,j-i}) \quad , \ i,j=1,\dots,n \quad .$$

Again we associate a polynomial with the matrix

$$A(\theta) = A(\theta_1,\theta_2) = \sum A_j(\theta_2)\exp(i\theta_1 j)$$

We shall assume that A and each A_k is symmetric. Then $A(\theta)$ is real and

$$A(\theta) = A_0(\theta_2) + 2 \Sigma A_j(\theta_2)\cos j\theta_1 \quad . \tag{2}$$

We shall make the corresponding assumption on $A(\theta)$ as in the one-dimensional case in order to consider Toeplitz matrices relatedto elliptic difference operators of order 2m. We thus suppose the existence of positive constants C_0 and C_1 such that

$$C_0|\theta|^{2m} \leq A(\theta) \leq C_1|\theta|^{2m} \quad , \; |\theta|^2 = \theta_1^2 + \theta_2^2 \leq 2\pi^2 \quad . \tag{3}$$

Furthermore we assume that the diagonal blocks are well conditioned, i.e. there is a positive constant d such that

$$A_0(\theta_1) > d \quad , \text{ for all } \theta_1 \; . \tag{4}$$

Let $x=(x_{jk})$ be a vector in R^{n^2} and introduce the corresponding function

$$x(\theta) = \Sigma \; x_{ij}\exp(i(\theta_1 j+\theta_2 k)) \quad .$$

Parseval´s equalty then reads

$$(x,Ax) = (2\pi)^{-2} \int_S A(\theta)| x(\theta)|^2 d\theta \quad ,$$

where S is a square in the θ_1-θ_2-plane with side 2π .

4.2. SSOR preconditioning of block matrices

In this section we suppose that the D matrix contains the diagonal blocks. The following lemma is crucial.

Lemma 4.

Let $x(\theta)$ be of the form

$$x(\theta) = \sum_{j,k=1}^{n} x_{jk} \exp(i(\theta_1 j+\theta_2 k)) \quad .$$

Then

$$\sup_x \int_S |\theta|^2 |x(\theta)|^2 d\theta \Big/ \int_S |\theta|^{2m} |x(\theta)|^2 \, d\theta \ < \ Cn^{2m-2}$$

<u>Proof.</u>

We may assume that

$$\Sigma |x_{jk}|^2 = 1$$

so that by Schwartz's inequality

$$|x(\theta)|^2 \le n^2$$

and by Parseval's equality

$$(2\pi)^{-2} \int_S |x(\theta)|^2 \, d\theta = 1 \quad .$$

If m=1 the lemma is true. Hence we suppose that m>1. Let

$$F(r) = (Cn)^{2m-2} r^{2m} - r^2 \quad , \quad r > 0 \quad ,$$

where C is some positive constant to be choosen later. Let r=a be the point where F(r) attains its minmal value. We find that

$$a = 1/[Cn \ m^{1/(2m-2)}]$$

and that F(a)<0.

Define b>0 by F(b)=-F(a) . Since F(2/(Cn)) > F(a) it follows that b < 2/(Cn).
We then have

$$(2\pi)^{-2} \int_{|\theta|<b} |x(\theta)|^2 d\theta < (2\pi)^{-2} n^2 \pi b^2 < [1/\pi C^2] < 1/2$$

if we choose C=1 . Then a < b < 1 and

$$\int_{S \setminus D} |x(\theta)|^2 d\theta > \int_D |x(\theta)|^2 d\theta \quad ,$$

where D is the disc with center at the origin and radius b. By the construction of b it now follows that

$$\int_S F(|\theta|) \, |x(\theta)|^2 d\theta \ > \ 0$$

and the proof is complete.

Lemma 5.

Suppose that $A(\theta)$ satisfies (3). Then there is a positive constant C such that

$$K < Cn^{2m-2} \quad , \text{ for all } n.$$

Proof.

First we observe that

$$(x,(D+2L)D^{-1}(D+2L^{*})x) \leq d^{-1}\|(D+2L^{*})x\|^{2} ,$$

where d is the positive constant that was introduced in section 4.1.

Using a two-dimensional version of Lemma 2 we get

$$\|(D+2L^{*})x\|^{2} \leq (2\pi)^{-2} \int_{S} | D(\theta) +2L(-\theta)|^{2} |x(\theta)|^{2} d\theta \quad .$$

The remainder of the of the proof closely follows the one-dimensional case:

$$\text{Re } (D(\theta)+2L(-\theta)) = A(\theta)$$

and

$$\text{Im } (D(\theta)+2L(-\theta)) = 2 \text{ Im } L(-\theta) \quad .$$

By (3) we obtain

$$\text{Im } L(-\theta) = -i\theta_{1}\{\Sigma\, ja_{j} + O(\theta_{2}^{2})\} + O(\theta_{1}^{3}) \quad ,$$

where $a_{j} = A_{j}(0)$.

We can now conclude that there is a positive constant C such that

$$| D(\theta)+2L(-\theta)|^{2} \leq C\theta^{2} .$$

Lemma 4 completes the proof.

We now state our main theorem for the block Toeplitz case.

Theorem 3.

Let A be a block Toeplitz matrix of order $n^2 * n^2$ with a characteristic function $A(\theta)$ satisfying (3-4) and let , K and w be defined as in Theorem 1. Then the spectral condition number of A with SSOR preconditioning is less than Cn^{2m-1} for some positive constant C independent of n.

It can be shown, [1], that we cannot reduce the power of n in the above result.

The same technique can be used to bound the condition number of point SSOR preconditioning of block Toeplitz matrices, cf [1]. Generalizations to higher dimensions are also obvious. In all cases the the bound of the last theorem hold.

REFERENCES

1. ANDERSSON, L. - SSOR Preconditioning of Toeplitz Matrices, Thesis, Chalmers University of Technology, 1976.

2. AXELSSON, O. - On Preconditioning and Convergence Acceleration in Sparse Matrix Problems, CERN Report 74-10, 1974.

3. AXELSSON, O., - Solution of linear systems of equations: iterative methods, Sparse Matrix Techniques, Ed. Dold, A. and ECKMANN, B., Lecture Notes in Mathematics no 572, Springer Verlag, 1977.

4. EVANS, D.J. - The Use of Pre-conditioning in Iterative Methods for Solving Linear Equations with Symmetric Positive Definite Matrices, JIMA Vol 4, pp 295-314, 1968.

5. PARTER, S. - Extreme Eigenvalues of Toeplitz Forms and Applications to Elliptic Difference Equations, Trans. AMS Vol. 99, pp. 153-192, 1961.

ON THE CONVERGENCE THEORY AND APPLICATION OF SYMMETRIC COMPACT PRECONDITIONED ITERATIVE METHODS

D.J. Evans & E.A. Lipitakis

Department of Computer Studies, University of Technology, Loughborough, Leicestershire, U.K.

ABSTRACT

Sparse symmetric approximate factorization procedures are presented in the form of compact preconditioned iterative methods. The definition of the acceleration parameters for the convergence of these semi-direct methods are shown to be based on lower and upper bounds of the extreme eigenvalues of the iteration matrix. Optimum values of these parameters are established for the real eigenvalues of the iteration matrix and estimates for the computational work required to reduce the L_2-norm of the error vector by a specified factor ε are presented. Finally, the application of the methods to both linear and non-linear problems is discussed and experimental results are given.

1. INTRODUCTION

In this paper the Extended to the Limit (EL) sparse LDL^T factorization solution methods for the large sparse system $A\underline{u}=\underline{s}$ derived from the Finite Difference (FD) discretization of self-adjoint elliptic P.D.E.s in 2 dimensions are introduced. The coefficient matrix is shown to be factorized exactly (the "Limit" case) into triangular systems (i.e. $A=LDL^T$, where A is a large, sparse, symmetric-quindiagonal (n×n) matrix, D is a diagonal matrix, L is a lower triangular matrix retaining (m-1) outermost off-diagonal entries and L^T denotes the transpose of L) from which direct methods of solution can be obtained. In addition, a sparse approximate factorization of the form $A \simeq L_r D_r L_r^T$, where L_r and D_r are appropriate sparse approximations to L and D respectively, is shown to yield implicit iterative schemes for the FD solution, which can then be obtained as the limit of a convergent sequence of vectors $\{u_{i+1}\}$ generated by the implicit iterative schemes, e.g.

$$L_r D_r L_r^T(\underline{u}_{i+1}-\underline{u}_i) = \alpha_i \underline{r}_i \ , \ i=0,1,2,\ldots, \qquad (1)$$

where $\underline{r}_i=\underline{s}-A\underline{u}_i$, $i=0,1,2,\ldots$, and the α_i are predetermined sequences of acceleration parameters. In an earlier paper [20], the authors have demonstrated that (1) can be regarded as belonging to the class of preconditioned iterative methods where the conditioning matrix $(L_r D_r L_r)^{-1}$ is derived from a factorisation rather than splitting process.

The necessary and sufficient conditions for the convergence of the iteration (1) are derived and the analysis for the selection of the optimum parameters $\alpha_i \equiv \alpha$ (where α is a constant) and sequence of acceleration parameters α_i, when the eigenvalues of $(L_r D_r L_r^T)^{-1}A$ are real, is presented. For a predetermined fixed choice of the "fill-in" parameter r in the symmetric factorisation procedure $L_r D_r L_r^T$ and the parameter α, an upper bound of $O(h^{-3} r\log\varepsilon^{-1})$ is obtained for the computational work required to reduce the L_2-norm of the error by a factor ε, where h denotes the mesh size of the uniform grid under consideration. Furthermore, a sequence of parameters α_i yields work estimates of $O(h^{-5/2} r^{3/2} \log\varepsilon^{-1})$.

Then, an implicit semi-direct method is introduced which is a combination of the EL symmetric root-free sparse factorization procedures and the conjugate gradient method. This leads to a powerful class of Compact Preconditioned iterative methods - the Symmetric Implicit Conjugate Gradient method - which is used to solve self-adjoint elliptic difference equations efficiently.

Finally, composite "inner-outer" iterative schemes are developed, which are combinations of the derived EL factorisation procedures as outlined earlier and standard non-linear methods, i.e. Picard/Newton, for solving non-linear boundary-value problems.

2. THE APPROXIMATE FACTORIZATION PROCEDURES

Several iterative procedures for solving large sparse symmetric structured linear systems based on sparse factorization counterparts of the Choleski Square root (i.e. $A=LL^T$), root free Choleski (i.e. $A=LDL^T$) and normalized symmetric factorization (i.e. $A=\bar{D}T^T T\bar{D}$) procedures have recently appeared in the literature [18],[3],[12],[5],[7],[4].

In this section the EL symmetric sparse factorization procedures are presented which are both the exact and Approximate triangular factorization methods for the solution of large sparse systems derived from the FD discretization of self-adjoint P.D.E.s by use of a matrix Root Free Choleski factorization "bordering" technique [6] (henceforth called the RFC and ARFC(r) algorithms respectively). Since the former factorization method can be considered a "Limit" case of the latter, as will be shown in the forthcoming presentation, only the approximate LDL^T factorization procedure is described here.

Let us consider the resulting linear system from the FD discretization of a 2 dimensional self-adjoint elliptic P.D.E., viz,

$$A\underline{u} = \underline{s} \quad , \tag{2}$$

where A is a large, sparse, symmetric, diagonally-dominant quindiagonal matrix of the following partitioned form:

$$A \equiv \left[\begin{array}{cccc|cccc} \multicolumn{4}{c|}{\longleftarrow m \longrightarrow} & & & & \\ b_1 & a_1 & & & c_1 & & & \\ a_1 & b_2 & a_2 & & & c_2 & & 0 \\ & \ddots & \ddots & \ddots & & 0 & \ddots & \\ & & a_{m-2} & b_{m-1} & a_{m-1} & & & \\ \hline c_1 & & & a_{m-1} & b_m & a_m & & c_{n-m+1} \\ & c_2 & & 0 & a_m & \ddots & \ddots & \\ & & \ddots & & & \ddots & \ddots & a_{n-1} \\ 0 & & & & c_{n-m+1} & & a_{n-1} & b_n \end{array}\right] \equiv \left[\begin{array}{c|c} A_{11} & A_{12} \\ \hline A_{12}^T & A_{22} \end{array}\right], \tag{3}$$

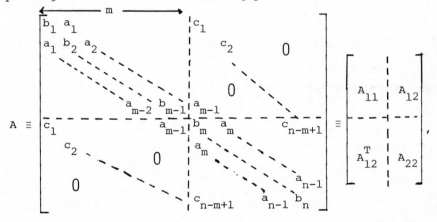

and $\underline{u}$ and $\underline{s}$ are n-column vectors, which consist respectively of the FD solution and the source term of the given P.D.E. plus the known boundary values.

The idea of the approximate factorization[18],[2],[13],[16],[21], is based on the simple replacement of the coefficient matrix A by a matrix (A+B) such that

$$A+B = L_r D_r L_r^T \quad , \tag{4}$$

where D_r is a diagonal matrix, L_r is a sparse strictly lower triangular matrix and L_r^T denotes the transpose of L_r. Obviously there is a large number of such matrices B, where the matrix (A+B) can be factored as in (4), leading to such "easily solvable" systems $L_r D_r L_r^T \underline{u}=\underline{s}$. In the following, we shall attempt to outline a strategy whereby L_r and D_r are easily determined and the approximate factorization (4) can be chosen to be as accurate as we require by an appropriate choice of r. We define the matrices L_r and D_r such that:

$L_r \equiv$ 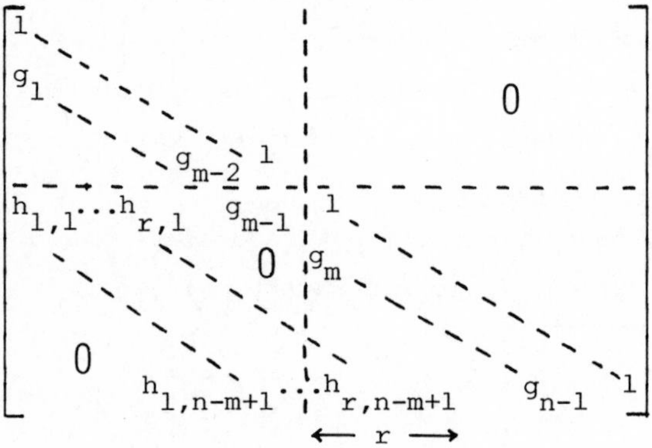

$\equiv \begin{bmatrix} L_{11} & 0 \\ L_{21} & L_{22} \end{bmatrix}$, (5)

and

$D_r \equiv$ 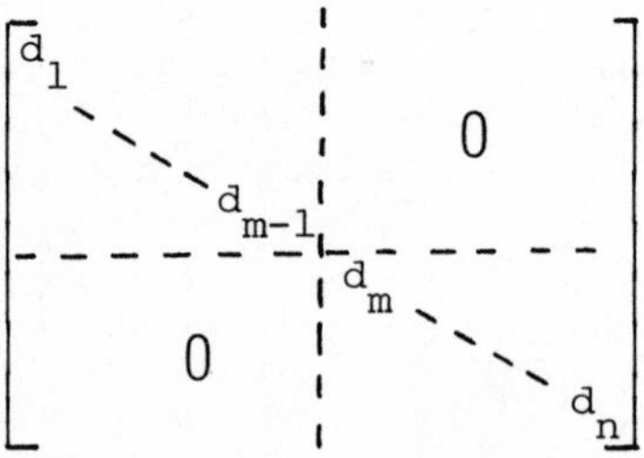

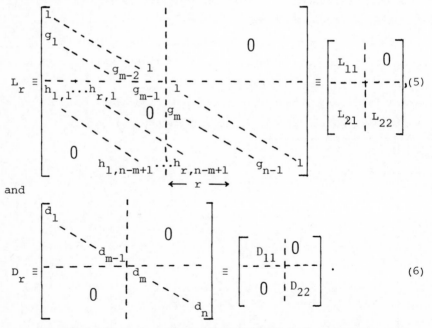

$\equiv \begin{bmatrix} D_{11} & 0 \\ 0 & D_{22} \end{bmatrix}$. (6)

It should be noted that if the "fill-in" parameter r is chosen to be r=m-1 (i.e. (m-1) outermost off-diagonal entries have been retained in L_r) then $L_r \equiv L$ and $D_r \equiv D$, where L is a strictly lower triangular matrix retaining (m-1) outermost off-diagonal elements and $D \equiv \text{diag}\{\tilde{d}_1, \tilde{d}_2, \ldots, \tilde{d}_m, \ldots, \tilde{d}_n\}$.

The coefficient matrix is then factorized exactly, i.e.

$$A = LDL^T , \tag{7}$$

to yield a direct algorithmic procedure for the FD solution. In this case from the relationships (5),(6) we have

$$LDL^T = \left[\begin{array}{c|c} L_{11} & 0 \\ \hline L_{12} & L_{22} \end{array}\right] \left[\begin{array}{c|c} D_{11} & 0 \\ \hline 0 & D_{22} \end{array}\right] \left[\begin{array}{c|c} L_{11}^T & L_{12}^T \\ \hline 0 & L_{22}^T \end{array}\right] =$$

$$= \left[\begin{array}{c|c} L_{11}D_{11}L_{11}^T & L_{11}D_{11}L_{12}^T \\ \hline L_{12}D_{11}L_{11}^T & L_{12}D_{11}L_{12}^T + L_{22}D_{22}L_{22}^T \end{array}\right] , \tag{8}$$

and a combination of (3),(7) and (8) leads to the identities:

$$A_{11} = L_{11}D_{11}L_{11}^T \quad ; \quad A_{12} = L_{11}D_{11}L_{12}^T ; \tag{9a}$$

$$A_{12}^T = L_{12}D_{11}L_{11}^T \quad ; \quad A_{22} = L_{12}D_{11}L_{12}^T + L_{22}D_{22}L_{22}^T . \tag{9b}$$

Since the exact factorization can be obtained from its approximate counterpart when $r \to m-1$ (i.e. $L_r^{-1}L = I + O(\varepsilon)$ and $D_r^{-1}D = I + O(\varepsilon)$) then the approximate $L_rD_rL_r^T$ factorization procedure is given in the following compact algorithmic form:

$$d_1 = b_1 \; ; \quad g_1 = a_1/d_1 , \tag{10}$$

<u>for i=2,3,...,m-2</u>

$$d_i = b_i - d_{i-1}g_{i-1}^2 \; ; \; g_i = a_i/d_i , \tag{11}$$

and

$$d_{m-1} = b_{m-1} - d_{m-2}g_{m-2}^2 . \tag{12}$$

<u>For j=1,2,...,n-m+1</u>, we have

$$h_{1,j} = c_j/d_j \; ; \; g_{m+j-2} = a_{m+j-2}/d_{m+j-2} , \tag{13}$$

whilst, <u>for j≤r-1</u>

$$h_{i,j} = -d_{i+j-2}g_{i+j-2}h_{i-1,j}/d_{i+j-1}, \tag{14}$$

for i=2,3,...,r-j+1

Then, for j>1 and r>1

$$\left[\begin{array}{l} h_{i,j} = (-d_{i+j-2}g_{i+j-2}h_{i-1,j} - \sum_{k=1}^{i-1} d_{k+j-1}h_{k,j}h_{k-i+r+1,i+j-r-1})/d_{i+j-1}, \quad (15) \\ \text{for either } i=(r-j+2),(r-j+3),\ldots,r \text{ and all } j\leq r, \\ \text{or} \quad i=2,3,\ldots,r \quad \text{for } j>r. \end{array}\right.$$

Then, for i=r

$$d_{m+j-1} = b_{m+j-1} - d_{i+j-1}g_{i+j-1}^2 - 2d_{i+j-1}g_{i+j-1}h_{i,j} - \sum_{k=1}^{i} d_{k+j-1}h_{k,j}^2 \quad . \quad (16)$$

An approximate solution of the linear system (1) can then be obtained by solving $L_r D_r L_r^T \underline{u} = \underline{s}$ and by setting $L_r^T \underline{u} = \underline{y}$ the problem reduces to solve the system $L_r D_r \underline{y} = \underline{s}$, i.e.

$$y_1 = s_1/d_1 \; ; \; y_i = (s_i - d_{i-1}g_{i-1}y_{i-1})/d_i, \text{ for } i=2,3,\ldots,m-1, \quad (17a)$$

and
$$y_i = (s_i - d_{i-1}g_{i-1}y_{i-1} - \sum_{k=i-m+1}^{i-m+r} d_k h_{k-i+m,i-m+1} y_k)/d_i, \text{ for } i=m,m+1,\ldots,n. \quad (17b)$$

The final solution is obtained from a back substitution process given by
$$u_n = y_n \; ; \; u_i = y_i - g_i y_{i+1} - \sum_{k=\tilde{p}}^{\tilde{q}} h_{i-k+m,k-m+1} y_k \, ,$$

where $\tilde{p}$ and $\tilde{q}$ are defined in [8], p.63. The memory requirement for this algorithm is approximately (r+4)n words. The amount of work involved for the factorization is $\simeq(r^2+4r+2)n$ multiplicative operations. Given that the forward-back substitution process requires $\simeq(3r+4)n$ operations, the total number of the multiplicative operations required is $\simeq(r^2+7r+6)n$. The choice of the optimum fill-in parameter r for the 2D-model problem has been discussed in [4]. It should be mentioned that if $c_i=0$, $i\in[1,n-m+1]$, cf. (3), the algorithm reduces to the $L_r D_r L_r^T$ form of the common tridiagonal system [19] which is encountered in solving two-point boundary value problems.

3. CONVERGENCE CRITERIA

In this section, by following the same approach as given for the convergence analysis of an approximate LU-factorization semi-direct method [11], we derive the necessary and sufficient conditions for the convergence of the iteration (1) and present the analysis for the selection of the optimum parameter α and sequence of parameters α_i, from the eigenvalues of $(L_r D_r L_r^T)A^{-1}$.

Let us consider the implicit iterative scheme

$$L_r D_r L_r^T(\underline{u}_{i+1}-\underline{u}_i) = \alpha\underline{r}_i, \quad i=0,1,2,..., \tag{18}$$

where α is a preconditioned acceleration parameter. Let $\underline{e}_i$ denote the error vector of (18) satisfying the relationship

$$\underline{e}_{i+1} = (I-\alpha\Omega^r)\underline{e}_i, \; i=0,1,2,... \tag{19}$$

or equivalently

$$\underline{e}_{i+1} = (I-\alpha\Omega^r)^{i+1}\underline{e}_0, \; i=0,1,2,... \tag{20}$$

where

$$\Omega^r \equiv (L_r D_r L_r^T)^{-1}LDL, \tag{21}$$

and I is the identity matrix. Let H_α^r be the iteration matrix of the stationary iterative scheme (18), i.e.

$$H_\alpha^r \equiv I-\alpha\Omega^r, \tag{22}$$

and $\lambda_j(H_\alpha^r)$, $j\in[1,n]$ be the eigenvalues of H_α^r, where $\lambda_{min}(H_\alpha^r)\equiv\lambda_1(H_\alpha^r)$ and $\lambda_{max}(H_\alpha^r)\equiv\lambda_n(H_\alpha^r)$ denote the eigenvalue of minimum and largest algebraic value of H_α^r respectively.

<u>Lemma 3.1</u>: Let LDL^T and $L_r D_r L_r^T$, $r\in[1,m-1]$, be positive definite matrices. Then, the iterative scheme (18) converges iff

$$0 < \alpha < 2/\lambda_{max}(\Omega^r). \tag{23}$$

<u>Proof</u>: Since the iterative scheme (18) converges iff the spectral radius of H_α^r is less than unity, i.e. $\rho(H_\alpha^r)<1$, or equivalently

$$|\lambda_{max}(I-\alpha\Omega^r)| < 1, \tag{24}$$

and all the eigenvalues of Ω^r are positive it can be seen that

$$-1 < 1-\alpha\lambda_{max}(\Omega^r) < 1, \tag{25}$$

from which the conclusion of the Lemma follows.

<u>Lemma 3.2</u>: Let LDL^T and $L_r D_r L_r^T$, $r\in[1,m-1]$, be positive definite

matrices. Then, the optimum choice of the parameter $\alpha\in(0,2/\lambda_{max}(\Omega^r))$ in the iterative scheme (18) is given by

$$\alpha_{opt} = \frac{2}{\lambda_{min}(\Omega^r)+\lambda_{max}(\Omega^r)} \quad . \tag{26}$$

<u>Proof</u>: The spectral radius of the iteration matrix H^r_α of (18) is minimized with respect to α when

$$|\lambda_{min}(H^r_\alpha)| = |\lambda_{max}(H^r_\alpha)| \ . \tag{27}$$

Since the eigenvalues of H^r_α are decreasing functions of α we get

$$|1-\alpha\lambda_{max}(\Omega^r)| = 1-\alpha\lambda_{min}(\Omega^r) \ , \tag{28}$$

or equivalently

$$\alpha(\lambda_{min}(\Omega^r)+\lambda_{max}(\Omega^r)) = 2 \ , \tag{29}$$

from which the conclusion of the Lemma follows.

It should be noted that from (29) and the definition (21) we have

$$\rho(H^r_\alpha) = 1 - \frac{2\lambda_{min}(\Omega^r)}{\lambda_{min}(\Omega^r)+\lambda_{max}(\Omega^r)} = \frac{\lambda_{max}(\Omega^r)-\lambda_{min}(\Omega^r)}{\lambda_{max}(\Omega^r)+\lambda_{min}(\Omega^r)} \ . \tag{30}$$

<u>Lemma 3.3</u>: Let LDL^T and $L_rD_rL_r^T$, $r\in[1,m-1]$, be positive definite matrices. Then, assuming that ξ_1,ξ_2 are positive numbers for all non-zero $\underline{x}$ we have that

$$\frac{(LDL^T\underline{x},\underline{x})}{(L_rD_rL_r^T\underline{x},\underline{x})}\in[\xi_1,\xi_2] \quad , \tag{31}$$

iff

$$\lambda_j(\Omega^r)\in[\xi_1,\xi_2] \ , \ j\in[1,n] \ . \tag{32}$$

<u>Proof</u>: It will be sufficient to prove that the extreme eigenvalues of Ω^r lie in the interval $[\xi_1,\xi_2]$. Since $\Omega^r\equiv(L_rD_rL_r^T)^{-1}LDL^T$ is similar to the Hermitian form $(LDL^T)^{\frac{1}{2}}(L_rD_rL_r^T)^{-1}(LDL^T)^{\frac{1}{2}}$ we obtain

$$\lambda_{max}(\Omega^r) = \lambda_{max}((LDL^T)^{\frac{1}{2}}(L_rD_rL_r^T)^{-1}(LDL^T)^{\frac{1}{2}}) = $$
$$= \max_{\underline{x}\neq 0} \frac{((LDL^T)^{\frac{1}{2}}(L_rD_rL_r^T)^{-1}(LDL^T)^{\frac{1}{2}})}{||\underline{x}||^2} \ . \tag{33}$$

Then, by choosing $\underline{x}=[(LDL^T)^{\frac{1}{2}}(L_rD_rL_r^T)^{-1}(LDL^T)^{\frac{1}{2}}]^{-\frac{1}{2}}(LDL^T)^{\frac{1}{2}}\underline{z}$ and using the "self-adjoint" property of $[(LDL^T)^{\frac{1}{2}}(L_rD_rL_r^T)^{-1}(LDL^T)^{\frac{1}{2}}]^{-\frac{1}{2}}$

and $(LDL^T)^{\frac{1}{2}}$ the relationship (33) leads to

$$\lambda_{max}(\Omega^r) = \max_{\underline{z}\neq\underline{0}} \frac{(LDL^T\underline{z},\underline{z})}{(L_rD_rL_r^T\underline{z},\underline{z})} \leq \frac{\lambda_{max}(LDL^T)}{\lambda_{min}(L_rD_rL_r^T)} . \tag{34}$$

In a similar way we can obtain the result

$$\lambda_{min}(\Omega^r) = \min \frac{(LDL^T\underline{z},\underline{z})}{(L_rD_rL_r^T\underline{z},\underline{z})} \geq \frac{\lambda_{min}(LDL^T)}{\lambda_{max}(L_rD_rL_r^T)} . \tag{35}$$

Finally, from (34) and (35) the conclusion of the Lemma follows.

Let us now consider the implicit iterative scheme

$$(L_rD_rL_r^T)(\underline{u}_{i+1}-\underline{u}_i) = \alpha_i\underline{r}_i \ , \ i=0,1,2,... \tag{36}$$

and the corresponding error vector $\underline{e}_\nu$ given by

$$e_\nu = \prod_{i=0}^{\nu-1}(I-\alpha_i\Omega^r)e_0 . \tag{37}$$

The rate of convergence of the iterative scheme (36) is maximized for that choice of α_i which minimizes the spectral radius of

$$H_\nu^r = \prod_{i=0}^{\nu-1}(I-\alpha_i\Omega^r) , \tag{38}$$

or equivalently, since the eigenvalues of H_ν^r are real (the eigenvalues of Ω^r are real) for those values of α_i which minimize the max $|F_\nu(x)|$, where

$$F_\nu(x) = \prod_{i=0}^{\nu-1}(1-\alpha_i x) , \quad x\in[\tilde{\lambda}_1,\tilde{\lambda}_n] , \tag{39}$$

with $\tilde{\lambda}_1,\tilde{\lambda}_n$ the extreme eigenvalues of Ω^r.

The polynomial $F_\nu(x)$ can be defined by the classical Chebyshev theory as

$$F_\nu(x) = \frac{T_\nu\left(\dfrac{\tilde{\lambda}_1+\tilde{\lambda}_n-2x}{\tilde{\lambda}_n-\tilde{\lambda}_1}\right)}{T_\nu\left(\dfrac{\tilde{\lambda}_1+\tilde{\lambda}_n}{\tilde{\lambda}_n-\tilde{\lambda}_1}\right)} , \quad x\in[\tilde{\lambda}_1,\tilde{\lambda}_n] , \tag{40}$$

where T_ν is the well-known Chebyshev polynomial of degree ν defined by

$$T_\nu(x) \equiv \begin{cases} \cos(\nu\cos^{-1}x), & |x|<1,\ \nu\geq 0, \\ \cosh(\nu\cosh^{-1}x), & x\geq 1,\ \nu\geq 0. \end{cases} \tag{41}$$

Furthermore, the maximum value of $F_\nu(x)$ as $\nu\to+\infty$ is given by

$$\max|F_\nu(x)| = \left\{T_\nu\left(\frac{\tilde{\lambda}_n+\tilde{\lambda}_1}{\tilde{\lambda}_n-\tilde{\lambda}_1}\right)\right\}^{-1}, \tag{42}$$

with the property: $F_\nu(0)=1$.

For the implementation of the sequence α_i Stiefel's scheme [15] can be used, i.e. assuming that $\underline{x}_k$ denotes the result of using a Chebychev sequence of length k, then calculate $\underline{x}_{k+1}$ by

$$\underline{x}_{k+1} = \underline{x}_k+\delta\underline{x}_k, \tag{43}$$

$$\delta\underline{x}_k = \frac{4T_k(z)(L_rD_rL_r^T)^{-1}\underline{r}_k}{(\tilde{\lambda}_n-\tilde{\lambda}_1)T_{k+1}(z)} + \frac{T_{k-1}(z)}{T_{k+1}(z)}\delta\underline{x}_{k-1}, \tag{44}$$

where

$$\delta\underline{x}_0 = \frac{2(L_rD_rL_r^T)^{-1}\underline{r}_0}{\tilde{\lambda}_1+\tilde{\lambda}_n} \quad\text{and}\quad z = \frac{\tilde{\lambda}_1+\tilde{\lambda}_n}{\tilde{\lambda}_n-\tilde{\lambda}_1}.$$

Alternatively, the sequence α_i can be derived by choosing a value of ν and computing the α_i, $i=0,1,\ldots,(\nu-1)$ as the reciprocal of the roots of $T_\nu\left(\frac{\tilde{\lambda}_n+\tilde{\lambda}_1-2x}{\tilde{\lambda}_n-\tilde{\lambda}_1}\right)$. Then, the so obtained sequence can be used repeatedly to reduce the norm of the residual r_i to the desired accuracy. It should be pointed out that since the extreme eigenvalues of Ω^r are generally unknown, the minimization of (40) can be achieved by considering the constants ξ_1,ξ_2 (instead of $\tilde{\lambda}_1,\tilde{\lambda}_n$) respectively such that $[\xi_1,\xi_2]\supset[\tilde{\lambda}_1,\tilde{\lambda}_n]$. Then, Stiefel's scheme can be defined by

$$\underline{x}_{k+1} = \underline{x}_k+\delta\underline{x}_k, \tag{45}$$

$$\delta\underline{x}_k = \frac{4T_k(\tilde{z})(L_rD_rL_r^T)^{-1}\underline{r}_k}{(\xi_2-\xi_1)T_{k+1}(\tilde{z})} + \frac{T_{k-1}(\tilde{z})}{T_{k+1}(\tilde{z})}\delta\underline{x}_{k-1}, \tag{46}$$

where
$$\delta\underline{x}_0 = \frac{2(L_rD_rL_r^T)^{-1}\underline{r}_0}{\xi_1+\xi_2} \quad \text{and} \quad \tilde{z} = \frac{\xi_2+\xi_1}{\xi_2-\xi_1}\,, \tag{47}$$

whilst the polynomial $F_\nu(x)$ is now given by

$$F_\nu(x) = \frac{T_\nu\left(\dfrac{\xi_1+\xi_2-2x}{\xi_2-\xi_1}\right)}{T_\nu\left(\dfrac{\xi_2+\xi_1}{\xi_2-\xi_1}\right)}\ . \tag{48}$$

An alternative form of the algorithmic scheme defined in (45)-(47) can be found in [19].

4. ESTIMATES OF THE COMPUTATIONAL WORK

In this final section we give estimates of the computational work required for the convergence of the iteration schemes (18) and (36) for the optimum parameter α and sequence of parameters α_i respectively assuming that the coefficient matrix A is positive definite. A similar approach has been given by Dupont et al [3] with the fill-in parameter r chosen to be r=1.

Since L_r and D_r are sparse approximations to L and D, cf. (5),(6), respectively, we define

$$\tilde{L}_r = L-L_r \quad \text{and} \quad \tilde{D}_r = D-D_r\,, \tag{49}$$

and assume throughout this section that

$$||A|| \equiv ||A||_1 = \max_{j\in[1,n]} \{\sum_{i=1}^{n} |A_{i,j}|\}\ . \tag{50}$$

Then, by combining (21),(49) we have

$$\begin{aligned}\Omega^r &= (L_rD_rL_r^T)^{-1}[(L_r+\tilde{L}_r)(D_r+\tilde{D}_r)(L_r^T+\tilde{L}_r^T)] = \\ &= I+(L_rD_rL_r^T)^{-1}(L_rD_r\tilde{L}_r^T+L_r\tilde{D}_rL_r^T+L_r\tilde{D}_r\tilde{L}_r^T+\tilde{L}_rD_rL_r^T+\tilde{L}_rD_r\tilde{L}_r^T+\tilde{L}_r\tilde{D}_rL_r^T+ \\ &\qquad \tilde{L}_r\tilde{D}_r\tilde{L}_r^T).\end{aligned} \tag{51}$$

By taking norms we obtain

$$\begin{aligned}||\Omega^r|| \leq 1 &+ ||(L_r^T)^{-1}||.||\tilde{L}_r^T|| + ||D_r^{-1}||.||\tilde{D}_r|| + ||(D_rL_r^T)^{-1}||.||\tilde{D}_r\tilde{L}_r^T|| \\ &+ ||L_r^{-1}||.||\tilde{L}_r|| + ||(L_rL_r^T)^{-1}||.||\tilde{L}_r\tilde{L}_r^T|| + ||(L_rD_r)^{-1}||.||\tilde{L}_r\tilde{D}_r|| \\ &+ ||(L_rD_rL_r^T)^{-1}||.||\tilde{L}_r\tilde{D}_r\tilde{L}_r^T||.\end{aligned} \tag{52}$$

It can be easily seen that for a real $(n\times n)$ non-singular matrix H we have

$$||H^{-1}||\leqslant||H||.||H^{-2}||\leqslant||H||^2||H^{-1}||^2 \cdot \frac{1}{||H||} = \frac{M_H^2}{n^2} \cdot \frac{1}{||H||}, \tag{52a}$$

where $M_H = n||H||.||H^{-1}||$ is the M-condition number of matrix H [17]. Then, by using (52a) the inequality (52) gives

$$||\Omega^r||\leqslant 1+\Lambda_1 \frac{||\tilde{L}_r^T||}{||L_r^T||} + \Lambda_2 \frac{||\tilde{D}_r||}{||D_r||} + \Lambda_3 \frac{||\tilde{D}_r\tilde{L}_r^T||}{||D_rL_r^T||} + \Lambda_4 \frac{||\tilde{L}_r||}{||L_r||} +$$
$$+\Lambda_5 \frac{||\tilde{L}_r\tilde{L}_r^T||}{||L_rL_r^T||} + \Lambda_6 \frac{||\tilde{L}_r\tilde{D}_r||}{||L_rD_r||} + \Lambda_7 \frac{||\tilde{L}_r\tilde{D}_r\tilde{L}_r^T||}{||L_rD_rL_r^T||} \leqslant$$
$$\leqslant 1+ \Lambda\left(\frac{||\tilde{L}_r^T||}{||L_r^T||} + \frac{||\tilde{D}_r||}{||D_r||} + \frac{||\tilde{D}_r\tilde{L}_r^T||}{||D_rL_r^T||} + \frac{||\tilde{L}_r||}{||L_r||} + \frac{||\tilde{L}_r\tilde{L}_r^T||}{||L_rL_r^T||} + \right.$$
$$\left. + \frac{||\tilde{L}_r\tilde{D}_r||}{||L_rD_r||} + \frac{||\tilde{L}_r\tilde{D}_r\tilde{L}_r^T||}{||L_rD_rL_r^T||}\right), \tag{52b}$$

where $\Lambda_i \equiv \frac{1}{n^2}M_i^2$, $i\in[1,7]$ (M_i, $i\in[1,7]$ are respectively the M-condition numbers of the matrices $L_r^T, D_r, D_rL_r^T, L_r, L_rL_r^T, L_rD_r$ and $L_rD_rL_r^T$) and $\Lambda\equiv \max_{i\in[1,7]}\{\Lambda_i\}$. Furthermore, recalling that an average value of the M-condition number of a real $(n\times n)$ non-singular matrix is given by $n^{\frac{1}{2}}\log n$, we obtain the result $\Lambda=\frac{1}{n}\log^2 n$.

Let us assume now that the elements $|h_{k,\ell}|$, $k\in[1,r]$, $\ell\in[1,n-m+1]$, cf. (5) are monotonically decreased along each column ℓ and let $\tilde{\rho}$ be the ratio of the monotonicity, i.e. $|h_{r,\ell}|=\tilde{c}_{r-1}\tilde{\rho}|h_{r-1,\ell}|=\ldots=(\prod_{i=0}^{r-1}\tilde{c}_i)\tilde{\rho}^{(r-1)}|h_{1,\ell}|$, where $\tilde{\rho}\in(0,1)$, $\tilde{c}_i$, $i\in[0,r-1]$ is a sequence of positive constants with $\tilde{c}_0\equiv 1$ and $\tilde{c}_i\geqslant 1$, $i\in[1,r-1]$. Then, it can be readily seen that

$$\sum_{k=1}^{r}|h_{k,\ell}| = [1+\tilde{c}_1\tilde{\rho}+\tilde{c}_2\tilde{c}_1\tilde{\rho}^2+\ldots+(\prod_{i=0}^{r-1}\tilde{c}_i)\tilde{\rho}^{(r-1)}]|h_{1,\ell}| = \sum_{\mu=0}^{r-1} P_\mu\tilde{\rho}^\mu|h_{1,\ell}|, \tag{53}$$

where $$P_\mu \equiv \prod_{i=0}^{\mu} \tilde{c}_i, \quad \mu\in[0,r-1]. \tag{53a}$$

The consideration of the parameters $\tilde{c}_i$, $i\in[0,r-1]$ is closely related with the monotonicity behaviour of the elements $|h_{i,j}|$ particularly when $r\to m-1$ (see remark in [4], p.7).

In order to derive bounds for ξ_1,ξ_2 we now proceed as follows. By definition we have

$$||L_r||= \max_{j\in[1,n]} \left\{\sum_{i=1}^{n} |(L_r)_{i,j}|\right\} = \max_{\ell\in[1,n-m+1]} \left\{ \sum_{k=1}^{n} |h_{k,\ell}|+|g_{m+\ell-2}|+1\right\}, \tag{54}$$

$$||D_r||= \max_{j\in[1,n]} \{|d_1|,|d_2|,\dots,|d_{m-1}|,|d_m|,|d_{m+1}|,\dots,|d_n|\}, \tag{55}$$

and

$$||\tilde{L}_r||= \max_{j\in[1,n]} \left\{\sum_{i=1}^{n} |(\tilde{L}_r)_{i,j}|\right\} = \max_{\ell\in[1,n-m+1]} \left\{ \sum_{k=m-r}^{m-1} |h_{k,\ell}|\right\}, \tag{56}$$

$$||\tilde{D}_r||= \max_{j\in[1,n]} \{0,0,\dots,0,|\tilde{d}_m-d_m|,|\tilde{d}_{m+1}-d_{m+1}|,\dots,|\tilde{d}_n-d_n|\}. \tag{57}$$

(Note that $\tilde{d}_i=d_i$, $i\in[1,m-1]$, cf. (8)-(12)).

Then, from the relationships (54)-(57) we obtain that

$$\frac{||\tilde{L}_r||}{||L_r||} = \max_{\ell\in[1,n-m+1]} \left\{ \frac{\sum_{k=m-r}^{m-1} |h_{k,\ell}|}{\sum_{k=1}^{r} |h_{k,\ell}|+|g_{m+\ell-2}|+1} \right\} \leqslant$$

$$\leqslant \max_{\ell\in[1,n-m+1]} \left\{ \frac{\sum_{k=1}^{m-1} |h_{k,\ell}|-\sum_{k=1}^{r} |h_{k,\ell}|}{\sum_{k=1}^{r} |h_{k,\ell}|} \right\} \leqslant \frac{\sum_{\mu=0}^{m-2} P_\mu \tilde{\rho}^\mu}{\sum_{\mu=0}^{r-1} P_\mu \tilde{\rho}^\mu} -1, \tag{58}$$

and

$$\frac{||\tilde{D}_r||}{||D_r||} = \frac{1}{|d_1|} \max_{j\in[m,n]} |\tilde{d}_j-d_j| \leqslant \frac{b_m}{b_1} . \tag{59}$$

An analogous relationship to (58) can be obtained for the ratio $\frac{||\tilde{L}^T||}{||L_r^T||}$ and then a combination of (52),(58),(59) leads to

$$0 \leq ||\Omega^r|| \leq -1+2\Lambda \frac{\sum_{\mu=0}^{m-2} p_\mu \tilde{\rho}^\mu}{\sum_{\mu=0}^{r-1} p_\mu \tilde{\rho}^\mu} + q < -1+2\Lambda\sigma\frac{m}{r} + q \ , \tag{60}$$

where
$$q \equiv \Lambda\left\{\frac{||\tilde{L}_r\tilde{D}_r\tilde{L}_r^T||}{||L_rD_rL_r^T||} + \frac{||\tilde{D}_r\tilde{L}_r^T||}{||D_rL_r^T||} + \frac{||\tilde{L}_r\tilde{L}_r^T||}{||L_rL_r^T||} + \frac{||\tilde{L}_r\tilde{D}_r||}{||L_rD_r||} + \frac{b_m}{b_1}\right\} - 2(\Lambda-1), \tag{61}$$

and σ is a positive constant depending on $\tilde{c}_i$, cf.(53a).

The upper right inequality in (60) can be easily verified by induction recalling that $r\in[1,m-1]$. Then, by defining

$$0 \leq \Lambda \frac{\sum_{\mu=0}^{m-2} p_\mu \tilde{\rho}^\mu}{\sum_{\mu=0}^{r-1} p_\mu \tilde{\rho}^\mu} + \frac{q}{2} \leq C_2^{(m,r)} \quad , \tag{62}$$

from (60) we obtain

$$0 < ||\Omega^r|| \leq -1+2C_2^{(m,r)} \quad , \tag{63}$$

where $C_2^{(m,r)}$ is a non-negative constant depending on m and r.

In order to obtain a lower bound for $||\Omega^r||$, from (49) we have

$$L_r = L-\tilde{L}_r \quad \text{and} \quad D_r = D-\tilde{D}_r \quad . \tag{64}$$

The relationships (51) can then be written as

$$\begin{aligned}\Omega^r &= [(L-\tilde{L}_r)(D-\tilde{D}_r)(L^T-\tilde{L}_r^T)]^{-1}LDL^T = \\ &= [LDL^T+L\tilde{D}_r\tilde{L}_r^T+\tilde{L}_rD\tilde{L}_r^T+\tilde{L}_r\tilde{D}_rL^T-(LD\tilde{L}_r^T+L\tilde{D}_rL^T+\tilde{L}_rDL^T+\tilde{L}_r\tilde{D}_r\tilde{L}_r^T)]^{-1} \\ &\qquad LDL^T. \end{aligned} \tag{65}$$

Since all the terms at the extreme right side of (65) are non-negative matrices then by taking norms we can obtain

$$\begin{aligned} ||\Omega^r|| &\geq \frac{||LDL^T||}{||LDL^T+L\tilde{D}_r\tilde{L}_r^T+\tilde{L}_rD\tilde{L}_r^T+\tilde{L}_r\tilde{D}_rL^T-(LD\tilde{L}_r^T+L\tilde{D}_rL^T+\tilde{L}_rDL^T+\tilde{L}_r\tilde{D}_r\tilde{L}_r^T)||} \geq \\ &\geq \frac{1}{1+\dfrac{||L\tilde{D}_r\tilde{L}_r^T+\tilde{L}_rD\tilde{L}_r^T+\tilde{L}_r\tilde{D}_rL^T||}{||LDL^T||}} \ . \end{aligned} \tag{66}$$

Then, by defining

$$0 \leq \frac{||L\tilde{D}_r\tilde{L}_r^T+\tilde{L}_rD\tilde{L}_r^T+\tilde{L}_r\tilde{D}_rL^T||}{||LDL^T||} \leq C_1^{(r)} , \tag{67}$$

where $C_1^{(r)}$ is a constant depending on r and also independent of the mesh size h, the inequalities (66) and (63) yield

$$\frac{1}{1+C_1^{(r)}} \leq ||\Omega^r|| \leq -1+2C_2^{(m,r)} . \tag{68}$$

Let us consider now the positive numbers ξ_1,ξ_2 such that

$$\xi_2 \leq -1+2C_2^{(m,r)} \quad \text{and} \quad \xi_1 \geq \frac{1}{1+C_1^{(r)}} . \tag{69}$$

Then, assuming that a uniform grid of mesh size h is considered (i.e. $h^{-1}=m$) it can be readily seen from (60),(62) and (68) that

$$\xi_2 = O(h^{-1}r^{-1}) . \tag{70}$$

It should be pointed out that if the "fill-in" parameter r is chosen such that r=m-1 then $L\equiv L_r$, $D\equiv D_r$ and from the relationships (64),(67) we get that $||L\tilde{D}_r\tilde{L}_r^T+\tilde{L}_rD\tilde{L}_r^T+\tilde{L}_r\tilde{D}_rL^T||\to 0$, i.e. $C_1^{(r)}\geq 0$, while from (62) we obtain that $C_2^{(m,r)}\geq 1$. In the case r=1, since $\frac{||L\tilde{D}_r\tilde{L}_r^T+\tilde{L}_rD\tilde{L}_r^T+\tilde{L}_r\tilde{D}_rL^T||}{||LDL^T||} \to \gamma$, where $\gamma\in [1,3]$.

Note that the elements of matrices $\tilde{L}_r$ and $\tilde{D}_r$ are identical to the elements of L and D respectively, except for those along the principal diagonal, the co-diagonal and m^{th}-diagonal) the inequality (67) yields $C_1^{(r)} \geq \gamma$, while from the relationship (62) we get that $C_2^{(m,r)}\geq\Lambda \sum_{\mu=0}^{m-2} P_\mu\tilde{\rho}^\mu + \frac{q}{2}$, with $P_\mu, \tilde{\rho}, \Lambda$ and q as defined earlier in this section. The behaviour of the values of the constants $C_1^{(r)}, C_2^{(m,r)}$ in connection with the inequalities (68), (69) is closely related to the eigenvalue bounds distribution (Fig.1) for the model problem (i.e. the Laplace equation on the unit square with zero boundary values) as r→m-1.

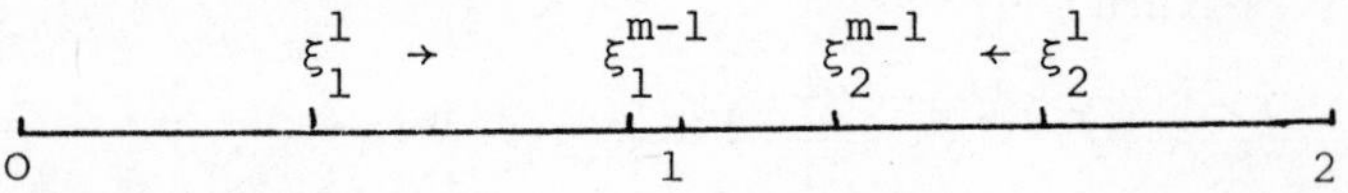

FIGURE 1: Distribution of the eigenvalue bounds ξ_1^r, ξ_2^r for the model problem as $r \to m-1$.

Then, we can propose the following theorem:

Theorem 4.1: Let LDL^T and $L_rD_rL_r^T$, $r \in [1,m-1]$, be positive definite matrices. Then, for an appropriate choice of α, cf.(26), the implicit iterative scheme

$$(L_rD_rL_r^T)\underline{u}_{i+1} = (L_rD_rL_r^T)\underline{u}_i - \alpha\underline{r}_i \ , \tag{71}$$

gives a sequence $\underline{u}_{i+1}$, which converges to $A^{-1}\underline{s}$ and the number of iterations required to reduce the L_2-norm of the error by a factor $\varepsilon > 0$ is given by

$$\nu = O(h^{-1}r^{-1}\log\varepsilon^{-1}) \ . \tag{72}$$

Furthermore, the total number of arithmetic operations for the calculation of u_ν is

$$O(h^{-3}r\log\varepsilon^{-1}) . \tag{73}$$

Proof: From Lemmas 3.2, 3.3 it can be seen that for the optimum value of α each iteration of (71) reduces the L_2-norm of the error by a factor

$$\frac{\xi_2 - \xi_1}{\xi_1 + \xi_2} = 1 - \frac{2}{1 + \xi_2/\xi_1} = 1 - O(hr) \ . \tag{74}$$

Consequently, the number of iterations of the iterative scheme (71) required to reduce the L_2-norm of the error by a factor $\varepsilon > 0$ is $\nu = O(h^{-1}r^{-1}\log\varepsilon^{-1})$. Since each iteration requires $O(h^{-2}r^2)$ arithmetic operations to be carried out, the total number of arithmetic operations required is bounded from above by $O(h^{-3}r\log\varepsilon^{-1})$.

Theorem 4.2: Let LDL^T and $L_rD_rL_r^T$, $r \in [1,m-1]$, be positive definite matrices. Suppose that there exist positive numbers ξ_1, ξ_2 such that

$$\frac{(LDL^T\underline{u},\underline{u})}{(L_rD_rL_r^T\underline{u},\underline{u})} \in [\xi_1, \xi_2] \ , \quad \forall \underline{u} \neq \underline{0} \ , \tag{75}$$

where ξ_1 is independent of the mesh size h and $\xi_2=O(h^{-1}r^{-1})$. Then, the number of iterations of the implicit iterative scheme

$$(L_r D_r L_r^T)\underline{u}_{i+1} = (L_r D_r L_r^T)\underline{u}_i - \alpha_i \underline{r}_i , \quad (76)$$

required to reduce the L_2-norm of the error by a factor $\varepsilon>0$ for proper values of $\{\alpha_i\}$, $i=1,2,\ldots,\nu-1$, cf.(42), is given by

$$O(h^{-\frac{1}{2}} r^{-\frac{1}{2}} \log\varepsilon^{-1}) . \quad (77)$$

Furthermore the total number of arithmetic operations required for the calculation of the solution $\underline{u}_\nu$ is given by

$$O(h^{-5/2} r^{3/2} \log\varepsilon^{-1}) . \quad (78)$$

Proof: From the hypothesis of the theorem the convergence factor, cf.(47), becomes

$$\frac{\xi_1+\xi_2}{\xi_2-\xi_1} = 1+2/(\frac{\xi_2}{\xi_1}-1) = 1+O(hr) . \quad (79)$$

The maximum absolute value of the polynomial $F_\nu(x)$ of degree ν on $[\xi_1,\xi_2]$, cf.(48), can be obtained in an analogous manner as in (42), i.e.

$$\max_{x\in[\xi_1,\xi_2]} |F_\nu(x)| = \left\{T_\nu(\frac{\xi_1+\xi_2}{\xi_2-\xi_1})\right\}^{-1} . \quad (80)$$

From the relationships (79),(48) and the definitions of

$$T_\nu(x) \equiv \cosh(\nu\cosh^{-1}x);\ \cosh^{-1}x\equiv\log(x+(x^2-1)^{\frac{1}{2}}),x\geq 1,$$

we obtain that the number of iterations of the iterative scheme (76) required to make (80) less than $\varepsilon>0$ is given by

$$\nu = O(h^{-\frac{1}{2}} r^{-\frac{1}{2}} \log\varepsilon^{-1}).$$

Since each iteration requires $O(h^{-2}r^2)$ arithmetic operations to be carried out we conclude that the total number of arithmetic operations involved for the determination of u_ν is given by

$$O(h^{-5/2} r^{3/2} \log\varepsilon^{-1}).$$

Numerical results for the model problem confirming the established estimates in Theorems 4.1 and 4.2 are given in Tables 1 and 2. The calculations were performed on an ICL 1904S at Loughborough University of Technology and the initial guess was taken to be the zero vector, while the right-hand side vector of the system $A\underline{u}=\underline{s}$ was chosen as the product of the coefficient matrix by the solution vector of which the components are n-pseudo random numbers chosen from a uniform distribution in the interval (0,1). The optimum values of the parameters α_i are calculated from the well known relationship [15],

$$\alpha_i=2/((\lambda_{max}+\lambda_{min})-(\lambda_{max}-\lambda_{min})\cos[\frac{(2i-1)\pi}{2\nu^*}]),\ i\in[1,\nu^*],$$

assuming that only a cycle of ν^*-iterations is to be performed. The cycle parameter ν^* was chosen to be proportional to $(h^{-1}r^{-1})$. The behaviour of the error measures $||r_i||_2$ and $||u-u_k||_2$ of the implicit methods (71) and (76) are illustrated in Fig. 2.

h^{-1}	r=1		r=2		r=3		r=4	
	α	$\tilde{\nu}$	α	$\tilde{\nu}$	α	$\tilde{\nu}$	α	$\tilde{\nu}$
10	0.50	38	0.85	17	0.95	14	1.05	14
15	0.35	65	0.70	27	0.90	17	0.98	17
20	0.25	100	0.55	39	0.80	23	0.95	21
30	0.20	139	0.40	63	0.55	39	0.80	31

TABLE 1: Experimental values of α and the number of iterations $\tilde{\nu}$ of the iterative scheme (18) which are required to reduce the L_2-norm of the error by a factor $\varepsilon=10^{-6}$ for several mesh sizes and values of the "fill-in" parameter r for the model problem.

h^{-1}	r=1		r=2		r=3		r=4	
	α_i	$\tilde{\nu}$	α_i	$\tilde{\nu}$	α_i	$\tilde{\nu}$	α_i	$\tilde{\nu}$
10	[1.568,...,0.305]	18	[1.509,...,0.927]	15	[1.395,...,1.629]	11	[1.423,...,1.634]	11
15	[1.598,...,0.239]	37	[1.849,...,0.426]	17	[1.926,...,1.014]	15	[1.716,...,1.716]	14
20	[1.089,...,0.569]	69	[1.712,...,0.331]	19	[1.692,...,0.557]	17	[2.376,...,1.419]	16
30	[0.256,...,0.240]	133	[1.742,...,0.242]	35	[1.711,...,0.328]	19	[2.290,...,0.524]	19

TABLE 2: Number of iterations $\tilde{\nu}$ of the iterative scheme (36) required to reduce the L_2-norm of the error by a factor of $\varepsilon=10^{-6}$ for several mesh sizes and values of the fill-in parameter r for the model problem. The first and last term of the sequence α_i are also given.

5. THE SYMMETRIC (ROOT-FREE) IMPLICIT CONJUGATE GRADIENT METHOD

A class of implicit semi-direct methods based on the combination of RFC sparse factorization procedures and first order implicit Simultaneous Displacement and Richardson methods were presented in Section 3. In this section we introduce the Symmetric (root-free) Implicit Conjugate Gradient method (henceforth called the SICG method) for solving large sparse symmetric linear systems of algebraic equations. Both the concept of the Extended to the Limit sparse factorization techniques [9],[4],[10] and the property of the eigenvalues of $(L_r D_r L_r^T)^{-1}A$ to be close to unity as $r \to m-1$, cf.(5), lead to fast convergence rates for the SICG method, which can be defined as follows:

Let $\underline{u}_0$ be an arbitrary initial approximation of the solution $\underline{u}$. Then, form the residual,

$$\underline{r}_0 = \underline{s} - A\underline{u}_0 \ , \tag{81}$$

solve
$$(L_r D_r L_r^T)\underline{r}_0^* = \underline{r}_0 \ , \tag{82}$$

and set
$$\underline{\sigma}_0 = \underline{r}_0^* \ . \tag{83}$$

Then, for i=0,1,2,... calculate the vectors $\underline{u}_{i+1}, \underline{r}_{i+1}, \underline{\sigma}_{i+1}$ and scalar quantities α_i, β_{i+1} as follows:

form
$$\underline{q}_i = A\underline{\sigma}_i \ , \tag{84}$$

set
$$p_i = (\underline{r}_i, \underline{r}_i^*), \text{ (only when } i=0) \ , \tag{85}$$

evaluate
$$\alpha_i = p_i/(\underline{\sigma}_i, \underline{q}_i) \ , \tag{86}$$

compute
$$\underline{u}_{i+1} = \underline{u}_i + \alpha_i \underline{\sigma}_i \ , \tag{87}$$

and
$$\underline{r}_{i+1} = \underline{r}_i - \alpha_i \underline{q}_i \ . \tag{88}$$

Then, solve
$$(L_r D_r L_r^T)\underline{r}_{i+1}^* = \underline{r}_{i+1} \ , \tag{89}$$

form
$$p_{i+1} = (\underline{r}_{i+1}, \underline{r}_{i+1}^*) \ , \tag{90}$$

evaluate
$$\beta_{i+1} = p_i/p_{i+1} \ , \tag{91}$$

and compute
$$\underline{\sigma}_{i+1} = \underline{r}_{i+1}^* + \beta_{i+1}\underline{\sigma}_i \ . \tag{92}$$

Let us assume that the SICG method converges, i.e. $\rho(I-(L_r D_r L_r^T)^{-1}LDL^T)<1$, where $\rho(H)$ denotes the spectral radius of the non-singular $(n\times n)$ matrix H, and let η be the number of iterations required for convergence to a specified accuracy ε. Then, by ignoring the single scalar operations, the SICG method at the ith stage, $i \in (1,n]$, requires $(3r+4)n$ multiplications for the solution of system (89), 5n mults for the product of A by a vector, 3n mults for the scalar by vector products, 3n additions and 2n mults for the vector inner products. Consequently, the convergence of the SICG method requires a total of $[(3r+14)n$ mults $+ 3n$ adds$]\eta+(3r+9)n$ mults $+ n$ adds.

For comparative reasons and in order to illustrate the effectiveness of the derived implicit semi-direct methods for solving elliptic boundary-value problems we consider the 2D-linear elliptic P.D.E., i.e.,

Problem I: $$\frac{\partial^2 U}{\partial x^2} + \frac{\partial^2 U}{\partial y^2} + \sigma U = f, \quad (x,y) \in R, \tag{93}$$

where R is the unit square, subject to the Dirichlet boundary conditions

$$U(x,y) \equiv \tilde{\gamma}, \quad (x,y) \in \Gamma, \tag{93a}$$

with Γ the exterior boundary of R. By covering the region under consideration with a rectilinear grid system with equal mesh spacings and with a column-wise ordering a set of n-simultaneous, in-homogeneous, linear difference equations, cf.(2), is formed, where the coefficient matrix is of the following block form:

$$A \equiv \begin{bmatrix} B_1 & C_1 & & & \\ C_1 & B_2 & C_2 & & 0 \\ & \ddots & \ddots & \ddots & \\ & & & & C_{m-2} \\ 0 & & & C_{m-2} & B_{m-1} \end{bmatrix}, \text{ where } B_i \equiv \begin{bmatrix} 4-h^2\sigma & -1 & & & \\ -1 & 4-h^2\sigma & -1 & & 0 \\ & \ddots & \ddots & \ddots & \\ & & & & -1 \\ 0 & & & -1 & 4-h^2\sigma \end{bmatrix},$$

$$\text{for } i \in [1,m-1], \tag{94}$$

and $$C_i \equiv \text{diag}\{-1,-1,\ldots,-1\}, \quad i \in [1,m-2]. \tag{95}$$

By choosing $f=1$, $\sigma=0.1$ and $\tilde{\gamma}=0$ the resulting linear systems were solved by using the semi-direct ISD, CG and SICG methods. Numerical results are given in Tables 3,4 and the behaviour of the error measures $||r_i||_2$ and $(r_i,(L_r D_r L_r^T)^{-1} r_i)^{\frac{1}{2}}$ of the CG and SICG methods is illustrated in Fig.3.

The initial guess was taken to be the zero vector, while the right side vector of the system $\underline{Au}=\underline{s}$ was chosen as the product of the coefficient matrix by the solution vector of which the components are n-pseudo-random numbers chosen from a uniform distribution in the interval (0,1).

h^{-1}	r=1		r=2		r=3		r=4	
	α	ν	α	ν	α	ν	α	ν
10	0.50	38	0.85	17	0.95	15	1.05	15
15	0.35	60	0.70	25	0.90	17	1.00	17
20	0.25	91	0.55	35	0.80	22	0.95	19
30	0.20	119	0.40	53	0.55	34	0.80	25

TABLE 3: Experimental optimum values of α and the number of iterations ν of the ISD method which are required to reduce the L_∞-norm of the residual by a factor $\varepsilon=10^{-6}$ for several mesh sizes and values of the "fill-in" parameter r for the problem (93).

h^{-1}	CG method	SICG method			
		r=1	r=2	r=3	r=4
10	26	14	10	9	9
15	39	18	12	10	10
20	52	21	14	11	11
30	75	27	18	14	13
40	100	32	22	17	15
50	120	36	24	19	17

TABLE 4: The number of iterations of the CG and SICG methods required to reduce the L_∞-norm of the residual by a factor $\varepsilon=10^{-6}$ for several mesh sizes and values of the "fill-in" parameter r for the model problem (the resulting sparse matrices are of order 81,196,361,841,1421 and 2401).

Figure 2: Behaviour of the error measures (i.e.) $\|r_i\|_2$, $\|u-u_k\|_2$) of the implicit methods (71) (ISD) and (76) (IR) when applied to the model problem coefficient matrix of order N=841 (grid 19x19).

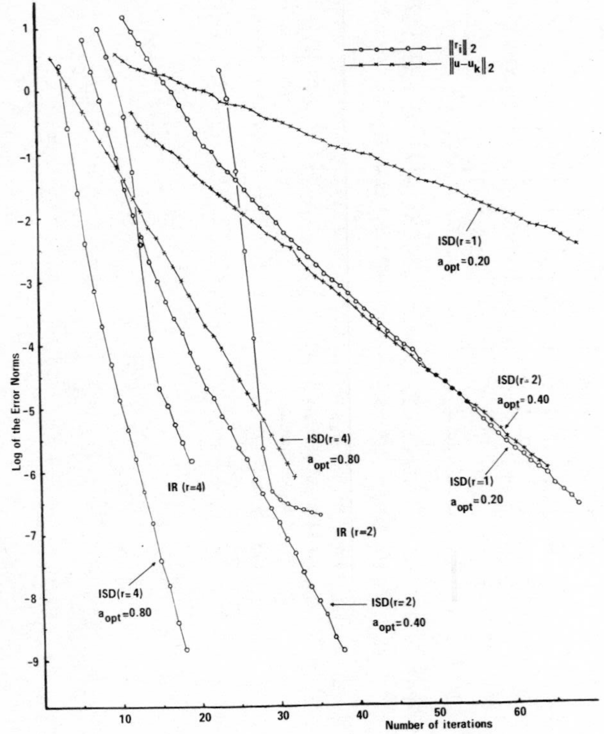

Figure 3: Behaviour of the error measures (i.e. $\|r_i\|_2$ and $(r_i, (L D L^T)^{-1} r_i)^{1/2}$ of CG and SICG methods applied to problem (I) with coefficient matrix of order N=2401 (grid 49x49).

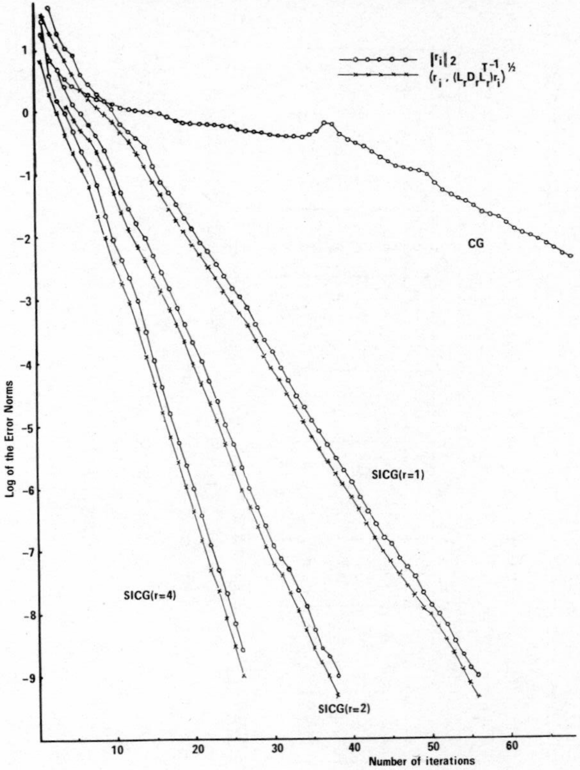

6. NON-LINEAR ELLIPTIC BOUNDARY VALUE PROBLEMS

We now consider a class of boundary-value problems defined by the non-linear elliptic P.D.E. in two space dimensions, viz.

$$\frac{\partial}{\partial x}\left(w\frac{\partial U}{\partial x}\right)+\frac{\partial}{\partial y}\left(\tilde{w}\frac{\partial U}{\partial y}\right) = f(x,y,U), \quad (x,y)\in R, \tag{96}$$

where $w,\tilde{w}$ are non-negative functions dependent on x,y and R is a given, finite, simply-connected, closed 2D-region subject to the boundary conditions

$$aU+\frac{\partial U}{\partial \zeta} = b \ , \quad (x,y)\in\Gamma \ , \tag{96a}$$

where Γ is the exterior boundary of R. On Γ, a and b are positive, piecewise continuous and ζ denotes the direction of the normal derivative. The coefficients of equation (96) are assumed to be given continuous functions of (x,y) in R and on Γ respectively and $w,\tilde{w}$ possess continuous first partial derivatives. The solution U(x,y) is also continuous and has continuous partial derivatives of the second order in R.

If a rectilinear network of mesh spacings h_x,h_y in the X,Y directions respectively is superimposed over the region R, then the central finite difference analogue of (96) at the point $(i,j)\equiv(ih_x,jh_y)$ is given by

$$[\delta_x(w_{i,j}\delta_x)+\delta_y(\tilde{w}_{i,j}\delta_y)]u_{i,j} = f(x_i,y_j,u_{i,j}) \ , \tag{97}$$

where δ_τ denotes the central difference operator with respect to τ. Then, the solution of (97) can be obtained by the Picard iteration, viz.

$$[\delta_x(w_{i,j}\delta_x)+\delta_y(\tilde{w}_{i,j}\delta_y)]u_{i,j}^{(\ell+1)} = f(x_i,y_j,u_{i,j}^{(\ell)}) \ , \tag{98}$$

and the Newton iteration, viz.

$$[\delta_x(w_{i,j}\delta_x)+\delta_y(\tilde{w}_{i,j}\delta_y)]u_{i,j}^{(\ell+1)}-f'(x_i,y_j,u_{i,j}^{(\ell)})u_{i,j}^{(\ell+1)} =$$

$$= f(x_i,y_j,u_{i,j}^{(\ell)})-f'(x_i,y_j,u_{i,j}^{(\ell)})u_{i,j}^{(\ell)} \ , \tag{99}$$

where f' denotes the Jacobian of f with respect to $u_{i,j}^{(\ell)}$. Both iterative schemes (98),(99) lead to large, sparse systems of the form

$$A_\ell \underline{u}^{(\ell+1)} = s(\underline{u}^{(\ell)}) \ , \tag{100}$$

(with $A_\ell \equiv A$ for the Picard iteration) which have to be solved.

Utilizing the derived algorithmic procedures for the given non-linear equation [14], we consider now the {Picard/Newton - RFC algorithm} and {Picard/Newton - (Implicit iteration (18)} compositive iterative schemes, where the Picard/Newton method is the outer iteration, while the inner iteration is carried out either directly by the RFC algorithm or by the ARFC(r) algorithm in conjunction with a standard iterative method, i.e. Simultaneous Displacement. It can be easily seen that the proposed composite "inner-outer" iterative scheme in the former case reduces to an equivalent one-level iteration, while for the latter, the usual two level iteration can be written as

$$L_r D_r L_r^T [\underline{u}_{(\eta+1)}^{(\ell+1)} - \underline{u}_{(\eta)}^{(\ell+1)}] = \alpha \underline{r}_{(\eta)}^{(\ell+1)} , \qquad (101)$$

where the superscript ℓ denotes the outer iteration index and the subscript η the inner iteration index, D_r and L_r are respectively diagonal and lower sparse triangular matrices appropriate to the particular method (Picard/Newton), $\underline{r}_{(\eta)}^{(\ell+1)}$ is the residual factor $\underline{r}_{(\eta)}^{(\ell+1)} = s(\underline{u}_{(\eta)}^{(\ell)}) - A_\ell \underline{u}_{(\eta)}^{(\ell+1)}$ and α is a pre-conditioned acceleration parameter.

Further details for the derivation of the composite iterative schemes and the choice of the fill-in parameter r can be found in [10].

Consider now a problem given by the non-linear elliptic P.D.E. i.e.,

<u>Problem II</u>:
$$\frac{\partial^2 U}{\partial x^2} + \frac{\partial^2 U}{\partial y^2} = e^U , \quad (x,y) \in R, \qquad (102)$$

where
$$R \equiv \left\{ (x,y), \begin{array}{c} 0 \leq x \leq x_{max} \\ 0 \leq y \leq y_{max} \end{array} \right\} , \qquad (102a)$$

subject to the Dirichlet boundary conditions

$$U(x,y) \equiv \tilde{\gamma}, \quad (x,y) \in \Gamma , \qquad (102b)$$

where Γ is the exterior boundary of R.

The equation (102) arises in Magnetic-hydrodynamics and specifically is of physical interest in diffusion-reaction, vortex problems and electric charge consideration [1]. The linearized

Picard and quasi-linearized Newton iterations are respectively outer iterative schemes of the form,

$$(\delta_x^2+\delta_y^2)u^{(\ell+1)} = e^{u^{(\ell)}} , \tag{103}$$

and

$$(\delta_x^2+\delta_y^2)u^{(\ell+1)}-[e^{u^{(\ell)}}u^{(\ell+1)}] = [1-u^{(\ell)}]e^{u^{(\ell)}} . \tag{104}$$

Assuming that $h_x=h_y=h$ and denoting the resulting set of $(x_{max}/h_x-1)(y_{max}/h_y-1)$ numbers $u_{i,j}^{(\ell)}$ (which are the approximations to the solution at (ih_x,jh_y) at the ℓth Picard/Newton iteration) by $u^{(\ell)}$, we respectively obtain the finite difference equations,

$$4u_{i,j}-u_{i+1,j}-u_{i-1,j}-u_{i,j+1}-u_{i,j-1} = -h^2e^{u^{(\ell)}} , \tag{105}$$

and

$$(4+h^2e^{u^{(\ell)}})u_{i,j}-u_{i+1,j}-u_{i-1,j}-u_{i,j+1}-u_{i,j-1} = -h^2(1-u^{(\ell)})e^{u^{(\ell)}} . \tag{106}$$

Then, a columnwise ordering is introduced and the resulting large sparse system is of the form,

$$\overbrace{}^{m}$$
$$\begin{bmatrix} b_1 & c_1 & \ddots & \tau_1 & \ddots & 0 \\ a_2 & \ddots & \ddots & 0 & \ddots & \tau_{n-m+1} \\ \ddots & \ddots & \ddots & \ddots & \ddots & \\ \nu_m & 0 & \ddots & \ddots & \ddots & c_{n-1} \\ 0 & \ddots & \nu_n & & a_n & b_n \end{bmatrix} \underline{u}^{(\ell+1)} = s(\underline{u}^{(\ell)}) , \tag{107}$$

where

$$c_{\theta(m-1)} = 0; \quad a_{\theta(m-1)+1} = 0, \quad \theta\in[1,m-2],$$

while the diagonal elements b_i, $i\in[1,n]$ are equal to unity. For the Picard iteration the elements of c,a,τ,ν are identically equal to $-1/4$ with $s(u_i^{(\ell)})=-h^2/4.e^{u_i^{(\ell)}}$, $i\in[1,n]$, while for the Newton iteration the elements of c,a,τ,ν are equal to $1/(4+h^2e^{u^{(\ell)}})$ with

$$s(\underline{u}_i^{(\ell)}) = -h^2e^{u_i^{(\ell)}}(1-u_i^{(\ell)})/(4+h^2e^{u_i^{(\ell)}}), \quad i\in[1,n].$$

The inner iteration was terminated when the following criteria was achieved, i.e.

$$\frac{|u_{i+1}-u_i|}{|u_{i+1}|+1} < \varepsilon_I , \tag{108}$$

where ε_I was initially taken as $\varepsilon_I=10^{-2}$ and then was decreased at each iterative step by $\varepsilon_I/10$ to 10^{-6} where it remained constant during the remaining iterative steps. The criterion for determination of the outer iteration was,

$$\max_j \left| \frac{u_{i,j}^{(\ell+1)}-u_{i,j}^{(\ell)}}{u_{i,j}^{(\ell+1)}} \right| \lesssim \varepsilon_O = 10^{-6} . \qquad (109)$$

Numerical experiments were carried out with $h=1/20$, $x_{max}=y_{max}=1$, and the results are given in Tables 5,6. The initial guesses $u^{(0)}=0$ and $u^{(0)}=6$ were chosen for the boundary conditions $U\equiv 0$ and $U\equiv 10$ respectively. The number of outer iterations in Table 6 where the values of the solution $u^{(\ell+1)}$ near the origin and the centre of the region R are given, denotes how many times the criterion (109) has been applied. The value of $u^{(\ell)}$ at the grid point $(\frac{1}{2},\frac{1}{2})$ is given in Table 6 as $V_1=\phi_1-10^{-7}x_1$, and $V_2=\phi_2+10^{-6}x_2$, where $\phi_1=-0.069768$ and $\phi_2=3.33684$.

Accuracy	Method-Outer Iter.	No.of iters.	Value of $u^{(\ell)}$ at $(\frac{1}{2},\frac{1}{2})$	Value of $u^{(\ell)}$ at (1/20,1/20)
$\varepsilon_O=10^{-6}$			B.C. $U\equiv 0$	
	Picard	6	-0.06976 83265 88	-0.00420 44278 91
	Newton	3	-0.06976 83275 65	-0.00420 44279 15
			B.C. $U\equiv 10$	
	Picard	(†)	-	-
	Newton	6	3.33684 36822	7.42077 39657
$\varepsilon_O=10^{-9}$			B.C. $U\equiv 0$	
	Picard	8	-0.06976 83275 72	-0.00420 44279 15
	Newton	4	-0.06976 83275 63	-0.00420 44279 15
			B.C. $U\equiv 10$	
	Picard	(†)	-	-
	Newton	7	3.33684 36821	7.42077 39657

TABLE 5: The performance of {Picard/Newton - RFC algorithm} composite schemes for $\nabla^2 U=e^U$ on a grid (19×19).

(†) The conditions for local convergence are not satisfied.

α	No.of Overall Iters.	No.of Outer Iters.	Value of $u^{(\ell)}$ at $(\frac{1}{2},\frac{1}{2})$ x_1	α	No.of Overall Iters.	No.of Outer Iters.	Value of $u^{(\ell)}$ at $(\frac{1}{2},\frac{1}{2})$ x_1
Picard			B.C.	$U\equiv 0$,		Newton	
1.25	120	52	335332	1.25	100	41	335025
1.20	65	29	320427	1.20	59	26	321270
1.15	43	20	320611	1.10	31	16	323599
1.10	32	16	332494	1.05	24	13	331438
1.05	25	13	323790	1.00	20	10	322849
1.00	20	11	329695	0.95	20	9	305785
0.95	17	10	348548	0.90	22	10	310573
0.90	18	9	309031	0.75	27	11	295296
0.85	21	10	347044	0.60	35	13	282748
0.80	20	9	316003	0.50	42	13	253770
			B.C.	$U\equiv 10$			
The conditions for local convergence are not satisfied				"	"	"	x_2
				1.45	75	6	38378
				1.35	55	6	36827
				1.30	48	6	37550
				1.20	38	6	37574
				1.15	33	6	35962
				1.10	31	6	37393
				1.05	31	6	38455
				1.00	33	6	39092
				0.95	36	6	39707
				0.90	39	6	40724

TABLE 6: The performance of {Picard/Newton - Implicit scheme (18)} composite schemes with "fill-in" parameter r=4 for $\nabla^2 U=e^U$ on a grid (19×19).

REFERENCES

1. Belman, R., M.L. Juncosa, R. Kalaba - "Some Numerical Experiments using Newton's Method for Nonlinear Parabolic and Elliptic Boundary Value Problems", C.A.C.M., 4, pp.187-191 1961.

2. Buleev, N.I. - "A Numerical Method for the Solution of Two-dimensional and Three-dimensional Equations of Diffusion", Math.Sbornik 51, No.2, pp.227-238, 1960.

3. Dupont, T., R.P. Kendall, H.H. Rachford - "An Approximate Factorization Procedure for Solving Self-Adjoint Elliptic Difference Equations", SIAM J.Numer.Anal.5, pp.559-573, 1968.

4. Evans, D.J., E.A. Lipitakis - "A Normalized Implicit Conjugate Gradient Method for the Solution of Large Sparse Systems of Linear Equations", Comp.Meth. in Appl.Mech. & Engng. 23, pp.1-19, 1979.

5. Gustafsson, I. - "A Class of First Order Factorization Methods", B.I.T. 18, pp.142-156, 1978.

6. Householder, A.S. - "The Theory of Matrices in Numerical Analysis", New York: Blaisdell, 1964.

7. Kershaw, D.S. - "The Incomplete Choleski-Conjugate Gradient Method for the Iterative Solution of Linear Equations", J. Comp.Phys. 26, pp.43-65, 1978.

8. Lipitakis, E.A. - "Computational and Algorithmic Techniques for the Solution of Elliptic and Parabolic Partial Differential Equations in Two and Three-Space Dimensions", Doctoral Thesis, Loughborough University of Technology, U.K., 1978.

9. Evans, D.J. E.A. Lipitakis - "A Sparse LU Factorization Procedure for the Solution of Parabolic Differential Equations" Proc.Intern.Conf. on Num. Methods in Thermal Problems, Swansea U.K. pp.954-966, 1979.

10. Lipitakis, E.A., D.J. Evans - "Solving Non-linear Elliptic Difference Equations by Extendable Sparse Factorization Procedures", Computing 24, pp.325-339, 1980.

11. Lipitakis, E.A., D.J. Evans, - "The Rate of Convergence of an Approximate Matrix Factorization Semi-Direct Method", Numerische Mathematik. 36, pp.237-252, 1981.

12. Meijering, J.A., H.A. Van der Vorst - "An Iterative Solution Method of Linear Systems of which the Coefficient Matrix is a Symmetric M-matrix", Math. of Comp.31, pp.148-162, 1977.

13. Oliphant, T.A. - "An Extrapolation Procedure for Solving Linear Systems", Quarterly of Appl.Math. 20, No.3, pp.257-267, 1962.

14. Ortega, J.M., W.C. Rheinboldt - "Iterative Solution of Non-Linear Equations in Several Variables", New York, Academic Press, 1970.

15. Stiefel, E.A. - "Kernel Polynomials in Linear Algebra and Their Numerical Applications", Nat.Bur.Standards Appl.Math. Ser. 49, pp.1-22, 1958.

16. Stone, H.L. - "Iterative Solution of Implicit Approximation of Multidimensional P.D.E.s", SIAM J.Numer.Anal. 5, pp.530-558, 1968.

17. Turing, A.M. - "Rounding-off Errors in Matrix Processes", Quart.J.Mech. and Appl.Math. 1, pp.287-308, 1948.

18. Varga, R.S. - "Factorization and Normalized Iterative Methods" in "Boundary Problems in Differential Equations", Ed. R.E. Langer, Univ. of Wisconsin Press, Madison, 1960.

19. Varga, R.S. - "Matrix Iterative Analysis", Englewood Cliffs, N.J., Prentice Hall, 1962.

20. Evans, D.J., Lipitakis E.L. and Missirlis, N.M. - "On Sparse and Compact Preconditioned Conjugate Gradient Methods for Partial Differential Equations", Int.J.Comput.Math. 9, pp.55-80, 1981.

21. Evans, D.J. - "Iterative Sparse Matrix Algorithms", in 'Software for Numerical Mathematics' (Ed. D.J. Evans) 49-84. Academic Press, 1974.

ANALYSIS OF INCOMPLETE FACTORIZATIONS WITH FIXED STORAGE ALLOCATION

O. AXELSSON
Mathematics Institute
University of Nijmegen
Toernooiveld, Nijmegen
The Netherlands

N. MUNKSGAARD
CE-DATA
Teknikerbyen 32
Virum
Denmark

For incomplete factorization of sparse symmetric and positive definite matrices we consider techniques for preserving positive definiteness during the factorization. Further, the problem of factorization in a limited storage environment is considered. The methods described have been implemented in a computer code and results from a large number of numerical experiments are displayed.

1. INTRODUCTION

When solving large and sparse linear equations

$$A\underset{\sim}{x} = \underset{\sim}{b}$$

using direct methods one is interested in obtaining a small fill-factor, i.e. a relative small number of fill-ins in relation to the number of non-zeros in the matrix A. It is possible for some problems with special structures to achieve this in a full factorization. For finite element problems for instance, the nested dissection factorization of George (1978) has experimentally shown to have a fill-factor of order O(log n) where n is the number of unknowns. This applies, however only for twodimensional problems. For cubic grids it is shown in Eisenstat et. al. (1976) that the fill-factor is at least $O(n^{2/3})$. Hence for more general problems, the fill-factor is of higher order for full factorizations. An alternative is to look at incomplete factorizations (see e.g. Varga (1960), Buleev (1960), Oliphant (1962), Stone

(1968), Dupont, Kendall and Rachford (1968), Axelsson (1972), Bracha-Barak and Saylor (1973), Jacobs (1973), Beauwens and Quenon (1976), Jennings and Malik (1977), Meijerink and van der Vorst (1977), Gustafsson (1978), Kershaw (1977), Axelsson and Munksgaard (1979), Munksgaard (1979), Manteuffel (1979). Note that in some of the above works only a special incomplete factorization was considered.)

In such methods one can control the fill-in by allowing only a few non-zero entries in each pivot row. There are two main strategies; either nonzero entries are confined to certain in advance chosen locations or the positions of the nonzero entries are determined during the factorization itself.

Obviously, one may also combine these two strategies, i.e. choose some non-zero positions which are fixed during the factorizations, for instance those which are non-zero in the given matrix, and in addition allow for fill-in in positions, which are determined during the factorization.

A simple criterion of neglecting fill-in entries during the factorization is based on comparison of the size of the absolute value of the entry and its corresponding diagonal entries. Small entries are then neglected. However, in this way we may destroy positive definiteness. In Example 1.1 the matrix is positive definite for $0 < \varepsilon \leq \frac{1}{2}$ but if the fill-in entries in positions (2.4), (4.2) after the first elimination step are neglected, the resulting matrix is indefinite.

We shall here consider mainly the modified incomplete factorization method where deleted entries are moved to some other location in the same row, usually to the diagonal (see Gustafsson (1978)).

If this is done in example 1.1, i.e. if the fill-in entry ε in positions (2.4), (4.2) is moved to the diagonal entries (2.2), (4.4), then the resulting matrix is positive definite. However in example 1.2 (where $0 < \varepsilon < 1/8$), the fill-in entry in position (2.4) is negative and we destroy in this example the positive definiteness due to the moving (and not just deleting) the fill-in entries in positions (2.4), (4.2).

It is wellknown that even if all pivot entries are non-zero, this situation may cause numerical instabilities, because at least in large finite element problems, a change of sign of pivot entries usually occurs through pivotentries of small absolute values.

In section 2 we discuss various problems in obtaining a good incomplete factorization algorithm for positive definite matrices, and we will discuss different possibilities of keeping the matrix positive definite during the factorization.

Since in each particular application of a sparse matrix solver, one would like to make use of the total available core storage, it is of interest to allow a fill-in, so that this storage is used in an efficient manner. This means that the incomplete factorization routine should not make use of the total allowable storage at a too early stage of the factorization, but also that it does not save too much of it to the last stage. In section 3 we discuss an algorithm for dynamic change in the size of the fill-ins in order to ensure a reasonable use of the allocated storage.

Finally in section 4 we show the results of numerical experiments carried out on a large number of testproblems.

$$\begin{bmatrix} 1 & -1 & 0 & \varepsilon & 0 \\ -1 & 2 & -1 & 0 & -\varepsilon \\ 0 & -1 & (1+\varepsilon/2) & -\varepsilon/2 & 0 \\ \varepsilon & 0 & -\varepsilon/2 & (1+\varepsilon^2) & -1 \\ 0 & -\varepsilon & 0 & -1 & (1+\varepsilon) \end{bmatrix} \qquad \begin{bmatrix} 1 & -1 & \varepsilon & -\varepsilon \\ -1 & (1+\varepsilon/2) & 0 & 0 \\ \varepsilon & 0 & 1 & 0 \\ -\varepsilon & 0 & 0 & 1 \end{bmatrix}$$

<u>EXAMPLE 1.1.</u> <u>EXAMPLE 1.2.</u>

2. ON STABLE INCOMPLETE FACTORIZATIONS

To find the solution $\underset{\sim}{x}$ of $A\underset{\sim}{x} = \underset{\sim}{b}$ using a preconditioned conjugate gradient solver, we repeatedly have to solve linear systems on the form

$$LU(\underset{\sim}{x}^{\ell+1} - \underset{\sim}{x}^{\ell}) = -\tau_{\ell}\cdot\underset{\sim}{d}^{\ell} \qquad \ell = 0,1,\ldots,$$

where LU is an incomplete sparse triangular factorization of the given matrix A, $\underset{\sim}{d}^{\ell}$ are searchdirections and τ_{ℓ} parameters, which are calculated by the conjugate gradient algorithm. For details regarding the preconditioned conjugate gradient method, see e.g. Axelsson (1976).

Initially, $\underset{\sim}{d}^0 = -\underset{\sim}{r}^0$, $\underset{\sim}{r}^0 = A\underset{\sim}{x}^0 - \underset{\sim}{b}$, where $\underset{\sim}{r}^0$ is the initial residual. In general, a good choice of an initial approximation $\underset{\sim}{x}^0$ (which otherwise can be chosen arbitrarily) is the solution of

$$LU\underset{\sim}{x}^0 = \underset{\sim}{b}.$$

The rate of convergence of the preconditioned conjugate gradient method is determined by the distribution of eigenvalues of the matrix $(LU)^{-1}A$. The best rate of convergence is obtained if the eigenvalues are clustered in one or a few groups and then the value of the condition number is of less importance (see Axelsson (1976) and Jennings (1977)). With some exceptions the theory of the method is however only applicable for symmetric and positive definite matrices A and LU but practice has shown good results also for unsymmetric and indefinite matrices (see Axelsson and Gustafsson (1979) and Axelsson and Munksgaard (1977)).

For results regarding convergence of conjugate gradient type methods for indefinite problems, see e.g. Paige and Saunders (1975) and Chandra (1978). For a generalized conjugate gradient method on unsymmetric problems, see e.g. Axelsson (1980) and the references quoted therein.

At each iterative step we have to solve two (triangular) systems and we do not want that rounding errors do influence the rate of convergence noticeably. Hence the condition numbers of L,U should not be too large and small pivots should not appear. However, the condition number of LU must be large if LU is going to mimic A and if A has a large condition number. In the incomplete factorization method this condition number is actually smaller than that of A but in the modified incomplete factorization it may be assymptotically somewhat larger for some problems.

The appearance of small pivot entries is more difficult to avoid. Small pivots may in particular be expected when the pivot entries change sign. Hence in the first place we should avoid nonpositive pivots.

If we have a symmetric positive definite matrix it is well-known that all pivots in a complete factorization are positive. For incomplete factorizations the situation is not as simple however, because it consists of the following two operations for each pivot

(i) the elimination step,

(ii) the modification step.

The elimination step is performed as usual in a Gaussian elimination method, and the symmetry and definiteness of the matrix A^0, given before the elimination step, is unchanged. As well known, this is so, since the matrix $A^{(1)}$ left after the first elimination defined by

$$L_1 A^{(0)} L_1^T = \begin{bmatrix} 1 & \underline{0}^T \\ \underline{0} & A^{(1)} \end{bmatrix}$$

is itself symmetric and positive definite.

The modification step may however destroy positive definiteness. In the modification step (ii) we consider the numerically small entries generated as fill-ins in the elimination step (i). The small elements can be determined either by a numerical test (see Section 3) or by their geometrical positions in the matrix as proposed by Meijerink and van der Vorst (1977) for the ICG-methods. We modify the matrix either by deleting the fill-in - which is small enough - or by moving it to some other location in the same row for instance to the diagonal. The latter technique is due to Gustafsson (1978) called the modified incomplete conjugate gradient method (MICG).

If we define the rowsum of a matrix row as the sum of all entries in the row, it is seen that for the MICG-type of modification the rowsum of all rows in $A^{(k)}$, $k = 0,1,\dots,n-1$ (the successive matrices we get in the process) will be unchanged by the modification step. This means in particular that if the rowsums of $A^{(0)}$ are non-negative and if $A^{(0)}$ is a positive definite Z-matrix, i.e. having all off-diagonal entries non-positive, then the modification step preserves both the positive definiteness and the Z-matrix property. In addition, it is easily seen that the rowsums do not decrease during the elimination process of a Z-matrix. Hence if we let L have diagonal entries = 1 then the pivot entries $a_{kk}^{(k)}$ are bounded below by the sum of the off-diagonal entries of U and the original rowsum, i.e.

$$a_{kk}^{(k)} \geq \sum_{j=k+1}^{n} (-u_{kj}) + s_k^{(0)}$$

where

$$s_k^{(0)} = \sum_{j=1}^{n} a_{kj}.$$

In particular, no pivot entry is small if we have at least one (not small) off-diagonal entry in every row of A (and hence of U) and if the final row of A has a rowsum which is not small. In such a case we are guaranteed to get a stable modified incomplete factorization.

However for positive definite matrices, which are not Z-matrices, one may loose positive definiteness due to the modification step. This is easily checked by testing the method on some simple matrices (see for instance Gustafsson (1979) and Example 1.2).

For this type of matrices, we have therefore considered an off-diagonal modification to try to preserve positive definiteness of the matrix $A^{(k)}$, $k = 1,\dots,n-1$.

In the diagonal modification it is only the negative entries that are causing problems because moving a positive entry to the diagonal only increases the dominance of the diagonal elements.

If however there is a positive off-diagonal in the row, it may be more satisfactory to move the negative element to this location. Fig. 2.2 a and 2.2 b show two variants of this kind of rowsum and symmetry preserving modification. In the full developed version (fig. 2.2 b) the modified off-diagonal elements have been decreased in size, while diagonal entries have only been increased. Hence later fill-in entries will not be larger than they would have been if this modification had not been done. In case the element to move is negative, and so are all other off-diagonal non-zero's, we use the usual diagonal modification to obtain the rowsum preservation. See fig. 2.1.

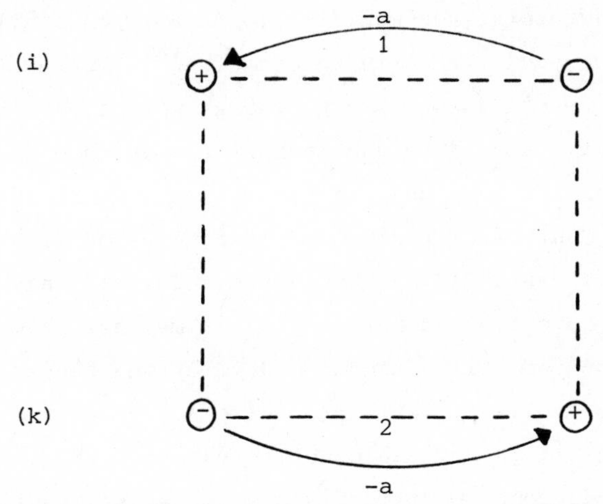

FIGURE 2.1. All off-diagonal entries in row (i) are negative. Small elements are moved to the diagonal $(a > 0)$.

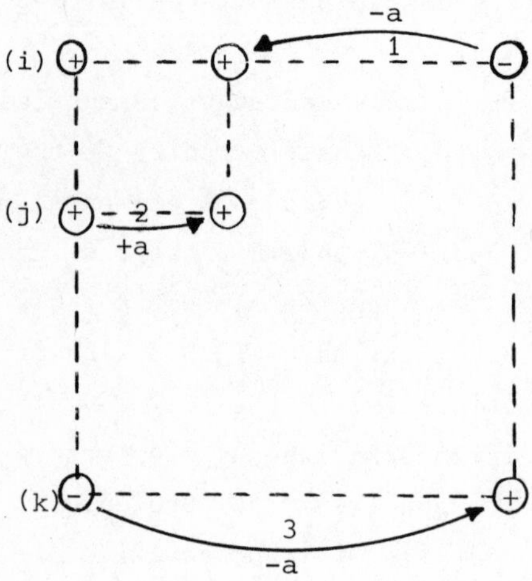

FIGURE 2.2 a
Semi off-diagonal modification

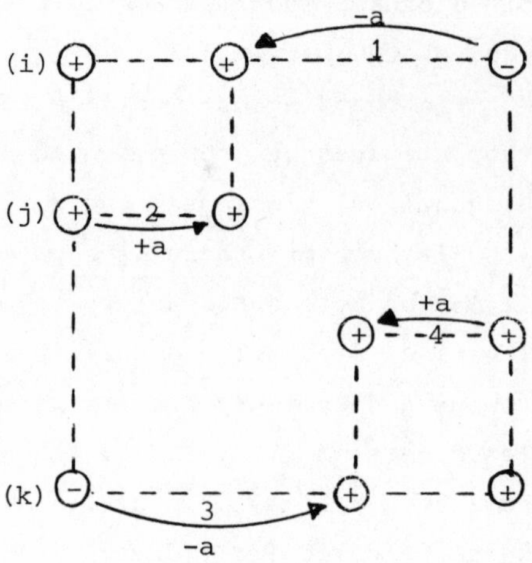

FIGURE 2.2 b
Complete off-diagonal modification

This way of performing the modification step does still not in general guarantee positive definiteness of $A^{(k)}$. There is of cause the alternative of adding a large enough positive number δ_i to the pivot entry, should this be nonpositive or too small. This latter approach has been used in connection with IC-methods in Kershaw (1978), Manteuffel (1980) and in connection with MIC-type methods in Munksgaard (1979). The disadvantage of this method is that the condition number of $(LU)^{-1}A$ may now be much larger than if we perform off-diagonal modifications, because we calculate LU from a different matrix $(A + \text{diag}(\delta_i))$ than the given one (A). Hence it is advisable not to use this final remedy as long as the off-diagonal modifications work satisfactorily. For discretized elliptic problems some results have been proven indicating that the condition number may be decreased asymptotically with respect to a problem parameter if one uses a MIC-method.

Such results are found in e.g. Gustafsson (1981). For some second order problems it means that the spectral condition number $\kappa(A) = O(h^{-2})$, where h is a stepsize-parameter, is decreased to $\kappa((LU)^{-1}A) = O(h^{-1})$, $h \to 0$, if LU is calculated from $A + \zeta_1 h^2 \operatorname{diag}(A) + \zeta_2 h \operatorname{diag}(A')$ where $\zeta_1, \zeta_2 > 0$ and

$$(A')_{ii} = \begin{cases} 0 & \text{if } \sum_{j<i} a_{ij} \geq \sum_{j>i} a_{ij} \\ (A)_{ii}, & \text{otherwise.} \end{cases}$$

It is easily seen that for the IC methods the spectral condition number is still $O(h^{-2})$. Hence at least for such problems where h is small enough it is essential to perform the modification step of moving entries (Gustafssons result is for the case where the entries are moved to the diagonal).

For other types of matrices however, it may not be the case.

3. FIXED SPACE FACTORIZATION

A major problem in factorizing sparse matrices is to determine the amount of storage needed to hold the factorized form. The incomplete factorizations exhibit the same problem if a numerical dropcriterion is used, but we have considered the design of algorithms, that perform an incomplete factorization in a fixed amount of storage locations and utilize the available space as efficient as possible.

We have worked with two different algorithms for dropping elements in a fixed space factorization. The first is based on a combination of a 'geometrical' and a numerical criterion, whereas the second tries to distribute the storage locations equally among the rows of the matrix.

In the first algorithm, the 'geometrical' criterion restricts all locations specified for the coefficientmatrix (as well zeros as non-zeros can be specified) to be maintained during the factorization. Only non-zeros which are not contained in this structure, are subject to the numerical dropcriterion. In the numerical criterion we compare the value of the potential fill-in with the corresponding diagonal entries, and we only accept the element in the matrix structure if

$$|a_{ij}^{(k)}| \geq c \sqrt{a_{ii}^{(k)} a_{jj}^{(k)}}. \tag{3.1}$$

Here the coefficient c determines the amount of fill-in for the factorization. If $c \leq 0$ we obtain a complete factorization, and in the other end $c \geq 1$ (for positive definite matrices) gives a factorization, where only elements specified in the structure of the coefficient matrix are involved. By changing c during the factorization we can adjust the amount of fill-in.

The simplest way to change c is to run with a fixed value, until the available space is used, and then change c to 1 so no more fill-in will be accepted. This factorization is illustrated graphically in Figure 3.1, which shows the mumber of fill-ins relative to the space allocated for the factorization (i_{fill}) as function of the relative pivot step number (i_p, $0 < i_p \leq 1$). Curve (1) shows the case where c has been chosen so small, e.g. 10^{-5}, that the allocated space has been filled after about 50% of the pivot steps. Consequently we must drop all further fill-ins of which several may be much greater than 10^{-5}, and hence there might be a better approximation to the coefficient matrix, in this amount of space.

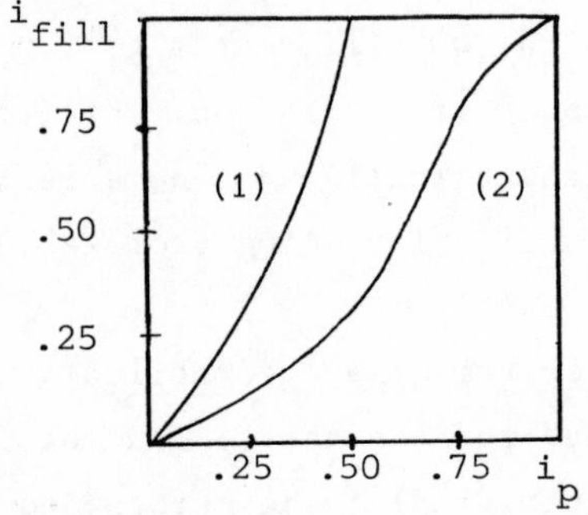

FIGURE 3.1 Relative fill-in as a function of relative pivot step.

To explain the way in which we change c we first describe how a sparse matrix normally fills. Usually pivots are chosen from rows with as few other elements as possible if we wish to minimize the number of fill-ins. There is only few elements in the beginning, and later on, when the density of the matrix increases, the number of fill-ins in each step is also growing. A typical elimination curve for a symmetric sparse matrix is shown as curve (2) in Figure 3.1. If we consider problems arising from the discretization of partial differential equations which is the main purpose of the algorithm we consider here, the curve is slightly more linear.

Our proposal for an algorithm which adaptively changes c is the following. We try to force the elimination curve to follow an ideal curve, which distributes the fill-ins uniformly between the first 90% of the pivot steps. We allow the actual elimination curve to deviate 3% from the ideal curve, but in case of larger deviations we increase the value of c, and redefine the ideal elimination curve to distribute the remaining storage uniformly on the pivot steps up to $i_p = 0.9$. If i_{pc} and i_{fc} are the values of i_p and i_{fill} in a pivot step where the 3% diviation is exceeded, the ideal curve is defined as

$$i_{fill}(i_p) = i_{fc} + \frac{1.0-i_{fc}}{0.9-i_{pc}}(i_p - i_{pc}) \qquad i_{pc} \le i_p < 0.9$$

$$i_{fill}(i_p) = 1.0 \qquad 0.9 \le i_p \le 1.0$$

The curve is initially defined for $(i_{pc}, i_{fc}) = (0.,0.)$.

This produces a piecewise linear and discontinuous ideal curve that is determined by the rate of fill-in and which always tends to distribute the available space uniformly among the remaining pivot steps with i_p values less than 0.9 (see Fig. 3.2).

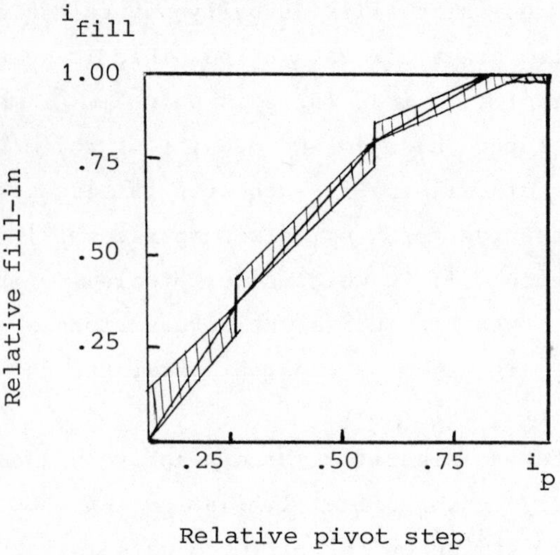

FIGURE 3.2 Elimination curve with modifying drop tolerance

Various techniques have been considered for decreasing the value of c when there is more space available for the factorization than it is possible to utilize with the initial specified value of c. If we decrease c during the factorization we accept fill-ins of greater magnitude than some of those we have dropped until we changed c. As regards clustering of eigenvalues of $(C^{-1}A)$ where C is a representation of the incomplete factorized matrix, our experiments indicate however, that using a fixed value of c gives just as good an incomplete factorization as if we decrease c at some pivot step during the factorization. The satisfactory way of decrasing c for obtaining a better clustering of the eigenvalues is to restart the factorization with a new c value, but that is considered to be too time consuming to incorporate in the algorithm.

The other algorithm considered, is adapted to suit factorization of equations with banded structure. It is well known, that the fill-ins in banded matrices often are equally distributed among the rows of the matrix. We have tried to design an algorithm that also distributes the available space equally among the rows.

If the available number of storage locations is IA, we initially allow each row to contain NN = IA/N non-zeros where N is the order of the matrix. After each row operation (i.e. after a multiple of the pivot row have been added to an other row) we drop the smallest elements of the row so that we end up with only NN elements. After each pivot step we ajust NN to regain space not used for fill-ins. If we for instance have NZ non-zeros after pivot step IP we reset NN to

$$NN = NZ/N + (IA-NZ)/(N-IP).$$

In this way we are able to increase NN if the initial pivot steps do not utilize the available space.

This algorithm has the advantage that we do not carry around with small elements and drop potential fill-ins of larger size. We are presently working on a variant of this algorithm, which actually perform the rowoperations quite normally. We only drop elements in the pivot row. When a row enters the pivot position we drop the smallest elements so we end up with NN non-zeros in the row. Using this algorithm on banded matrices, we need to reserve some extra space (apart from the NN $*$ N locations used to hold the factorized matrix), for fill-ins which we will drop when they enters the pivot row. Looking at a banded matrix (fig. 3.3) it is obvious, that the space needed is $\frac{1}{2} * r_m^2$ where r_m is the maximum bandwidth.

By using this algorithm we note, that if location a_{ij} receive <u>small</u> contributions from elimination step k,k+1,...,i-1, each of which might have been dropped if the other algorithm was used, it is here allowed to accumulate and $|a_{ij}^{(i)}|$ may be large.

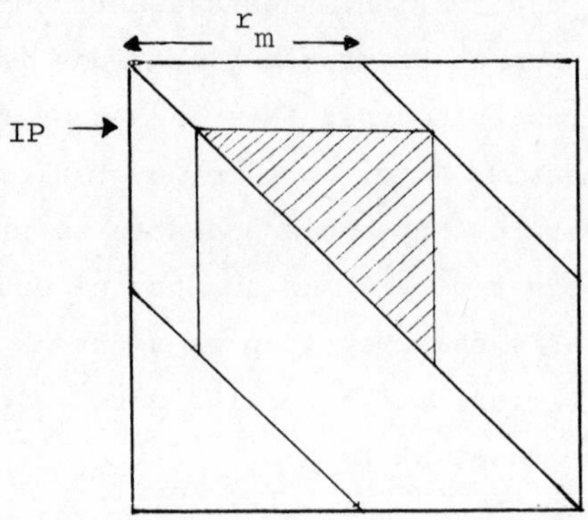

FIGURE 3.3. Influence area at pivot step IP in bandmatrix with maximum bandwidth r_m.

4. NUMERICAL RESULTS

A large number of testruns have been performed with the Harwell MA31 code Munksgaard (1979) which implements the diagonal modification factorization considered in section 2 and also contains the possibility of using the first of the fixed space factorization algorithms described in section 3.

We have worked with the following versions of this code:

MA31 A: MA31 with fixed value of c (3.1) during the factorization.

MA31 B: The factorization is performed with off-diagonal modification as described in section 2. c is fixed during the factorization.

MA31 C: MA31 with fixed space factorization using the variable c-method (section 3).

MA31 D: MA31 with fixed space factorization using the second method described in section 3.

The testruns have been carried out on two different computers which will explain inconsistencies between the tables. The results reported in tables 4.2, 4.3 and 4.5 are performed on an IBM 370/168 and for table 4.4 is used a DEC 2020, which is considerably slower than the IBM-machine.

We have used the following testmatrices to illustrate the behaviour of the algorithms.

(i) FPFD and SPFD: five and seven point discretizations of Laplace's equation in two and three dimensions. The regions considered is the unit square and cube with natural boundary conditions.

(ii) FPI1 and FPI2: five point discretization in an irregular mesh of Laplace equation with natural (1) and Dirichlet boundary conditions (2) (see fig. 4.1).

(iii) BIHA: 13 point discretization of biharmonic plate problem of Stiefel (1959) (see fig. 4.2).

(iv) GRDL: graded L-problem of George (1977). Linear triangulars finite element discretization of Laplace equation (see fig. 4.3).

Table 4.1 gives some data on the matrices used for the numerical experiments.

Matrix	n	nz	Matrix	n	nz	Matrix	n	nz
FPFD 16	256	480	FPI1 16	465	1349	BIHA 16	270	1341
FPFD 31	1024	1984	FPI1 32	1953	5765	BIHA 32	1054	5943
SPFD 8	512	1344	FPI2 16	480	1393	GRDL 5	406	1155
SPFD 10	1000	2700	FPI2 32	1984	5857	GRDL 8	1009	2928

TABLE 4.1. Testmatrices n = order; nz = non-zeros in upper triangular part.

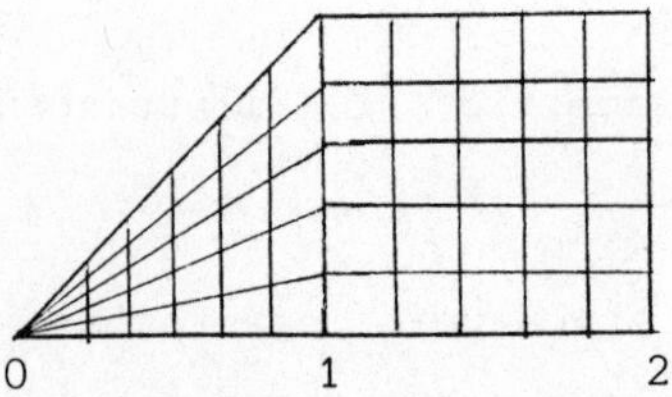

Fig. 4.1 FPI1 and FPI2 discretizations.

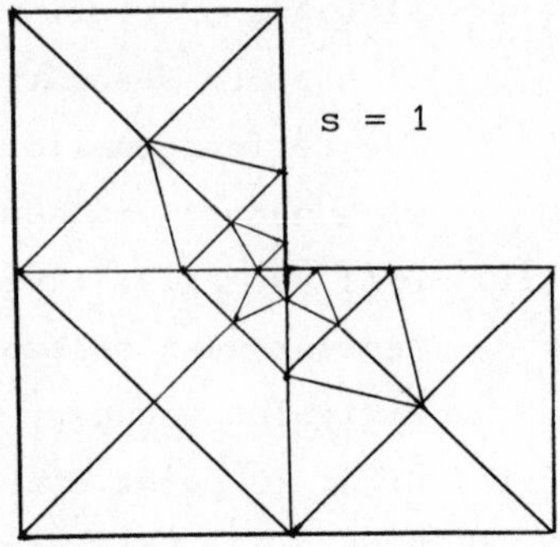

Fig. 4.3 GRDL. Graded L discretization.

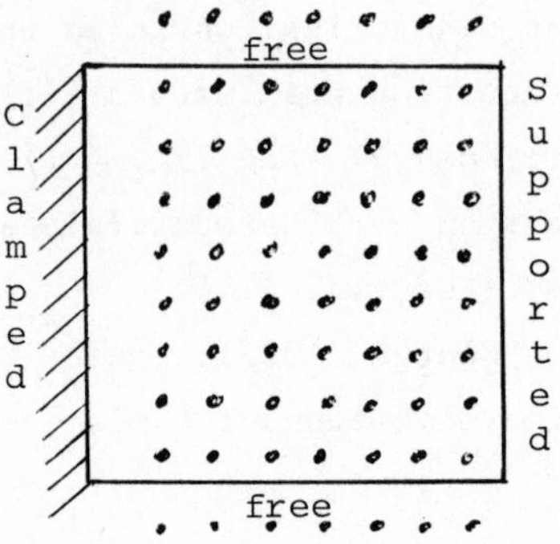

Fig. 4.2 BIHA. 13-point discretization.

MA31 uses the numerical dropcriterion and the relative drop coefficient c can be specified by the user. For the following comparisons we have run the code with three different values of c, namely $c = 1$ for which no fill-in is accepted, $c = 10^{-2}$, and $c = 0$ for which the code converts to a direct method.

As we discussed in section 2, with the preservation of the rowsum we guarantee a low order growth in the condition number of $C^{-1}A$ when the meshsize of the discretization is decreased. If $\mathcal{H}(C^{-1}A) = O(h^{-1})$ for second order problems, we can expect the number of iterations to grow with a factor $\sqrt{2}$ when h is halved. Table 4.2 shows the number of iterations required to reduce the residual norm to 10^{-8} using MA31B.

Matrix	Iterations $c=1$	$c=10^{-2}$	Matrix	Iterations $c=1$	$c=10^{-2}$	Growth factor $c=1$	$c=10^{-2}$
FPFD 16	15	5	FPFD 32	19	6	1.3	1.2
FPI1 16	17	6	FPI1 32	25	9	1.5	1.5
FPI2 16	19	6	FPI2 32	27	9	1.4	1.5
BIHA 16	41	27	BIHA 32	78	53	2.0	2.0

TABLE 4.2. Iterations required to make $\|\underset{\sim}{r}\|_2 < 10^{-8}$ where $\underset{\sim}{r} = A\underset{\sim}{x} - \underset{\sim}{b}$, $\|\underset{\sim}{r}_0\|_2 = 1$.

The theory for preconditioned conjugate gradient methods shows that for positive definite matrices the error in norm $\|\cdot\|_A$ is decreasing monotonically. In practice one often finds that $\|\underset{\sim}{r}\|_2$ also decreases. Since it is easier to test on this latter quantity we have chosen it in the stop criterion for the iterations.

From the table it is seen, that the growth in iterations is in accordance with the theory for second order problems. Further, we see, that the growth of $H(C^{-1}A)$ for the biharmonic problem seems to be $O(h^{-2})$.

When factorizing the biharmonic equation matrix, it was not possible to keep the active matrix positive definite when the diagonal modification with $c = 10^{-2}$ is used. This problem is solved in MA31 by adding the maximum element of the row of a negative pivot to the diagonal. However, as we see in the actual case (table 4.3), this results in a large number of extra iterations.

It is therefore a much better solution to implement the off-diagonal modification algorithm described in section 2. This algorithm is a bit more complicated than the diagonal modification, and it might be expected to be more timeconsuming in the factorization step. This is true for the no-fill-in factorization, where we observe a 5% increase in the factorization time, which is regarded as being acceptable. For $c = 10^{-2}$ however, the time is decreased with 20% because the number of fill-in, using the off-diagonal modification is reduced significantly.

		Diagonal modif. MA31A			Off-diagonal modif. MA31B		
	c	Fact. time	Iter. time	Iter. nr.	Fact. time	Iter. time	Iter. nr.
BIAH 16	1	.248	.819	41	.257	.908	46
BIAH 16	10^{-2}	.594	>5.50*	>200*	.477	.661	27
BIAH 32	1	1.052	6.681	78	1.102	7.109	83
BIAH 32	10^{-2}	2.519	>27.00*	>200*	2.017	5.382	53

TABLE 4.3. BIHA-factorization and iteration times in sec.

Let us finally consider the fixed-space factorizations described in section 3. The first algorithm implementated in MA31C uses the dropcriterion (3.1) where c is increased if it is impossible to fit the factorization into the (IA3) available storage locations with a fixed value of c.

In table 4.4 is shown a number of runs with MA31C for different start values of c but with the same IA3 value. It is seen, that if the start value of c is so large, that the available space can not be utilized, or if c have been chosen so small that it has to be increased several times to limit the amount of fill-in then the number of iterations required to obtain the desired accuracy is usually a little larger than for the case where the start value of c is well estimated in relation to the available space. Hence we conclude that the algorithm can be sensitive to bad estimations of the start value of c.

The second algorithm implemented in MA31D is independent of c since it always drops the smallest elements in the row. From the runs shown in table 4.4 it is seen, that this algorithm results in a factorization, that is quite competitive to the best factorizations performed with MA31C. Thus in the case of bounded structured matrices, there is no doubt that we will prefer to use this algorithm since it is independent of initial parameter choices. The algorithm have not yet been tested on other types of matrices so it is hard to draw any conclusions on its behaviour on general sparse matrix structures, where the first algorithm (MA31C) seems to be quite efficient (see Munksgaard 1979).

For both algorithms it is seen, that about 92-96% of the available storage locations is used for the factorizations, so there is a relative small waste of storage in both algorithms.

Problem	CODE	c	Fac.time sec.	non-zeros in L	No. of iterations
FPFD 16	MA31C	10^{-1}	1.41	592	10
n=256		10^{-2}	1.56	760	7
		10^{-3}	1.74	735	10
IA3=764		10^{-4}	1.62	759	11
	MA31D		1.63	706	8
FPFD 32	MA31C	10^{-1}	6.05	2631	16
n=1024		10^{-2}	7.10	3305	13
		10^{-3}	7.20	3338	13
IA3=3492		10^{-4}	7.29	3299	17
	MA31D		7.60	3312	11
SPFD 8	MA31C	10^{-1}	3.11	1344	9
n=512		10^{-2}	3.85	1784	8
		10^{-3}	4.24	2005	10
IA3=2144		10^{-4}	4.27	2004	9
	MA31D		4.87	2058	8
BIHA 16	MA31C	10^{0}	3.16	1425	33
n=270		10^{-1}	3.56	1472	32
		10^{-2}	6.16	2430	35
IA3=2570		10^{-3}	6.96	2470	36
	MA31D		6.90	2508	32

TABLE 4.4. Fixed space factorizations. IA3 is the available space in L. Runs made on DEC2020.

Let us finally consider how the preconditioned methods compare with a direct method. As mentioned MA31 converts to a direct method when $c = 0$.

	Factor time			Solve time		
Matrix	c=1	$c=10^{-2}$	c=0	c=1	$c=10^{-2}$	c=0
FPFD 32	.441	.696	2.337	1.077	.637	.067
SPFD 10	.482	1.134	11.505	.752	.571	.153
GRDL 8	.615	1.158	4.401	1.834	1.321	.123

TABLE 4.5. Factor and solve times in seconds for MA32 with different relative dropcoefficients c. $\|\underset{\sim}{r}_0\|_2 = 1$.

Whether to use a direct, or a preconditioned iterative method, depends first of all on the number of righthand sides involved in the actual problem, but as regards "one-off" problems (i.e. with only one righthand side) it is seen from table 4.5 that the preconditioned methods is preferable for the partial differential equation problems considered here.

From Table 4.5 we find that the factor plus solution times for $c = 1$ (no fill-in) and for $c = 10^{-2}$ (medium fill-in) are almost equal. But since the solution time is smaller for $c = 10^{-2}$, the medium fill-in algorithm will be more efficient for "many-off" problems, in particular for large problems. The numerical results indicates that c should decrease with increasing number of righthand sides.

Storage problems, which we do not consider here, may however also have an influence on the choice of solution method. The direct methods often require much more storage than the preconditioned methods, because all fill-ins have to be maintained.

5. CONCLUSION

MA31 has by Munksgaard (1979) been compared with an other direct method, The Yale Sparse Matrix Package (Eisenstat et al (1977)) which has a good reputation for solving sparse positive definite problems efficiently. The solution times for MA31 has been found to be comparable with the Yale code over a wide range of problems of different origin. Though an improved version of the Yale code have been developed, we can conclude, that the algorithm implemented by MA31 completed with the off-diagonal modification algorithm described in this paper, supplies a solution technique which may be an important alternative to the traditional methods.

REFERENCES

1. AXELSSON, O. - A generalized SSOR method. BIT 13, pp. 443-457, 1972.
2. AXELSSON, O. - A class of iterative methods for finite element equations. Comp. Meth. Appl. Mech. Eng., 9, pp. 123-137, 1976.
3. AXELSSON, O. - Solution of linear systems of equations. Iterative methods in sparse matrix techniques, Ed. V.A. Barker, Lecture Notes in Mathematics 572, Springer-Verlag, 1977.
4. AXELSSON, O. - Conjugate gradient type methods for unsymmetric and inconsistent systems of linear equations. Linear Algebra and its Applications, 29, pp. 1-16, 1980.
5. AXELSSON, O. and GUSTAFSSON, I. - A modified upwind scheme for convective transport equations and the use of a conjugate gradient method for the solution of non-symmetric systems of equations. J. Inst. Maths. Applics, 23, pp. 321-337, 1979.
6. AXELSSON, O. and MUNKSGAARD, N. - A class of preconditioned conjugate gradient methods for the solutions of a mixed finite-element discretization of the biharmonic operator. Int. J. Numer. Meth. Eng. 14, pp. 1001-1019.
7. BEAUWENS, R. and QUENON, L. - Existence criteria for partial matrix factorizations in iterative methods. SIAM J. Numer. Anal., 13, pp. 615-643, 1976.
8. BRACHA-BARAK, A. and SAYLOR, P.E. - A symmetric factorization procedure for the solution of elliptic boundary value problems. SIAM J. Numer. Anal., 10, pp. 190-206, 1973.
9. BULEEV, N.I. - A numerical method for the solution of two-dimensional and three-dimensional equations of diffusion. Mat. Sb. 51, pp. 227-238, 1960; English transl. Rep.BNL-TR-551, Bookhaven National Laboratory, Upton, New York, 1973.

10. CHANDRA, R. - Conjugate Gradient Methods for Partial Differential Equations, Ph.D. Dissertation. Report 129, Department of Computer Science, Yale University, 1978.
11. DUPONT, T., KENDALL, R.P. and RACHFORD, H.H. - An approximate factorization procedure for solving self-adjoint elliptic difference equations. SIAM Journal on Numerical Analysis, 5, pp. 559-573, 1968.
12. EISENSTAT, S.C., GURSKY, M.C., SCHULTZ, M.H. and SHERMAN, A.H. - Yale sparse matrix package I. The symmetric codes. Report 112. Dept. of Computer Science, Yale University. 1977.
13. EISENSTAT, S.C., SCHULTZ, M.H. and SHERMAN, A.H. - Applications of an element model for Gaussian elimination, in J.R. Bunch and D.J. Rose (editors), Sparse Matrix computations, Academic Press 1976.
14. GEORGE, A. - An automatic nested dissection algorithm for finite element problems. SIAM Journal on Numerical Analysis, 15, pp. 1053-1069, 1978.
15. GUSTAFSSON, I. - A class of first order factorization methods. BIT, 18, pp. 142-156, 1978.
16. GUSTAFSSON, I. - Stability and rate of convergence of modified incomplete Cholesky factorization methods 79.02 R (thesis), Department of Computer Science, Chalmers University of Technology, Gothenburg, Sweden, 1979.
17. HESTENESS, M.R. and STIEFEL, E. - Methods of conjugate gradients for solving linear systems. NBS J. Res., 49, pp. 409-436, 1952.
18. JACOBS, D.A.H. - The strongly implicit procedure for Biharmonic Problems. J. Comp. Phys., 13, pp. 303-315, 1973.
19. JENNINGS, A. - Influence of the eigenvalue spectrum on the convergence rate of the conjugate gradient method. J. Inst. Math. Applics., 20, 61-72, 1977.
20. JENNINGS, A. and MALIK, G.A. - Partial elimination. J. Inst. Math. Applics., 20, pp. 307-316, 1977.
21. KERSHAW, D.S. - The incomplete Cholesky-conjugate gradient method for the iterative solution of systems of linear equations. J. Comp. Phys., 26, pp. 43-65, 1978.
22. MENTEUFFEL, T.A. - An incomplete factorization technique for positive definite linear systems. Math. Comp., 34, pp. 473-497, 1980.
23. MEIJERINK, J.A. and VAN DER VORST, H.A. - An iterative solution method for linear systems of which the coefficient matrix is a symmetric M-matrix. Math. Comp., 31, pp. 148-162, 1977.
24. MUNKSGAARD, N. - New factorization codes for sparse symmetric and positive definite matrices. BIT, 19, pp. 43-52, 1978.
25. MUNKSGAARD, N. - Solving sparse symmetric sets of linear equations by preconditioned conjugate gradients. AERE-Harwell report. CSS 67, 1979.
26. OLIPHANT, T.A. - An extrapolation process for solving linear systems. Quarterly on Applied Mathematics, 20, pp. 257-267,

1962.
27. PAIGE, C.C. and SAUNDERS, M.A. - Solution of sparse indefinite systems of linear equations. SIAM Journal on Numerical Analysis, 12, pp. 617-629, 1975.
28. STIEFEL, E. - The self adjoint boundary value problems. Refined iterative methods for computation of the solution and the eigenvalues of self adjoint boundary value problems, eds. M. Engeli, Th. Ginsburg, H. Rutishauser and E. Stiefel. Der Eidgenössischen Technischen Hochschule, Zürich, 1959.
29. STONE, H.L. - Iterative solution of implicit approximations of multidimensional partial differential equations. SIAM Journal on Numerical Analysis, 5, pp. 530-558, 1968.
30. VARGA, R.S. - Factorization and normalized iterative methods. Boundary Problems in Differential Equations, ed. R.E. Langer, University of Wisconsin Press, Madison, pp. 121-142, 1960.

STABILIZED INCOMPLETE LU-DECOMPOSITIONS AS PRECONDITIONINGS FOR THE TCHEBYCHEFF ITERATION

HENK A. VAN DER VORST

Academisch Computer Centrum Utrecht

Budapestlaan 6

Utrecht - de Uithof

the Netherlands

ABSTRACT

The usefulness of simple incomplete LU-decompositions as preconditionings for the Tchebycheff iteration, in order to solve certain linear systems Ax=b with a non-symmetric matrix, has already been reported. In this chapter we consider certain stability problems that may occur in the construction of the preconditioning and show how they can be overcome. Further we consider the question whether it is whorthwhile to seek for more complicated decompositions.

1. INTRODUCTION

An important class of problems consists of the linear equations Ax=b that arise from discretisation of second order partial differential equations that include first order derivative terms. Discretisation of this type of equation yields a linear system with a non-symmetric nonsingular matrix. These problems are in general more difficult to solve than linear systems with a symmetric matrix A useful method to solve nonsymmetric linear systems is Manteuffel's variant of the Tchebycheff iteration [1]. Application of this method is restricted to situations where all the eigenvalues

---

The research reported in this chapter has been supported in part by the European Research Office, London, through Grant DAJA-37-80-C-0243.

of the matrix A are in the right half plane. Meijerink and Van der Vorst [2] introduce the idea of incomplete LU-decomposition and Van der Vorst [3] reports how this idea can be used as an effective preconditioning for the Tchebycheff iteration. Unfortunately it is not always possible to construct an incomplete decomposition in a straight-forward manner and some steps have to be taken to prevent instability.

In this chapter we consider in more detail the use of a parameter in order to master the instability problems without anihilating the efficiency of the algorithm. It appears (by accident?) that stability as well as optimal efficiency is achieved for almost the same value of the parameter. Van der Vorst [3] considers the use of incomplete decompositions, and their stabilized variants, that have a similar sparsity structure as the matrix A has. Here we will consider the usefulness of more complicated incomplete decompositions. In contrast to the situation for symmetric linear systems (see for instance [4]) they seem to be of little practical value.

2. MANTEUFFEL'S ADAPTIVE VERSION OF TCHEBYCHEFF ITERATION

In this section a short presentation of the Tchebycheff iteration is given. For more details the reader is referred to Manteuffel [1]. For the solution of the linear system $Ax=b$ the following two step iterative method can be used

$$(2.1) \qquad x_{k+1} = -\alpha_k A x_k + (1 + \beta_k) x_k - \beta_k x_{k-1} + \alpha_k b$$

It is shown by Manteuffel [1] that this method converges to the solution of $Ax=b$ if the eigenvalues of A are enclosed in an ellipse, with focii $d-c$ and $d+c$, in the right half plane and if α_k and β_k are defined in terms of parameters of that ellipse as follows.

$$(2.2) \qquad \alpha_k = \frac{2}{c}\,\frac{T_k(\frac{d}{c})}{T_{k+1}(\frac{d}{c})} \qquad \beta_k = \frac{T_{k-1}(\frac{d}{c})}{T_{k+1}(\frac{d}{c})}$$

where $T_k(z)=\cos(k \arccos(z))$, for z complex. The constants d and c should be chosen to define a family of ellipses which contains the ellipse with focii d-c and d+c (c complex eventually) that encloses the spectrum of A and for which the convergence factor r_c (see (2.4)) is minimal. For computational purposes the following scheme is convenient:

Given x_0 , define $x_1=x_0+p_0$, where $p_0=\frac{1}{d}r_0$

$r_0=b-Ax_0$

$\alpha_1=2d/(2d^2-c^2)$

$\beta_1=d\alpha_1-1$

then

$$(2.3) \qquad \left.\begin{aligned} r_i &= b - Ax_i \\ p_i &= \alpha_i r_i + \beta_i p_{i-1} \\ x_{i+1} &= x_i + p_i \\ \alpha_{i+1} &= (d - (c/2)^2\alpha_i)^{-1} \\ \beta_{i+1} &= d\alpha_{i+1} - 1 \end{aligned}\right\} \text{for } i=1,2,.....$$

The asymptotic value for the convergence factor r_c of this Tchebycheff iteration is given by

$$(2.4) \qquad r_c = \frac{a+\sqrt{a^2-c^2}}{d+\sqrt{d^2-c^2}}$$

where d is the center of the ellipse, c is the focal distance and a is the length of the axis in the x-direction (see figure 1). Manteuffel [1] has provided an algorithm for estimating the parameters a, d and c adaptively.

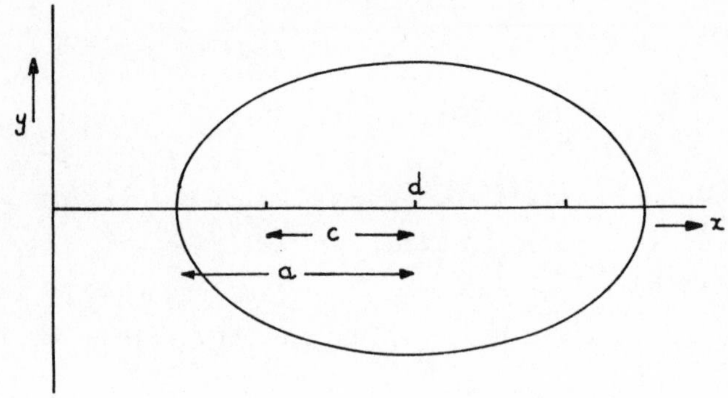

FIGURE 1. Manteuffel parameters

In order to improve the speed of convergence of the algorithm (2.3) one may use a well-chosen preconditioning matrix K and solve the equation $K^{-1}Ax=K^{-1}b$. In that case the scheme (2.3) has only to be changed with respect to the formula for r_i which should be replaced by

$$r_i = K^{-1}(b - Ax_i) \tag{2.5}$$

For successful application of the Manteuffel algorithm it is necessary for $K^{-1}A$ to have all its eigenvalues in the right half plane. For a number of situations this property is guaranteed in [3]. In the coming section we will consider the use of incomplete LU-decompositions as preconditionings.

3. INCOMPLETE LU-DECOMPOSITIONS

The idea of incomplete LU-decompositions has been described by Meijerink and Van der Vorst [2]. They also proved that an incomplete LU-decomposition of the matrix A exists when A is an M-matrix. For a large number of practical problems the property of being an M-matrix is not valid or cannot easily be verified.

Various difficulties may arise in the construction of an incomplete decomposition and it is often not easy a task to tackle those problems without detoriating the efficiency of the ultimate preconditioned process to much.
In this section we present one particular possibility to overcome the stability problems that may occur.

3.1 Simple decompositions

We first consider an incomplete decomposition that has a similar sparsity structure as the matrix A has. This decomposition will be denoted as

$$L_1 D_1^{-1} U_1$$

The factors L_1, D_1 and U_1 are defined by the following relations.

(3.1)

a. $D_1^{-1} = \mathrm{diag}(L_1) = \mathrm{diag}(U_1)$

b. if the elements of A, L_1 and U_1 are denoted by a_{ij}, l_{ij}, u_{ij} resp, then

$l_{ij}=0$ for $i<j$, $l_{ij}=a_{ij}$ for $i>j$

$u_{ij}=a_{ij}$ for $i<j$, $u_{ij}=0$ for $i>j$

c. the elements of the diagonal matrix D_1 are defined by the relation $\mathrm{diag}(L_1 D_1^{-1} U_1) = \mathrm{diag}(A)$

For the pentadiagonal block M-matrices, arising from discretisation of elliptic pde's, this is the nonsymmetric variant of the ICCG(1, 1)-decomposition described by Meijerink and Van der Vorst[4], who also prove its existence for general structured M-matrices. We will now prove here the existence of the incomplete decomposition defined by (3.1) for an extended class of matrices.

Theorem 1. If A=M+N, M a symmetric M-matrix, $N=-N^T$, then the incomplete decomposition $L_1 D_1^{-1} U_1$ of A exists. If $\bar{L}_1 \bar{D}_1^{-1} \bar{U}_1$ is the incomplete decomposition of M, then the diagonal elements of D_1 are not smaller than the corresponding ones of $\bar{D}_1$.

Proof. The elements of D_1, $\bar{D}_1$, A, M and N are denoted by d_i, $\bar{d}_i$,

a_{ij}, m_{ij} and n_{ij} resp.. From the definition of the d_i's it follows that $d_1=\bar{d}_1$ and $d_2 \geq \bar{d}_2$. Now suppose that $d_i \geq \bar{d}_i$ for $i=1,2,\ldots,k-1$, then

$$\begin{aligned} d_k &= a_{kk} - \sum_{j=1}^{k-1} a_{jk} a_{kj} / d_j = \\ &= m_{kk} - \sum_{j=1}^{k-1} (m_{jk}^2 - n_{jk}^2)/d_j \\ &\geq m_{kk} - \sum_{j=1}^{k-1} m_{jk}^2 / d_j \geq m_{kk} - \sum_{j=1}^{k-1} m_{jk}^2/\bar{d}_j = \bar{d}_k \end{aligned}$$

Thus $d_k \geq \bar{d}_k > 0$, from which the theorem follows by an induction argument. //

This by itself is a useful result, but now it appears that depending on the size of the elements of N the factors L_1 and U_1 may be severely ill-conditioned (see Van der Vorst [3]). Several ways can be followed to overcome this problem. We will focus our attention to the construction of an incomplete decomposition of the matrix $A+(\sigma-1)\mathrm{diag}(A)$, $\sigma \leq 1$, instead of A. By choosing σ large enough the ill-conditioning will be prevented. This decomposition will be denoted by

(3.2) $\quad L_\sigma D_\sigma^{-1} U_\sigma$

<u>Theorem 2.</u> If $A=M+N$, M a symmetric M-matrix, $N=-N^T$, then the incomplete decomposition $L_\sigma D_\sigma^{-1} U_\sigma$ of $A+(\sigma-1)\mathrm{diag}(A)$ exists for $\sigma \geq 1$. If $\bar{L}_1 \bar{D}_1^{-1} \bar{U}_1$ is the incomplete decomposition of M, then the diagonal elements of D_σ are not smaller than the corresponding ones of $\bar{D}_1$.

<u>Proof</u>. We have $A+(\sigma-1)\mathrm{diag}(A) = M_\sigma + N$, where $M_\sigma = M+(\sigma-1)\mathrm{diag}(A) = M+(\sigma-1)\mathrm{diag}(M)$. From Varga[5],th.3.11, it follows that M_σ is an M-matrix. Applying theorem 1 gives the existence of $L_\sigma D_\sigma^{-1} U_\sigma$. Following the proof of theorem 1 and inserting σm_{kk} for m_{kk} completes the proof. //

An attractive parameterless variant has been described by Van der Vorst [3]. The reasons that we consider here the $L_\sigma D_\sigma^{-1} U_\sigma$-decomposition are the following:

a. if σ is well-tuned, a higher efficiency of the preconditioned Tchebycheff iteration can be achieved, specially for near-symmetric matrices.

b. the σ-strategy can also be employed for M-matrices. In that case σ has to be chosen less than 1.0 .

c. Manteuffel [6] shows that this variant can also be used effectively for the construction of an incomplete decomposition of a positive definite matrix.

d. Some interesting effects can be analysed easier for the σ-variant. This may lead eventually to better parameterless strategies for the construction of well-conditioned factors.

For convergence of the Tchebycheff iteration it is necessary that the preconditioned matrix has all its eigenvalues in the right half plane. This is valid for the $(L_1 D_1^{-1} U_1)^{-1}$-preconditioning when A is an M-matrix (see [3]). If A is not an M-matrix we have the following result.

Theorem 3. If $A=M+N$, where M is a symmetric M-matrix, $N=-N^T$ and if $L_\sigma D_\sigma^{-1} U_\sigma$ is the incomplete decomposition of $A+(\sigma-1)\text{diag}(A)$, $\sigma\geq 1$, then there is a $\sigma_0 \geq 1$ such that the eigenvalues of $(L_\sigma D_\sigma^{-1} U_\sigma)^{-1}A$ have all positive real part for $\sigma\geq\sigma_0$.

Proof. We have

$$A+(\sigma-1)\text{diag}(A)=L_\sigma D_\sigma^{-1} U_\sigma-R \tag{3.3}$$

The strictly lowertriangular part of A is denoted by L and the strictly uppertriangular part by U. This gives $L+U+\sigma\text{diag}(A)=(L+D_\sigma)D_\sigma^{-1}(D_\sigma+U)-R$ or $D_\sigma+LD_\sigma^{-1}U-R=\sigma\text{diag}(A)$. By definition it follows that R is equal to the off-diagonal part of $LD_\sigma^{-1}U$. Since L and U do not depend on σ and since by theorem 2 the elements of D_σ^{-1} are bounded by those of $\bar{D}_1$, we have that R is bounded for $\sigma\geq 1$. From (3.3) it follows that $L_\sigma D_\sigma^{-1} U_\sigma=A+(\sigma-1)\text{diag}(A)+R$ or

$$L_\sigma D_\sigma^{-1} U_\sigma = \sigma(\text{diag}(A) + \frac{1}{\sigma}(-\text{diag}(A) + A + R)) \tag{3.4}$$

Using that $-\text{diag}(A)+A+R$ is bounded we have

$$(3.5)\quad (L_\sigma D_\sigma^{-1} U_\sigma)^{-1} A = (\sigma(\text{diag}(A)+O(\tfrac{1}{\sigma})))^{-1} A = \tfrac{1}{\sigma}((\text{diag}(A))^{-1} A + O(\tfrac{1}{\sigma}))$$

By a theorem of Van der Vorst[3] we know that all eigenvalues of $(\text{diag}(A))^{-1}A$ have positive real part. Thus for large σ we have the result to be proven. //

From relation (3.5) we see that if σ grows to infinity the effectiveness of the preconditioning reduces to that of simple diagonal scaling of the matrix A by its diagonal elements.

3.2 More complicated decompositions

More complicated incomplete decompositions arise if we allow for more non-zero fill-in in the factors. In this case it is difficult to give general formulas and therefore we will present a special incomplete decomposition for a frequently occurring type of matrices. If the pde

$$(3.6)\qquad -(Au'_x)'_x - (Bu'_y)'_y + Gu'_x + Hu'_y + Cu = F$$

is discretised using central differences over a rectangular grid imposed on a rectangular region then the resulting discretisation matrix has the nonzero structure as given in figure 2.

A =
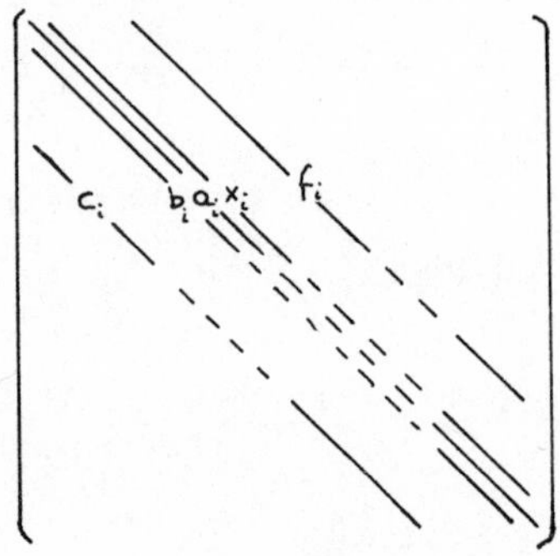

FIGURE 2.

L_{13} =

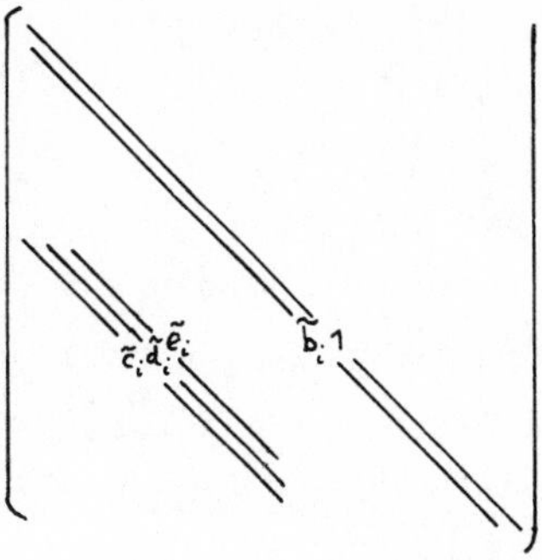

U_{13} =

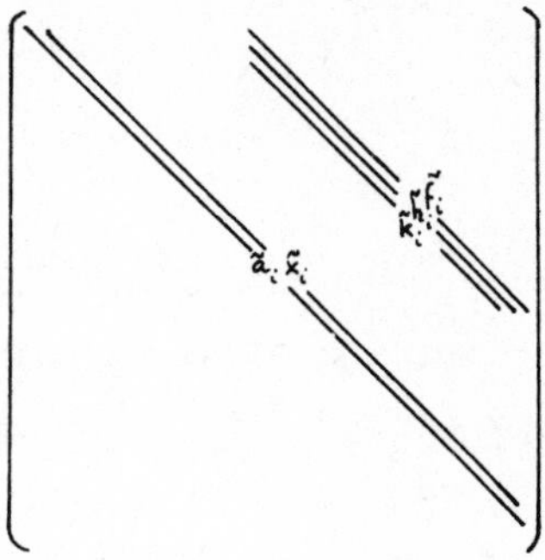

FIGURE 3.

In ref.[4] it is concluded that the incomplete Choleski-decomposition for symmetric M-matrices with 2 extra non-zero diagonals is in general a good (optimal?) choice. In analogy we present here the $L_{13}U_{13}$-decomposition, where the components are given in figure 3. The elements of L_{13} and U_{13} can be computed rowwise from:

$$\begin{aligned}
\tilde{c}_i &= c_i/\tilde{a}_{i-m} \\
\tilde{d}_i &= -\tilde{x}_{i-m}\tilde{c}_i/\tilde{a}_{i-m+1} \\
\tilde{e}_i &= -\tilde{x}_{i-m+1}\tilde{d}_i/\tilde{a}_{i-m+2} \\
\tilde{b}_i &= (b_i - \tilde{h}_{i-m}\tilde{c}_i - \tilde{k}_{i-m+1}\tilde{d}_i)/\tilde{a}_{i-1} \\
\tilde{a}_i &= a_i - \tilde{f}_{i-m}\tilde{c}_i - \tilde{h}_{i-m+1}\tilde{d}_i - \tilde{k}_{i-m+2}\tilde{e}_i - \tilde{x}_{i-1}\tilde{b}_i \\
\tilde{x}_i &= x_i - \tilde{f}_{i-m+1}\tilde{d}_i - \tilde{h}_{i-m+2}\tilde{e}_i \\
\tilde{k}_i &= -\tilde{h}_{i-1}\tilde{b}_i \\
\tilde{h}_i &= -\tilde{f}_{i-1}\tilde{b}_i \\
\tilde{f}_i &= f_i
\end{aligned} \tag{3.7}$$

Again this decomposition may suffer from ill-conditioning when the matrix is very non-symmetric, i.e. $(A-A^T)/2$ is large as compared to $(A+A^T)/2$. In a similar way as described in the previous section this problem can be overcome by computing the factors of the matrix $A+(\sigma-1)\mathrm{diag}(A)$. That means that in (3.7) the formula for a_i has to be replaced by

(3.8) $$\tilde{a}_i = \sigma a_i - \tilde{f}_{i-m}\tilde{c}_i - \tilde{h}_{i-m+1}\tilde{d}_i - \tilde{k}_{i-m+2}\tilde{e}_i - \tilde{x}_{i-1}\tilde{b}_i$$

where $\sigma \geq 1$ is again a suitable chosen problem dependent constant. In our numerical experiments it appeared that the best performance was achieved for smaller values of σ then in the $L_\sigma D_\sigma^{-1} U_\sigma$-case.

3.3 Some comments on the implementation

Before considering the numerical experiments in the coming section some aspects can be discussed in advance. Eisenstat [7] has shown that the ICCG(1,1)-decomposition, which is the symmetric equivalent of the $L_\sigma D_\sigma^{-1} U_\sigma$-decomposition, can be implemented very efficiently as a preconditioning for the conjugate gradient algorithm. It is shown that the computational cost of a preconditioned iteration-step can be reduced almost to the cost of an iterationstep of the unpreconditioned process. Van der Vorst [3] shows that Eisenstat's implementation ideas can also be applied to the preconditioned Tchebycheff iteration, when $L_\sigma D_\sigma^{-1} U_\sigma$ is used as a preconditioning. A similar reduction in the amount of work cannot be achieved for the $L_{13}U_{13}$-decomposition.

Since the $L_{13}U_{13}$-decomposition requires also quite more computer storage and is far less easier to program in general, the choice for the $L_\sigma D_\sigma^{-1} U_\sigma$-decomposition or alike will be obvious in most cases. After the presentation of the numerical experiments we will again compare both decompositions in light of the previously made remarks.

4. NUMERICAL EXPERIMENTS

The numerical experiments described in this section have all been carried out on a CDC-Cyber 175/100 computer of the Academisch Computer Centrum Utrecht in 48 bits relative working precision. The residuals in all experiments, as far as listed, have been computed as $||Ax_i-b||_2$, where x_i is the i-th iterand in the iterative solution process for Ax=b.

No efforts have been made to start the Tchebycheff iteration with good parameters d and c. This means that in most cases the first 20 iterationsteps were only used to get estimates for d and c, while sometimes the iterationprocess even diverged during those 20 steps. Therefore the numbers of iterations give mostly an unfavourable impression of the actual convergence behaviour. We have chosen for this approach because we believe that it is in most practical situations the only choice. However if one has to solve a set of similar problems, a proper determination of d and c beforehand can improve the efficiency significantly.

4.1 Model equation

All the numerical experiments have been done with linear systems arising from standard 5-point finite difference discretisation of partial differential equations of the form

$$(4.1) \qquad -(Du'_x)'_x - (Eu'_y)'_y + Gu'_x + Hu'_y + Cu = F$$

defined over a rectangular region, with the restriction that $D(x,y)>0$, $E(x,y)>0$ and $C(x,y)\geq 0$. Along the boundaries we have Dirichlet boundary conditions and equidistant rectangular grids are used for the discretisation. The terms u'_x and u'_y are discretised by central differences too. For more details on this point see Van der Vorst[3]. The matrix A of the resulting linear system Ax=b then has the nonzero structure as in figure 2. Also in this case the matrix A can be splitted as A=M+N, where M is a symmetric M-matrix coming from the second order derivative and linear terms and N comes from the first order terms. If G=G(y) and H=H(x) we have $N=-N^T$.

4.2 $-u''_{xx}-u''_{yy}+\beta(u'_x+u'_y)+u=f$

A number of effects can be demonstrated for this relatively simple equation. We consider the matrix A that results when discretisation takes place over an equidistant grid with meshwidth 1.0 in both directions. Typical values for the elements of A (see figure 2) then are

$$\begin{aligned} &c_i = -1-\beta/2 \qquad && b_i = -1-\beta/2 \\ (4.2)\qquad &a_i = 5 && \\ &x_i = -1+\beta/2 && f_i = -1+\beta/2 \end{aligned}$$

In figure 4 we give the conditionnumber $C = \sqrt{\lambda_{max}(A^TA)/\lambda_{max}(AA^T)}$ as a function of β.

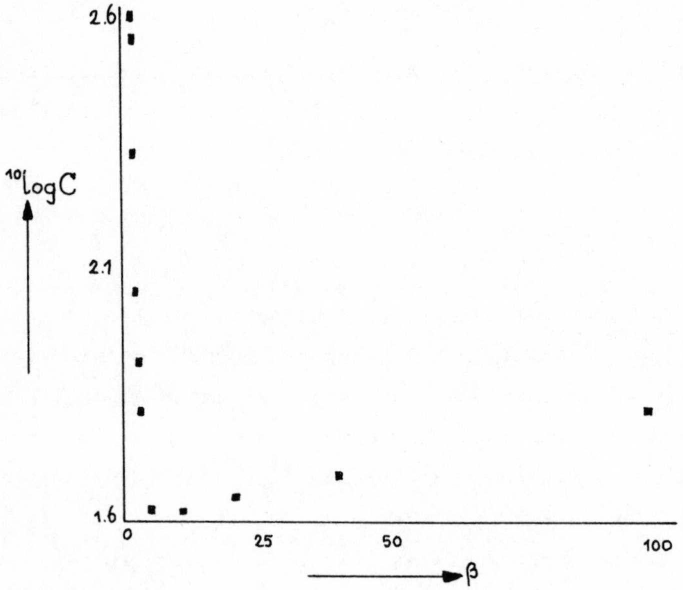

FIGURE 4. Conditionnumber as function of β

We see that even for large β, the conditionnumber C is quite reasonable. When an $L_\sigma D_\sigma^{-1} U_\sigma$-decomposition is constructed then the factors L_σ and U_σ are well-conditioned as soon as the diagonal elements Δ_i of D_σ are comparable to the sum of the off-diagonal elements in L_σ and U_σ. This implies that for the modelequation:

(4.3) $$\Delta_i \simeq 2 + \beta$$

For the pentadiagonal matrix A the Δ_i can be computed from

(4.4) $$\Delta_i = \sigma a_i - b_i x_{i-1}/\Delta_{i-1} - c_i f_{i-m}/\Delta_{i-m}$$

where m is the number of unknowns in the x-direction (the bandwidth of A). The values of Δ_i converge quickly to the largest root of the equation

(4.5) $$x = 5\,\sigma - 2(1 - \beta^2/4)/x$$

that results from (4.4) when the Δ_i are replaced by x and the values of the elements of A are inserted. The largest root has to be equal to $2+\beta$, according to (4.3), therefore σ can be computed from

(4.6) $$2+\beta = 5\sigma - 2(1- \beta^2/4)/(2 + \beta)$$

which gives

(4.7) $$\sigma = (3 + \beta/2)/5$$

Up until now we have considered the question how to choose σ in order to have well-conditioned factors. A more interesting point is how the convergence of the preconditioned Tchebycheff iteration depends on the choice of σ. In order to get some insight at this point we have for a small-sized problem, N=64 (=8^2), computed all the eigenvalues of $(L_\sigma D_\sigma^{-1} U_\sigma)^{-1}A$. Then the ellipse containing all these eigenvalues and for which the rate of convergence r_c is minimal is constructed. We will now present the results for different values of β.

a. β=100

The decomposition is well-conditioned, according to (4.7) when σ=10.6 . In figure 5 the value of r_c has been plotted as a function of σ from which we see that r_c is minimal for a value of σ slight-

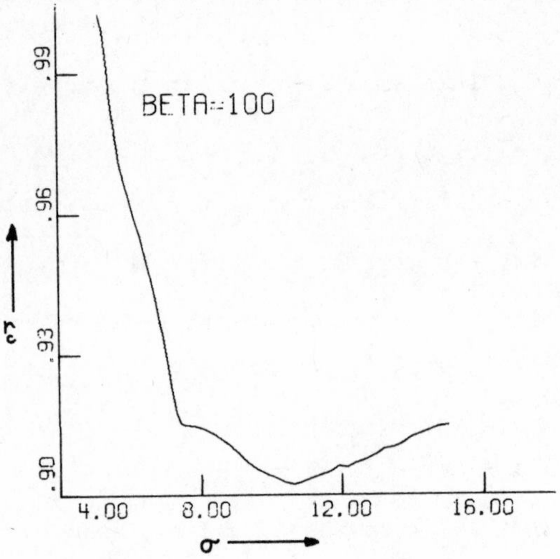

FIGURE 5. r_c as function of σ, for β=100

ly larger than 10.5 .

b. β=20.

The factors L_σ and U_σ are well-conditioned when σ=2.6 and from figure 6 it can be seen that r_c is minimal for σ≃2.5 .

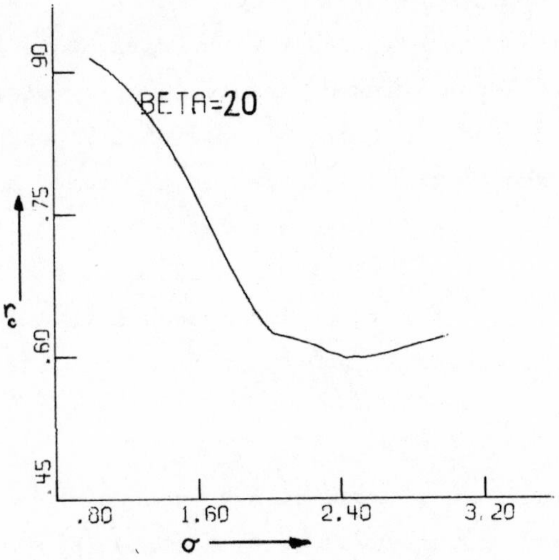

FIGURE 6. r_c as function of σ, for β=20

These two examples are typical for a large number of examples, from which we conclude that the value of σ for which well-conditioning occurs is pretty close to the value of σ for which the rate of convergence r_c is minimal when the matrix A is very non-symmetric. In the following example we have chosen a value for β for which the resulting matrix A is a diagonally dominant (near-symmetric) M-matrix. In this case the factors of the standard incomplete $L_1D_1^{-1}U_1$-decomposition are well-conditioned.

c. β=0.1

From figure 7 we see that the rate of convergence r_c is minimal for $\sigma \simeq 0.9$, whereas application of (4.7) learns that the factors L_σ and U_σ are well-conditioned for $\sigma \geq 0.61$.

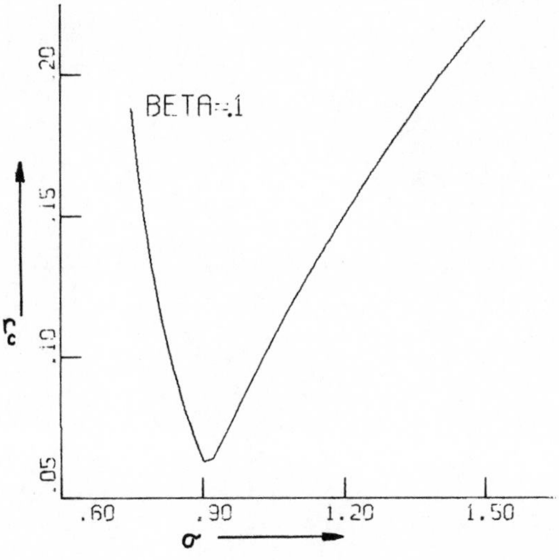

FIGURE 7. r_c as function of σ, for β=0.1

The use of a value σ<1.0 for M-matrices is not a new result. Manteuffel [6] shows that for the preconditioned conjugate gradient algorithm the highest efficiency is achieved for a value of σ less than 1.0. The result is also related to the idea behind the MICCG-process for symmetric matrices, described by Gustaffson [9], in which actually the diagonal elements of the matrix are decreased in value although not all by the same factor. We are able to prove that all eigenvalues of $(L_\sigma D_\sigma^{-1} U_\sigma)^{-1}A$ tend to lie on a circle for $\sigma\to\infty$. In our experiments it appeared that the ellipse that contains all the eigenvalues, for which r_c is minimal, is very close to a circle as soon as σ approaches the value for which r_c is minimal. As a typical example we show the ellipses for the case $\beta=20$ for the 64-th order matrix described previously, for $\sigma=1.0$ in figure 8 and for $\sigma=2.5$ in figure 9.

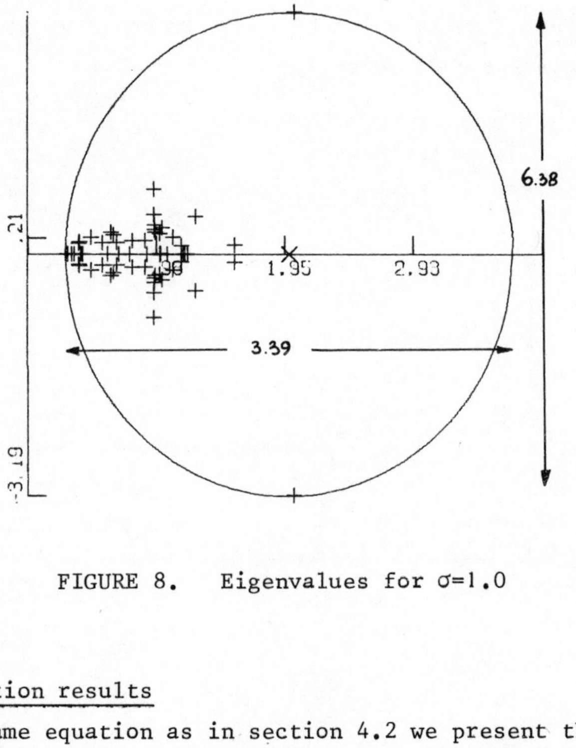

FIGURE 8. Eigenvalues for $\sigma=1.0$

4.3 Iteration results

For the same equation as in section 4.2 we present the iteration results for the Tchebycheff iteration without preconditioning and with the preconditionings $L_\sigma D_\sigma^{-1} U_\sigma$ and $L_{13}U_{13}$.

For the unpreconditioned process the iteration parameters have been chosen initially as d=5, c=0, while for the preconditioned iteration they were initialised to d=1.0, c=0. The value of σ for the $L_\sigma D_\sigma^{-1} U_\sigma$-preconditioning has been determined by (4.7) and the value

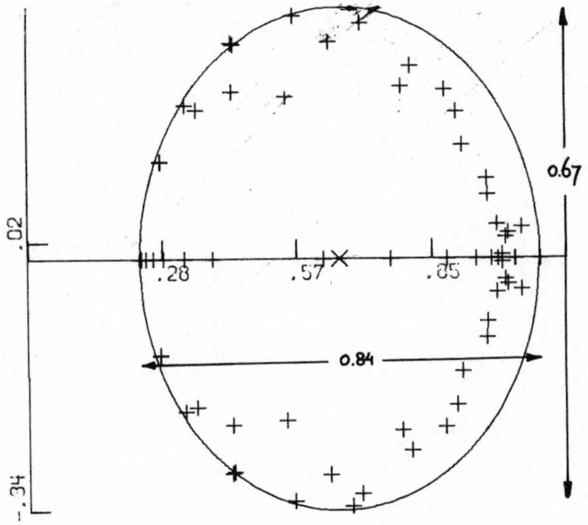

FIGURE 9. Eigenvalues for $\sigma=2.5$

σ for the $L_{13}U_{13}$-decomposition has been determined experimentally as to give the fastest convergence. The results are given in Table I for $\beta=4.0$ and in Table II for $\beta=20.0$.

Method	Number of iterations	Final residual / Initial residual
Manteuffel algorithm without preconditioning	120	$1.9_{10}{-8}$
$L_\sigma D_\sigma^{-1} U_\sigma$-preconditioning with $\sigma=1.0$	25	$9.1_{10}{-9}$
$L_{13}U_{13}$-preconditioning with $\sigma=1.0$	9	$1.3_{10}{-8}$

TABLE I. Iteration results for $\beta=4.0$

Method	Number of iterations	$\frac{\text{Final residual}}{\text{Initial residual}}$
Manteuffel algorithm without preconditioning	200	$6.5_{10}{-7}$
$L_\sigma D_\sigma^{-1} U_\sigma$-preconditioning with $\sigma=2.6$	40	$6.4_{10}{-9}$
$L_{13}U_{13}$-preconditioning with $\sigma=1.75$	25	$4.3_{10}{-9}$

TABLE II. Iteration results for $\beta=20.0$

4.4 $-u''_{xx}-u''_{yy}+((au)'_x+au'_x)/2=f$

This equation has been considered by Widlund [8] who gives results for the generalized Lanczos method and numerical results of the Tchebycheff iteration for the discretised system have been presented by Van der Vorst[3]. Since first derivative terms in only one direction (the x-direction) are present, it can be shown that the incomplete decomposition has well-conditioned factors (see [3]), which implies that $\sigma=1.0$ can safely be used in the $L_\sigma D_\sigma^{-1} U_\sigma$- decomposition. No efforts have been made in this case to find a more optimal value.

The equation is discretised over a rectangular equidistant grid with equal gridspacing in both x- and y-direction over the region $[0,1] * [0,1]$. The Dirichlet boundary conditions as well as the right hand side are chosen in such a way that

$$u = \sin \pi x \sin \pi y \; e^{(x/2+y)^3}$$

is a solution to the partial differential equation. The iteration parameters for the Tchebycheff iteration are initialized to d=1.0, c=0.0 . The results are given in Table III.

If we take into account the efficiency that can be achieved for the

a	Method	No. of unknowns	No. of iter.	Final residual / Initial residual
$2e^{3.5(x^2+y^2)}$	no prec.	225	200	$4.9_{10}{-2}$
,,	$L_1D_1^{-1}U_1$	225	29	$2.1_{10}{-8}$
,,	$L_{13}U_{13}$	225	16	$2.1_{10}{-8}$
,,	no prec.	961	200	$5.2_{10}{-2}$
,,	$L_1D_1^{-1}U_1$	961	47	$3.1_{10}{-8}$
,,	$L_{13}U_{13}$	961	29	$2.4_{10}{-8}$
$20e^{3.5(x^2+y)^2}$	no prec.	225	200	$2.9_{10}{-1}$
,,	$L_1D_1^{-1}U_1$	225	14	$2.0_{10}{-9}$
,,	$L_{13}U_{13}$	225	10	$1.9_{10}{-9}$
,,	no prec.	961	200	$6.5_{10}{-1}$
,,	$L_1D_1^{-1}U_1$	961	26	$2.2_{10}{-9}$
,,	$L_{13}U_{13}$	961	14	$1.5_{10}{-9}$

TABLE III. Iterationresults for the equation from section 4.4

$L_\sigma D_\sigma^{-1} U_\sigma$-preconditioning (see 3.3), this in contrast with the situation for the $L_{13}U_{13}$-preconditioning, then it appears that also in this case the $L_{13}U_{13}$ gives only a minor saving in computertime over the $L_\sigma D_\sigma^{-1} U_\sigma$-preconditioning.

5. CONCLUSIONS

In ref.[3] it is shown that the incomplete Crout decompositions are worthwhile to be considered as a preconditioning for the Tchebycheff iteration in order to solve Ax=b, where A is a nonsymmetric matrix. In the same paper it is demonstrated that in general for very nonsymmetric matrices one can avoid to determine an optimal

value for the parameter σ for stabilization of the decomposition. Instead of this one can use the $L_{EQ}D_{EQ}^{-1}U_{EQ}$-decomposition. The material presented in this chapter admits two major conclusions. The first one is that it doesn't seem to pay off, for the linear systems considered here, to construct more complicated decompositions that is decompositions which allow for more non-zero fill-in.
The second conclusion is that the value of σ for which the $L_\sigma D_\sigma^{-1} U_\sigma$ preconditioning is well-conditioned is close to the value for which the best convergence behaviour is achieved. Specially for matrices that are near-symmetric it therefore seems to be worthwhile to use the parameter version and to determine a value of σ for wich good efficiency is obtained. In practice such a value might be less than 1.0.
It should be mentioned here that in most situations we are not able to prove that all the eigenvalues of $(L_\sigma D_\sigma^{-1} U_\sigma)^{-1}A$ have positive real part, which is essential for the convergence of the Tchebycheff iteration.

6. REFERENCES

1. MANTEUFFEL, T.A. - The Tchebycheff Iteration for Non-symmetric linear systems. Num. Math., Vol. 28, pp. 307-327, 1977.
2. MEIJERINK, J.A. & VAN DER VORST, H.A. - An Iterative Solution Method for Linear Systems of which the Coefficient Matrix is a Symmetric M-Matrix. Math. of Comp., Vol. 31, No. 137, pp. 148-162, 1977.
3. VAN DER VORST, H.A. - Iterative Solution Methods for Certain Sparse Linear Systems with a Non-symmetric Matrix arising from PDE-Problems. accepted for pub. in Journal of Comp. Phys.
4. MEIJERINK, J.A. & VAN DER VORST, H.A. - Guidelines for the Usage of Incomplete Decompositions in Solving Sets of Linear Equations as Occur in Practical Problems. accepted for pub. in Journal of Comp. Phys.
5. VARGA, R. S. - Matrix Iterative Analysis, Prentice-Hall, Englewood Cliffs N.J., 1962.
6. MANTEUFFEL, T.A. - The Shifted Incomplete Cholesky Factorization. Sandia Laboratories Report SAND 78-8226, 1978
7. EISENSTAT, S.C. - Efficient implementation of a Class of Pre-

conditioned Conjugate Gradient Methods. Research Report No. 185, Yale University, 1980.

8. WIDLUND, O. - A Lanczos Method for a Class of Non-symmetric Systems of Linear Equations. Siam J. Num. Anal., Vol. 15, No. 4, pp. 801-812, 1978.
9. GUSTAFFSON, I. - A Class of First Order Factorization Methods. BIT, Vol. 18, pp. 142-156, 1977.

MODIFIED INCOMPLETE CHOLESKY (MIC) METHODS

IVAR GUSTAFSSON
Mathematical Institute, Catholic University,
Nijmegen, The Netherlands

A class of modified incomplete factorization methods for the solution of large, sparse systems of equations, arising from discretized models for boundary value problems of partial differential equations, is studied. Asymptotic results for the computational complexity and for the storage requirement are developed, results from numerical experiments are presented and comparisons with other iterative and with direct methods are carried out.

1. PRELIMINARIES

We shall study the numerical solution of large, linear systems of equations arising from finite element (f.e.) or finite difference (f.d.) approximations of linear elliptic partial differential equations on the form

$$-\nabla(a\nabla u) + bu = f \text{ in } \Omega \tag{1.1}$$

with suitable boundary conditions on $\partial\Omega$. We assume that $a \geq a_0 > 0$, $b \geq 0$ $\forall x \in \overline{\Omega}$. As we shall see the methods to be presented are also appropriate (with minor modifications) for other types of problems, for instance special kinds of non-self-adjoint problems and fourth order problems.

We solve the sparse system of N linear equations

$$Au = f$$

by a preconditioned iterative process, which basic form is

$$C(u^{\ell+1}-u^{\ell}) = -\beta_{\ell}(Au^{\ell}-f), \quad \ell = 0,1,\ldots, \tag{1.2}$$

where C is the preconditioning matrix and β_{ℓ} an iteration parameter.

It is well known that if A and C are symmetric and positive definite then the Chebyshev method or a conjugate gradient (CG) type at method can be used to accelerate (1.2). As we shall see later this is also possible for some special kinds of non-symmetric problems. In this contribution the CG method will be used. In the symmetric case it converges to a relative residual error ε in at most

$$\mathrm{int}[\tfrac{1}{2}\sqrt{\kappa_1}\ \ln {}^{2}/_{\varepsilon} + 1] \tag{1.3}$$

iterations, where κ_1 is the effective spectral condition number of $\tilde{A} = C^{-\frac{1}{2}}AC^{-\frac{1}{2}}$, i.e. κ_1 is the quotient between the largest and smallest positive eigenvalue of $\tilde{A}$ [3].

Let $R = C-A$. Then $A = C-R$ is called a splitting of A. The following particular kind of splitting is discussed in e.g. [35] and [12].

Definition 1.1. The splitting $A = C-R$ is a *regular splitting* if C^{-1} and R have non-negative entries i.e. if $C^{-1} \geq 0$, $R \geq 0$.

It is shown in [36] that for a regular splitting the spectral radius ρ of $C^{-1}R$ satisfies $\rho < 1$ if and only if $A^{-1} \geq 0$. Hence, for a regular splitting of a monotone matrix, (1.2) converges with $\beta_\ell = 1$, $\ell = 0,\ldots$ In the CG method the rate of convergence is determined by κ_1 instead of by ρ and hence the splitting does not have to be regular.

In order to get an efficient method, the preconditioning matric C has to have the following properties.

(a) It should be easily calculated.

(b) It should not need too much storage.

(c) Systems with the matrix C should be easily solved. Typically, C = LU with sparse lower and upper triangular factors L and U.

(d) The effective spectral condition number κ_1 of $C^{-\frac{1}{2}}AC^{-\frac{1}{2}}$ should be much smaller than $\kappa(A)$, the effective spectral condition number of A.

The choice of C presented here is based on modified incomplete (Cholesky) (MIC) factorization (defined in the following section) of A and fulfilles the desired properties stated above. For instance, the sparsity complexity of C will be $O(N), N \to \infty$. This means that the construction of C as well as each iteration can be carried out in $O(N)$ operations (multiply-adds). For wide classes of f.e. matrices we show that κ_1 is of order $O(\sqrt{\kappa(A)})$, $N \to \infty$ and hence the required number of iterations will be considerably smaller than for the unpreconditioned method corresponding to $C = I$, the unity matrix.

In the following, we shall consider some model problems and model matrices due to the following definitions.

Definition 1.2. The problem (1.1) with $\Omega = \{(x,y);\ 0 < x < 1,\ 0 < y < 1\} \subset R^2$ is denoted the *model problem*. If in addition $a \equiv 1$, $b \equiv 0$ then we have the *restricted model problem*. By chosing Dirichlet and Neumann boundary conditions we obtain the *(restricted) model Dirichlet problem* and the *(restricted) model Neumann problem*, respectively.

Definition 1.3. The matrix corresponding to the (restricted) model Dirichlet (Neumann) problem, discretized by standard (5-point) f.d. approximations will be denoted the *(restricted) model Dirichlet(Neumann)matrix*. The same (or a similar) matrix arises in f.e. approximations based on a uniform rightangled triangulation and linear Lagrangian basis functions.

2. INCOMPLETE FACTORIZATIONS

It is well known that a complete Gaussian or Cholesky factorization of a sparse matrix produces fill-in within the band in the upper and lower triangular factors. This leads to a relatively high computational complexity for the factorization and to considerably large storage requirement. By using an incomplete factorization we keep the sparsity and hence reduce the computational cost and amount of storage.

Let $A = (a_{ij})$ be the $N \times N$ matrix to be factored and let $P^* = \{(i,j);\ a_{ij} \neq 0\}$. Further, let P be positions, where we allow non-zero entries in the triangular factors $L = (\ell_{ij})$ and $U = (u_{ij})$ of C, i.e. $(i,j) \notin P \Rightarrow \ell_{ij} = u_{ij} = 0$.

For well structured matrices such as f.e. matrices, the index set P can be chosen in advance. More generally, P can be determined during the approximate factorization, depending for instance on the size of the produced entries, see e.g. [30], [34].

We always assume that $(i,i) \in P$, $i = 1,\ldots,N$ and in particular we shall study factorizations for which $P^* \subseteq P$. Such factorizations are said to be at least as dense as A.

In the incomplete (IC) factorization methods we just drop entries appearing outside P during the elimination while in the modified (MIC) factorization methods these entries are added to the diagonal.

We have the following particular classes of approximate factorizations.

Definition 2.1. If $P = P^*$, (that is if we allow no fill-in,) then we use the notations *IC(0)* and *MIC(0)*. Further, if we in the (M)IC(0) algorithm disregard (or move to the diagonal) corrections to off-diagonal entries, then we obtain the *(M)IC(0)* factorization*.

We note that the MIC(0)* method in its symmetric version is of type $C = (\tilde{D}+L_A)\tilde{D}^{-1}(\tilde{D}+L_A^T)$, where L_A is the strictly lower triangular part of A and $\tilde{D}$ is a positive (in case the factorization is stable, see Section 3) diagonal matrix.
This method can be regarded as a generalized SSOR method [37], cf. also [2].

Example 2.1. We describe the $IC(0) \equiv IC(0)^*$ and $MIC(0) \equiv MIC(0)^*$ factorizations for the following simple example.

$$\begin{bmatrix} 4 & -1 & 0 & -2 \\ -1 & 4 & -1 & 0 \\ 0 & -1 & 4 & -1 \\ -2 & 0 & -1 & 4 \end{bmatrix} \xrightarrow{\text{elimination}} \begin{bmatrix} 4 & -1 & 0 & -2 \\ 0 & 3\frac{3}{4} & -1 & -\frac{1}{2} \\ 0 & -1 & 4 & -1 \\ 0 & -\frac{1}{2} & -1 & 3 \end{bmatrix}$$

$$\xrightarrow{\text{IC(0)}} \begin{bmatrix} 4 & -1 & 0 & -2 \\ 0 & 3\frac{3}{4} & -1 & 0 \\ 0 & -1 & 4 & -1 \\ 0 & 0 & -1 & 3 \end{bmatrix} \qquad \xrightarrow[D=0]{\text{MIC(0)}} \begin{bmatrix} 4 & -1 & 0 & -2 \\ 0 & 3\frac{1}{4} & -1 & 0 \\ 0 & -1 & 4 & -1 \\ 0 & 0 & -1 & 2\frac{1}{2} \end{bmatrix}$$

In general a MIC factorization can be stated $C = LU = A+D+R$ where the positive diagonal matrix D includes one or two preconditioning parameters, which will be determined in Section 4. Obviously, the defect matrix $R = (r_{ij})$ satisfies the rowsum criterion $\sum_j r_{ij} = 0$, $i = 1,2,\ldots,N$ and we simply write rowsum$(R) = 0$.

Definition 2.2. Let P_1 and P_2 be two index sets such that $P_1 \subset P_2$. Then the factorization corresponding to P_2 is said to be *more accurate* than that corresponding to P_1 (if they are not identical). In particular we also say that MIC(0) is more accurate than MIC(0)*.

Using the following strategy we can successively derive more and more accurate factorizations:

1. At first let $P = P^*$. This gives L and U with the same sparsity pattern as A and represents the MIC(0) method.
2. Form $C = LU$ and $R = C-A$.
3. Re-define L and U in such a way that these matrices are allowed to contain nonzero entries in positions where R has nonzero entries as well. This defines a new index set P.
4. If you are not satisfied with the given degree of accuracy, repeat from stage 2.

For well structured matrices, this strategy can be carried through formally, that is, the *structure* of L,U and R can be derived without computational effort. In special simple cases we derive MIC(d) algorithms, where $d > 0$ indicates that L contains d more nonzero subdiagonals than the lower part of A. We then also get particularly simple recursion formulas for the calculation of the entries of L and U, see the following example. For more general matrices some computational work may be needed to obtain the fill-in actually produced in each cycle of the above strategy. Obviously, since we are only interested in the sparsity pattern P (in each cycle) we don't perform the factorization before we have derived the final structure. Practical experiments indicate that we get the most efficient method (from the point at view of computational cost) after one or two cycles.

Example 2.1. Let the model matrix A and the matrices $C = LL^T$ and R in the MIC(0) method be defined by the following graphs, where as usual graph nodes coincide with f.d. nodes and m is the half band-width of A.

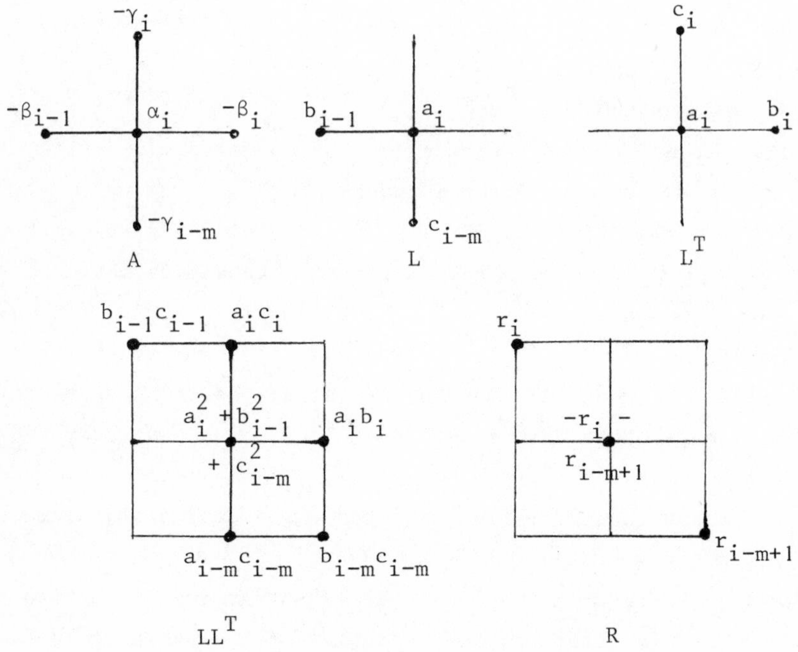

The relation $C = LL^T = A+D+R$ gives the following recursion formulas for the entries of L:

$$a_i^2 = \alpha_i + d_i - b_{i-1}^2 - c_{i-m}^2 - r_i - r_{i-m+1} =$$
$$= \alpha_i + d_i - b_{i-1}(b_{i-1}+c_{i-1}) - c_{i-m}(b_{i-m}+c_{i-m})$$
$$b_i = -\beta_i / a_i$$
$$c_i = -\gamma_i / a_i$$
$$r_i = b_{i-1}c_{i-1}.$$

The strategy to obtain a more accurate factorization now leads to the MIC(1) method defined by

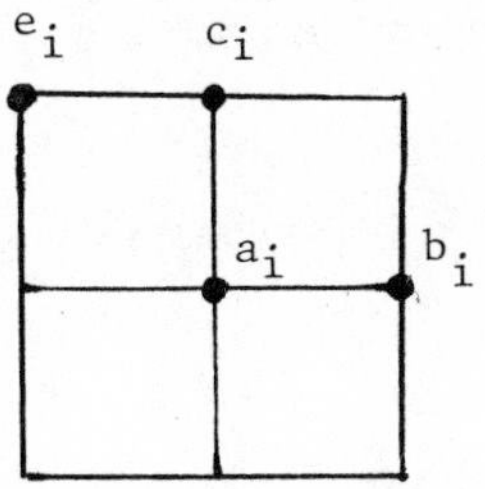

L^T

and the formulas:

$$a_i^2 = \alpha_i + d_i - b_{i-1}(b_{i-1}+e_{i-1}) - c_{i-m}^2 - e_{i-m+1}(b_{i-m+1}+e_{i-m+1})$$
$$b_i = -(\beta_i + c_{i-m+1}e_{i-m+1})/a_i$$
$$c_i = -\gamma_i / a_i$$
$$e_i = -b_{i-1}c_{i-1}/a_i.$$

Continuing in this way we get a sequence of more and more accurate factorizations; MIC(0), MIC(1), MIC(2), MIC(4), MIC(7), MIC(12) etc.

We note that we may avoid the square-roots by making a MIC factorization of the type $C = (L+D_1)D_1^{-1}(L^T+D_1)$, D_1 a diagonal matrix and L strictly lower triangular.

We also remark that for the five-point f.d. matrix, the

$MIC(0)^* \equiv MIC(0)$ method is identical with the Dupont-Kendall-Rachford method [14], [15] as well as with (in case of a Dirichlet problem) the generalized SSOR method in [2], apart from a minor difference in the choice of D.

The IC(0) method for the model Dirichlet matrix was introduced in [35], cf. also [31] and [32]. In a more general sense the IC methods are presented in [29], where also the powerful combination with the CG method is used (ICCG-methods). For the generalized SSOR method this combination was introduced in [2].

3. STABILITY OF MODIFIED INCOMPLETE FACTORIZATIONS

In this section we shall consider modified incomplete LU factorizations with diag(L) = I, of not necessarily symmetric matrices. We need some definitions.

Definition 3.1. A matrix $A = (a_{ij})$ is an *L-matrix* if $a_{ij} \leq 0$, $i \neq j$. Further A is an *M-matrix* if it is an L-matrix and if $A^{-1} \geq 0$.

For symmetric matrices, an equivalent definition of the notion M-matrix is that A is a positive definite L-matrix, see e.g. [17].

Definition 3.2. A matrix $A = (a_{ij})$ is said to be *weakly (strictly) diagonally dominant* if $a_{ii} \geq \sum_{j \neq i} |a_{ij}|$ $(a_{ii} > \sum_{j \neq i} |a_{ij}|)$, $i = 1,\ldots,N$.

Definition 3.3. An incomplete LU factorization algorithm is said to be *stable* if the entries of diag(U) are positive. Otherwise, that is, if any entry of diag(U) is non-positive, the algorithm is said to be *unstable*. Further, a factorization algorithm is *at least as stable* as another algorithm if the entries of diag(U) in the former algorithm are not smaller than the corresponding entries in the latter algorithm.

It is a well known fact that a complete LU decomposition is stable (if we disregard rounding errors) for e.g. symmetric, positive definite matrices and for strictly diagonally dominant matrices.

It is shown in [29] that an arbitrary IC factorization of an M-matrix is a regular splitting (Definition 1.1.) and furthermore it is at least as stable as the complete factorization. For a general positive definite matrix, however, the IC algorithms may be unstable, see [27] and [24].

For the MIC factorzations we state the following results.

Theorem 3.1. Let $A = (a_{ij})$ be a weakly diagonally dominant matrix. Then any MIC factorization of A is stable.

Proof. In a MIC factorization, the matrix A+D, where D is a positive diagonal matrix, is factored approximately. Since A is weakly diagonally dominant, the matrix actually factored is strictly diagonally dominant. For simplicity we therefore (in this proof) assume that A itself is strictly diagonally dominant. We also assume that (in case $P \subset P^*$) the first moving step is already performed. Apparently, the resulting matrix which we still denote A is strictly diagonally dominant as well.

Let $\tilde{A} = (\tilde{a}_{ij})$, $i,j \geq 2$ be the matrix that results when the first column of A is eliminated using the first row and let $\hat{A} = (\hat{a}_{ij})$ be the corresponding matrix after one MIC step, compare with Example 2.1. It suffices to show that $\hat{A}$ is strictly diagonally dominant (in particular $\hat{a}_{22} > 0$), since the argument can then be repeated for each step of the elimination.

Consider row k, $2 \leq k \leq N$. We have for $j = 1,\ldots,N$

$$\tilde{a}_{kj} = a_{kj} - a'_{kj}, \tag{3.1}$$

where $a'_{kj} = a_{k1}a_{1j}/a_{11}$ and hence, from the diagonal dominance of the first row of A,

$$\sum_{j>1} |a'_{kj}| \leq |a_{k1}|. \tag{3.2}$$

Further from the diagonal dominance of row k of A and (3.2) we get

$$a_{kk} > \sum_{j\neq k} |a_{kj}| \geq \sum_{\substack{j>1\\ j\neq k}} |a_{kj}| + \sum_{j>1} |a'_{kj}|.$$

Thus, for the diagonal entries of $\tilde{A}$ we have from (3.1)

$$\tilde{a}_{kk} = a_{kk} - a'_{kk} > \sum_{\substack{j>1\\ j\neq k}} \{|a_{kj}| + |a'_{kj}|\}. \tag{3.3}$$

Now consider the modification step determined by the index set P. From (3.3) we obtain

$$\hat{a}_{kk} = \tilde{a}_{kk} + \sum_{(k,j)\notin P} \tilde{a}_{kj} \geq \tilde{a}_{kk} - \sum_{(k,j)\notin P} |\tilde{a}_{kj}| >$$

$$> \sum_{\substack{j>1\\ j\neq k}} \{|a_{kj}| + |a'_{kj}|\} - \sum_{(k,j)\notin P} |\tilde{a}_{kj}| \geq \sum_{j\neq k} |\tilde{a}_{kj}| - \sum_{(k,j)\notin P} |\tilde{a}_{kj}|$$

$$= \sum_{\substack{(k,j)\in P\\ j\neq k}} |\tilde{a}_{kj}| = \sum_{j\neq k} |\hat{a}_{kj}| \geq 0.$$

Hence, $\hat{A}$ is strictly diagonally dominant and the theorem is proved.

Corollary 3.1. For the MIC(0)* factorization of a weakly diagonally dominant matrix we have

$$u_{ii} > \sum_{j>i} |a_{ij}| \tag{3.4}$$

Proof. This follows immediately from the fact that in the MIC(0)* method off-diagonal entries remain unchanged.

Corollary 3.2. Let $C = A + D + R$ be a MIC factorization of a symmetric, weakly diagonally dominant matrix. Then C is positive definite if $D > 0$ and positive semidefinite if $D = 0$.

Proof. For $D > 0$ the result follows directly from Theorem 3.1. Furthermore, this theorem can be carried out with weak inequalities when $D = 0$. Hence $\mathrm{diag}(U) \geq 0$ and C is positive semidefinite.

Although Theorem 3.1 ensures $\mathrm{diag}(U) > 0$ for an arbitrary MIC factorization of a weakly diagonally dominant matrix one may fear numerical instability if the diagonal entries become too small. However, in the same way as for a complete factorization, the diagonally dominance of $\hat{A}$ (in each step) implies numerical stability.

We also remark that, since the MIC factorizations are stable for *weakly* diagonally dominant matrices, they are stable for f.d. and (certain) f.e. approximations of (1.1) with pure Neumann boundary conditions and $b \equiv 0$, see examples in [24].

For f.e. approximations with linear basis functions over a quadrilateral or triangular mesh with all angles $\leq \pi/2$ A is a (weakly) diagonally dominant L-matrix. For higher order f.e. equations the MIC as well as the IC factorizations may be unstable.

In [27] Kershaw has successfully used a simple modification of the IC algorithms to avoid instability, see also [28]. Whenever an entry in the diagonal of U becomes negative, it is replaced by a positive number and the process is continued. This idea can obviously be used even in the MIC algorithms. Then, however, the property rowsum(R) = 0 is lost for rows corresponding to modified diagonal entries. If this modification is not needed too often this will likely have only a slight influence on the rate of convergence of the corresponding MICCG method. In Section 5 we shall describe other methods to overcome instability and to keep the fast rate of convergence for more general systems of f.e. equations.

4. RATE OF CONVERGENCE OF MICCG METHODS

In this section we shall consider symmetric MIC factorizations

$$C = LL^T = A+D+R \tag{4.1}$$

of a symmetric, positive semidefinite matrix A. If the factorization is stable then C is symmetric and positive definite and the effective spectral condition number κ_1 of $C^{-\frac{1}{2}}AC^{-\frac{1}{2}}$ determines the rate of convergence of the MICCG method, see (1.3).

In our applications $\kappa(A) = O(h^{-2})$, $h \to 0$. We will show that for wide classes of weakly diagonally dominant f.d. or f.e. matrices, the result $\kappa_1 = O(h^{-1})$, $h \to 0$ is valid. For the IC methods we have $\kappa_1 = O(h^{-2})$, $h \to 0$ i.e. actually the same order as for $\kappa(A)$ [24]. Here h is a normalized step-size parameter defined as follows. Assume that we have a coarse mesh, representing $h = 1$, to start with. We then subdivide the domain in a uniform way by intersecting each side of the elements into (say) k equal parts. The finer mesh corresponding to $h = 1/k$ is now derived by connecting the intersection points cf. also [8]. We also note that since a more general, smoothly varying mesh, is equivalent to smoothly varying coefficients in (1.1) this case is included by the results to be stated in this section.

Lemma 4.1. Let A be a weakly diagonally dominant L-matrix. Then any MIC factorization of A yields a defect matrix $R = C-A-D$ which is negative semidefinite.

Proof. This is immediately seen from the fact that the entries r_{ij} of R satisfy $r_{ij} \geq 0$, $j \neq i$ and $\Sigma_j r_{ij} = 0$, $j = 1,\ldots,N$.

In the following k, m_j, $j = 1,2,\ldots$ stand for positive constants independent of h, not necessarily the same at different instances.

The definition of the positive diagonal matrix D will (for instance) depend on the boundary conditions and on a possible discontinuity of the material coefficient a in (1.1), see below. In all applications it will be chosen in such a way that the relation

$$0 \leq (Dx,x)/(Ax,x) \leq m_1, \quad x \notin N(A) \tag{4.2}$$

holds, where $N(A)$ is the nullspace of A and $(x,y) = x^T y$. In our applications we have $N(A) = 0$ (Dirichlet problems) or $N(A)$ is spanned by the constant vector (pure Neumann problems). In both cases $Rx = 0$, $x \in N(A)$ and since we are considering the *effective* condition number we may restrict the analysis to $x \notin N(A)$. In fact, since $D > 0$, any zero-eigenvalue at A corresponds to a zero-eigenvalue of $C^{-\frac{1}{2}}AC^{-\frac{1}{2}}$. For an application where a pure Neumann problem is solved see [23].

In the following lemma we state a sufficient condition for asymptotically fast convergence.

Lemma 4.2. Let A be a weakly diagonally dominant L-matrix. Then, for an arbitrary MIC factorization with D satisfying (4.2), the result $\kappa_1 = O(h^{-1})$, $h \to 0$ holds *if*

$$-(Rx,x) \le (1+kh)^{-1}(Ax,x). \tag{4.3}$$

Proof. From (4.1), (4.3) and Lemma 4.1 follows,
$(1-\frac{1}{1+kh})(Ax,x) \le (Cx,x) \le (Ax,x)+(Dx,x) \le (1+m_1)(Ax,x)$.
Hence, $(1+m_1)^{-1} \le (Ax,x)/(Cx,x) \le 1+1/kh$ and
$\kappa_1 \le (1+m_1)(1+1/kh) = O(h^{-1})$, $h \to 0$.

For a certain class of matrices, we shall show that (4.3) is valid already for the MIC(0)* method, the least accurate factorization for which $P \supseteq P^*$. For the restricted model problem this is shown for more general factorizations of type MIC(d), $d \ge 0$ in [24], cf. also [21].

We recall that for the MIC(0)* method we have
$C = LL^T = (\tilde{D}+L_A)\tilde{D}^{-1}(\tilde{D}+L_A^T)$, where L_A is the strictly lower part of $A = (a_{ij})$. Hence, $L^T = \tilde{D}^{\frac{1}{2}} + \tilde{D}^{-\frac{1}{2}}L_A^T$ and we immediately get the following relation for the entries of $L^T = (\ell_{ij})$;

$$\ell_{ij} = a_{ij}/\ell_{ii}, \quad j > i. \tag{4.4}$$

Let n be the space dimension. We assume that

(i) A is a symmetric, weakly diagonally dominant L-matrix of order $N = O(h^{-n})$, $h \to 0$,

(ii) $a_{ii} \ge -\sum_{j>i} a_{ij}(2-m_i h)$, $i \in \mathcal{N}_1$, where $\mathcal{N}_1 \subseteq \mathcal{N} = \{i;\ 1 \le i \le N\}$ and the number of indices in $\mathcal{N}_2 = \mathcal{N}\setminus\mathcal{N}_1$ is $O(h^{-n+1})$, $h \to 0$.

For the definition of $\mathcal{N}_2$ in some applications, see below. We now define $D = \mathrm{diag}(d_i)$ in a proper way namely

(iii) $d_i = \begin{cases} \xi_1 h^2 a_{ii}, & i \in \mathcal{N}_1 \\ \xi_2 h a_{ii}, & i \in \mathcal{N}_2, \end{cases}$

where $\xi_i > 0$, $i = 1,2$ are parameters.

Lemma 4.3. For the MIC(0)* factorization, with D satisfying (iii), of A satisfying (i) and (ii) the entries ℓ_{ii} of diag(L^T) are bounded below by

$$\ell_{ii}^2 \geq - \sum_{j>i} a_{ij}(1+kh), \quad i = 1,\dots,N, \tag{4.5}$$

where $k = \min\{h^{-1}, \sqrt{\xi_1}, \xi_1/m_0, \xi_2\}$, $m_0 = \max_i m_i$.

Proof. Let the diagonal matrix $\hat{D} = \mathrm{diag}(\hat{d}_i)$ be defined by $\hat{d}_i = \sum_j a_{ij}$, $i = 1,\dots,N$. Obviously $\hat{D} \geq 0$. The proof of (4.5) is now carried out by induction on i. We show the induction step. From the definition of the MIC(0)* factorization and (4.4) we have

$$\begin{aligned}\ell_{ii} &= a_{ii}(1+d_i) - \sum_{p<i} \ell_{pi} \sum_{j>p} \ell_{pj} \\ &= a_{ii}(1+d_i) - \sum_{p<i} a_{pi}/\ell_{pp} \sum_{j>p} a_{pj}/\ell_{pp} \\ &\geq a_{ii}(1+d_i) + (1+kh)^{-1} \sum_{p<i} a_{pi} \\ &= (\hat{d}_i - \sum_{j>i} a_{ij})(1+d_i) - \{1+d_i-(1+kh)^{-1}\} \sum_{j<i} a_{ij}.\end{aligned}$$

If $i \in N_2$, then

$$\ell_{ii} \geq (\hat{d}_i - \sum_{j>i} a_{ij})(1+\xi_2 h) \geq - \sum_{j>i} a_{ij}(1+kh)$$

with $k \leq \xi_2$.

On the other hand, if $i \in N_1$, then

$$\ell_{ii} \geq (\hat{d}_i - \sum_{j>i} a_{ij})(1+\xi_1 h^2) - (\xi_1 h^2 + kh - k^2h^2) \sum_{j<i} a_{ij}. \tag{4.6}$$

From (ii) follows that

$$- \sum_{j>i} a_{ij}(2-m_i h) \leq a_{ii} = \hat{d}_i - \sum_{j>i} a_{ij} - \sum_{j<i} a_{ij}$$

and thus

$$- \sum_{j<i} a_{ij} \geq - \sum_{j>i} a_{ij}(1-m_i h) - \hat{d}_i. \tag{4.7}$$

From (4.6) and (4.7) we now get

$$\ell_{ii}^2 \geq - \sum_{j>i} a_{ij}(1+kh)+\hat{d}_i(1-kh+k^2h^2)$$

$$- \sum_{j>i} a_{ij}\{(1-m_i h)(\xi_1 h^2-h^2k^2)+\xi_1 h^2-m_i kh^2\}.$$

We consider two cases. At first, if $m_i = 0$, then

$$\ell_{ii}^2 \geq - \sum_{j>i} a_{ij}(1+kh) \text{ with } k \leq \min\{h^{-1},\sqrt{2\xi_1}\}. \tag{4.8}$$

Secondly, if $m_i > 0$, then

$$\ell_{ii}^2 \geq - \sum_{j>i} a_{ij}(1+kh) \text{ with } k \leq \min\{h^{-1},\sqrt{\xi_1},\xi_1/m_i\}.$$

Hereby the lemma is proved.

Lemma 4.4. Let $\hat{A} = A\text{-diag}(\hat{d}_i)$, where $\hat{d}_i = \sum_j a_{ij}$, $i = 1,\ldots,N$ and A satisfies (i). Then for the MIC(0)* factorization $\hat{C} = \hat{L}\hat{L}^T = \hat{A}+\hat{R}$ (with $D = 0$) of $\hat{A}$ we have

$$\hat{\ell}_{ii}^2 = - \sum_{j>i} a_{ij}, \quad i = 1,\ldots,N.$$

Proof. The proof is similar to the previous one. Apparently, $\text{Rowsum}(\hat{A}) = 0$. The induction step reads

$$\ell_{ii}^2 = \hat{a}_{ii} - \sum_{p<i} \hat{\ell}_{pi} \sum_{j>p} \hat{\ell}_{pj} = \hat{a}_{ii} - \sum_{p<i} \hat{a}_{pi}/\hat{\ell}_{pp} \sum_{j>p} \hat{a}_{pj}/\hat{\ell}_{pp}$$

$$= \hat{a}_{ii} - \sum_{j>p} \hat{a}_{pj} = \hat{a}_{ii} - \sum_{j<i} \hat{a}_{ij} = - \sum_{j>i} \hat{a}_{ij} = - \sum_{j>i} a_{ij}.$$

We are now able to state the main result.

Theorem 4.1. Assume that A satisfies (i) and (ii) and that D in the MIC(0)* factorization of A satisfies (iii) and (4.2). Then $\kappa_1 = O(h^{-1})$, $h \to 0$ and the computational complexity of the corresponding MICCG method is $O(N^{1+1/2n} \ln N)$, $N \to \infty$, where n is the space dimension.

Proof. Consider the MIC(0)* factorizations $C = LL^T = A+D+R$ of A and $\hat{C} = \hat{L}\hat{L}^T = \hat{A}+\hat{R}$ of $\hat{A}$ defined in Lemma 4.3. We note that the elements r_{ij} of R are calculated from

$$r_{ij} = \sum_{p<i} \ell_{pi}\ell_{pj} = \sum_{p<i} \frac{a_{pi}a_{pj}}{\ell_{pp}^2}, \quad j > i$$

and similarly for $\hat{R} = (\hat{r}_{ij})$,

$$\hat{r}_{ij} = \sum_{p<i} \frac{\hat{a}_{pi}\hat{a}_{pj}}{\hat{\ell}_{pp}^2} = \sum_{p<i} \frac{a_{pi}a_{pj}}{\hat{\ell}_{pp}^2}, \quad j > i.$$

From Lemma 4.3 and Lemma 4.4 we then get

$$0 \le r_{ij} \le (1+kh)^{-1} \hat{r}_{ij}, \; j > i. \tag{4.9}$$

From Corollary 3.2 follows that $\hat{C}$ is positive semidefinite or equivalently,

$$-(\hat{R}x,x) \le (\hat{A}x,x), \qquad \forall x \in R^N. \tag{4.10}$$

We have, since R is symmetric and rowsum(R) = 0,

$$-(Rx,x) = \sum_i \sum_{j>i} r_{ij}(x_i-x_j)^2 \tag{4.11}$$

and similarly for $\hat{R}$.

Using (4.9), (4.10) and (4.11) we obtain

$$-(Rx,x) \le -(1+kh)^{-1}(\hat{R}x,x) \le -(1+kh)^{-1}(\hat{A}x,x) \le (1+kh)^{-1}(Ax,x).$$

Hence, the assumptions of Lemma 4.2 are fulfilled and $\kappa_1 = O(h^{-1})$, $h \to 0$.

The second part of the theorem follows from (1.3) and the fact that each iteration can be carried out in $O(N)$, $N \to 0$ arithmetic operations. In a n-dimensional problem we have $N = O(h^{-n})$ and the $\ln(N)$ factor in the expression for the complexity comes from the fact that we let $\varepsilon = O(h^{\sigma})$, $\sigma > 0$ in order to make the iterations error of the same order as the discretization error.

What is stated in Theorem 4.1 can be proved in a similar way for more general MIC factorizations with $P \supseteq P^*$ [24]. A quite general proof, however, is much more complicated since it has to take care of the fill-in actually produced by the incomplete factorization.

The relation (i) is fulfilled for linear f.e. approximations over quadrilateral and triangular meshes having all angles $\leq \pi/2$. In addition, (ii) is fulfilled if the mesh points are ordered in a natural (rowwise or similar) way for the following types of selfadjoint elliptic pde problems;

a) Dirichlet problems with constant coefficients (Laplace equation). In this case $N_1 = N$ and $m_i = 0$, $i = 1,\ldots,N$.

b) Dirichlet problems with Lipschitz continuous material coefficients. Here $N_1 = N$.

c) Neumann problems. Then, N_2 represents points on and/or near a Neumann boundary.

d) Problems with discontinuous material coefficients. Here, N_2 represents points on and/or near an interface over which the coefficients are discontinuous.

For a more detailed discussion of the various types of problems, see [24].

It remains to show that (4.2) is satisfied for the different definitions of D. This is trivially seen for Dirichlet problems. Further, for a pure Neumann problem this is shown in [15] and for a problem with mixed boundary conditions in [24], where also the ideas for a general proof is presented. The crucial point is of course that the number of indices in N_2 and N_1 is $O(h^{-n+1})$ and $O(h^{-n})$, respectively.

We note that the ordering of the unknows is only essential for the definition of D. Any ordering for which (ii) is valid is appropriate. For instance, the points on a Neumann boundary may be numbered first. For the SSOR method the ordering of the unknows is significant for Neumann problems, see [1].

<u>Example 4.1</u>. Consider the model Neumann matrix and the model Dirichlet matrix arising from discontinuous coefficient a in (1.1 In the following figures we indicate how to define D. We assume a rowwise, upwards numbering of meshpoints. Then the indexset N_2 is represented by the star-marked meshpoints.

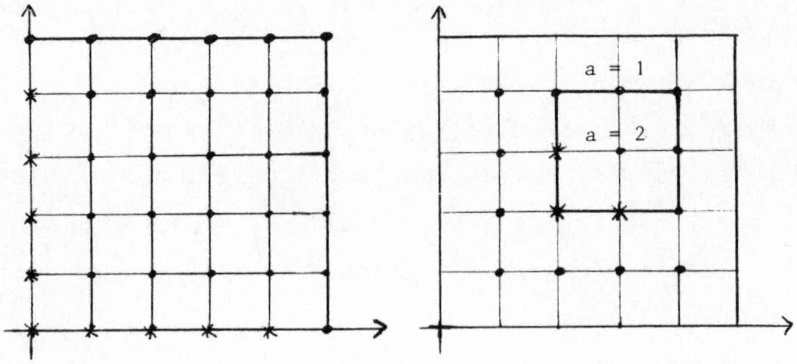

a Neumann problem a discont. Dirichlet problem

For special (model) matrices it is possible to calculate optimal values of the preconditioning parameters ξ_i, $i = 1,2$. Here we consider the Dirichlet problem and hence we have one parameter $\xi = \xi_1$ and $D = \xi h^2 D_A$, where $D_A = \text{diag}(A)$. If A is a weakly diagonally dominant L-matrix, then by Lemma 4.2,

$$\kappa_1 \le (1+m_1)(1+1/kh) = (1+m_1)+m_2 h^{-1},$$

where $m_2 = m_2(\xi) = (1+m_1)/k$. We want to minimize m_2 with respect to ξ.

Let $h^2\lambda_0$ be the smallest eigenvalue of $D_A^{-\frac{1}{2}} A D_A^{-\frac{1}{2}}$. Then by (4.2), $m_1 = \xi/\lambda_0$. Hence, $m_2(\xi) = (1+\xi/\lambda_0)/k$. For the MIC(0)* method, k is given in Lemma 4.2. For instance, in the constant coefficient case, $k = \sqrt{2\xi}$, see (4.8), and we readily find that $m_2(\xi)$ is minimized for $\xi = \xi_{opt} = \lambda_0$ and that

$$\kappa_1(\xi_{opt}) \le 2 + \sqrt{2}\lambda_0^{-\frac{1}{2}} h^{-1} . \tag{4.12}$$

<u>Example 4.2</u>. For the restricted model Dirichlet matrix we have $\lambda_0 = \pi^2/2$ (+ $O(h^2)$) and hence for the MIC(0)* factorization we get by (4.12)

$$\xi_{opt} = \pi^2/2, \quad \kappa_1(\xi_{opt}) \le 2(1+\pi^{-1}h^{-1}).$$

For the MIC(d), d > 0 methods we have $k = \alpha\xi^{-\frac{1}{2}}$ (where α

depends on d) and thus the same optimal value of ξ.

We remark that numerical experiments (see e.g. [20],) show that the number of iterations is fairly insensitive to the choice of the parameters ξ_1,ξ_2. For many problems the choice $D = \xi h^2 \operatorname{diag}(A)$, $\xi > 0$ (say $\xi = 5$), or even $D = 0$, has turned out to be quite satisfactory independent of e.g. the type of boundary conditions or the possible presence of a discontinuity of the material coefficients.

5. SOLUTION OF MORE GENERAL F.E. EQUATIONS

Let us at first consider the solution of higher order f.e. equations. We assume that Ω is a polygonal domain in (for simplicity) R^2 and that the discrete mesh Ω_h is derived by uniform meshrefinement of a coarse mesh Ω_1 (,which does not have to be uniform,) as described in Section 4. We assume that all angles in the triangulation Ω_1 (and hence also in Ω_h) are $\le \pi/2$. Let $V_h^{(p)}$ denote the space of continuous piecewise polynomials of degree p over Ω_h. For $p = 1$, the results of the previous section are valid. In particular, due to the assumptions made on Ω_h, the f.e. matrix, denoted $A^{(1)}$, is a weakly diagonally dominant L-matrix. For $p \ge 2$ we present two different methods which generalize these results.

5.1. Method 1; Spectral equivalence

Let $A^{(p)}$ (of the same order (N) as $A^{(1)}$) be the f.e. matrix obtained by using standard Lagrangian basis functions of degree $p > 1$. Then one can show (see [8]) that $A^{(1)}$ and $A^{(p)}$ are spectrally equivalent i.e.

$$m_0(A^{(1)}x,x) \le (A^{(p)}x,x) \le m_1(A^{(1)}x,x), \quad \forall x \in R^N. \tag{5.1}$$

This follows from the fact that the corresponding element stiffnes matrices have the same nullspace (actually, for elements not meeting a Dirichlet boundary, spanned by the constant vector) and that the mesh is refined in a uniform way.

Let C be a MIC factorization of $A^{(1)}$. Then $\kappa(C^{-\frac{1}{2}}A^{(1)}C^{-\frac{1}{2}}) = O(h^{-1})$, $h \to 0$ and from (5.1) we conclude that $\kappa(C^{-\frac{1}{2}}A^{(p)}C^{-\frac{1}{2}}) = O(h^{-1})$, $h \to 0$ as well. Hence we can use an incomplete factorization of the simpler matrix $A^{(1)}$ as a preconditioning to $A^{(p)}$ and thereby obtain the same asymptotic rate of convergence as when solving the system with matrix $A^{(1)}$.

5.2. Method 2; Block-factorization

This method is based on a special choice of basis functions. Here, we present it in its simplest form, for a more general and often more efficient approach see [10], cf. also [11].

Let ϕ_1, ϕ_2, ϕ_3 be linear basis functions associated with corner nodes of the triangulation and let $\phi_4, \ldots, \phi_q$, $q = (p+1)(p+2)/2$ the basis functions of degree p associated with edge and interior nodes. It is easily seen that these basis functions form a basis for $V_h^{(p)}$. The element stiffness matrices have the form

$M_e = \begin{bmatrix} A_e & C_e^T \\ C_e & B_e \end{bmatrix}$, where the 3×3 matrix A_e corresponds to the linear basis functions and B_e is positive definite. By making an over all numbering of mesh points such that corner nodes are numbered first we get a global N×N matrix of the form $M = \begin{bmatrix} A & C^T \\ C & B \end{bmatrix}$, where $A = A^{(1)}$ is of order $O(N/p^n)$ in an n dimensional problem. Since B_e is positive definite we have $\kappa(B) = O(1)$, $h \to 0$ as well as $\kappa(D_B^{-1}B) = O(1)$, $h \to 0$, where $D_B = \text{diag}(B)$.

We consider preconditioning matrices $\tilde{M}$ to M on the form

$$\tilde{M} = \begin{bmatrix} L_A L_A^T & 0 \\ 0 & L_B L_B^T \end{bmatrix} \quad \text{(diagonal block factorizations)}$$

and

$$\tilde{M} = \begin{bmatrix} L_B & 0 \\ C^T L_B^{-T} & L_A \end{bmatrix} \begin{bmatrix} L_B^T & L_B^{-1}C \\ 0 & L_A^T \end{bmatrix} = \begin{bmatrix} L_B L_B^T & C \\ L^T & L_A L_A^T + C^T (L_B L_B^T)^{-1} C \end{bmatrix},$$

(full-block factorizations),

where in the latter case we have reordered the system so that

$M = \begin{bmatrix} B & C \\ C^T & A \end{bmatrix}$. Here $L_A L_A^T$ and $L_B L_B^T$ are (approximate) factorizations of A and B, respectively.

It follows (see [10]) that if $L_A L_A^T$ is a complete factorization of A and $L_B L_B^T$ is an incomplete factorization (e.g. an IC factorization) of B, then $\kappa_1 = \kappa(\tilde{M}^{-\frac{1}{2}} M \tilde{M}^{-\frac{1}{2}}) = O(1)$, $h \to 0$, for both types of block factorizations. In fact $L_B L_B^T$ may even be equal to D_B. For $L_A L_A^T$ being a MIC factorization of A we get $\kappa_1 = O(h^{-1})$, $h \to 0$. However, since A is of small size compared to M, κ_1 is essentially O(1) for $h \geq h_0$, if we choose a sufficient accurate MIC factorization, for model problems a MIC(d) algorithm, where apparently h_0 depends on d.

Similar block factorization techniques were used in [4] for the solution of Navier equations of elasticity.

5.3. A class of non-symmetric problems

Consider differential equations on the form

$$Lu = -\nabla(a\nabla u) + \vec{v}\cdot\nabla u + bu = f, \quad x \in \Omega \tag{5.2}$$

with appropriate boundary conditions on $\partial\Omega$.

Assume that we use a positive difference scheme (e.g. upwind, modified upwind [5] or Il'in scheme [26]) for the discretization of (5.2). Then the associated matrix A is diagonally dominant and hence any MIC factorization $C = A+D+R$ is stable.

Let A_0 be the matrix corresponding to $\vec{v} \equiv 0$ and let $A_s = \frac{1}{2}(A+A^T)$ be the symmetric part of A. We assume that $(A_s x,x) \geq (A_0 x,x)$, $\forall x \in R^N$. A sufficient condition for this to be true is that $\mathrm{div}(\vec{v}) \leq 0$, since $\frac{1}{2}(L+L^*)u = -\nabla(a\nabla u) - \frac{1}{2}\mathrm{div}(\vec{v})u + bu$.

Under the assumptions above the eigenvalues of $C^{-1}A$ are sitated in an ellipse in the complex plane, centered at (1,0) and with axes $\alpha = 1-m_1 h$ and $\beta = m_2 h$ (after normalization and for h sufficiently small). Then the number of iterations, p, in the Chebyshev acceleration method (or a generalized conjugate direction method [9]) is of order $p = O([\ln(1+\sqrt{1-\alpha^2+\beta^2}) - \ln(\alpha+\beta)]^{-1}) = O(h^{-\frac{1}{2}})$, $h \to 0$, i.e. the same order as for a corresponding symmetric problem [24].

5.4. Fourth order problems

For some discretizations of the Biharmonic operator an idea similar to that in Method 1 (Section 5.1) above can be used; One makes a MIC factorization of a matrix corresponding to a related second order problem. We assume that this matrix is a weakly diagonally dominant L-matrix (, which is true under assumptions stated earlier). The square of this MIC preconditioning matrix is now used as preconditioning for the fourth order problem. One finds that the condition number is $O(h^{-2})$, $h \to 0$, while for the unpreconditioned matrix it is $O(h^{-4})$, $h \to 0$. Hence, as in second order problems we gain a square root in the order of the condition number by using this kind of preconditioning. For a detailed discussion of this method, for instance, for which kind of discretizations it is appropriate, see [9] cf. also [6].

6. NUMERICAL EXPERIMENTS

We present the number of iterations needed for different (M)ICCG methods to solve some typical test problems. For further numerical experiments we refer to [14] and the references therein. As initial approximation u_0 to the conjugate gradient method we have used $u_0 = C^{-1}f$. Since C resembles A this is always a good choice. For more accurate initial approximations see [24]. The iterations were stopped when the residual error (in ℓ_2-norm) was reduced by a factor ε, that is, when $\|r^{\ell}\| \le \varepsilon\|r^0\|$. We note that since we use a *relative* stopping criterion the number of iterations is almost independent of the initial approximation (except for very special ones).

Example 6.1. The restricted model Dirichlet matrix, $\varepsilon = 10^{-4}$.

N	IC(0)	MIC(0)	IC(1)	MIC(1)	IC(2)	MIC(2)	IC(4)	MIC(4)
49	6	6	4	4	4	3	3	3
225	9	8	7	6	6	5	5	4
961	16	12	12	9	10	8	8	6
3969	30	17	26	13	21	11	15	9

It is confirmed that the rate of convergence of the ICCG and MICCG methods is $O(N^{\frac{1}{2}})$ and $O(N^{\frac{1}{4}})$, respectively. By considering the total number of operations (multiply-adds) we find that the MIC(2) method is the most efficient for N = 3969.

Example 6.2. The restricted model Dirichlet problem discretized by bilinear finite elements, $\varepsilon = 10^{-6}$.

N	MIC(0)	MIC(0)*	SSOR	IC(0)	IC(0)*	MIC(2)
100	8	8	9	8	9	5
400	11	12	12	13	15	7
1600	16	17	18	23	27	10

We note that all of these methods except MIC(2) have the same sparsity pattern and hence need the same work per iteration (in the direct application, see however [16]). The MIC(2) method needs a bit more storage and work per iteration. Nevertheless, the smaller number of iterations leeds to an operation count about 35% smaller than for MIC(0) for N = 1600.

Example 6.3. The discontinuous problem

$$-\nabla\cdot(a\nabla u) = f \equiv 1 \text{ on } \Omega = \{(x,y); 0 < x < 1,\ 0 < y < 4/3\},$$

with mixed boundary conditions, see the figure, where d is a parameter to vary the degree of discontinuity. The five-point difference method was used and $\varepsilon = 10^{-6}$.

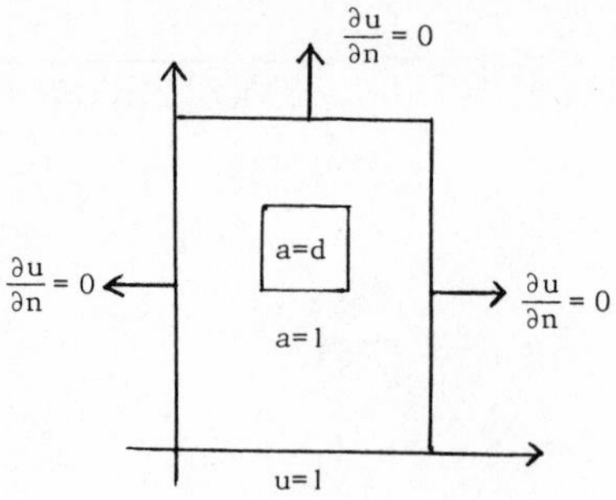

d	1			100			10 000		
h^{-1}	SSOR	MIC(0)	MIC(2)	SSOR	MIC(0)	MIC(2)	SSOR	MIC(0)	MIC(2)
6	11	10	6	13	12	7	15	13	7
12	16	14	8	21	17	10	26	18	11
24	26	19	12	34	23	14	42	25	15
27	29	21	13	37	25	15	46	27	16

Example 6.4. The restricted model Dirichlet problem discretized by quadratic f.e. over a uniform rightangled triangular mesh. Method 1 is the direct application of MIC on $A^{(2)}$, Method 2 is the method based on spectral equivalence (Method 1 in Section 5.1) with factorizations of $A^{(1)}$ and Methods 3 and 4 are the diagonal block and full block factorization methods (see Method 2 in Section 5.2) with $L_B L_B^T = \text{diag}(B)$ and $L_B L_B^T$ an IC(0) factorization of B, respectively and with different MIC factorizations of $A = A^{(1)}$, $\varepsilon = 10^{-4}$.

N	Method 1		Method 2			Method 3			Method 4		
	MIC(0)	MIC(4)	MIC(0)	MIC(1)	MIC(2)	MIC(0)	MIC(2)	MIC(4)	MIC(0)	MIC(2)	MIC(4)
49	6	4	7	6	6	13	12	12	6	5	5
225	9	5	9	7	6	14	12	12	7	5	5
961	13	7	13	10	8	17	13	12	8	5	5
3969	19	10	19	14	11	22	15	12	10	6	5

We remark that in Methods 3 and 4, the MIC(4) factorizations of $A^{(1)}$ give the same number of iterations as a complete factorization of $A^{(1)}$ (for these values of N). For N = 3969, Method 4 needs about 170 operations per unknown (for MIC(4)) while the other methods need about 250 operations per unknown (with the optimal MIC method).

Example 6.5. The Biharmonic problem

$$\begin{cases} \Delta^2 u = f & \text{in } K_e \\ u = u_n & \text{on } \partial K_e, \end{cases}$$

where K_e is the unit square *and* u_n *is the outward pointing normal derivative*, discretized by the 13-point f.d. approximation. Method 1 is the application of MIC directly on the derived matrix A and Method 2 is the method presented in Section 5.4 with MIC factorizations of A_1, a matrix corresponding to a second order problem, in this case the restricted model Dirichlet matrix, $\varepsilon = 10^{-3}$.

h^{-1}	Factorization of A_1	Factorizations of A	
	MIC(2)	MIC(0)	MIC(4)
5	4	4	2
10	6	7	3
20	10	16	7
40	17	37	14

The work per unknown for $h^{-1} = 40$ is for the MIC(2) factorization of A_1, the MIC(0) and MIC(4) factorizations of A about 630, 1210 and 620 operations, respectively. One realizes that one needs a quite accurate and hence more complicated factorization of A to reach the same efficiency as for a simpler factorization of A_1.

Example 6.6. The non-symmetric problem

$$-\Delta u + \gamma(u'_x - u'_y) = 1 \text{ in } K_e, \quad u = 0 \quad \text{on } \partial K_e,$$

where $\gamma \geq 0$ is used to vary the degree of non-symmetry. As discretization method we have used standard 5-point approximation for

Δu and upwind differences for the derivatives of first order. The minimum residual CG method [7] was used and $\varepsilon = 10^{-3}$.

h	C \ γ	0	5	10	100	10^5
1/8	I	8	18	19	18	18
	MIC(0)	4	5	5	7	7
	MIC(2)	3	3	3	3	3
1/16	I	18	34	34	34	35
	MIC(0)	6	6	7	10	11
	MIC(2)	4	4	4	4	4
1/32	I	36	72	72	65	62
	MIC(0)	9	9	10	12	13
	MIC(2)	6	6	6	5	5

Observe that, when one uses a more accurate MIC factorization, the number of iterations is almost independent on γ even for fairly large values of γ, that is, for singularly perturbed problems.

7. CONCLUSIONS AND COMPARISONS WITH OTHER METHODS

For wide classes of f.e. problem, including nonsymmetric and discontinuous problems, we have shown that the computational complexity of MICCG methods is of order $O(N^{1+1/2n} \ln N)$, $N \to \infty$ in an n-dimensional problem. The sparsity complexity and hence the need of storage is of optimal order $O(N)$, $N \to \infty$.

Besides the comparisons with the IC and SSOR methods presented in Section 6, the MIC methods have been compared with other factorization methods, for instance so called second order (or almost second order) methods e.g. the symmetric Stone's method [33], [13], [32], [22] and with factorizations where fill-in is moved to other subdiagonals than the main diagonal [24]. From these comparisons, the MIC methods turn out to be preferable with respect to rate of convergence, need of storage and simplicity

In [24] the MICCG methods are compared with some direct

methods e.g. methods of nested dissection type [18], [19]. It turns out that for not too small problems and not too many right-hand sides (with the same coefficient matrix) the iterative methods are faster. In particular this is true for three-dimensional problems. Furthermore, this author feels that the MICCG methods are easier to implement on computers. A general non-symmetric MIC(0) program is presented in [24]. For the MIC(0)* method one can use a special implementation to decrease the number of operations [16]. Moreover, the methods presented in this contribution can be made stable for rounding errors [8] and can take advantage of a good initial approximation, present for instance in non-linear and time-dependent problems. For a special choice of initial approximation see [24] and [8].

Finally, we recall that the Method 2 in Section 5.2 is essentially of optimal order, $O(N \ln N)$, $N \to \infty$, for not too fine meshes. The computational complexity of this method (for numerical tests see [10] and Example 6.4) is comparable to that for the best implementations of multigrid (multilevel) methods, cf. [25] and references therein.

REFERENCES

1. ANDERSSON, L. - SSOR preconditioning of Toeplitz matrices, Report 76.02R, Department of Computer Sciences, Chalmers University of Technology, Göteborg, Sweden, 1976.
2. AXELSSON, O. - A generalized SSOR method, BIT, Vol.12, pp. 443-467, 1972.
3. AXELSSON, O. - A class of iterative methods for finite element equations, Comp. Meth. Appl. Mech. Eng., Vol.9, pp. 123-137, 1976.
4. AXELSSON, O. and GUSTAFSSON, I. - Iterative methods for the solution of the Navier equations of elasticity, Comp. Meth. Appl. Mech. Eng., Vol.15, pp. 241-258, 1978.
5. AXELSSON, O. and GUSTAFSSON, I. - A modified upwind scheme for convective transport equations and the use of conjugate gradient method for the solution of non-symmetric systems of equations, J. Inst. Math. Applics., Vol. 23, pp. 321-337, 1979.
6. AXELSSON, O. and GUSTAFSSON, I. - An iterative solver for a mixed variable formulation of the biharmonic problem, Comp. Meth. Appl. Mech. Eng., Vol. 20, pp. 9-16, 1979.

7. AXELSSON, O. - A generalized conjugate direction method and its application on a singular perturbation problem, 8th Biennial Numerical Analysis Conference, Dundee, Ed. Watson G.A., Lecture Notes in Mathematics #773, Springer, Berlin, 1980.
8. AXELSSON, O. and GUSTAFSSON, I. - A preconditioned conjugate gradient method for finite element equations, which is stable for rounding errors, INFORMATION PROCESSING 80, Ed. Lavington, S.H., pp. 723-728, North Holland, 1980.
9. AXELSSON, O. and GUSTAFSSON, I. - A note on the solution of the Biharmonic problem by Modified Incomplete Factorizations, (in progress).
10. AXELSSON, O. and GUSTAFSSON, I. - An essentially optimal method for finite element equations, (in progress).
11. BANK, R. and DUPONT, T. - Analysis of a two-level scheme for solving finite element equations, Report CNA-159, Center for Numerical Analysis, The University of Texas at Austin, U.S.A., 1980.
12. BEAUWENS, R. - Factorization Iterative Methods, M-operators and H-operators, Numer. Math., Vol. 31, pp. 335-351, 1979.
13. BRACHA-BARAK, A. and SAYLOR, P. - A symmetric factorization procedure for the solution of elliptic boundary value problems, SIAM J. Numer. Anal., Vol.10, pp. 190-206, 1973.
14. DUPONT, T., KENDALL, R. and RACHFORD, H.H., Jr. - An approximate factorization procedure for solving self-adjoint elliptic difference equations, SIAM J. Numer. Anal., Vol.5, pp. 559-573, 1968.
15. DUPONT, T. - A factorization procedure for the solution of elliptic difference equations, SIAM J. Numer. Anal., Vol.5, pp. 753-782, 1968.
16. EISENSTAT, S. - Efficient Implementation of a class of Preconditioned Conjugate Gradient Methods, Research Report #185, Department of Computer Sciences, Yale University, U.K. 1980.
17. FIEDLER, M. and PTAK, V. - On matrices with non-positive off-diagonal elements and positive principal minors, Czech. Math J., Vol.12, pp. 382-400, 1962.
18. GEORGE, A. - Nested dissection of a regular finite element mesh, SIAM J. Numer. Anal., Vol. 10, pp. 345-363, 1973.
19. GEORGE, A. - Numerical experiments using dissection methods to solve n by n grid problems, Research Report CS-75-07, University of Waterloo, Canada, 1975.
20. GUSTAFSSON, I. - A class of first order factorization methods, Report 77.04 R, Department of Computer Sciences, Chalmers University of Technology, Göteborg, Sweden, 1977.
21. GUSTAFSSON, I. - A class of first order factorization methods, BIT, Vol. 18, pp. 142-156, 1978.
22. GUSTAFSSON, I. - On first and second order symmetric factorization methods for the solution of elliptic difference equations, Report 78.01R, Department of Computer Sciences, Chalmers University of Technology, Göteborg, Sweden, 1978.
23. GUSTAFSSON, I. - On modified incomplete Cholesky factorization methods for the solution of problems with mixed boundary con-

ditions and problems with discontinuous material coefficients, Int. J. Numer. Meth. Engng., Vol.14, pp. 1127-1140, 1979.

24. GUSTAFSSON, I. - Stability and rate of convergence of modified incomplete Cholesky factorization methods, Report 79.02R, Department of Computer Science, Chalmers University of Technology, Göteborg, Sweden, 1979.
25. HEMKER, P.W. - Introduction to multigrid methods, Colloquium Numerical Solution of Partial Differential Equations, Ed. Verwer, J.G., MC SYLLABUS 44, pp. 59-67, Mathematisch Centrum, Amsterdam, 1980.
26. IL'IN, A.M. - Differencing scheme for a differential equation with a small parameter affecting the highest derivative, Mat. Zametki, Vol. 6, pp. 237-248; Math. Notes, Vol. 6, pp. 596-602, 1969.
27. KERSHAW, D. - The incomplete Cholesky conjugate gradient method for the iterative solution of systems of linear equations, J. Comput. Phys., Vol. 26, pp. 43-65, 1978.
28. MANTEUFFEL, T.A. - The shifted incomplete Cholesky factorization, Technical report, Appl. Math. Division 8325, Sandia Laboratories, Livermore, California, USA, 1978.
29. MEIJERINK, J.A. and VAN DER VORST, H.A. - An iterative solution method for linear systems of which the coefficient matrix is a symmetric M-matrix, Math. Comp., Vol. 31, pp.148-162, 1977.
30. MUNKSGAARD, N. - Solution of general sparse symmetric sets of linear equations, Report No. NI-78-02, Inst. for Num. Anal., Technical Univ. of Denmark, Lyngby, Denmark, 1978.
31. OLIPHANT, T.A. - An extrapolation process for solving linear systems, Quart. Appl. Math., Vol. 20, pp. 257-267, 1962.
32. SAYLOR, P. - Second order strongly implicit symmetric factorization methods for the solution of elliptic difference equations, SIAM J. Numer. Anal., Vol. 11, pp. 894-908, 1974.
33. STONE, H.L. - Iterative solution of implicit approximations of multidimentional partial differential equations, SIAM J. Numer. Anal., Vol. 5, pp. 530-558, 1968.
34. TUFF, A.D. and JENNINGS, A. - An iterative method for large systems of linear structural equations, Int. J. Numer. Meth. Engng., Vol. 7, pp. 175-183, 1973.
35. VARGA, R.S. - Factorization and normalized iterative methods, Boundary problems in differential equations, Ed. Langer, R.E., pp. 121-142, Madison, University of Wisconsin Press, 1960.
36. VARGA, R.S. - Matrix iterative analysis, Englewood Cliffs, New York, Prentice Hall, 1962.
37. YOUNG, D. - Iterative solution of large linear systems, Academic Press, New York and London, 1971.

PRECONDITIONING SYMMETRIC INDEFINITE MATRICES

PAUL E. SAYLOR
Department of Computer Science
The University of Illinois U-C
1304 West Springfield
Urbana, IL 61801

Abstract

Let Ax = b be a linear system where A is a symmetric matrix with positive and negative eigenvalues. The basic conjugate gradient method assumes A is positive definite. A preconditioning method is discussed that yields a symmetric positive definite preconditioned matrix of the form Q(A)A, where Q is a polynomial, to which the conjugate gradient method may be applied. The solution method is a generalization of the cgT method of Rutishauser and uses Richardson's method. If an incomplete factorization is used the preconditioned matrix is nonsymmetric and the conjugate gradient method not directly applicable. The preconditioned system may be solved by Richardson's method if the eigenvalues of the preconditioned matrix are real.

1 INTRODUCTION

The problem considered in this paper is the solution of a symmetric indefinite system of linear equations, Ax = b. Here, indefinite means that there are eigenvalues of opposite sign. This condition either interferes with or renders impossible the solution by standard methods: Gaussian elimination requires pivoting for stability, which may destroy sparsity; the conjugate gradient method is unstable; and the successive-over-relaxation method diverges. Because symmetric indefinite matrices result from the solution of important scientific and engineering problems, efficient solution methods are significant. Examples of problems from which symmetric indefinite matrices arise are the numerical solution of Stokes flow, elasticity and constrained minimization problems (references in [5]). They appear also in less familiar areas such as the numerical simulation of step motors [6]. (A step motor is an electric motor used, for example, to control printers, teletypes, disk and tape drives, and machine tools.)

If the iterative solution of a linear system converges slowly, then preconditioning would be appropriate but the existence of positive and negative eigenvalues also interferes with preconditioning. To explain, some notation and definitions are useful. Let K be a <u>preconditioning matrix</u>, used to transform a given system $Ax = b$ into a <u>preconditioned system</u> $K^{-1}Ax = K^{-1}b$. The most familar preconditioning matrices result from an incomplete factorization that produces the lower and upper triangular factors of K such that $K = LU$. The basic conjugate gradient method cannot be applied to $K^{-1}A$ since this matrix is nonsymmetric. If A is symmetric positive definite, the usual algorithms yield $U = L^T$, and the conjugate gradient method may be applied to the alternative preconditioned system

$$L^{-1}A(L^T)^{-1}L^Tx = L^{-1}b \tag{1}$$

If A is symmetric indefinite then the preconditioning matrix may be assumed symmetric but is not usually positive definite, and therefore cannot be factored as $K = LL^T$. Thus the form (1) does not exist and the conjugate gradient method is not applicable. Nonsymmetric extensions of the conjugate gradient method could be applied to $K^{-1}Ax = K^{-1}b$, but those will not be considered here. Instead preconditioning by a polynomial in A,

$$Q(A)Ax = Q(A)b$$

will be discussed. Since $Q(A)A$ is symmetric, the basic conjugate gradient method applies if the eigenvalues are positive. A simple way to assure positive eigenvalues is to let Q(A) = A, but this vilipends the intuitive idea of preconditioning that Q(A) should approximate A^{-1}. In Section 4, algorithms to yield a polynomial approximation to A^{-1} are described. The resulting solution method, that is, the conjugate gradient method applied to Q(A)Ax = Q(A)b, is called the conjugate gradient Richardson's method (cgRm) iteration and is a generalization of the cgT method of Rutishauser [11], which is described in Section 2. Richardson's method is another iterative method to solve symmetric indefinite systems [10]. Its role in the cgRm iteration is this. The practical implementation of cgRm is as an inner - outer iteration in which the inner iteration is Richardson's method and the outer iteration is the conjugate gradient method (Subsection 4.2).

Incomplete factorizations give extraordinary leverage to an iterative solution process and can be used straightforwardly with Richardson's method if the eigenvalues of $K^{-1}A$ are real (Subsection 3.5). No incomplete factorization is described here; if one is available this paper focuses only on a way in which it may be used.

It is essential for Richardson's method that eigenvalue bounds of A be known approximately. These may be computed dynamically, and the ideas are outlined in Section 5.

A survey of other solution methods for symmetric indefinite matrices appears in a paper of Fletcher [5] in which the author also develops a stabilized form of the conjugate gradient method. Another solution method is discussed in the thesis of Chandra [3].

I am indebted to Carl de Boor and John R. Rice for sending a copy of their report [4] and to Olaf Widlund for a copy of the report of Atlestam [2]. R. Roloff pointed out to me the inner outer formulation of the cgT method, as well as other aspects of linear algebra, provided the examples for Figures 4 and 5, and served as interlocutor. John Taylor and Daeshik Lee helped in the preparation. Support from NSF grant number MCS-7906123 is gratefully acknowledged.

2 PRELIMINARIES

In this section, definitions and notations are given and Richardson's method introduced for the symmetric positive definite case. The cgT method of Rutishauser is then described.

2.1 Richardson's Method

Let $t_0, \ldots$ be a set of iteration parameters, and $x^{(0)}$ an initial guess. Richardson's method is

$$r^{(j)} = b - Ax^{(j)} \tag{2a}$$

$$x^{(j+1)} = x^{(j)} + t_j r^{(j)} \tag{2b}$$

In practice, the parameters are not distinct but repeat in cycles, i.e., $t_j = t_{j-k}$, for $j \geq k$ where k is called the period of the iteration. Richardson's method is an example of a gradient method. It is distinguished from other gradient methods such as the method of steepest descent or the conjugate gradient method by the use of Chebyshev analysis to select the parameters, and is also referred to as a Chebyshev iteration [9]. As a term "Richardson's method" conforms to usage in many references [1,7,13] and avoids confusion with cyclic Chebyshev iterative methods.

2.1(a) Residual Polynomials

Let

$$e^{(j)} = x - x^{(j)}$$

be the error and

$$r^{(j)} = b - Ax^{(j)}$$

be the residual error, or, simply, the residual. Since

$$r^{(j)} = Ae^{(j)}$$

any statement about the residual error also applies to the error.

The purpose of an iteration parameter sequence is to minimize (approximately) the residual error, $r^{(k)}$, after one period of the iteration, and so the discussion begins with an expression for the residual in terms of the parameters. Such an expression is easily derived from the relation

$$r^{(k)} = [I - t_{k-1}A]r^{(k-1)}, \tag{3}$$

which is a consequence of multiplying (2b) by A then subtracting from b = b. From (3) it follows that

$$r^{(k)} = R_k(A)r^{(0)}$$

where

$$R_k(\lambda) = \prod_{j=0}^{k-1} (1-t_j\lambda).$$

Any polynomial R_k of degree k such that $R_k(0) = 1$ is called a residual polynomial [11], and $R_k(A)$ is called a residual polynomial matrix. Iteration parameters are thus reciprocals of the roots of a residual polynomial.

2.1(b) Optimum Iteration Parameters

The technique to determine iteration parameters for Richardson's method will now be sketched. It is based on Chebyshev polynomials, and is due not to Richardson but to Lanczos [8]. Shortley [12] and Young [14] also discussed the idea. It does not attain the goal of a minimum residual; only the conjugate gradient method, among the class of gradient methods, succeeds in minimizing the residual with respect to a vector norm.

Let (β_3, β_4) be an interval of positive numbers, $0<\beta_3<\beta_4$. (The odd notation is consistent with that used in a later section.) Suppose

$$P_k(\lambda;\beta_3,\beta_4) = \frac{T_k\left[\dfrac{\beta_3+\beta_4-2\lambda}{\beta_4-\beta_3}\right]}{T_k\left[\dfrac{\beta_3+\beta_4}{\beta_4-\beta_3}\right]}$$

where $T_k(\mu) = \cos(k\cos^{-1}\mu)$ for $|\mu|\leqslant 1$, is the Chebyshev polynomial of degree k. Since $P_k(0;\beta_3,\beta_4) = 1$, it is a residual polynomial, called the <u>Chebyshev residual polynomial</u>. It may be proved that

$$\max_{\beta_3\leqslant\lambda\leqslant\beta_4} |P_k(\lambda;\beta_3,\beta_4)| \leqslant \max_{\beta_3\leqslant\lambda\leqslant\beta_4} |R_k(\lambda)|$$

where R_k is any degree k residual polynomial. Let

$$M_{\beta_3,\beta_4} = \max_{\beta_3\leqslant\lambda\leqslant\beta_4} |P_k(\lambda;\beta_3,\beta_4)|.$$

It follows from properties of the Chebyshev polynomial that

$$M_{\beta_3,\beta_4} < 1.$$

Therefore, Richardson's method converges if the eigenvalues of A lie in (β_3,β_4). This condition may be relaxed. Richardson's method also converges if the eigenvalues are less than β_3, since the Chebyshev residual polynomial is less than 1 (and positive) in $(0,\beta_3)$.

Any Chebyshev residual polynomial is <u>optimum</u> in the sense of having minimum uniform norm on (β_3,β_4) subject to the constraint

$P_k(0;\beta_3,\beta_4) = 1$. See Figure 1. It would be more correct to state that a residual polynomial is optimum <u>on</u> (β_3,β_4). This qualification can be omitted with no loss of clarity.

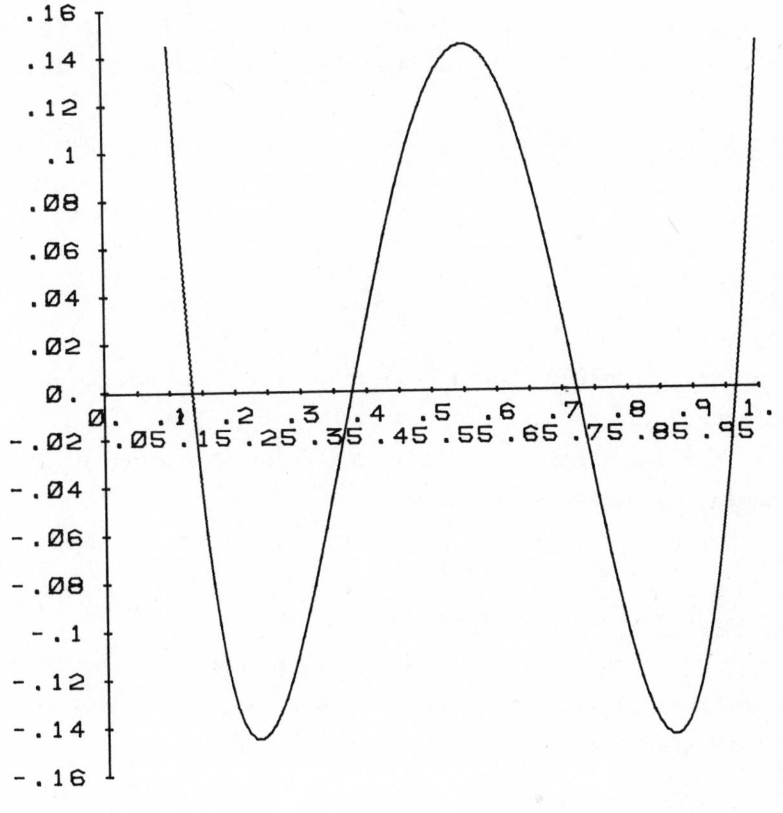

FIGURE 1 The optimum residual polynomial for $(\beta_3,\beta_4) = (.1,1)$.

Among the class of optimum residual polynomials there is only one that is best for a given matrix. To explain, let λ_N and λ_1 be the smallest and largest eigenvalues of A. The parameters from $P_k(\lambda;\lambda_N,\lambda_1)$ yield the best rate of convergence in the sense that

$$M_{\lambda_N,\lambda_1} \leqslant \max_{\lambda_N \leqslant \lambda \leqslant \lambda_1} P_k(\lambda;\beta_3,\beta_4).$$

The residual polynomial $P_k(\lambda;\lambda_N,\lambda_1)$ will be said to be A-optimum; the parameters are also called A-optimum. Observe that this residual polynomial and corresponding iteration parameters are A-optimum for any matrix of any order with the same smallest and largest eigenvalues as A. Richardson's method depends only on the bounds of an eigenvalue distribution, not on the distribution.

The spectral radius of any residual polynomial matrix, $R_k(A)$, is bounded by

$$\max_{\lambda_N \leq \lambda \leq \lambda_1} |R_k(\lambda)|.$$

The optimum Chebyshev residual polynomial is not optimum in the sense that the spectral radius of $P_k(A;\lambda_N,\lambda_1)$ is minimized, but only in the sense that a bound on the spectral radius of residual polynomial matrices is minimized.

Ordering the iteration parameters is important: see [1,14].

2.2 The cgT Method of Rutishauser

The conjugate gradient Tchebyshev (cgT) method of Rutishauser for a symmetric positive definite linear set, $Ax = b$, consists of the conjugate gradient method applied to

$$A^{-1}[I - P_k(A;\beta_3,\beta_4)]Ax = A^{-1}[I - P_k(A;\beta_3,\beta_4)]b, \tag{4}$$

where (β_3,β_4), $0<\beta_3$, is any interval containing the eigenvalues. This definition is due to Rutishauser [11].

Henceforth the dependence of P_k on (β_3,β_4) will not be shown unless necessary for clarity.

Rutishauser's equivalent system (4) is a preconditioning of $Ax = b$, in which there appears to be an unreasonable requirement for the inverse of A,

$$K = A^{-1}[I - P_k(A)].$$

It is easily seen that K reduces to a polynomial in A from the fact that $1 - P_k(\lambda)$ has a root at zero, which implies

$$Q_{k-1}(\lambda) = [1 - P_k(\lambda)]/\lambda$$

is a polynomial in λ of degree k-1, called a <u>preconditioning polynomial</u>. Therefore $K = Q_{k-1}(A)$.

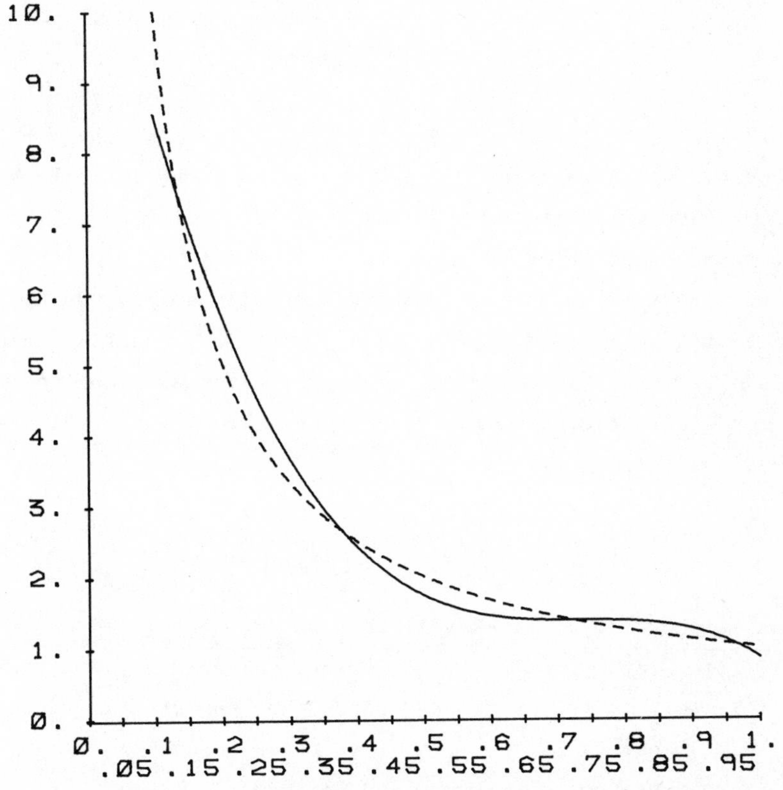

FIGURE 2 Optimum degree 3 polynomial approximation to $1/\lambda$ on $(\beta_3, \beta_4) = (.1,1)$ as derived from the optimum residual polynomial of Figure 1. The dashed line shows $1/\lambda$.

Observe that Q_{k-1} is such that $\lambda Q_{k-1}(\lambda)$ is the optimum uniform approximation to 1 on (β_3,β_4) over all polynomials of degree k with a root at zero. In this sense, Q_{k-1} is an optimum approximation to $1/\lambda$. See Figure 2. It is not the optimum uniform approximation in the sense that the absolute error is uniformly smallest over (β_3,β_4) but is in the sense that the relative error is uniformly smallest, which follows from the fact that the relative error is

$$\frac{|1/\lambda - Q_{k-1}(\lambda)|}{\lambda} = |1 - \lambda Q_{k-1}(\lambda)|$$

$$= |P_k(\lambda)|.$$

In this sense, if $(\beta_3,\beta_4) = (\lambda_N,\lambda_1)$, $Q_{k-1}(A)$ is the optimum polynomial approximation to A^{-1}, and will be said to be A-optimum. Otherwise if $(\beta_3,\beta_4) \neq (\lambda_N,\lambda_1)$ it will be called an optimum preconditioning polynomial.

It should be noted that convergence follows from using any (β_3,β_4) if k is odd. To prove this it is sufficient that $Q_{k-1}(A)A$ be symmetric definite, since definiteness implies that the conjugate gradient method converges. Now, (positive) definiteness holds if

$$Q_{k-1}(\lambda_i)\lambda_i > 0 \tag{5}$$

for any eigenvalue of A. Recall that $Q_{k-1}(\lambda)\lambda = 1 - R_k(\lambda)$, and $|R_k(\lambda)| < 1$ for $0 < \lambda \leq \beta_4$. Therefore (5) holds if $\lambda \leq \beta_4$. If $\beta_4 < \lambda$ it follows that $R_k(\lambda) < 0$ since the number of roots is odd, and (5) holds.

These facts on convergence and optimum polynomial preconditioning are summarized in

Theorem 1. (I) If Q_{k-1} is the A-optimum preconditioning polynomial, the relative error in the approximation of $1/\lambda$ by

$Q_{k-1}(\lambda)$ on (λ_N, λ_1) is uniformly smallest among all polynomials of degree k-1.

(II) If Q_{k-1} is any optimum preconditioning polynomial for which k is odd, then the cgT method converges.

3 RICHARDSON'S METHOD FOR SYMMETRIC INDEFINITE MATRICES

Richardson's method will be applied in this section to the solution of a linear set, Ax = b, for which the eigenvalues are real and contained in the union, S, of two disjoint intervals,

$$S = (\beta_1, \beta_2) \cup (\beta_3, \beta_4),$$

where $\beta_2 < 0 < \beta_3$. The eigenvalues must be real, but the matrix need not be symmetric. This fact is essential if an incomplete factorization is used to precondition a symmmetric matrix since an incomplete factorization transforms symmetry into nonsymmetry, but, under certain conditions, transforms real eigenvalues into real eigenvalues. See Subsection 4.3. Statements in Section 2 on Richardson's method also hold if the matrix is nonsymmetric but the eigenvalues are positive real.

A restriction to two disjoint intervals is not characteristic of approximation theory, but shows only the rudimentary stage of parts of the theory of iterative methods.

3.1 Effects of Positive and Negative Eigenvalues

Both positive and negative eigenvalues are assumed to exist. Some of the ways in which this fact impinges on Richardson's method are compiled below.

- The optimum residual polynomials on S are not orthogonal.
- An analytic representation of optimum residual polynomials is due to Atlestam [2] but is not obviously helpful for computing

roots or parameters; parameters may be computed from an algorithm due to de Boor and Rice [4]. The results of Atlestam may be useful in showing how to determine the degree of the residual polynomial so as to attain a given bound on the uniform norm, which is needed to determine the number of iterative steps necessary to reduce the residual error by a given amount.

• One (and only one) root of the optimum residual polynomial may lie outside S, a troublesome property to contend with in the dynamic computation of eigenvalues.

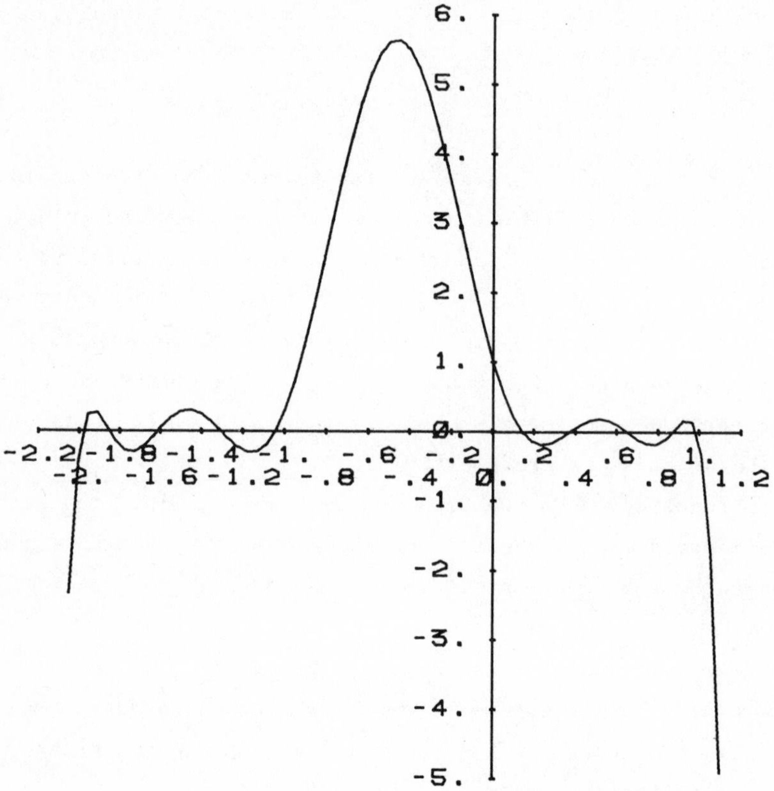

FIGURE 3 The cr-optimum residual polynomial for S = (-2,-1)$\cup$(.1,1) with five negative and five positive roots: -1.98, -1.79, -1.42, -.891, .904, .125, .296, .550,

3.2 Constraining the Roots to S

Another approach to residual polynomials is due to Roloff [10], who constrains the roots to S. Consider the class of residual polynomials with J_L negative and J_R positive roots in S, and degree $k = J_L + J_R$. The polynomial from this class with the smallest uniform norm on S will be called the cr-optimum residual polynomial (for constrained root). It will be denoted by $P_{cr,k}$, with a special partition of k,

$$k = J_L + J_R,$$

understood. It is equioscillating on (β_1, β_2) and (β_3, β_4), as in Figure 3. To explain the relation between optimum and cr-optimum let

$$\Delta_L = \max_{\lambda \varepsilon (\beta_1, \beta_2)} |P_{cr,k}(\lambda)|$$

and

$$\Delta_R = \max_{\lambda \varepsilon (\beta_3, \beta_4)} |P_{cr,k}(\lambda)|.$$

If $\Delta_L = \Delta_R$, then the cr-optimum residual polynomial is optimum. Roloff showed for sufficiently large k that the roots of the optimum residual polynomial lie in S. He also showed that if J_L is the number of negative roots of the optimum residual polynomial then J_L/k is independent of k for sufficiently large k. Roloff observed that this fact is helpful in the following way. Computing the optimum residual polynomial is more efficient if the number of positive or negative roots is known in advance. To determine either J_L or J_R for large k, it is sufficient to do so for small k.

3.3 Examples

Let $S = (-11,-10) \cup (10,12)$. Roots of the optimum residual polynomial of degree 3 are -10.5174, 11, and 32.5174. See Figure 5. The cr-optimum residual polynomial with one negative and two positive roots is shown in Figure 4.

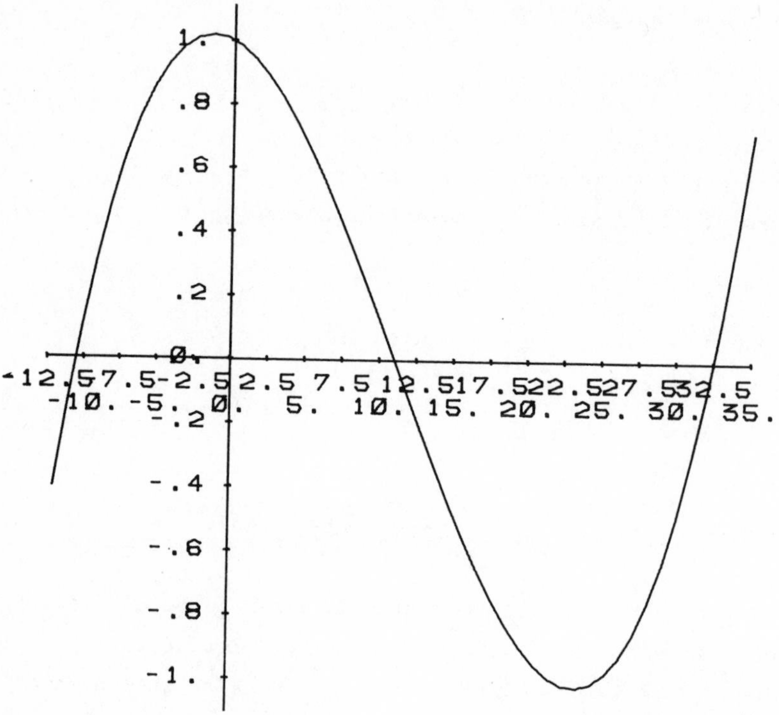

FIGURE 4 The cr-optimum residual polynomial for the same S as in Figure 4, $\Delta_L = .1816$, $\Delta_R = .00846$. The roots are -10.5233, 10.3050, and 11.7183.

The algorithm of Roloff [10] is used here and below to compute cr-optimum residual polynomials.

A cr-optimum preconditioning polynomial is shown in Figure 6. It is derived from the residual polynomial in Figure 3.

4 PRECONDITIONING SYMMETRIC INDEFINITE MATRICES

4.1 The cgRm Iteration

The conjugate gradient Richardson's method (cgRm) iteration consists of the conjugate gradient method applied to the preconditioned set

$$Q_{k-1}(A)Ax = Q_{k-1}(A)b$$

where $Q_{k-1}(\lambda) = [1 - R_k(\lambda)]/\lambda$ and R_k is an optimum or cr-optimum residual polynomial. The corresponding preconditioning polynomial will be called optimum or cr-optimum, respectively.

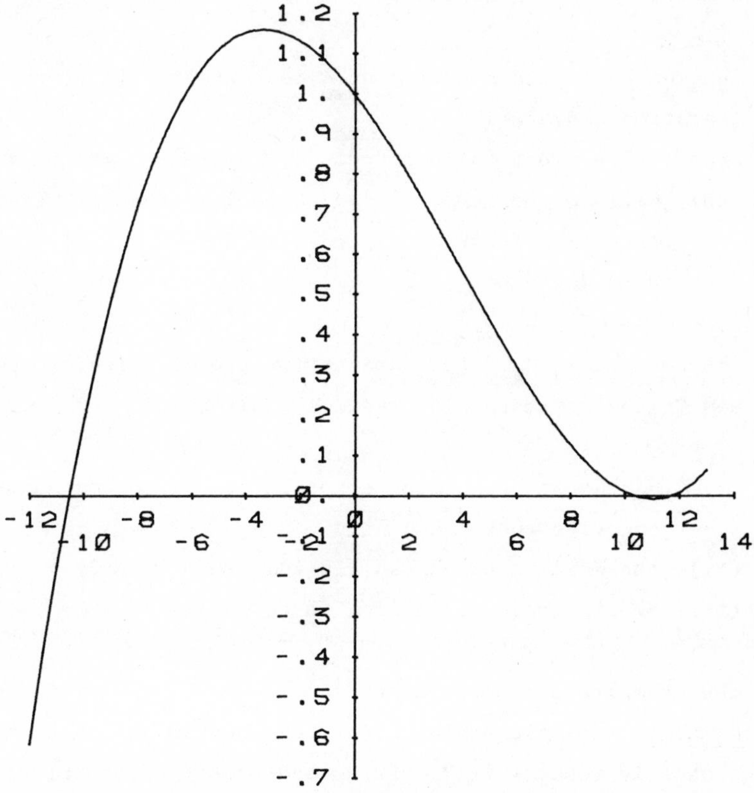

FIGURE 5 The optimum residual polynomial of degree 3 on S = (-11,-10) (10,12). The uniform norm on S is .1228.

The conjugate gradient method converges if $Q_{k-1}(A)A$ is symmetric positive definite. In the case A is symmetric positive definite, $Q_{k-1}(A)A$ is positive definite if k is odd for any (β_3, β_4); see Theorem 1. The proof depended on the property that $R_k(\lambda)$ is negative if $\beta_4 < \lambda$. No such general convergence statement is possible in the indefinite case for two reasons: (i) a root of R_k may lie outside S; and (ii) the maximum of R_k in (β_2, β_4) may be greater than 1. Either reason may cause $1 - R_k(\lambda)$ to be negative if λ is an eigenvalue of A outside S. Convergence still results under rather general conditions if more information is assumed for the eigenvalues.

<u>Theorem</u> <u>2</u> If the spectrum of A is contained in S, then the cgRm iteration converges.

Proof: The proof follows from [4, Theorem 1], which implies that the maximum of $|R_k(\lambda)|$ on S is less than 1. Therefore $1 - R_k(\lambda_i)$ is positive for any eigenvalue and $Q_{k-1}(A)A$ is symmetric positive definite.

<u>Theorem</u> <u>3</u> Let Q_{k-1} be a cr-optimum preconditioning polynomial and R_k the corresponding residual polynomial.

If:

(i) R_k has an odd number of positive and an odd number of negative roots; and

(ii) the maximum of $|R_k(\lambda)|$ is less than 1 on S;

(iii) no eigenvalue of A lies in (β_2, β_3);

then:

the cgRm iteration converges.

<u>Remark</u>. A simple statement of the theorem is that convergence usually results if R_k has an odd number of positive and an odd number of negative roots.

<u>Proof</u> of Theorem 3: The objective is to show that $1 - R_k(\lambda_i)$ is positive for any eigenvalue of A. By (iii) this holds if λ_i is in S, and by (ii) it holds if λ_i is outside S.

4.2 Algorithm for the Conjugate Gradient Richardson's Method Iteration

The preconditioned system

$$Q_k(A)Ax = Q_k(A)b$$

though a deft algebraic relation is not practical to compute with since optimization algorithms produce the roots of R_k rather than Q_k. An inner/outer iteration is recommended in which the inner iteration is Richardson's method and the outer is the conjugate gradient method. This is the form of the algorithm described by Rutishauser [11].

A statement of the algorithm follows. Let x_0 be an initial guess, and $t_0,\ldots,t_{k-1}$ be iteration parameters for one period.

(I) Set $\tilde{x}^{(0)} := x_0$.

(II) Initial inner loop for Richardson's method.

For $i = 0,\ldots,k-1$:

Set $\tilde{r}^{(i)} := b - A\tilde{x}^{(i)}$.

Set $\tilde{x}^{(i+1)} := \tilde{x}^{(i)} + t_i\tilde{r}^{(i)}$.

(III) Set $j := 0$.

(IV) Prepare for outer loop.

Set $q^{(j)} := \tilde{x}^{(k)} - x^{(j)}$.

(V) Conjugate gradient segment.

Set $\alpha_j := (r^{(j)},q^{(j)})/(d^{(j)},Ad^{(j)})$.

Set $x^{(j+1)} := x^{(j)} + \alpha_j d^{(j)}$.

Set $r^{(j+1)} := r^{(j)} - \alpha_j Ad^{(j)}$.

(VI) Richardson's method inner loop.

Set $\tilde{x}^{(0)} := x^{(j+1)}$.

Set $\tilde{r}^{(0)} := r^{(j+1)}$.

For $i = 0,\ldots,k-1$:

set $\tilde{x}^{(i+1)} := \tilde{x}^{(i)} + t_i\tilde{r}^{(i)}$.

Set $\tilde{r}^{(i)} := b - A\tilde{x}^{(i)}$.

(VII) Set $q^{(j+1)} := \tilde{x}^{(k)} - x^{(j+1)}$.

(VIII) Conjugate gradient segment.

$$\text{Set } \gamma_j := (q^{(j+1)}, r^{(j+1)})/(q^{(j)}, r^{(j)}).$$

$$\text{Set } d^{(j+1)} := r^{(j+1)} + \gamma_j d^{(j)}.$$

(IX) Set $j := j + 1$. Return to (V).

4.3 Incomplete Factorizations

If K is an incomplete factorization of A such that K is symmetric positive (or negative) definite, then the preconditioned matrix

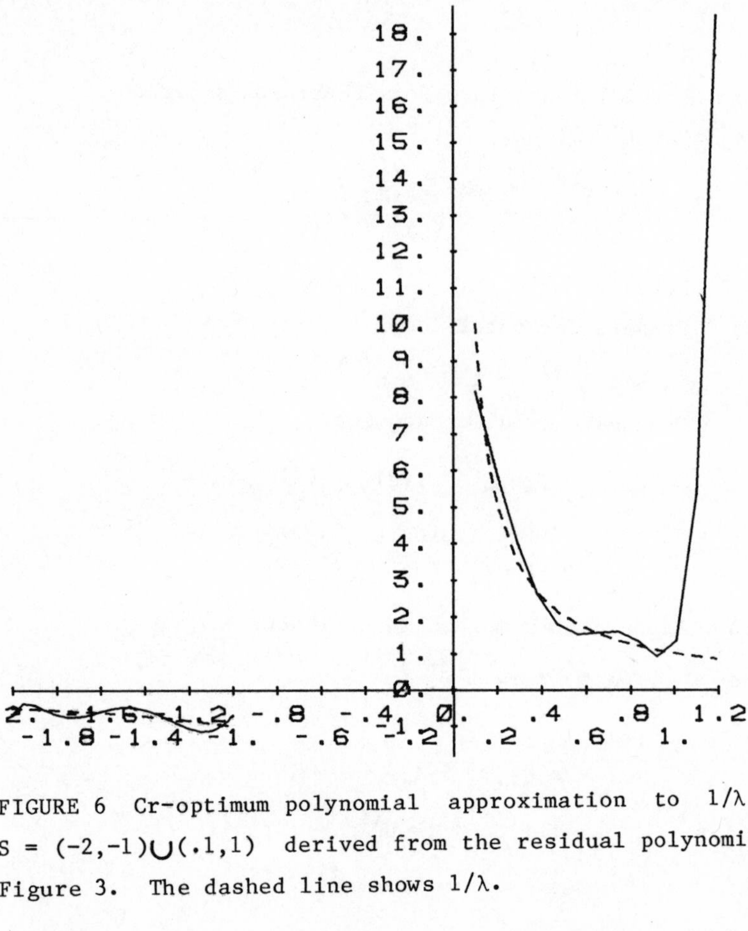

FIGURE 6 Cr-optimum polynomial approximation to $1/\lambda$, on $S = (-2,-1) \cup (.1,1)$ derived from the residual polynomial of Figure 3. The dashed line shows $1/\lambda$.

$K^{-1}A$ is similar to the positive definite matrix $K^{-1/2}AK^{1/2}$. Richardson's method may therefore be applied to $K^{-1}Ax = K^{-1}b$ since the eigenvalues are real. If K is symmetric but not positive definite then $K^{-1}A$ may still have real eigenvalues, in which case Richardson's method is applicable. If nonreal eigenvalues exist, for which each real part is positive then various methods are possible including [9]. In the general case in which there are eigenvalues with a positive real part and eigenvalues with a negative real part, there are few novel solution techniques.

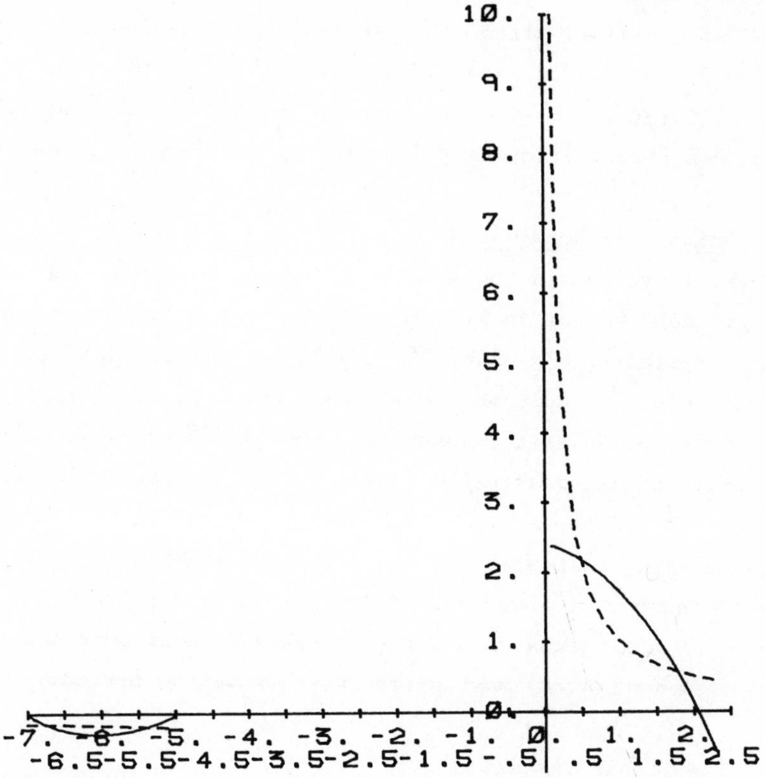

FIGURE 7 The cr-optimum polynomial approximation to $1/\lambda$ derived from the residual polynomial of Subsection 4.4a.

4.4 Examples

The tests described here employ diagonal matrices, an indication of the undeveloped state of numerical experiments. However, algebraic properties of the conjugate gradient method as well as many roundoff properties are independent of the (orthogonal) coordinate system, facts that justify the convenient ease of diagonal matrices. A more incisive criticism is that the eigenvalues are known. In practice they must be computed by a dynamic algorithm, which is discussed in the next section.

4.4a The Residual Polynomial

In both examples, the period of Richardson's method is 4. Parameters are derived from the residual polynomial with roots -6.765816, -5.380278, .4562219 and 1.775159;, it is cr-optimum on $S = (-7,-5) \cup (.1,2)$ with $\Delta_L = .842$ and $\Delta_R = .762$. The derived cr-optimum preconditioning polynomial is plotted in Figure 7.

4.4b Arithmetic Work Units

The cgRm iteration is compared to two other methods, the conjugate gradient method applied to $Ax = b$ and the conjugate gradient method applied to $A^2x = Ab$, which will be called cgA and cgA^2 respectively. A unit of work is that required for one conjugate gradient step, under the assumption that $Ax^{(j)}$ takes 5N multiples as would be necessary if A resulted from a partial differential equation on a 2D-domain. A work unit is therefore defined to be the number of multiples for one step of cgA, which is 10N, where N is the number of unknowns. The cgA^2 method takes 1.5 work units. Since Richardson's method takes .6 work units, one step of cgRm takes (1+.6k) work units, where k is the period.

4.4c The Linear Systems

Let $\lambda_1,\ldots,\lambda_n$ be the negative and $\lambda_{n+1},\ldots,\lambda_N$ the positive eigenvalues. In each example $\lambda_1 = -7$, $\lambda_n = -5$, $\lambda_{n+1} = .1$ and $\lambda_{N-1} = 2$, which are the boundary points of S. The remaining eigenvalues, except λ_N, are uniformly distributed in S by setting $\lambda_i = -7 + 2i/(n-1)$, $0 \leqslant i \leqslant n-1$, and $\lambda_i = .1 + 1.9i/(N-n-2)$,

$0 \leq i \leq N-n-2$. The components of the starting vector are each set to 1 and the right side taken to be zero. Eigenvalue λ_N is choosen such that the initial residual is 10^{-10}, which causes the conjugate gradient method to oscillate, as shown by the solid curve in Figures 8 and 9.

In the example of Figure 8, N = 50 and n = 5. This yields λ_N = 10.1004, to six places. In Figure 8 the vertical scale is the Euclidean norm of the residual. The dashed line is a plot of cgA^2 and the hatched a plot of cgRm. The dot to the left of the vertical scale (halfway up) marks 10^{-4}. The marks reading upward occur at $2^3\times10^{-4}$, $3^3\times10^{-4}$, etc. The number of cgA^2 steps suffi-

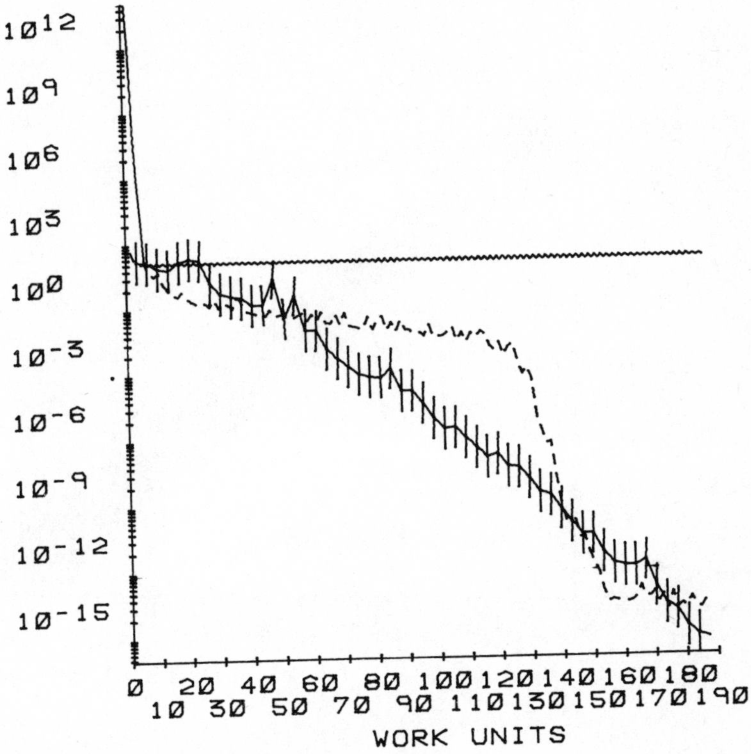

FIGURE 8 Euclidean norm of the residual plotted against work for the cgA (line), cgA^2 (dashed line) and cgRm (hatched line) iterations for a 50 by 50 matrix.

cient for the exact solution if there were no roundoff would be the same as the number of unknowns which is 50, or 75 work units. The exact value of the Euclidean norm of the residual at 50 cgA^2 steps is .0121 which is marked on the vertical scale as $4.95^3 \times 10^{-4}$. The solution then oscillates while the residual slowly declines for 75 more work units before converging precipitously in a manner characteristic of the conjugate gradient method. The performance of cgRm, shown by the hatched line is superior and smoother but only after 55 work units.

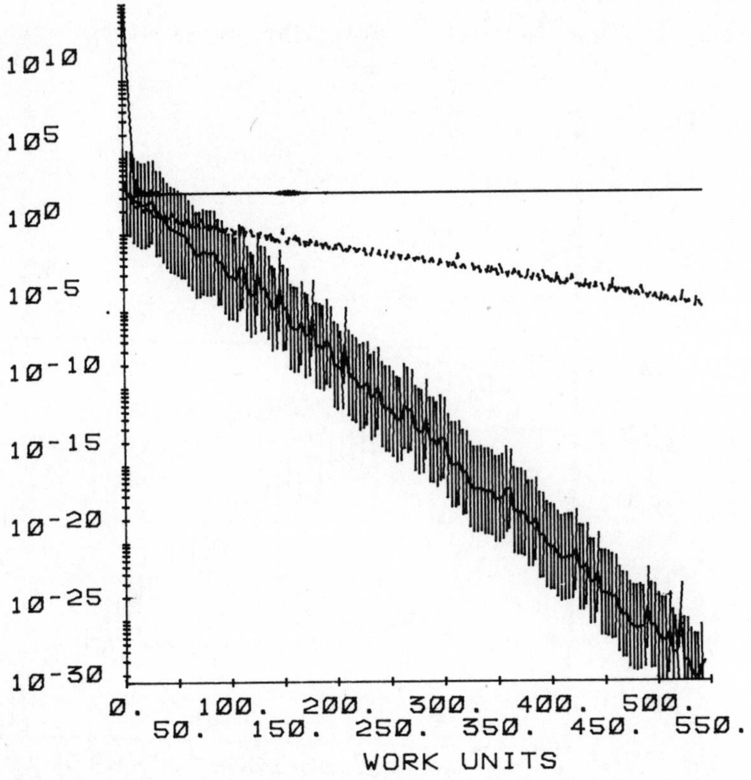

FIGURE 9 Euclidean norm of the residual plotted against work for cgA (line), cgA^2 (dashed line) and cgRm (hatched line) for a 500 by 500 matrix. Dot to left of vertical axis is placed at 10^{-6}.

In the example of Figure 9, $n = 50$ and $N = 500$, and $\lambda_N = 21.6623$. The percentage of negative and positive eigenvalues is the same as in the first example. The cgRm iteration after about 160 work units reduces the error to 10^{-6}, an average reduction per step equal to .917, i.e., $.917^{160} = 10^{-6}$, considered large for the error reduction per step in the iterative solution of symmetric positive definite systems.

4.4d Remark

Observe that Richardson's method used alone is superior since the error reduction per step is .84, which is the larger of Δ_L and Δ_R, and it takes less work per step. An advantage of the cgRm iteration is that with an odd number of negative and an odd number of positive roots, it generally converges (Theorem 3). If Richardson's method were used alone for a matrix with eigenvalues less than β_1 and greater than β_2, it would diverge.

5 ADAPTIVE PARAMETER STRATEGY

An adaptive parameter strategy, which computes parameters dynamically, during execution, is necessary for the practical use of Richardson's method. Ideally $S = (\beta_1, \beta_2) \cup (\beta_3, \beta_4)$ should contain the eigenvalues and β_i should be an eigenvalue. In practice the β_i's are input parameters for a routine. A parameter strategy would allow the routine to compute eigenvalues autonomously, refining each β_i until it approximated an eigenvalue.

The strategy described here depends on the use of cr-optimum residual polynomials which are monotone when the argument is less than β_1 or greater than β_4. It is convenient to assume no eigenvalue lies in (β_2, β_3) but this is not essential. To illustrate the technique assume $\lambda_1 < \beta_1$ and $|R_k(\lambda_i)| < |R_k(\lambda_1)|$ for each other eigenvalue. In this section R_k is assumed to be a cr-optimum residual polynomial. It follows that $r^{(k)}$ is approximately the eigenvector of A corresponding to λ_1, and therefore λ_1 is approximately $(Ar^{(k)},r^{(k)})/||r^{(k)}||^2$, which will be the new value of β_1. The iteration continues and at each succeeding step another β_i is improved.

If $\lambda_1 < \beta_1$ it may be expected that $1 < |R_k(\lambda_1)|$, in which case Richardson's method is diverging. If the cgRm method is used with an odd number of negative roots and an odd number of positive roots of the residual polynomial then it generally converges. Eigenvalue estimates may be computed from the Richardson's method inner iteration described above. Information may also be obtained as follows. Let B be the tridiagonal matrix,

$$B = \left(\dots - \frac{\gamma_{i-1}}{\alpha_{i-1}}, \frac{1}{\alpha_i} + \frac{\gamma_{i-1}}{\alpha_{i-1}}, - \frac{1}{\alpha_i} \dots \right)$$

where the parameters are defined in Subsection 4.2, and the first row is $1/\alpha_0$, $-1/\alpha_0$, 0,... It is a well-known property of the conjugate gradient method, which shows it is equivalent to the Lanczos algorithm, that the eigenvalues of B are approximate eigenvalues of the system matrix, in this case $Q_{k-1}(A)A = I - R_k(A)$. Let μ_i denote an eigenvalue of B. Suppose $\Delta_R < \Delta_L$. Exclude all μ_i for which λ_i would lie in S by requiring that $1 + \Delta_L < |\mu_i|$. Next solve $1 - R_k(\lambda_i) = \mu_i$ for λ_i, which is feasible since $R_k(\lambda)$ is monotonic to the left of β_1 and to the right of β_2. There are two roots of opposite sign that are candidates for the value of λ_i since $R_k(\lambda) < 0$ for λ not in (β_1,β_4). The correct sign is given by the sign of $(A\tilde{r}^{(k)},\tilde{r}^{(k)})$ where $\tilde{r}^{(k)}$ is the Richardson's method residual from the inner iteration.

REFERENCES

1. ANDERSON, R.S. and GOLUB, G.H. - Richardson's Nonstationary Matrix Iterative Procedure, Dept. of Computer Science,

Stanford, CA, Report No. 304, Aug., 1972.

2. ATLESTAM, B. - Tschebycheff Polynomials for Sets Consisting of Two Disjoint Intervals with Application to Convergence Estimates for the Conjugate Gradient Method, Dept. of Computer Science, Chalmers Institute of Technology and the University of Göteborg, Göteborg, Sweden, Report No. 77.06R.
3. CHANDRA, RATI - Conjugate Gradient Methods for Partial Differential Equations, Dept. of Computer Science, Yale University, New Haven, Conn., Research Report No. 129, Jan., 1978.
4. DE BOOR, CARL and RICE, JOHN R. - Extremal Polynomials with Application to Richardson's Iteration for Indefinite Linear Systems, Mathematics Research Center, Report No. 2107, University of Wisconsin, June, 1980. To appear in SIAM Journal on Scientific and Statistical Computing.
5. FLETCHER, R. - Conjugate Gradient Methods for Indefinite Systems, Symposium on Numerical Analysis, Ed. Dold, A. and Eckmann, B., Springer-Verlag, New York, 1975.
6. FOLKERTS, C.H. - Development of Linear and Nonlinear Magnetic Circuit Models for the Prediction of Step Motor Static Performance, Ph.D. thesis, Dept. of Electrical Engineering, The University of Illinois at Urbana-Champaign, Urbana, Ill., Aug., 1978.
7. FORSYTHE, GEORGE E. and WASOW, WOLFGANG R. - Finite Difference Methods for Partial Differential Equations, Wiley, New York, 1960.
8. LANCZOS, C. - Solution of Systems of Linear Equations by Minimized Iterations. J. Res. Nat. Bur. of Standards, Vol. 49, pp. 33-53, 1952.
9. MANTEUFFEL, THOMAS A. - The Tchebyshev Iteration for Nonsymmetric Linear Systems, Numer. Math. Vol. 28, pp. 307-327, 1977.
10. ROLOFF, R.R. - Iterative Solution of Matrix Equations for Symmetric Matrices Possessing Positive and Negative Eigenvalues, Dept. of Computer Science, University of Illinois, 1304 W. Springfield, Urbana, IL 61801, Report No. UIUCDCS-R-79-1018, Oct., 1979.
11. RUTISHAUSER, H. - Theory of Gradient Methods, Refined Iterative Methods for Computation of the Solution and the Eigenvalues of Self-Adjoint Boundary Value Problems, Engli, M., Ginsburg, Th., Rutishauser, H. and Stiefel, E., Springer-Verlag, Heidelburg, 1959.
12. SHORTLEY, G. - Use of Tschebyscheff Polynomial Operators in the Numerical Solution of Boundary Value Problems. J. Appl. Phys., Vol. 24, pp. 392-396, 1953.
13. YOUNG, D.M. - Iterative Solution of Large Linear Systems, Academic Press, New York, 1971.
14. YOUNG, D.M. - On Richardson's Method for Solving Linear Systems with Positive Definite Matrices. J. Math. Phys., Vol. XXXII, pp. 243-255, 1954.

THE PRECONDITIONED CHEBYSHEV ITERATIVE METHOD FOR UNSYMMETRIC LINEAR SYSTEMS OF EQUATIONS

D.J. Evans & M.A. Kammoonah

Department of Computer Studies, University of Technology, Loughborough, Leicestershire, U.K.

ABSTRACT

In this paper, the application of preconditioning to the Chebyshev iterative method of solution is presented. Large,sparse unsymmetric linear systems derived from the finite difference discretization of a 2nd order elliptic partial differential equations over a rectangular domain are obtained and solved by a second order preconditioned iterative method based on Chebyshev polynomials.

Further, by use of an adaptive procedure to determine the optimum preconditioning parameter, a numerical example is described and experimental results obtained which confirm the theory.

1. INTRODUCTION

In this paper, the use of the preconditioning strategy is extended to systems of equations with unsymmetric coefficient matrix A. Thus, we have to solve,

$$A\underline{\phi} = \underline{b} , \qquad (1.1)$$

with $A = I-L-U$,

and is assumed to be unsymmetric (i.e. $U \neq L^T$).

Then, by applying the preconditioning technique, Evans (1968) we have to solve the alternative system of equations:

$$A_\omega \underline{\phi}_\omega = \underline{b}_\omega , \qquad (1.2)$$

with $A_\omega = (I-\omega L)^{-1} A (I-\omega U)^{-1}$,

$$\underline{\phi}_\omega = (I-\omega U)\underline{\phi}, \text{ and } \underline{b}_\omega = (I-\omega L)^{-1}\underline{b} .$$

Then, equation (1.2) can be written in the form:

$$FAG\underline{\phi}_\omega = \underline{b}_\omega , \tag{1.3}$$

where, $G = (I-\omega U)^{-1}$, $F = (I-\omega L)^{-1}$, and $F \neq G^T$.

The inverse of the matrices G and F with respect to a right hand side vector can still be obtained easily by a backward and forward substitution process, (Evans, 1968). Also, G and F are simply the upper and lower triangular copies of the coefficient matrix A so that the sparseness of A is again preserved.

The matrix A_ω is also unsymmetric and probably has complex eigenvalues.

In the next section we apply the preconditioning technique to modify the standard Simultaneous Displacement iterative method, (Evans, 1968).

Whilst the application of the preconditioning technique to the Chebyshev semi-iterative method, (Stiefel, 1958) will be presented in Section 3.

2. THE PRECONDITIONING SIMULTANEOUS DISPLACEMENT ITERATIVE METHOD

To solve the system of equations (1.2) we consider the iteration method, given by

$$\underline{\phi}_\omega^{(n+1)} = \underline{\phi}_\omega^{(n)} + \alpha(\underline{b}_\omega - A_\omega\underline{\phi}_\omega^{(n)}) , \tag{2.1}$$

where α is the acceleration parameter, to be discussed later.

If we now premultiply (2.1) by $(I-\omega L)$, we obtain the following result,

$$(I-\omega L)(I-\omega U)\underline{\phi}^{(n+1)} = (I-\omega L)(I-\omega U)\underline{\phi}^{(n)} + \alpha(\underline{b} - A\underline{\phi}^{(n)}) . \tag{2.2}$$

If we denote

$$G_\omega = (I-\omega L(I-\omega U) ,$$

then (2.2) becomes,

$$G_\omega\underline{\phi}^{(n+1)} = G_\omega\underline{\phi}^{(n)} + \alpha(\underline{b} - A\underline{\phi}^{(n)}) . \tag{2.3}$$

To derive a more compact form we add and subtract L and U on both sides of equation (2.3) to obtain the result

$$(A+B_\omega)\underline{\phi}^{(n+1)} = (A+B_\omega)\underline{\phi}^{(n)} + \alpha(\underline{b} - A\underline{\phi}^{(n)}) , \tag{2.4}$$

where, $A = I-L-U$, and $B_\omega = L+U-\omega L-\omega U+\omega^2 LU = (1-\omega)(L+U)+\omega^2 LU$.

A convergence criteria for equation (2.4) can be derived from the following.

Remark (2.1): By a comparison of (2.4) and (2.3) we obtain

$$G_\omega = A + B_\omega \quad .$$

Remark (2.2): By a similarity transformation, we can show that

$$\begin{aligned} A_\omega &\equiv (I-\omega L)(I-\omega L)^{-1} A(I-\omega U)^{-1}(I-\omega L)^{-1} \\ &= A((I-\omega L)(I-\omega U))^{-1} \\ &= A(A+B_\omega)^{-1} \quad . \end{aligned}$$

Remark (2.3): Since

$$A_\omega = (I-\omega U)(I-\omega U)^{-1}(I-\omega L)^{-1}A(I-\omega U)^{-1},$$

then by a similarity transformation, we have the result

$$\begin{aligned} A_\omega &\equiv ((I-\omega L)(I-\omega U))^{-1}A \\ &= (A+B_\omega)^{-1}A. \end{aligned}$$

Now, we can say that the matrices A_ω, $(A+B_\omega)^{-1}A$ and $A(A+B_\omega)^{-1}$ all have the same eigenvalues, but different eigenvectors, i.e., they are similar.

Lemma (2.1): The iteration (2.4) is convergent if, and only if,

$$\mathrm{Re}(\lambda_i(A^{-1}B_\omega)) > \frac{\alpha}{2} - 1 \ ,$$

where A and B_ω are arbitrary matrices defined in (2.4).

Proof: Let $\underline{\Phi}$ be the exact solution of (1.1), then the error of the nth iterate is given by:

$$\underline{E}^{(n)} = \underline{\Phi} - \underline{\phi}^{(n)} \quad . \tag{2.5}$$

Then, by subtracting (2.4) from the equation

$$(A+B_\omega)\underline{\phi} = (A+B_\omega)\underline{\phi} + \alpha(\underline{b}-A\underline{\Phi}) \ ,$$

yields, $(A+B_\omega)(\underline{\Phi}-\underline{\phi}^{(n+1)}) = (A+B_\omega)(\underline{\Phi}-\underline{\phi}^{(n)})-\alpha(A(\underline{\Phi}-\underline{\phi}^{(n)}))$

or by use of (2.5) is

$$\underline{E}^{(n+1)} = (A+B_\omega)^{-1}(A+B_\omega-\alpha A)\underline{E}^{(n)} .$$

Thus, the matrix

$$(A+B_\omega)^{-1}(A+B_\omega-\alpha A) = (I-\alpha(A+B_\omega)^{-1}A) = (I-\alpha A_\omega) , \qquad (2.6)$$

is the iteration matrix (error operator) of (2.4).

It is well known that, the iteration (2.4) is convergent if, and only if,

$$\rho(I-\alpha A_\omega) < 1 , \qquad (2.7)$$

where $\rho(M)$ denotes the spectral radius of matrix M.

Equation (2.7) can be written as

$$\lambda_i(I-\alpha A_\omega) \in \delta(0,0;1) , \qquad (2.8)$$

where δ is defined to be the circular region as:

$$\delta(u,v;R) = \{(u,v):(u-x)^2+(v-y)^2<R^2\} .$$

Then, by division we have the result

$$\lambda_i(\frac{1}{\alpha} I-A_\omega) \in \delta(0,0;\frac{1}{\alpha}) , \qquad (2.9)$$

and hence,

$$\lambda_i(A_\omega) \in \delta(\frac{1}{\alpha},0;\frac{1}{\alpha}) , \qquad (2.10)$$

i.e. $\lambda_i(A_\omega^{-1})$ has a real part greater than $\frac{\alpha}{2}$, or,

$$Re(\lambda_i((A+B_\omega)^{-1}A)^{-1}) > \frac{\alpha}{2} .$$

Since,

$$((A+B_\omega)^{-1}A)^{-1} = (I+A^{-1}B_\omega),$$

then we have the result,

$$Re(\lambda_i(I-A^{-1}B_\omega)) > \frac{\alpha}{2} .$$

Thus, it follows that the iteration (2.4) is convergent if, and only if, (iff)

$$Re(\lambda_i(A^{-1}B_\omega)) > \frac{\alpha}{2} - 1 . \qquad (2.11)$$

Corollary (2.1): If A is a positive definite matrix, and B_ω is an Hermitian matrix, then the iteration (2.4) is convergent if, and only if, $(A+B_\omega - \frac{\alpha}{2} A)$ is positive definite.

Proof: If A and B_ω are of Hermitian form, then

$$\lambda_i(A^{-1}B_\omega) \text{ is real (for all i)},$$

and (2.11) becomes,

$$\lambda_i(A^{-1}B_\omega) > \frac{\alpha}{2} - 1 ,$$

which is equivalent to

$$\lambda_i(I(1-\frac{\alpha}{2}) + A^{-1}B_\omega) > 0.$$

But we have

$$(I(1-\frac{\alpha}{2}) + A^{-1}B_\omega) = A^{-1}A(I(1-\frac{\alpha}{2}) + A^{-1}B_\omega)$$
$$= A^{-1}(A+B_\omega - \frac{\alpha}{2}A).$$

Thus (2.11) is equivalent to

$$\lambda_i(A^{-1}(A+B_\omega - \frac{\alpha}{2}A)) > 0 .$$

By similarity transformation, we can show that:

$$(A^{-1}(A+B_\omega - \frac{\alpha}{2}A)) \equiv A^{\frac{1}{2}}(A^{-1}(A+B_\omega - \frac{\alpha}{2}A))A^{-\frac{1}{2}}$$
$$= A^{-\frac{1}{2}}(A+B_\omega - \frac{\alpha}{2}A)A^{-\frac{1}{2}} ,$$

which may be restated as:

$$< A^{-\frac{1}{2}}(A+B_\omega - \frac{\alpha}{2}A)A^{-\frac{1}{2}}\underline{\phi},\underline{\phi}>>0, \text{ (for all } \underline{\phi}\neq\underline{0}) .$$

Therefore, we have the result,

$$< (A+B_\omega - \frac{\alpha}{2}A)A^{-\frac{1}{2}}\underline{\phi},A^{-\frac{1}{2}}\underline{\phi}>>0, \text{ (for all } \underline{\phi}\neq\underline{0}) .$$

If we now let

$$\underline{x} = A^{-\frac{1}{2}}\underline{\phi},$$

then, we obtain the simpler result

$$< (A+B_\omega - \frac{\alpha}{2}A)\underline{x},\underline{x}>>0, \quad \text{(for all } \underline{x}\neq\underline{0}) .$$

Hence, the iteration (2.4) is convergent if, and only if the matrix $(A+B_\omega - \frac{\alpha}{2}A)$ is positive definite.

Remark (2.4): For $\alpha=1$, corollary (2.1) implies: the iteration (2.4) is convergent if, and only if, $(A+2B_\omega)$ is positive definite.

Lemma (2.2): If A and $(A+B_\omega)$ are positive definite, then the iteration (2.4) is convergent if, and only if,

$$0 < \alpha < \frac{2}{\lambda_{max}(A_\omega)} .$$

Proof: Since A and $(A+B_\omega)$ are positive definite, then they both have real eigenvalues. Then A_ω has real eigenvalues, and

$\lambda_i(A_\omega)>0$, (Remark (2.3)).

From equation (2.7) we have that the iteration (2.4) is convergent if, and only if,

$$|\lambda_i(I-\alpha A_\omega)| < 1,$$

i.e. $$-1<\lambda_i(I-\alpha A_\omega) < 1 .$$

If we let $$-1 < 1 -\alpha\lambda_{max}(A_\omega) ,$$

then

$$\alpha < \frac{2}{\lambda_{max}(A_\omega)},$$

and if $$1 - \alpha\lambda_i(A_\omega) < 1,$$

then

$$\alpha>0.$$

Hence, the iteration (2.4) is convergent if, and only if,

$$0 < \alpha < \frac{2}{\lambda_{max}(A_\omega)} . \tag{2.12}$$

Lemma (2.3): If A and $(A+B_\omega)$ are positive definite, then:

$$\alpha_B = \frac{2}{\lambda_{max}(A_\omega)+\lambda_{min}(A_\omega)} ,$$

is the best choice of the acceleration parameter α in the iteration (2.4).

Proof: From (2.6) we have the iteration matrix (2.4) is similar to the matrix $$M = (I-\alpha A_\omega) .$$

Thus, since all the eigenvalues of A_ω are positive, then the eigenvalues of M are decreasing functions of α. Clearly, the spectral radius of M is minimized with respect to α, when

$$|\lambda_{min}(M)| = |\lambda_{max}(M)| ,$$

i.e. when, $$1-\alpha\lambda_{min}(A_\omega) = |1-\alpha\lambda_{max}(A_\omega)| .$$

Thus, the best choice is given by

$$1-\alpha_B\lambda_{min}(A_\omega) = -1 + \alpha_B\lambda_{max}(A_\omega) ,$$

which gives the result,

$$\alpha_B = \frac{2}{\lambda_{max}(A_\omega)+\lambda_{min}(A_\omega)}\ , \tag{2.13}$$

as the best choice of α in the iteration (2.4).

Remark (2.5): As

$$\lambda_{min}(A_\omega) \to 0, \text{ then,}$$

$$\alpha_B = \frac{2}{\lambda_{max}(A_\omega)}\ ,$$

is the best choice of α in the iteration (2.4).

3. THE PRECONDITIONED SEMI-ITERATIVE METHOD

To solve the system of equations (1.1) using preconditioning first, we have to solve the system of equations,

$$A_\omega \underline{\phi}_\omega = \underline{b}_\omega\ ,$$

as defined in (1.2).

If we let $\underline{\phi}_\omega^{(n)}$ be the nth iterative preconditioned solution vector of (1.2), and $\underline{\Phi}$ be the exact solution vector, then, the error vector at n^{th} step is obtained by,

$$\underline{E}_\omega^{(n)} = \underline{\Phi} - \underline{\phi}_\omega^{(n)}\ .$$

If we let,

$$\underline{r}_\omega^{(n)} = \underline{b}_\omega - A_\omega \underline{\phi}_\omega^{(n)}\ ,$$

be the residual at the nth step, then the $(n+1)^{th}$ step solution vector approximation can be obtained by the formula,

$$\underline{\phi}_\omega^{(n+1)} = \underline{\phi}_\omega^{(n)} + \sum_{i=1}^{n} \gamma_{n,i} \cdot \underline{r}_\omega^{(i)}\ ,$$

i.e. at each step, we add a linear combination of the previous residuals, where the γ's are constant numbers, and the error vector at $(n+1)^{th}$ step can be obtained by,

$$\underline{E}_\omega^{(n+1)} = \underline{E}_\omega^{(n)} - \sum_{i=1}^{n} \gamma_{n,i} \underline{r}_\omega^{(i)}$$

$$= \underline{E}_\omega^{(n)} - A_\omega \sum_{i=1}^{n} \gamma_{n,i} \underline{E}_\omega^{(i)}\ .$$

Also, we can prove by induction, see Stiefel (1958), that if $P_n(z)$ is a polynomial of degree n such that

$$P_n(0) = 1,$$

then the error at n^{th} step is given by,

$$\underset{\sim}{E}_{\omega}^{(n)} = P_n(A_{\omega}).\underset{\sim}{E}_{\omega}^{(0)} \quad . \tag{3.1}$$

Since, any sequence of polynomials, $\{P_n(z)\}$, can be generated by choosing a sequence of constant numbers, $\{\gamma_{n,i}\}$, then we would like to choose the sequence of polynomials so that

$$||\underset{\sim}{E}_{\omega}^{(n)}|| \leq ||P_n(A_{\omega})|| \, ||\underset{\sim}{E}_{\omega}^{(0)}|| \, ,$$

is small.

Further, if we examine $P_n(A_{\omega})$ then it is well known, (Birkhoff, 1953), that:

1. If A_{ω} is diagonalizable then,

   $$||P_n(A_{\omega})|| \to 0 \quad \text{as } n \to \infty,$$

   if, and only if,

   $$P_n(\lambda_{\omega i}) \to 0 \quad \text{as } n \to \infty,$$

   for all eigenvalues $\lambda_{\omega i}$ of A_{ω}.

2. If A_{ω} is not diagonalizable and if $\lambda_{\omega i}$ is an eigenvalue of A_{ω}, with multiplicity m_i, then

   $$||P_n(A_{\omega})|| \to 0 \text{ as } n \to \infty,$$

   if, and only if,

   $$P_n^{(j)}(\lambda_{\omega i}) \to 0 \text{ as } n \to \infty,$$

   for all $j < m_i$, for each eigenvalue $\lambda_{\omega i}$.

Then, to choose the sequence of poynomials, $\{P_n(z)\}$, we have to consider following three important criteria:

1. $P_n(z)$ must have some recursive properties so that there is no need to store all the previous residuals.
2. We must choose $P_n(z)$ among polynomials of like degree such that

   $$P_n(0) = 1,$$

   to be as small as possible on the spectrum of A_{ω}.
3. If A_{ω} has non-linear elementary divisors, then the $P_n(z)$ derivatives must be small on the spectrum of A_{ω}.

Manteuffel (1975), has recently shown that the best polynomial which fits the previous criteria is the scaled and translated Chebyshev polynomials.

4. THE SCALED AND TRANSLATED CHEBYSHEV POLYNOMIALS

Let $F(0,1)$ be the family of ellipses centered at the origin (0), with foci at (+1) and (-1).

Let $F_i(0,1)$, $F_j(0,1) \in F(0,1)$ be members of the F family.

Suppose $z_i \in F_i(0,1)$, $z_j \in F_j(0,1)$, then we can say that:

$$\mathrm{Re}(\cosh^{-1}(z_i)) \leq \mathrm{Re}(\cosh^{-1}(z_j)),$$

if and only if,

$$F_i(0,1) \subseteq F_j(0,1),$$

(i.e. the ellipse $F_i(0,1)$ is inside or equal to the ellipse $F_j(0,1)$).

Now consider the "scaled Chebyshev polynomials",

$$C_n(z) = \frac{T_n(z)}{T_n(z_0)}, \quad z_0 \notin [-1,1]. \tag{4.1}$$

These polynomials exhibit an asymptotic behaviour.

For large n, we can prove that

$$C_n(z) = e^{n[\cosh^{-1}(z) - \cosh^{-1}(z_0)]}, \tag{4.2}$$

where e denotes the exponential function.

Then if $F_0(0,1)$ is a member of the family $F(0,1)$ passing through z_0, it can be shown that

$$\lim_{n\to\infty} C_n(z) = \begin{Bmatrix} 0, & \text{if } z \in F_0(0,1) \\ \infty, & \text{if } z \notin F_0(0,1) \end{Bmatrix}. \tag{4.3}$$

Definition 4.1: Let

$$r(z) = \lim_{z\to\infty} \left|C_n(z)^{\frac{1}{n}}\right| ,$$

be the "asymptotic convergence factor" of $C_n(z)$ at the point z. The asymptotic convergence factor, which will be referred to as the convergence factor, is related to the rate of convergence, which is defined as,

$$R_{conv.} = -\ell n(r(z)),$$

where,

$$r(z) = |e^{\cosh^{-1}(z)-\cosh^{-1}(z_O)}|$$

$$= e^{\mathrm{Re}[\cosh^{-1}(z) - \mathrm{Re}\cosh^{-1}(z_O)]} .$$

From the discussion above we can prove that if an ellipse is a subset of another ellipse, of the same family, then its asymptotic convergence factor is less than the asymptotic convergence factor of the larger ellipse.

In general, let us consider, $F(d,c)$ to be the family of ellipses centered at (d), with foci at $(d+c)$ and $(d-c)$, and $F_i(d,c), F_j(d,c) \in F(d,c)$ are members of the family, and $F_O(d,c)$ is the member of the family passing through the origin.

Let,

$$z = \frac{d-\lambda_\omega}{c} , \quad z_O = \frac{d}{c} ,$$

where d and c are any complex numbers.

Also, we can take,

$$P_n(\lambda_\omega) = \frac{T_n(\frac{d-\lambda_\omega}{c})}{T_n(\frac{d}{c})} , \tag{4.4}$$

to be "the scaled and translated Chebyshev polynomials", and

$$P_n(O) = 1.$$

As before, we can define,

$$r(\lambda_\omega) = \lim_{n\to\infty} |P_n(\lambda_\omega)^{\frac{1}{n}}| , \tag{4.5}$$

as the asymptotic convergence factor of $P_n(\lambda_\omega)$, at the point λ_ω, and

$$r(\lambda_\omega) = e^{\mathrm{Re}[\cosh^{-1}(\frac{d-\lambda_\omega}{c})]-\mathrm{Re}[\cosh^{-1}(\frac{d}{c})]} .$$

The relationship between the members of $F(d,c)$ and the convergence properties of the polynomial $P_n(\lambda_\omega)$ is:

1. $$\lim_{n\to\infty} P_n(\lambda_\omega) = \begin{Bmatrix} O, \text{ if } \lambda_\omega \in F_O(d,c) \\ \infty, \text{ if } \lambda_\omega \notin F_O(d,c) \end{Bmatrix} .$$

2. If $\lambda_{\omega i} \in F_i(d,c)$ and $\lambda_{\omega j} \in F_j(d,c)$ then,

$$r(\lambda_{\omega i}) \leq r(\lambda_{\omega j}) \text{ if, and only if,}$$

$$F_i(d,c) \subset F_j(d,c), \text{ and}$$

$$r(\lambda_\omega) = 1 \text{ if, and only if, } \lambda_\omega \in F_O(d,c).$$

3. We can easily prove that:

$$\lim_{c \to 0} P_n(\lambda_\omega) = \left(\frac{d-\lambda_\omega}{d}\right)^n = \left(1 - \frac{\lambda_\omega}{d}\right)^n .$$

4. If A_ω has non-linear elementary divisors, then the derivatives of the sequence of polynomials, $\{P_n(\lambda_\omega)\}$, must converge to zero on the eigenvalues of A_ω, and if each sequence of $P_n(\lambda_\omega)$ has the same region of convergence as $P_n(\lambda_\omega)$ does, i.e. if $\lambda_\omega \in F_O(d,c)$, which is a member of the family $F(d,c)$ passing through the origin, then,

$$\lim_{n \to \infty} P_n^{(j)}(\lambda_\omega) = 0, \text{ for all } j,$$

(see Manteuffel (1975)).

It is likely that we can choose polynomials that are small on the eigenvalue spectrum of the matrix A_ω, and since the spectrum is seldom known, then it is more practical to choose polynomials that are small on the region containing the spectrum of A_ω.

4.1 The Preconditioned Chebyshev Iterative Method

To solve the system of equations (1.2) we know that the error vector at the nth iterative step is given by the equation

$$\underline{E}_\omega^{(n)} = P_n(A_\omega)\underline{E}_\omega^{(0)} ,$$

which is based on the scaled and translated Chebyshev polynomials,

$$P_n(\lambda_\omega) = \frac{T_n\left(\frac{d-\lambda_\omega}{c}\right)}{T_n\left(\frac{d}{c}\right)} .$$

Let, $$D\underline{\phi}_\omega^{(n)} = \underline{\phi}_\omega^{(n+1)} - \underline{\phi}_\omega^{(n)} , \qquad (4.6)$$

be the nth difference vector, which can be written as

$$D\underline{\phi}_\omega^{(n)} = (P_n(A_\omega) - P_{n+1}(A_\omega))\underline{E}_\omega^{(O)} , \qquad (4.7)$$

since the Chebyshev polynomials $P_n(A_\omega)$, satisfy a recurrence relation then:

<u>for n>O,</u> we have:

$$P_{n+1}(\lambda_\omega) = \frac{T_{n+1}(\frac{d-\lambda_\omega}{c})}{T_{n+1}(\frac{d}{c})} = \frac{2(\frac{d-\lambda_\omega}{c})T_n(\frac{d-\lambda_\omega}{c}) - T_{n-1}(\frac{d-\lambda_\omega}{c})}{T_{n+1}(\frac{d}{c})}$$

$$= -\frac{2}{c}\left[\frac{T_n(\frac{d}{c})}{T_{n+1}(\frac{d}{c})}\right]\lambda_\omega P_n(\lambda_\omega) + \left[\frac{2d}{c}\frac{T_n(\frac{d}{c})}{T_{n+1}(\frac{d}{c})}\right]P_n(\lambda_\omega)$$

$$- \left[\frac{T_{n-1}(\frac{d}{c})}{T_{n+1}(\frac{d}{c})}\right]P_{n-1}(\lambda_\omega) ,$$

then,

$$(P_n(\lambda_\omega) - P_{n+1}(\lambda_\omega)) = \left[\frac{2}{c}\ \frac{T_n(\frac{d}{c})}{T_{n+1}(\frac{d}{c})}\right]\lambda_\omega P_n(\lambda_\omega) + \left[\frac{T_{n-1}(\frac{d}{c})}{T_{n+1}(\frac{d}{c})}\right](P_{n-1}(\lambda_\omega) - P_n(\lambda_\omega)),$$

and, by using equation (4.7) we obtain:

$$D\underline{\phi}_\omega^{(n)} = \left[\frac{2}{c}\ \frac{T_n(\frac{d}{c})}{T_{n+1}(\frac{d}{c})}\right] A_\omega P_n(A_\omega)\underline{E}_\omega^{(O)} + \left[\frac{T_{n-1}(\frac{d}{c})}{T_{n+1}(\frac{d}{c})}\right](P_{n-1}(A_\omega) - P_n(A_\omega))\underline{E}_\omega^{(O)} , \qquad (4.8)$$

Now since by definition the preconditioned residual vector $\underline{r}_\omega$ can be defined as

$$\begin{aligned} \underline{r}_\omega^{(n)} &= \underline{b}_\omega - A_\omega \underline{\phi}_\omega^{(n)} \\ &= A_\omega \underline{E}_\omega^{(n)} \\ &= A_\omega P_n(A_\omega)\underline{E}_\omega^{(O)} , \end{aligned}$$

then

$$D\underline{\phi}_\omega^{(n)} = \alpha_n . \underline{r}_\omega^{(n)} + \beta_n D\underline{\phi}_\omega^{(n-1)} , \qquad (4.9)$$

where

$$\alpha_n = \frac{2}{c}\frac{T_n(\frac{d}{c})}{T_{n+1}(\frac{d}{c})} \quad \text{and} \quad \beta_n = \frac{T_{n-1}(\frac{d}{c})}{T_{n+1}(\frac{d}{c})} .$$

Similarly, for n<0, by using equation (4.7) we have:

$$D\underline{\phi}_{\omega}^{(0)} = (P_0(A_{\omega})-P_1(A_{\omega}))\,\underline{E}_{\omega}^{(0)} ,$$

since,

$$P_0(\lambda_{\omega}) = \frac{T_0\left(\frac{d-\lambda_{\omega}}{c}\right)}{T_0\left(\frac{d}{c}\right)}$$

$$= 1,$$

and

$$P_1(\lambda_{\omega}) = \frac{T_1\left(\frac{d-\lambda_{\omega}}{c}\right)}{T_1\left(\frac{d}{c}\right)}$$

$$= \frac{\left(\frac{d-\lambda_{\omega}}{c}\right)}{\left(\frac{d}{c}\right)}$$

$$= \left(1 - \frac{\lambda_{\omega}}{d}\right) ,$$

then,

$$P_0(\lambda_{\omega}) - P_1(\lambda_{\omega}) = \frac{\lambda_{\omega}}{d} .$$

Hence,

$$D\underline{\phi}_{\omega}^{(0)} = \frac{1}{d} A_{\omega}\underline{E}_{\omega}^{(0)}$$

$$= \frac{1}{d}\,\underline{r}_{\omega}^{(0)} . \qquad (4.10)$$

Hence, the computational procedure for the preconditioned Chebyshev iterative method is as follows:

To solve the system of equations (1.2):

1. Input the column vector $\underline{b}$.
2. Give an initial guess for the preconditioned solution vector, $\underline{\phi}_{\omega}^{(0)}$, and an initialvalues for the parameters d,c and ω.
3. Transform the column vector, $\underline{b}$, into the preconditioned column vector $\underline{b}_{\omega}$, by using the backward substitution process $(I-\omega L)^{-1}$ on $\underline{b}$.
4. Obtain the initial residual vector,

$$\underline{r}_{\omega}^{(0)} = \underline{b}_{\omega} - A_{\omega}\underline{\phi}_{\omega}^{(0)} ,$$

where the matrix-vector multiplication $A_{\omega}\underline{\phi}_{\omega}^{(0)}$, is performed by using the $(I-\omega L)^{-1}A$ and $(I-\omega U)^{-1}$ operators respectively.

5. Find the initial difference vector,

$$D\underline{\phi}_{\omega}^{(0)} = \frac{1}{d}\underline{r}_{\omega}^{(0)} ,$$

and then obtain the new solution vector,

$$\underline{\phi}_{\omega}^{(1)} = \underline{\phi}_{\omega}^{(0)} + D\underline{\phi}_{\omega}^{(0)} .$$

6. For $n>0$, repeat as in 4 and 5 to find the nth residual vector, $\underline{r}_{\omega}^{(n)}$, using the equation,

$$\underline{r}_{\omega}^{(n)} = \underline{b}_{\omega} - A_{\omega}\underline{\phi}_{\omega}^{(n)} ,$$

where the matrix-vector multiplication, $A_{\omega}\underline{\phi}_{\omega}^{(n)}$ is performed by the given procedure as for $A_{\omega}\underline{\phi}_{\omega}^{(0)}$.

7. Find the nth difference vector,

$$D\underline{\phi}_{\omega}^{(n)} = \alpha_n.\underline{r}_{\omega}^{(n)} + \beta_n.D\underline{\phi}_{\omega}^{(n-1)} ,$$

and the $(n+1)^{th}$ solution vector,

$$\underline{\phi}_{\omega}^{(n+1)} = \underline{\phi}_{\omega}^{(n)} + D\underline{\phi}_{\omega}^{(n)} ,$$

respectively where the coefficients α_n and β_n are defined in equation (4.9) and can be recursively generated.

8. Repeat 6 and 7, keeping the values of the last $\underline{\phi}_{\omega}^{(n)}, \underline{r}_{\omega}^{(n)}$ and $D\underline{\phi}_{\omega}^{(n)}$ in storage, until the solution vector $\underline{\phi}_{\omega}^{(n+1)}$ satisfy the convergence criterion such as either

$$\frac{|\underline{\phi}_{\omega}^{(n+1)} - \underline{\phi}_{\omega}^{(n)}|}{\underline{\phi}_{\omega}^{(n+1)}} < \varepsilon,$$

or

$$RSD = ||\underline{\phi}_{\omega}^{(n+1)}|| < \varepsilon , \tag{4.11}$$

for a small value of ε, $\varepsilon>0$.

9. Finally, transform the final $(n+1)^{th}$ approximation preconditioned solution vector, $\underline{\phi}_{\omega}^{(n+1)}$ onto the solution vector of the system of equations (1.1), $\underline{\phi}^{(n+1)}$, by using the forward substitution process, $(I-\omega L)^{-1}$.

By using the scaled and translated Chebyshev polynomials, $P_n(A_{\omega})$, in the previous iterative procedure the three criteria, mentioned in section 3 were satisfied, where these polynomials have minimal and maximal modulus properties on ellipses, their derivative sequence is convergent and the iteration can be carried out recursively.

4.2 Convergence of the Preconditioned Chebyshev Iteration Method

In any given family of ellipses, $F(d,c)$ there are some members contained in its interior of the spectrum of A_ω. If the spectrum of A_ω lies on the interior of the member of the family passing through the origin, $F_O(d,c)$, then the preconditioned Chebyshev iteration will converge. Now, we have to choose the parameters d and c, to optimise the rate of convergence, which can be found as the solution of a "min-max problem". This is given as follows:

Suppose d and c have been chosen where d and c are real numbers since A_ω is a real valued matrix.

Let $F_s(d,c)$ be a smallest member of the family, $F(d,c)$, containing the spectrum of A_ω in the closure of its interior and it is convex. Then there are some eigenvalues $\lambda_{\omega s} \in F_s(d,c)$ on the boundary of the convex hull of the spectrum of A_ω, and all other eigenvalues are inside or on $F_s(d,c)$ which is associated with the convergence factor,

$$r(\lambda_{\omega s}) = \max_{\lambda_{\omega i}} [r(\lambda_{\omega i})] \ . \tag{4.12}$$

Since,

$$r(\lambda_\omega) = \left| e^{\cosh^{-1}\left(\frac{d-\lambda_\omega}{c}\right) - \cosh^{-1}\left(\frac{d}{c}\right)} \right| \ ,$$

and by using the log form of the $\cosh^{-1}$ function, we can prove that,

$$r(\lambda_\omega) = \left| \frac{\left(\frac{d-\lambda_\omega}{c}\right) + \left[\left(\frac{d-\lambda_\omega}{c}\right)^2 - 1\right]^{\frac{1}{2}}}{\left(\frac{d}{c}\right) + \left[\left(\frac{d}{c}\right)^2 - 1\right]^{\frac{1}{2}}} \right|$$

$$= \left| \frac{(d-\lambda_\omega) + [(d-\lambda_\omega)^2 - c^2]^{\frac{1}{2}}}{d + [d^2 - c^2]^{\frac{1}{2}}} \right|$$

$$= [P_n(\lambda_\omega)]^{1/n} \ . \tag{4.13}$$

Since $r(\lambda_\omega)$ is a function of d and c, then in order to optimise the choice of d and c we must make $r(\lambda_\omega)$ as small as possible.

Also, since A_ω is a real-valued matrix, then the "min-max problem" can be restricted such that:

The maximum is taken over a subset of eigenvalues of A_ω and the minimum is taken over d and c such that d and c^2 are real, i.e.

$$\min_{(d,c)} \max_{\lambda_{\omega i}} (r(\lambda_{\omega i})) .$$

Let

$$H = \{\lambda_{\omega i} / \lambda_{\omega i} \text{ is a vertex of the smallest convex polygon enclosing the spectrum of } A_\omega\},$$

i.e. H is considered as the hull of the spectrum and its elements determine the "min-max problem".

Let $R = \{(d,c2)/d>0,\ \ c2=c^2 \text{ and } c2<d^2\}$,

$H^+ = \{\lambda_{\omega j} \in H/\lambda_\omega > 0\}$ be the positive hull of the spectrum of A_ω,

and $K = \{\lambda_{\omega k} \in H^+$/there exist $(d,c2) \in R$ such that

$$r(\lambda_{\omega k}) = \max_{\lambda_{\omega i} \in H} [r(\lambda_{\omega i})] \quad \},$$

where $k<j<i$.

Now, the parameters d and c, which yield the optimum rate of convergence, will satisfy

$$\begin{aligned}
\min_{(d,c)} \max_{\lambda_{\omega i}} (r(\lambda_{\omega i})) &= \min_{(d,c)} \left(\max_{\lambda_{\omega i} \in H} (r(\lambda_{\omega i}))\right) \\
&= \min_{(d,c2) \in R} \left(\max_{\lambda_{\omega j} \in H^+} (r(\lambda_{\omega j}))\right) \\
&= \min_{(d,c2) \in R} \left(\max_{\lambda_{\omega k} \in K} \left| \frac{(d-\lambda_{\omega k}) + [(d-\lambda_{\omega k})^2 - c^2]^{\frac{1}{2}}}{d + [d^2 - c^2]^{\frac{1}{2}}} \right| \right).
\end{aligned} \tag{4.14}$$

Then the "min-max problem" solution, which has been proved by Manteuffel (1975) is as follows:

If the eigenvalues, $\lambda_{\omega j}$, in the positive hull, H^+, are known, then the optimal parameters d and c can be found at a point that minimizes the maximum of a finite number of real valued functions of two-real variables.

By considering that each function,

$$r(\lambda_{\omega j}) = r(\lambda_{\omega j},d,c2) ,$$

be a surface above the (d,c2)-plane, then the "min-max problem" solution is: either a local minimum of one of $r(\lambda_{\omega j},d,c2)$ surfaces in R, which is found explicitly or a local minimum along the intersection of two $r(\lambda_{\omega j},d,c2)$ surfaces, which is found as a root of a fifth degree polynomial, or the unique intersection point of three-$r(\lambda_{\omega j},d,c2)$ surfaces. This "min-max problem" solution then yields the smallest convergence factor,

$$r(\lambda_{\omega},d,c2) = r(\lambda_{\omega}) .$$

4.3 The Choice of Optimal Parameters

From the previous discussion, we can see that there are three important parameters d,c and ω, where,

$$d = \frac{\lambda_{max}+\lambda_{min}}{2}$$, which represent the centre of the ellipse,

$$c = \frac{\lambda_{max}-\lambda_{min}}{2}$$, which represent the focal length,

and ω is the preconditioning parameter to be discussed later. (Missirlis & Evans, 1981).

In this section we will consider only d and c, which have been optimised by Manteuffel (1975).

Now, since the parameters d and c are restricted such that

$$d > 0 , \quad c^2 < d^2 , \qquad (4.15)$$

then, by using the cosh definition on the scaled and translated Chebyshev polynomials, we have

$$P_n(\lambda_{\omega}) = \frac{e^{(n \cosh^{-1}(\frac{d-\lambda_{\omega}}{c}))} + e^{(-n \cosh^{-1}(\frac{d-\lambda_{\omega}}{c}))}}{e^{(n \cosh^{-1}(\frac{d}{c}))} + e^{(-n \cosh^{-1}(\frac{d}{c}))}}$$

$$= \left[\frac{e^{\cosh^{-1}\left(\frac{d-\lambda_\omega}{c}\right)}}{e^{\cosh^{-1}\left(\frac{d}{c}\right)}}\right]^n \left[\frac{1+e^{\left(-2n\cosh^{-1}\left(\frac{d-\lambda_\omega}{c}\right)\right)}}{1+e^{\left(-2n\cosh^{-1}\left(\frac{d}{c}\right)\right)}}\right] . \tag{4.16}$$

By use of equation (4.16) we have

$$P_n(\lambda_\omega) = [S(\lambda_\omega)]^n Q_n(\lambda_\omega) , \tag{4.17}$$

where,

$$S(\lambda_\omega) = \frac{e^{\left(\cosh^{-1}\left(\frac{d-\lambda_\omega}{c}\right)\right)}}{e^{\left(\cosh^{-1}\left(\frac{d}{c}\right)\right)}} = \frac{(d-\lambda_\omega)+((d-\lambda_\omega)^2-c^2)^{\frac{1}{2}}}{d+(d^2-c^2)^{\frac{1}{2}}},$$

and

$$Q_n(\lambda_\omega) = \frac{1+e^{\left(-2n\cosh^{-1}\left(\frac{d-\lambda_\omega}{c}\right)\right)}}{e^{\left(-2n\cosh^{-1}\left(\frac{d}{c}\right)\right)}} .$$

Since the branch of $\cosh^{-1}$ with non-negative real part is used and since (4.15) ensures that

$$\left(\frac{d}{c}\right) \notin [-1,1],$$

then,

$$\lim_{n\to\infty} Q_n = 1, \quad \text{for } \left(\frac{d-\lambda_\omega}{c}\right) \notin [-1,1].$$

Also, if we let

$$\gamma = e^{\left(-\mathrm{Re}\left(\cosh^{-1}\left(\frac{d}{c}\right)\right)\right)},$$

then,

$$0 \leq |Q_n(\lambda_\omega)| \leq \frac{2}{1-\gamma^n} , \text{ for } \left(\frac{d-\lambda_\omega}{c}\right) \in [-1,1]. \tag{4.18}$$

Notice, that for some λ_ω values Q_n does not approach unity quickly, i.e. those for which

$$\left|e^{\left(\cosh^{-1}\left(\frac{d-\lambda_\omega}{c}\right)\right)}\right| \text{ is nearly unity.}$$

For those which

$$[S(\lambda_\omega)]^n \to 0 \text{ most rapidly,}$$

then it is possible to say that

$$P_n(\lambda_\omega) = [S(\lambda_\omega)]^n, \text{ for large n.} \tag{4.19}$$

Also, from Dunford and Schwartz (1958), we have:

Let $S=S(A_\omega)$ be the operator such that

$$P_n(A_\omega) = S^n, \quad \text{for large } n, \tag{4.20}$$

Then according to (3.1) it is desirable to make the spectral radius of S as small as possible.

Since the parameters d and c represent the center and focal length of a family of ellipses in the complex plane C, with the foci (d+c) and (d-c), then each point $\lambda_\omega \in C$ is associated with an asymptotic convergence factor of the form,

$$r(\lambda_\omega) = |S(\lambda_\omega)| = \frac{(d-\lambda_\omega)+((d-\lambda_\omega)^2-c^2)^{\frac{1}{2}}}{d+(d^2-c^2)^{\frac{1}{2}}}, \tag{4.21}$$

and the member of this family of ellipses passing through the origin determines the region of convergence, i.e. if the spectrum of A_ω lies inside this ellipse then,

$$r(\lambda_{\omega i}) < 1 \text{ for each eigenvalue } \lambda_{\omega i},$$

and the iteration (1.2) will converge.

Since $r(\lambda_\omega)$ is a function of d and c, then $r(\lambda_\omega)$ must be as small as possible to optimise the choice of d and c, i.e. the optimisation of the rate of convergence can be found as the solution of a "min-max problem". From equations (3.1) and (4.19) it follows that:

For large n, each step of the iteration (4.9) will cause an error in the direction of the eigenvector associated with the eigenvalue $\lambda_{\omega i}$ to be multiplied by $r(\lambda_{\omega i})$.

If A_ω has a complete set of eigenvectors, then after a sufficiently large number of steps of the iteration (4.9), the error can be expressed as a linear combination of those eigenvectors associated with the eigenvalues with the largest convergence factor, i.e. the eigenvalues of the outermost members of the family of ellipses determined by d and c.

Hence, the d and c values can be optimised by using a four-step adaptive procedure, as follows:

1. Choose initial values of d and c, based on a prior knowledge of the matrix A_ω.

2. Perform a number of iteration steps, based on d and c, and store the last five residuals.
3. Obtain an estimate of the eigenvalue by a modified power method as given by Manteuffel (1975), and derive the positive hull, by adding the new eigenvalue estimates to any of the previously obtained ones.
4. Find the optimal values of d and c for the new set of eigenvalue estimates, as a solution of "min-max problem". Repeat 2,3 and 4 until the new solution vector satisfies the given convergence criterion (4.11).

5. APPLICATIONS TO PARTIAL DIFFERENTIAL EQUATIONS

In scientific applications such as oil reservoir and numerical weather forecasting problems, boundary value problems are encountered involving partial differential equations. They can be solved by approximating the solution of the differential equation at a discrete set of grid points on the domain by the solution of a linear system,

$$A\underline{\phi} = \underline{b} \ . \tag{5.1}$$

Here, let us consider the second order linear differential equation on a rectangular domain

$$[-\frac{\partial}{\partial x}(B_1(x,y)\frac{\partial}{\partial x}) - \frac{\partial}{\partial y}(B_2(x,y)\frac{\partial}{\partial y}) + C_1(x,y)\frac{\partial}{\partial x} + C_2(x,y)\frac{\partial}{\partial y} + Q(x,y)]\underline{\Phi} + \underline{b} = 0 \ , \tag{5.2}$$

where the functions B_1, B_2, C_1, C_2 and Q are continuous and differentiable, $\underline{\Phi}$ is the exact solution of the P.D.E. (5.2). The P.D.E. operator in equation (5.2) can be separated into two parts:

I. a self-adjoint operator:

$$[-\frac{\partial}{\partial x}(B_1(x,y)\frac{\partial}{\partial x}) - \frac{\partial}{\partial y}(B_2(x,y)\frac{\partial}{\partial y}) + Q(x,y)] ,$$

and

II. a non-self adjoint operator:

$$[C_1(x,y)\frac{\partial}{\partial x} + C_2(x,y)\frac{\partial}{\partial y}] ,$$

with the required boundary conditions.

By approximating these two operators by a finite difference equation on a regular grid, where h is the grid size and (x_i, y_j) are the grid points then with the standard row ordering of node points, left to right and bottom to top, the operator I gives a symmetric matrix and the operator II gives a non-symmetric matrix.

Let A be the five-point difference matrix associated with the differential operator (5.2) by using central difference approximations and the given Dirichlet boundary conditions. Hence we can divide the matrix A into two sub-matrices B and C such that

$$A = B + C \ , \tag{5.3}$$

where B is symmetric with block form,

$$B = \begin{bmatrix} B_1 & D_1 & & & & \\ D_1 & B_2 & D_2 & & 0 & \\ & \ddots & \ddots & \ddots & & \\ & & & & & \\ & 0 & & D_{k-2} & B_{k-1} & D_{k-1} \\ & & & & D_{k-1} & B_k \end{bmatrix} , \tag{5.4a}$$

where k is the number of grid points/column, and each of the B_i are tridiagonal matrices, defined as,

$$B_i = \begin{bmatrix} b_{1i} & a_{1i} & & & & \\ a_{1i} & b_{2i} & a_{2i} & & 0 & \\ & a_{2i} & b_{3i} & a_{3i} & & \\ & & \ddots & \ddots & \ddots & \\ & 0 & & a_{\ell-2,i} & b_{\ell-1,i} & a_{\ell-1,i} \\ & & & & a_{\ell-1,i} & b_{\ell,i} \end{bmatrix} ,$$

and ℓ is the number of grid points/row.

The elements of the matrix B can be shown to be given by

$$b_{i,j} = \frac{1}{h^2}[B_1(x_i - \tfrac{h}{2}, y_j) + B_1(x_i + \tfrac{h}{2}, y_j) + B_2(x_i, y_j - \tfrac{h}{2}) + B_2(x_i, y_j + \tfrac{h}{2})] + Q(x_i, y_j) \ ,$$

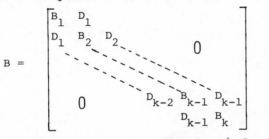

$$a_{i,j} = \frac{1}{h^2}[B_1(x_i + \frac{h}{2}, y_j)],$$

and the D_i are diagonal matrices defined as:

$$D_i = \begin{bmatrix} d_{1i} & & & 0 \\ & d_{2i} & & \\ & & \ddots & \\ 0 & & & d_{\ell i} \end{bmatrix},$$

with

$$d_{i,j} = \frac{1}{h^2}[B_2(x_i, y_j + \frac{h}{2}] .$$

The non-self adjoint operator when discretised using central difference approximations yields the anti-symmetric matrix C and its block form is given by

$$C = \begin{bmatrix} C_1 & F_1 & & & & 0 \\ -F_2 & C_2 & F_2 & & & \\ & -F_3 & C_3 & F_3 & & \\ & & \ddots & \ddots & \ddots & \\ 0 & & & -F_{k-1} & C_{k-1} & F_{k-1} \\ & & & & -F_k & C_k \end{bmatrix}, \qquad (5.4b)$$

where C_i are the tridiagonal matrices, defined as:

$$C_i = \begin{bmatrix} 0 & c_{1i} & & & & 0 \\ -c_{2i} & 0 & c_{2i} & & & \\ & -c_{3i} & 0 & c_{3i} & & \\ & & \ddots & \ddots & \ddots & \\ 0 & & & -c_{\ell-1,i} & 0 & c_{\ell-1,i} \\ & & & & -c_{\ell i} & 0 \end{bmatrix},$$

with

$$c_{i,j} = \frac{1}{2h}[C_1(x_i, y_j)],$$

and F_i are diagonal matrices, defined as:

$$F_i = \begin{bmatrix} f_{1i} & & & 0 \\ & f_{2i} & & \\ & & \ddots & \\ 0 & & & f_{\ell i} \end{bmatrix} ,$$

with $\quad f_{i,j} = \frac{1}{2h}[C_2(x_i,y_j)].$

The system (5.1) is a linear system of equations, where

$$A = I-L-U ,$$

is a real valued, unsymmetric, large, sparse, (N×N) matrix, with eigenvalues λ_i in the open right half complex plane, I is the identity matrix, L and U are respectively the lower and upper triangular matrices.

Such a system arises from applications of partial differential equations especially in conjunction with finite element methods in engineering.

The preconditioned form of (5.1) can be written as

$$A_\omega \underline{\phi}_\omega = \underline{b}_\omega , \qquad (5.5)$$

which is faster to converge, where,

$A_\omega = (I-\omega L)^{-1}A(I-\omega U)^{-1}$, is the preconditioned coefficient matrix,

$\underline{\phi}_\omega = (I-\omega U)\underline{\phi}$, is the preconditioned solution vector,

and $\quad \underline{b}_\omega = (I-\omega L)^{-1}\underline{b}$, is the preconditioned column vector.

We now wish to establish the bounds of the spectrum of the preconditioned coefficient matrix A_ω.

Since, we assumed that the matrix,

$$A = B + C ,$$

such that

$$B = \frac{A+A^T}{2} , \qquad C = \frac{A-A^T}{2} , \qquad (5.6)$$

B is symmetric and C is anti-symmetric.

From the preconditioned coefficient matrix

$$A_\omega = (I-\omega L)^{-1}A(I-\omega U)^{-1} ,$$

we have the transpose matrix defined as

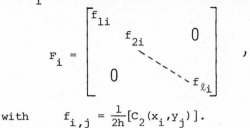

$$[A_\omega]^T = [(I-\omega U)^{-1}]^T[A]^T[(I-\omega L)^{-1}]^T$$
$$= (I-\omega L)^{-1}A^T(I-\omega U)^{-1} \ . \qquad (5.7)$$

Then, we can define the following matrices,

$$\frac{A_\omega+A_\omega^T}{2} = \frac{(I-\omega L)^{-1}(A+A^T)(I-\omega U)^{-1}}{2}$$
$$= (I-\omega L)^{-1}B(I-\omega U)^{-1}$$
$$= B_\omega \ , \qquad (5.8)$$

and similarly,

$$\frac{A_\omega-A_\omega^T}{2} = (I-\omega L)^{-1}C(I-\omega U)^{-1}$$
$$= C_\omega \ . \qquad (5.9)$$

Hence, from equations (5.8) and (5.9), we have the preconditioned coefficient matrix,

$$A_\omega = B_\omega + C_\omega \ ,$$

such that,

$$B_\omega = \frac{A_\omega + A_\omega^T}{2} \ , \qquad C_\omega = \frac{A_\omega - A_\omega^T}{2} \ ,$$

where, B_ω is symmetric and C_ω is anti-symmetric.

Suppose now λ_ω is an eigenvalue of the matrix A_ω corresponding to the eigenvector,

$$v = x+i.y \ ,$$

then from the definition of Rayleigh Quotient, we have,

$$\lambda_\omega = \frac{\langle A_\omega v,v\rangle}{\langle v,v\rangle}$$
$$= \frac{\langle B_\omega v,v\rangle}{\langle v,v\rangle} + \frac{\langle C_\omega v,v\rangle}{\langle v,v\rangle} \ .$$

Now because B_ω is symmetric, i.e.,

$$\langle B_\omega x,y\rangle = \langle B_\omega y,x\rangle \ ,$$

then we have the result,

$$\langle B_\omega v,v\rangle = \langle B_\omega(x+iy),(x+iy)\rangle$$
$$= \langle B_\omega x,x\rangle+i\langle B_\omega y,x\rangle-i\langle B_\omega x,y\rangle+\langle B_\omega y,y\rangle$$
$$= \langle B_\omega x,x\rangle+\langle B_\omega y,y\rangle \ .$$

In a similar manner because C_ω is anti-symmetric, i.e. we have

$$\langle C_\omega x,x\rangle = \langle C_\omega y,y\rangle = 0 \ ,$$
$$\langle C_\omega y,x\rangle = -\langle C_\omega x,y\rangle \ ,$$

then finally we have,

$$\langle C_\omega v,v\rangle = \langle C_\omega x,x\rangle + i\langle C_\omega y,x\rangle - i\langle C_\omega x,y\rangle + \langle C_\omega y,y\rangle$$

$$= 2i\langle C_\omega y,x\rangle .$$

Thus, combining the two previous results we have

$$\mathrm{Re}(\lambda_\omega) = \frac{\langle B_\omega x,x\rangle + \langle B_\omega y,y\rangle}{||x||^2+||y||^2} ,$$

and

$$|\mathrm{Im}(\lambda_\omega)| = \frac{2|\langle C_\omega y,x\rangle|}{||x||^2+||y||^2}$$

$$\leqslant \frac{2||C_\omega||_2||x||.||y||}{||x||^2+||y||^2}$$

$$\leqslant ||C_\omega||_2 , \tag{5.11}$$

where $||C_\omega||_2$ is equal to the spectral radius of C_ω. Since B_ω is symmetric, its eigenvalue spectrum is contained in some interval $[a_\omega, b_\omega]$ on the real line.

Since C_ω is anti-symmetric, its spectrum lies along the imaginary axis, i.e.,

$$\mathrm{Re}(\lambda_\omega) \in [a_\omega, b_\omega] ,$$
$$|\mathrm{Im}(\lambda_\omega)| \leqslant ||C_\omega||_2 .$$

If B_ω is positive definite (in particular when A_ω is diagonally dominant, with positive real diagonal elements), then the spectrum of A_ω lies in the right half complex plane.

By using Gersgorin's Theorem, the bounds for the interval,

$$[a_\omega, b_\omega] \text{ and } ||C_\omega||_2,$$

can be found in terms of the elements of the matrices B_ω and C_ω giving a rectangle known to contain the spectrum of the operator A_ω which is more closely approximated by an ellipse rather than a circle.

6. NUMERICAL EXAMPLE

In this section, we consider a particular numerical problem which is the solution of the partial differential equation,

$$[-\frac{\partial}{\partial x}(B_1(x,y)\frac{\partial}{\partial x}) - \frac{\partial}{\partial y}(B_2(x,y)\frac{\partial}{\partial y}) + C_1(x,y)\frac{\partial}{\partial x} + C_2(x,y)\frac{\partial}{\partial y} + Q(x,y)]\underline{\Phi}+\underline{b} = 0 \ ,$$

which is defined and approximated by a finite difference operator on the regular grid discussed in section 5.

If we let

$$B_1(x,y),\ B_2(x,y) > 0,$$

and
$$Q(x,y) \geqslant 0 \ ,$$

then the matrix B is positive definite, and if we take

$$C_1(x,y) \text{ and } C_2(x,y) \ ,$$

to be constant functions, then the matrix C is anti-symmetric, where,

$$A = B + C \ ,$$

is the coefficient matrix of the system of equations

$$A\underline{\phi} = \underline{b} \ .$$

Also, the eigenvalues of the matrix A lie in the right half plane.

Let us take
$$B_1(x,y) = B_2(x,y) = 1/4 \ ,$$

$$Q(x,y) = 0 \ ,$$

and
$$C_1(x,y) = C_2(x,y) = \beta/4 \ ,$$

(where β is a real number, to be taken in the range $0 \leqslant \beta \leqslant 2$).

If we now choose the square region such that:

$$0 \leqslant x \leqslant 41, \qquad 0 \leqslant y \leqslant 41 \ ,$$

with grid length h=1, then

$$A = B + C \ ,$$

is a (1600×1600) matrix where

$$B_i = \begin{bmatrix} 1 & -\frac{1}{4} & & & 0 \\ -\frac{1}{4} & 1 & -\frac{1}{4} & & \\ & \ddots & \ddots & \ddots & \\ & & -\frac{1}{4} & 1 & -\frac{1}{4} \\ 0 & & & -\frac{1}{4} & 1 \end{bmatrix} , \qquad D_i = \begin{bmatrix} -\frac{1}{4} & & & 0 \\ & -\frac{1}{4} & & \\ & & \ddots & \\ 0 & & & -\frac{1}{4} \end{bmatrix} ,$$

$$C_i = \begin{bmatrix} 0 & \frac{\beta}{8} & & & & \\ -\frac{\beta}{8} & 0 & \frac{\beta}{8} & & 0 & \\ & -\frac{\beta}{8} & 0 & \frac{\beta}{8} & & \\ & & \ddots & \ddots & \ddots & \\ & 0 & & -\frac{\beta}{8} & 0 & \frac{\beta}{8} \\ & & & & -\frac{\beta}{8} & 0 \end{bmatrix}, \text{ and } F_i = \begin{bmatrix} \frac{\beta}{8} & & & \\ & \frac{\beta}{8} & & 0 \\ & & \ddots & \\ & 0 & & \frac{\beta}{8} \end{bmatrix}.$$

The eigenvalues of the positive definite matrix B lie in the open interval (0,2) while the eigenvalues of the matrix C lie in the $(-\frac{\beta i}{2}, +\frac{\beta i}{2})$ along the imaginary axis.

If λ is an eigenvalue of the coefficient matrix A, then it is contained in the rectangular region:

$$\mathrm{Re}(\lambda) \in (0,2) ,$$

$$\mathrm{Im}(\lambda) \in (-\frac{\beta}{2}, +\frac{\beta}{2}) .$$

After the application of the preconditioning technique on the system of equations

$$A\underline{\phi} = \underline{b} ,$$

we have a new system of equations

$$A_\omega \underline{\phi}_\omega = \underline{b}_\omega ,$$

defined in (1.2), where the preconditioning parameter ω to be determined is such that $0 \leq \omega \leq 2$.

The preconditioning parameter ω, will for large values of ω (i.e. when $\omega \to 2$) cause the $\mathrm{Re}(\lambda_\omega)$ to exceed the open interval (0,2). Similarly for $\mathrm{Im}(\lambda_\omega)$ as shown in Figures 3 and 5.

For $\beta=0$, the matrix A_ω is symmetric, which yields the eigenvalue convex hull on the real axis only, i.e.,

$$\mathrm{Im}(\lambda_\omega) = 0 .$$

For small values of β, the matrix A_ω is nearly symmetric, which yields a hull close to the real axis. For large values of β, the matrix A_ω is unsymmetric, which yields a nearly vertical spectrum (Figure 3).

Tables 1,2 and 3 contain, for each value of β=0, 1 and 2 respectively, the initial as well as the computed optimal values of the center of the ellipse, d, and the focal length, c, in addition; the convergence factor associated with the best ellipse, and the number of the iterations needed to satisfy the convergence criteria as given in (4.11), for different values of the preconditioning parameter ω, $0 \leq \omega < 2$. As can be seen from these results the application of preconditioning produces a large increase in the convergence rate (i.e. the no. of iterations is decreased) in the region of the optimum preconditioning parameter.

Figures 1,3 and 5 show the work required to satisfy the convergence criterion, given in (4.11), for the considered cases of β=0, 1 and 2 respectively. For each β value it shows the best value of ω determined by an adaptive algorithm (Missirlis & Evans, 1981), which minimizes the number of iterations required to satisfy the convergence criterion.

It is obvious that when ω=0, the method becomes the usual Chebyshev iterative method, without the application of preconditioning.

Figures 2 and 4 show the hull of the eigenvalue estimates computed by the preconditioned Chebyshev iterative process and the best ellipse enclosing the approximate hull.

The initial choice of the parameters d and c, was based on the rectangle known to contain the spectrum of the matrix A_ω.

For these tests, the mesh size used was h=1. The symmetric part of A_ω is multiplied by a factor of $1/h^2$, and the anti-symmetric part, associated with the first order terms, is multiplied by a factor of 1/h. For smaller mesh sizes h, the matrix A would be more nearly symmetric, i.e. the type of system for which the preconditioned Chebyshev algorithm holds the greatest advantage.

Preconditioning parameter ω	Initial		Optimal		Convergence factor	No. of Iters.
	d	c	d	c		
0.0	1.0000	.9675	.9964	.9935	.92745	220
0.4	.7106	.7044	.7099	.7051	.89013	140
0.6	.5995	.5906	.5981	.5921	.86734	120
0.8	.5264	.5151	.5249	.5166	.83561	90
1.0	.5088	.4912	.5059	.4941	.80348	75
1.2	.5506	.5081	.5385	.5201	.76685	60
1.4	.6419	.5496	.6119	.5795	.71692	50
1.6	.9005	.6914	.8362	.7676	.65717	45
1.8	1.6745	1.2121	1.5610	1.3256	.55578	35
1.9	2.8847	2.3733	2.8786	2.3828	.53025	30
1.95	5.2782	4.6484	5.2532	4.6734	.61072	40

TABLE 1

where β = 0.0

Preconditioning parameter ω	Initial		Optimal		Convergence factor	No. of Iters
	d	c	d	c		
0.0	1.0000	.8765	.9991	.8791	.89752	150
0.4	.7240	.5491	.7324	.5381	.88502	120
0.8	.5572	.4087	.5831	.1015	.78699	60
1.0	.5457	.3243	.5570	.2848	.75346	50
1.2	.5733	.3610	.6270	.3164	.58453	30
1.4	.7827	.6141	.8709	.6191	.69695	50
1.6	1.2094	.9007	1.3545	.6349	.73570	50
1.8	3.4032	2.0660	3.1209	.8763	.74019	110

TABLE 2

where β = 1.0

Preconditioning parameter ω	Initial		Optimal		Convergence factor	No. of Iters
	d	c	d	c		
0.0	.9953	.8654	.9953	.1999	.86662	180
0.4	.7162	.5496	.7796	.1996i	.79386	120
0.6	.6264	.4434	.7194	.1367i	.7460	80
0.8	.5831	.1015	.7268	.2891i	.51150	40
1.0	.5570	.8093	.6150	.1725i	.50498	35
1.2	.6270	1.0004	1.5108	.2880i	.76354	60
1.4	.8709	.6191	1.3652	.6763i	.78308	126
1.6	1.3545	.6349	2.0995	1.6446i	.86028	160

TABLE 3

where $\beta = 2.0$

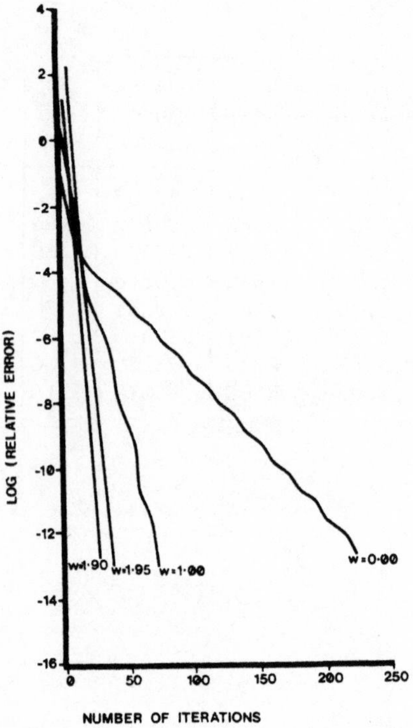

FIGURE 1

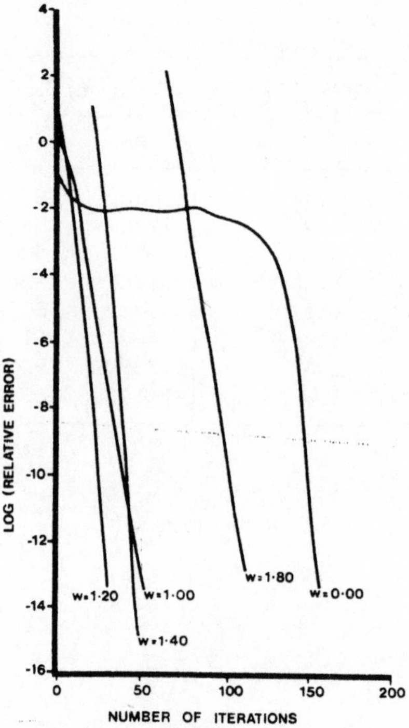

FIGURE 2

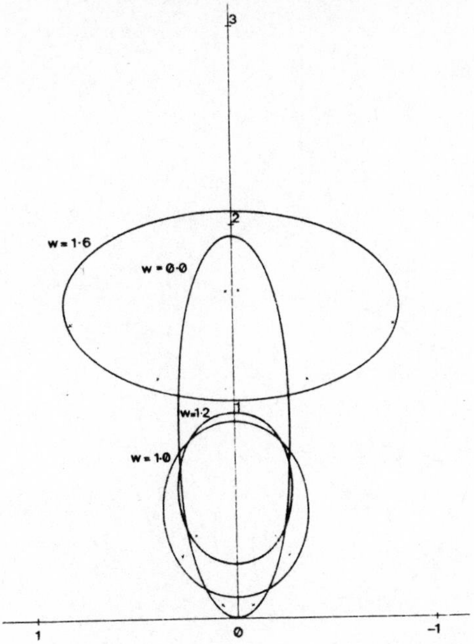

Figure 3, Best Ellipse for Indicated Value of w and β = 1·0

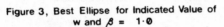

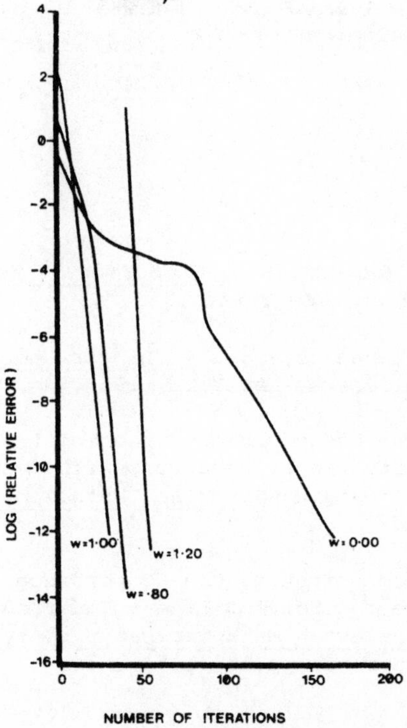

FIGURE 4

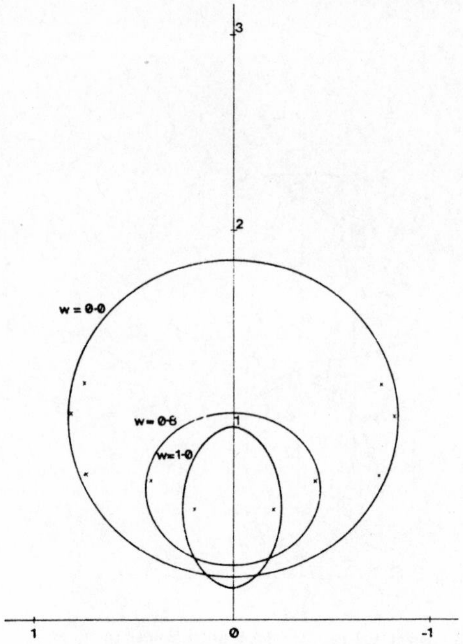

Figure 5, Best Ellipse for Indicated Value of w and $\beta = 2{\cdot}0$

REFERENCES

1. Birkhoff, G. and Maclane, S. - "A Survey of Modern Algebra", MacMillan, New York, 1953.

2. Dunford, N. and Schwartz, J.L. - "Linear Operator", Interscience Publishers, New York, 1958.

3. Evans, D.J. - "The Use of Preconditioning in the Iterative Methods for Solving Linear Equations with Symmetric, Positive Definite Matrices", J.Inst.Math.Applics., 4, pp.295-314, 1968.

4. Evans, D.J. - "Comparison of the Convergence Rates of Iterative Methods for Solving Linear Equations with Pre-Conditioning", The Greek Mathematical Society, C. Carathéodory Symposium, pp.106-135, 1973.

5. Kammoonah, M.A.M.S. - "A Preconditioned Chebyshev Iterative Method for Solving Symmetric and Unsymmetric Linear Systems", M.Phil. Thesis, Loughborough University of Technology, 1978.

6. Manteuffel, T.A. - "An Iterative Method for Solving Non-symmetric Linear Systems with Dynamic Estimation of Parameters", Ph.D. Thesis, Univ. of Illinois at Urbana-Champaign, 1975.

7. Missirlis, N.M. and Evans, D.J. - "On the Dynamic Acceleration of the Preconditioned Simultaneous Displacement (PSD) Method", Int.Jour.Comp.Math. 10, pp. 153-176, 1981.

8. Stiefel, E.L. - "Kernal Polynomials in Linear Algebra and Their Applications", U.S. N.B.S., Applied Math. Series, Vol.49, pp.1-22, 1958.

9. Varga, R.S. - "Matrix Iterative Analysis", Prentice-Hall, Englewood Cliffs, N.J. 1962.

10. Wrigley, H.W. - "Accelerating the Jacobi Method for Solving Simultaneous Equations by Chebyshev Extrapolation When the Eigenvalues of the Iteration Matrix are Complex", Computer J., Vol.6, 1963.

11. Young, D.M. - "Iterative Solution of Large Linear Systems", Academic Press, New York & London, 1971.

THE PRECONDITIONING BY DIRECT FACTORIZATION METHOD FOR SOLVING SELF-ADJOINT PARTIAL DIFFERENTIAL EQUATIONS

D.J. Evans & I.C. Demetriou

Department of Computer Studies, University of Technology, Loughborough, Leicestershire, U.K.

ABSTRACT

Preconditioning by direct factorization is a new acceleration strategy for solving the large, sparse systems of linear algebraic equations derived from elliptic partial differential equations, based on the preconditioning concept. The conditioning matrix which we consider is factorized into two rectangular easily invertible matrices. The PDF method can be seen as a fractional step iterative method which has three parts, a backward and a forward substitution followed by a simple Gaussian type elimination process. The method requires the selection of a preconditioning and an acceleration parameter to be applied. The number of iterations required for convergence varies approximately as h^{-1}, where h is the net mesh size. Since the iteration matrix has positive eigenvalues the semi-iterative technique applied resulted in a $O(h^{\frac{1}{2}})$ acceleration of the method. The numerical results show a substantial improvement of the convergence rate of the PDF method as compared to the SOR and SSOR methods.

1. THE PRECONDITIONING BY DIRECT FACTORIZATION METHOD (PDF METHOD)

Let us consider obtaining the solution of the system of linear algebraic equations

$$Au = b \ , \qquad (1)$$

where A is a $(v \times v)$ positive definite matrix derived from applying the five-point difference approximation formula on a rectangular grid to an elliptic partial differential equation.

Assuming the splitting

$$A = D-DB = D-C_L-C_U \ , \tag{2}$$

where C_L and C_U are the strictly lower and upper triangular parts of A and D=diag(A), let us define ([1]) the conditioning matrix M in the following form[(1)]

$$M = D[I-\omega U \mid \omega\sqrt{\bar{b}}e_v] \begin{bmatrix} I-\omega L \\ \hline \omega\sqrt{\bar{b}}e_v^T \end{bmatrix} = D[I-\omega(U+L)+\omega^2(UL+\bar{b}e_v e_v^T)], \tag{3}$$

where $L=D^{-1}C_L$, $U=D^{-1}C_U$, e_v the vth unit vector of the space $\mathbb{R}^{v\times v}$, $\bar{b}\gtrsim S(UL)$ and ω a real parameter which is chosen on the basis of á-priori information about the spectra of the operators involved in the algorithm.

Now because of the ill-conditioned character of A it is advisable to turn (1) into the better conditioned system,

$$A_\omega u = b_\omega \ , \tag{4}$$

where $A_\omega = M^{-1}A$ is the preconditioned matrix ([2],[5]) and $b_\omega = M^{-1}b$.

The preconditioning by direct factorization method (PDF method) is defined as follows:

$$u^{(n+1)} = (I-\tau M^{-1}A)u^{(n)} + M^{-1}b \ , \ n=0,1,\ldots \ , \tag{5}$$

where τ is a positive acceleration parameter to be chosen later.

We note that the scheme (5) can be seen as a fractional step scheme[(2)] ([10]), i.e. if we choose

$$\xi^{(n+1/3)} = \omega U\xi^{(n+1/3)} + r^{(n)} \ , \tag{6a}$$

---

(1) The idea behind this strategy has been applied previously by Evans [1972] in a direct method for solving the finite difference approximations to the partial differential equations, as well as by Evans and Hadjidimos [1978] in a factoring method described for the numerical solution of constant five diagonal linear systems (vd. [4] and [6]).

(2) A detailed analysis is given in [1] for the arithmetic operations to execute our algorithm by the Neithammer's [11] scheme in order to obtain computational efficiency. A saving of 20% is realized when Neithammer's scheme is applied to the PDF method.

$$\xi^{(n+2/3)} = \omega L\xi^{(n+2/3)} + \xi^{(n+1/3)} , \tag{6b}$$

$$(I+F)\xi^{(n+1)} = \xi^{(n+2/3)} , \tag{6c}$$

$$u^{(n+1)} = u^{(n)} - \tau\xi^{(n+1)} ,$$

where $r^{(n)} = Au^{(n)} - b$,

and[1] $F = \omega^2 b(I-\omega L)^{-1}(I-\omega U)^{-1}e_v$,

and where the third step (6c) is a simple Gaussian elimination process.

Since M is similar to a positive definite matrix the preconditioned matrix A_ω is also similar to the matrix

$$\begin{aligned}\bar{A}_\omega &= M^{\frac{1}{2}}A_\omega M^{-\frac{1}{2}} ,\\ &= M^{\frac{1}{2}}(M^{-1}A)M^{-\frac{1}{2}} ,\\ &= M^{-\frac{1}{2}}AM^{-\frac{1}{2}} .\end{aligned} \tag{7}$$

But since A is positive definite then $\bar{A}_\omega$ is positive definite which implies that A_ω is similar to a positive definite matrix.

Hence the eigenvalues of A_ω belong to the segment $[m(A_\omega), M(A_\omega)]$, where $m(A_\omega)$ and $M(A_\omega)$ are the minimal and maximal eigenvalues of A_ω respectively.

2. THE CHOICE OF $m(A_\omega)$ and $M(A_\omega)$

Our aim now is to evaluate the spectrum of the matrix A_ω, in order to find the optimal process (5) for minimizing the number of iterations. By [7] we can determine a lower bound on the minimal eigenvalue of A_ω. The technique followed is due to [9].

Lemma 2.1: Let A be a real symmetric positive definite matrix and $\bar{b}$,M and m numbers such that [14]

$$\begin{aligned}&-2\sqrt{\bar{b}} \leq m \leq m(B) ,\\ &M(B) \leq M \leq \min(1, 2\sqrt{\bar{b}}) ,\\ &S(UL+\bar{b}e_v e_v^T) \leq \bar{b} ,\end{aligned} \tag{8}$$

where m(B) and M(B) denote the minimal and maximal eigenvalues of B=U+L respectively. Then, for any ω in the range $1<\omega<2$, a

---

(1) By F we symbolize the matrix where only the last column is different than zero. For our algorithmic construction of F, we consider its last column.

lower bound on $m(A_\omega)$ is given by

$$m(A_\omega) \geq \begin{cases} \dfrac{1-M}{1-\omega M+\omega^2\bar{b}} & \text{, if } \bar{b}\geq\frac{1}{4} \text{ or if } \bar{b}<\frac{1}{4} \text{ and } \omega\leq\omega^*, \\ \dfrac{1-m}{1-\omega m+\omega^2\bar{b}} & \text{, if } \bar{b}<\frac{1}{4} \text{ and } \omega>\omega^* , \end{cases} \tag{9}$$

where for $\bar{b}<\frac{1}{4}$ we define ω^* by

$$\omega^* = \frac{2}{1+\sqrt{1-4\bar{b}}} \quad . \tag{10}$$

Before we estimate an upper bound for the maximal eigenvalue $M(A_\omega)$ of A_ω, we state a lower bound on the Rayleigh quotient of the conditioning matrix M. The matrix M given by (3) can be written in the form

$$M = \omega(2-\omega)A+D(C+E) \quad , \tag{11}$$

where
$$C = [(1-\omega)I+\omega U][(1-\omega)I+\omega L] \quad , \tag{12}$$

and $E=\omega^2\bar{b}e_v e_v^T$. Since C and E are symmetric matrices, exact symmetry is preserved in the matrix

$$C' = C+E \quad . \tag{13}$$

Thus, we can seek relations between the eigenvalues of C and E or between the eigenvalues of C', by applying a classical technique given in [12].

Since E is of rank unity, C can be partitioned in a similar form to E. Then, there is a real unitary matrix P of order v-1 such that

$$P^T C_1 P = \text{diag}(C_1) \quad , \tag{14}$$

where C_1 is the first minor matrix[(1)] of the partitioned form of C. Further the real unitary matrix R=diag{p,1} implies

$$R^T C' R = \left[\begin{array}{c|c} \text{diag}(C_1) & \ell \\ \hline \ell^T & \bar{C} \end{array}\right] + \left[\begin{array}{c|c} 0 & 0 \\ \hline 0 & \omega^2\bar{b} \end{array}\right] , \tag{15}$$

and the eigenvalues of C' and C satisfy the relations ([12] p.98)

---

(1) We mean the matrix arising by deleting the last column and row from a matrix.

$$\begin{cases} \lambda_i' = \lambda_i + \rho_i \omega^2 \bar{b} \\ 0<\rho_i \leq 1, \quad \sum \rho_i = 1 \end{cases} , \tag{16}$$

where λ_i' and λ_i are the eigenvalues of C' and C, respectively, in decreasing order.

Since now, C' is a symmetric matrix, then for any vector $v \neq 0$, its Rayleigh quotient satisfies the relations

$$\min_{v\neq 0} \frac{<v,C'v>}{<v,v>} \leq \frac{<v,C'v>}{<v,v>} \leq \max_{v\neq 0} \frac{<v,C'v>}{<v,v>} , \tag{17}$$

or in view of (16), relations (17) are written as,

$$m(C) + \omega^2 \rho_i \bar{b} \leq \frac{<v,C'v>}{<v,v>} \leq M(C') . \tag{18}$$

Further, since $m(C)=0$ (for C is positive semi-definite), we have

$$\omega^2 \rho_i \bar{b} \leq \frac{<v,DC'v>}{<v,Dv>} \leq M(C') . \tag{19}$$

We can now obtain a bound on $M(A_\omega)$ using the relation (19).

<u>Lemma 2.2:</u> Let A be a positive definite matrix and $\omega \in (1,2)$, then

$$M(A_\omega) \leq \frac{1}{\omega(2-\omega)+\omega^2 \frac{\bar{b}}{2}\rho} , \tag{20}$$

where $\rho \in (0,1]$.

<u>Proof</u>: Let μ be an eigenvalue of A_ω and v an associated eigenvector. Then, $A_\omega v=\mu v$ or $\mu = \frac{<v,Av>}{<v,Mv>}$ and by (11) and (19) we have

$$\mu \leq \frac{<v,Av>}{\omega(2-\omega)<v,Av>+\omega^2 \rho_i \bar{b}} . \tag{21}$$

Since now the right hand side of (21) is an increasing function of $<v,Av> \in [1-M(B),1-m(B)]$, the maximum value of that quantity with respect to $<v,Av>$ occurs when $<v,Av>=1-m(B)$. Hence,

$$\mu \leq \frac{1}{\omega(2-\omega)+\omega^2 \frac{\bar{b}}{2} \rho} , \tag{22}$$

is valid for any eigenvalue of A_ω when $-m(B) \to 1-$ and ρ has been chosen from the set $\{\rho_i : \sum\rho_i=1, 0<\rho_i \leq 1\}$. Thus, (20) is valid.

It is well known that the iterative scheme (5) is convergent if and only if

$$0<\tau<\frac{2}{M(A_\omega)} .$$

After obtaining the evaluation of $m(A_\omega)$ and $M(A_\omega)$ we can quote a theorem which establishes the condition for the minimization of the P-condition number of A_ω, i.e. $P(A_\omega) = \frac{M(A_\omega)}{m(A_\omega)}$, with respect to ω. Theorem 2.3 provides a theoretical foundation for the estimated parameters we use, i.e. ω_1, $P(A_{\omega_1})$ and τ_1, in the optimal process of the PDF method.

<u>Theorem 2.3</u>: Let $\bar{b}$,M and m be numbers such that

$$\begin{aligned} & m(B) \geq m \geq -2\sqrt{\bar{b}} , \\ & M(B) \geq M \geq 2\sqrt{\bar{b}} , \\ & M<1 , \\ & S(UL+\bar{b}e_v e_v^T) \leq \bar{b} , \end{aligned} \tag{23}$$

then

$$P(A_\omega) \leq \begin{cases} \dfrac{1-\omega M+\omega^2\bar{b}}{(1-M)\,[\omega(2-\omega)+\omega^2\frac{\bar{b}}{2}\rho]} , & \text{if } \bar{b}\geq\frac{1}{4} \text{ or if } \bar{b}<\frac{1}{4} \text{ and } \omega\leq\omega^* , \\ \dfrac{1-\omega m+\omega^2\bar{b}}{(1-m)\,[\omega(2-\omega)+\omega^2\frac{\bar{b}}{2}\rho]} , & \text{if } \bar{b}<\frac{1}{4} \text{ and } \omega>\omega^* , \end{cases} \tag{24}$$

where for $\bar{b}<\frac{1}{4}$, ω^* is defined by (10).

Moreover, a bound on $P(A_\omega)$ may be given if we let

$$\omega_1 = \begin{cases} \dfrac{2}{1+\sqrt{1-2M+4\bar{b}}} = \omega_M , & \text{if } M\leq 4\bar{b} , \\ \dfrac{2}{1+\sqrt{1-4\bar{b}}} = \omega^* , & \text{if } M\geq 4\bar{b} , \end{cases} \tag{25}$$

and therefore the corresponding bound of $P(A_\omega)$ is given by

$$P(A_{\omega_1}) \leq \frac{P(C_{\omega_1})}{1+kP(C_{\omega_1})} , \tag{26}$$

where [7]

$$P(C_{\omega_1}) \leq \begin{cases} \dfrac{1}{2}\left(1+\dfrac{\sqrt{1-2M+4\bar{b}}}{1-M}\right) = \dfrac{1}{2}\,\dfrac{(2-M\omega_M)}{(1-M)\,\omega_M} , & \text{if } M\leq 4\bar{b} , \\ \dfrac{1}{2}\left(1+\dfrac{1}{\sqrt{1-4\bar{b}}}\right) = \dfrac{1}{2-\omega^*} , & \text{if } M\geq 4\bar{b} , \end{cases} \tag{27}$$

and k is a positive constant lying in the interval

$$J = \begin{cases} \dfrac{(1-M)\omega_M^2\bar{b}}{(2-\omega_M M)(2-\omega_M)} & , \text{ if } M \leq 4\bar{b} , \\ \left(0, \dfrac{\omega^*\bar{b}}{2}\right) & , \text{ if } M \geq 4\bar{b}. \end{cases} \tag{28}$$

Thus,

$$P(A_{\omega_1}) \leq \begin{cases} \dfrac{(2-\omega_M M)}{(1-M)\omega_M} \cdot \dfrac{(2-\omega_M)}{4-\omega_M(2-\bar{b}\rho)} & , \text{ if } M \leq 4\bar{b} , \\ \dfrac{1}{2-\omega^*(1-\frac{\bar{b}}{2}\rho)} & , \text{ if } M \geq 4\bar{b} , \end{cases} \tag{29}$$

where $\rho \in (0,1]$.

From [7]we see that the preconditioned matrix C_ω of the Preconditioned Simultaneous Displacement method is in the form

$$C_\omega = (I-\omega U)^{-1}(I-\omega L)^{-1}D^{-1}A. \tag{30}$$

Proof: From Lemmas (2.1) and (2.2) the validity of (24) is evident.

Now the bound (24) on the P-condition number, $P(A_\omega)$ can be given in the form

$$P(A_\omega) \leq \frac{1}{\dfrac{(1-a)\omega(2-\omega)}{(1-\omega a+\omega^2\bar{b})} + \dfrac{1}{2}\dfrac{(1-a)\omega^2\bar{b}\rho}{(1-\omega a+\omega^2\bar{b})}} , \tag{31}$$

where

$$a = \begin{cases} M, \text{ if } \bar{b} \geq \frac{1}{4} \text{ or if } \bar{b} < \frac{1}{4} \text{ and } \omega \leq \omega^*, \\ m, \text{ if } \bar{b} < \frac{1}{4} \text{ and } \omega > \omega^* \end{cases} \tag{32}$$

Let $F_1(\omega;a,\bar{b})$ and $F_2(\omega;a,\bar{b})$ be the first and the second terms respectively of the denominator of (31).

The function $F_2(\omega;a,\bar{b})$ is increasing with respect to ω, in the range (1,2), then in a subinterval of ω, say I, $F_2(\omega;a,\bar{b})$ has a minimum at $\omega=\omega_1$, where $\omega_1 = \inf_{\omega\in I}\omega$.

$$F_2(\omega_1;a,\bar{b}) = \min_{\omega\in I} F_2(\omega;a,\bar{b}) = k . \tag{33}$$

Furthermore, in order to find a minimum on the bound of $P(A_\omega)$ we have,

$$\min_\omega P(A_\omega) \leq \frac{1}{\max_\omega (F_1+F_2)} ,$$

$$\leq \frac{1}{\max\limits_{\omega}\ (F_1 + \min\limits_{\omega\in I} F_2)} ,$$

$$\leq \frac{1}{\max\limits_{\omega} F_1 + \min\limits_{\omega\in I} F_2} ,$$

$$\leq \frac{1}{\max\limits_{\omega}\left[\frac{1}{P(C_\omega)}\right] + \min\limits_{\omega\in I} F_2} , \tag{34}$$

$$\leq \frac{P(C_{\omega_1})}{1+k\ P(C_{\omega_1})} ,$$

where $P(C_\omega)$ is the minimum bound on the P-condition number of the matrix C_ω, by [7] (vd. (30)) and where,

$$\omega_1 = \begin{cases} \dfrac{2}{1+\sqrt{1-2M+4\bar{b}}} = \omega_M , \text{ if } M\leq 4\bar{b} , \\ \dfrac{2}{1+\sqrt{1-4\bar{b}}} = \omega^* , \text{ of } M\geq 4\bar{b}. \end{cases} \tag{35}$$

From (27) concerning $P(C_\omega)$, the interval I may lie in the following interval

$$I \subseteq \begin{cases} (1,\omega^*] & , \text{ if } \bar{b}<\frac{1}{4} \text{ and } \omega_1=\omega_M , \\ [\omega^*,2) & , \text{ if } \bar{b}<\frac{1}{4} \text{ and } \omega_1=\omega^* , \\ (1,2) & , \text{ if } \bar{b}\geq\frac{1}{4} . \end{cases} \tag{36}$$

Thus, a realistic choice of I is

$$I = \begin{cases} (\omega_M,\omega^*] & , \text{ if } M\leq 4\bar{b}, \\ [\omega^*,2) & , \text{ if } M\geq 4\bar{b}. \end{cases} \tag{37}$$

By (33) we have that

$$k = \begin{cases} \dfrac{(1-M)\omega_M^2\bar{b}\rho}{(2-\omega_M M)(2-\omega_M)} & , \text{ if } M\leq 4\bar{b} , \\ \dfrac{\omega^*\bar{b}\rho}{2} & , \text{ if } M\geq 4\bar{b}, \end{cases} \tag{38}$$

and (28) and (29) are valid since $\rho\in(0,1]$.

We can modify the bound on $P(A_{\omega_1})$ given by (29) to yield

$$P(A_{\omega_1}) \leqslant \begin{cases} \dfrac{\frac{1}{2}\left(1+\frac{1}{\sqrt{1-M}}\right)}{1+\frac{M}{8\sqrt{1-M}}\rho}, & \text{if } \bar{b}\leq\frac{M}{4}, \\[2ex] \dfrac{\frac{1}{2}\left(1+\sqrt{\frac{2}{1-M}}\right)}{1+\frac{1}{8\sqrt{2(1-M)}}\rho}, & \text{if } \frac{M}{4}\leq\bar{b}\leq\frac{1}{4}, \\[2ex] \dfrac{\frac{1}{2}\left(1+\gamma^{-1}\sqrt{\frac{2}{1-M}}\right)}{1+\frac{\bar{b}\gamma}{4}\rho}, & \text{if } \bar{b}>\frac{1}{4}, \end{cases} \tag{39}$$

where $\rho\in(0,1]$ and $\gamma=\left(1+\frac{2(\bar{b}-1/4)}{1-M}\right)^{-\frac{1}{2}}$.

The optimal choice $\tau_1 = \frac{2}{m(A_{\omega_1})+M(A_{\omega_1})}$ of the acceleration parameter τ, is derived from the condition for obtaining

$$\inf_{\tau>0} \sup_{\mu\in[m(A_{\omega_1}),M(A_{\omega_1})]} |1-\tau\mu|. \tag{40}$$

3. THE RATE OF CONVERGENCE

For the PDF method with the optimal choice τ_1 of τ, the asymptotic rate of convergence is given by

$$R_\infty(H_{\omega_1}) \doteq \frac{2}{P(A_{\omega_1})}, \tag{41}$$

where $H_\omega=I-\tau A_\omega$ is the iteration matrix of the method. Also, the reciprocal rate of convergence is defined by

$$RR(H_{\omega_1}) = \frac{1}{R_\infty(H_{\omega_1})} \doteq P(A_{\omega_1})/2, \tag{42}$$

and indicates that the number of iterations required for convergence is approximately proportional to the P-condition number of the matrix A_ω.

The á-priori knowledge of the eigenvalues of A_ω implies a dependence upon ρ of the estimated rate of convergence. Thus, comparing the bounds on $RR(H_{\omega_1})$ with $RR(B_{\bar{\rho}})$, where $B_{\bar{\rho}}$ is the iteration matrix of the Jacobi over-relaxation method [13] and taking into account that $\rho\in(0,1]$ we have,

$$\frac{RR(H_{\omega_1})}{\sqrt{RR(B_{\bar{\rho}})}} \leqslant \begin{cases} \frac{1}{2\sqrt{2}}(1-k_1) & , \text{ if } \bar{b}\leqslant\frac{M}{4} , \\ \frac{1}{2}(1-k_2) & , \text{ if } \frac{M}{4}\leqslant\bar{b}\leqslant\frac{1}{4} , \\ \frac{1}{2}\gamma^{-1}(1-k_3) & , \text{ if } \bar{b}>\frac{1}{4} , \end{cases} \tag{43}$$

where $k_1,k_2,k_3\in(0,1)$ depending on M,ρ and $\bar{b}$.

Since now $RR(L_{\omega_b})\leqslant\sqrt{RR(B_{\bar{\rho}})}$ where L_ω is the iteration matrix of the SOR method we establish that

$$\frac{RR(H_{\omega_1})}{RR(L_{\omega_b})} \doteq \begin{cases} \frac{1}{\sqrt{2}}(1-k_1) , & \text{ if } \bar{b}\leqslant\frac{M}{4} , \\ (1-k_2) , & \text{ if } \frac{M}{4}\leqslant\bar{b}\leqslant\frac{1}{4} , \\ \gamma^{-1}(1-k_3) , & \text{ if } \bar{b}>\frac{1}{4} , \end{cases} \tag{44}$$

where $k_i\in(0,1)$ and A is a consistently ordered matrix.

Moreover, asymptotically for the SSOR method we have by [14] that

$$\frac{RR(H_{\omega_1})}{RR(G_{\omega_1})} \doteq \frac{1}{2}(1-k), \text{ where } k\in(0,1). \tag{45}$$

However, we are not able to specify exactly an á-priori asymptotic number to the above ratios since it is not possible to have á-priori knowledge of ρ (1), but the numerical results of the next section justify our assertion of the existence of k.

4. NUMERICAL EXPERIMENTS

In order to test the theoretical results obtained in Section 3, six problems were carried out involving the Dirichlet problem with the differential equation

$$\frac{\partial}{\partial x}(A\frac{\partial u}{\partial x}) + \frac{\partial}{\partial y}(C\frac{\partial u}{\partial y}) = 0 , \tag{46}$$

on the unit square $D=\{(x,y): 0\leqslant x\leqslant 1, 0\leqslant y\leqslant 1\}$, with zero boundary conditions.

(1) An approximate approach gives as ρ one of the elements of the set $\{c,ch,ch^2\}$, where h is the net mesh size and c a constant.

Before we can compute ω_1, $P(A_{\omega_1})$, we note that an upper bound M for S(B) is found by [13], and an upper bound $\bar{b}$ for S(UL) is determined by a similar process as in [14]. Thus,

$$S(UL)\leq||UL||_\infty = \max_{(x,y)\in D_h} [b_1(x,y)[b_3(x+h,y)+b_4(x+h,y)] + b_2(x,y)[b_3(x,y+h)+b_4(x,y+h)]] = \bar{b}\ , \qquad (47)$$

where $b_1 = \frac{A(x+\frac{h}{2},y)}{S(x,y)}$, $b_2 = \frac{C(x,y+\frac{h}{2})}{S(x,y)}$, $b_3 = \frac{A(x-\frac{h}{2},y)}{S(x,y)}$,

$$b_4 = \frac{C(x,y-\frac{h}{2})}{S(x,y)},\quad S(x,y) = A(x+\tfrac{h}{2},y)+A(x-\tfrac{h}{2},y)+C(x,y+\tfrac{h}{2})+C(x,h-\tfrac{h}{2}), \qquad (48)$$

and D_h is the totality of points $x_i=ih$, $y_j=jh$, where $h=\frac{1}{N}$, $1\leq i\leq N-1$, $1\leq j\leq N-1$, $i,j \in (N,N)$.

Further, it is easy to show [1] that if A and $C\in C^2(D+\partial D)$ then $S(UL)\leq\frac{1}{4}+O(h^2)$, as $h\to 0$. This result establishes the improvement of an order of magnitude, of the PDF-SI (vd. next section) over the PDF method.

By (39) given that $M=1-ch^2+O(h^4)$ we have that $P(A_{\omega_1})=C_1h^{-1}$. Thus by (46) we obtain the expected result for the asymptotic convergence rate of the PDF method, i.e. $R_\infty(H_{\omega_1})\doteq O(h)$. The numerical results given in Table (1) however, indicate that we can obtain O(h) convergence for the PDF method with estimated parameters. In this way, we see that for the SOR, SSOR and PDF method the number of iterations varies as h^{-1}.

We conclude this section by presenting in Table (1) the six problems with their estimated parameters ω_1 and $P(A_{\omega_1})$. The upper bound M for S(B) is given in Young [1977]; the upper bound $\bar{b}$ for S(UL) is computed by (48). Note that in Problems II and VI we can replace M by $2\sqrt{\bar{b}}$ and since $M>2\sqrt{\bar{b}}$. The estimated parameters ω_1,τ_1 and $P(A_{\omega_1})$ are computed by (25),(40) and (39)(1), respectively.

---

(1) In formula (39) we took ρ equal to h^2, i.e. $\rho=h^2$, where h the mesh size of the net.

The scheme (5) is then applied with a guess vector $u^{(0)}=(1,1,\ldots,1)$ and the procedure was terminated when the inequality $||u^{(n)}||_{\infty}\leq 10^{-6}$ was satisfied. The number of iterations, n_{IE}, required to obtain the solution is presented together with the number of iterations of the SSOR with estimated parameters and the SOR method with optima values.

In Table (1) it is shown that the SOR method with optimum parameters requires at least 33%, 150%, 26%, 32%, 17% and 128% more iterations than the PDF method with estimated values, respectively to the six problems in increasing order. However, it should be noted that the PDF method requires more work per iteration than the SOR method. The SSOR method, asymptotically requires about 63%, 25%, 77%, 53%, 95% and 153% more iterations as compared to the PDF method respectively.

5. TWO POINT BOUNDARY VALUE PROBLEM

We will consider a simple but typical problem of Mathematical Physics and use it to illustrate the effectiveness of the preconditioning scheme which we have applied earlier.

Let us consider the problem of obtaining the solution of the differential equation

$$-\frac{d^2u}{dx^2}+\gamma u=\beta(x) \text{ in the range } (0\leq x\leq 1), \tag{49}$$

with boundary conditions

$$u(0) = a \ , \quad u(1) = b,$$

where β represents the source term and a and b are given constants.

The difference analogue of (49) using the usual second order approximation, by Taylor series can be written as,

$$u(x_i)-\alpha u(x_{i-1})-\alpha u(x_{i+1}) = h^2\beta(x_i)/(2+\gamma h^2),\ i=1,2,\ldots,n, \tag{50}$$

at the grid points given by $x_i\equiv ih$, $1\leq i\leq n$ and $h=1/(n+1)$. Grouping the equations (50) into matrix form, we have the solution of (49) is given by

$$Au = g \ , \tag{51}$$

where

$$A \doteq \begin{bmatrix} 1 & -\alpha & & & 0 \\ -\alpha & 1 & -\alpha & & \\ & \ddots & \ddots & \ddots & \\ & & & & -\alpha \\ 0 & & & -\alpha & 1 \end{bmatrix}, \quad g = \begin{bmatrix} h^2\beta(x_1)/(2+\gamma h^2)+\alpha a \\ h^2\beta(x_2)/(2+\gamma h^2) \\ \cdot \\ \cdot \\ \cdot \\ \cdot \\ h^2\beta(x_N)/(2+\gamma h^2)+\alpha b \end{bmatrix}.$$

with $\alpha=1/(2+\gamma h^2)$.

Assuming the splitting

$$A = I-B = I-L-U \quad , \tag{52}$$

where L and U are strictly the lower and upper triangular parts of A and I the unit matrix, let us define [4] the conditioning matrix M in the following form:

$$M = [I-\omega U \mid -\omega\alpha e_v] \begin{bmatrix} I-\omega L \\ \hline -\omega\alpha e_v^T \end{bmatrix} \quad , \tag{53}$$

where e_v is the vth unit vector of the space $\mathbb{R}^{v\times v}$ and ω a real parameter.

Then it is advisable to turn (50) into the well conditioned system, i.e. the preconditioned system [2]

$$\mathcal{B}_\omega u = g_\omega \quad , \tag{54}$$

where

$$\mathcal{B}_\omega = M^{-1}A \quad , \tag{55}$$

is the preconditioned matrix [2].

System (54) has the optimistic p-condition number which can be easily ascertained from the following proposition.

<u>Proposition 5.1</u>: If A has the form (51) then,

$$P(\mathcal{B}_\omega) \leq \begin{cases} 1, & \text{if } \alpha^2 \leq \frac{1}{4} \\ k\ P(A), & \text{if } \alpha^2 \geq \frac{1}{4} \end{cases} \quad , \tag{56}$$

where k is a positive constant such that $k<1$ and where $\mathcal{B}_\omega=M^{-1}A$ is the preconditioned matrix of (54).

<u>Proof</u>:

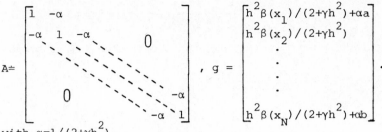

Let μ be an eigenvalue of the matrix $\mathcal{B}_\omega$ and v an associated normalized eigenvector. Then $\mathcal{B}_\omega v=\mu v$ and by taking inner products on both sides with respect to v we have

$$\mu = \frac{\langle v,Av\rangle}{\langle v,v\rangle-\omega\langle v,Bv\rangle+\omega^2\langle v,ULv\rangle+\omega^2\alpha^2\langle \ell v,\ell v^T\rangle} \tag{57}$$

Since A and M are positive definite matrices, then (57) becomes

$$\mu = \frac{1-a(v)}{1-\omega a(v)+\omega^2\alpha^2}, \tag{58}$$

where $a(v)=\langle v,Bv\rangle$ and $\alpha^2=\langle v,ULv\rangle$.

Let μ_1, μ_2 be two eigenvalues of B_ω and let v_1, v_2 be two normalized eigenvectors associated with them respectively. Then, if we suppose that,

$$\mu_1 = \mu_2 \text{ while } \langle v_1,Bv_1\rangle = a(v_1) \neq a(v_2) = \langle v_2,Bv_2\rangle, \tag{59}$$

by means of the relation (58) we have,

$$\frac{1-a(v_1)}{1-\omega a(v_1)+\omega^2\alpha^2} = \frac{1-a(v_2)}{1-\omega a(v_2)+\alpha^2\omega^2}, \tag{60}$$

where equivalently implies

$$[a(v_2)-a(v_1)]\ (\omega^2\alpha^2-\omega+1) = 0, \tag{61}$$

or

$$\omega^* = \frac{2}{1+\sqrt{1-4\alpha^2}}, \quad \text{if } \alpha^2\leq\frac{1}{4}, \tag{62}$$

where $\omega^*\in(0,2)$. Therefore (60) is valid for (62) and subsequently $P(B_{\omega^*})=1$.

On the other hand, if $\alpha^2\geq\frac{1}{4}$, since (58) with $b(v)=\alpha^2$ is a decreasing function with respect to a(v), then any eigenvalue μ of B_ω lies on

$$\left[\frac{1-M}{1-\omega M+\omega^2\alpha^2}, \frac{1-m}{1-\omega m+\omega^2\alpha^2}\right], \tag{63}$$

and hence the P-condition number of B_ω is minimized at $\omega_M=\frac{1}{\sqrt{\alpha^2}}$. Therefore,

$$P(B_{\omega_M}) \leq P(A)\left(\frac{2\sqrt{\alpha^2}-M}{2\sqrt{\alpha^2}-m}\right), \tag{64}$$

where the quantity $\frac{2\sqrt{\alpha^2}-M}{2\sqrt{\alpha^2}-m}$ is less than 1.

We have therefore proved (56).

6. THE P.D.F. METHOD WITH σ_1-ORDERING

In this section we again seek a solution of the self-adjoint elliptic equation of the 2nd order in a certain bounded plane region with specified values on the boundary.

We also assume that a σ_1-ordering has been applied to the grid points which gives the derived matrix A from the discretised elliptic partial differential equation to have the form (65).

Actually, the finite difference equations are ordered by choosing all the points in which (i+j) is even, and then all the points in which (i+j) is odd so that the coefficient matrix A has the simple partitioned form,

$$A = \begin{bmatrix} I & -U^* \\ -L^* & I \end{bmatrix} , \tag{65}$$

where $L^*=U^{*T}$.

Let

$$Au = b , \tag{66}$$

be the linear analogue to the elliptic partial differential equation, where without loss of generality we assume that,

$$A = I-B = I-L-U ,$$

where L and U the lower and upper triangular parts of A, respectively.

Clearly it can be seen that A is a positive definite matrix.

Now we consider the conditioning matrix ([2],[4]),

$$M_2 = (I-\omega\tilde{U})(I-\omega\tilde{L}) , \tag{67}$$

where

$$I-\omega\tilde{U} = \begin{bmatrix} I & -\omega U^* & \\ & I & -\omega U^* \end{bmatrix} , \tag{68}$$

and

$$I-\omega L^* = \begin{bmatrix} I & \\ -\omega L^* & I \\ & -\omega L^* \end{bmatrix} . \tag{69}$$

Therefore M_2 has the form

$$M_2 = \begin{bmatrix} I+\omega^2 U^*L^* & -\omega U^* \\ -\omega L^* & I+\omega^2 U^*L^* \end{bmatrix} . \tag{70}$$

The preconditioned matrix $B_\omega^{(2)}$, [2] is given by

$$B_\omega^{(2)} = M_2^{-1}A , \tag{71}$$

whereas the iterative scheme has the form

$$u^{(n+1)} = u^{(n)} - \tau M_2^{-1} A u^{(n)} + M_2^{-1} b. \tag{72}$$

It is easy to show that the matrices M_2 and $B_\omega^{(2)}$ are similar to positive definite matrices.

Lemma 6.1: Let A have the form (65). If μ is an eigenvalue of the preconditioned matrix $B_\omega^{(2)}$ such that

$$\mu \neq \frac{\omega-1}{\omega} + \sqrt{\left(\frac{\omega-1}{\omega}\right)^2 + \frac{1}{\omega^2}} , \tag{73}$$

and if $$\delta[\mu^2\omega^2+2\omega\mu(1-\omega)-1] = (\mu-1)^2 , \tag{74}$$

then δ is an eigenvalue of B^2.

Proof:

Let us assume that A has the form, A=I-L-U, where

$$L = \begin{bmatrix} O & O \\ L^* & O \end{bmatrix} \quad \text{and} \quad U = \begin{bmatrix} O & U^* \\ O & O \end{bmatrix} . \tag{75}$$

Let v be an associated eigenvector to the eigenvalue μ of $B_\omega^{(2)}$. Then,

$$B_\omega^{(2)} v = \mu v , \tag{76}$$

or

$$Av = \mu M_2 v , \tag{77}$$

or

$$Av = [I-\omega(U+L)+\omega^2 UL+\omega^2 UL]v ,$$

which implies

$$(\omega\mu-1)Bv-2\omega^2\mu ULv = (\mu-1)v . \tag{78}$$

Since A has the form (65) then,

$$B = \begin{bmatrix} O & U^* \\ L^* & O \end{bmatrix} , \tag{79}$$

and

$$UL = \begin{bmatrix} U^*L^* & O \\ O & O \end{bmatrix} . \tag{80}$$

Thus (78) becomes

$$\begin{bmatrix} -2\omega^2\mu U^*L^* & (\omega\mu-1)U^* \\ (\omega\mu-1)L^* & O \end{bmatrix} \begin{bmatrix} v_1 \\ v_2 \end{bmatrix} = (\mu-1) \begin{bmatrix} v_1 \\ v_2 \end{bmatrix}, \tag{81}$$

where we assume that $v=[v_1,v_2]^T$ is similarly partitioned to A. Furthermore, we have

$$\left.\begin{aligned} -2\omega^2\mu U^*L^*v_1+(\omega\mu-1)U^*v_2 &= (\mu-1)v_1 \quad (a) \\ (\omega\mu-1)L^*v_1 &= (\mu-1)v_2 \quad (b) \end{aligned}\right\} . \tag{82}$$

Multiplying (82a) by $(\mu-1)$ and (82b) by $(\omega\mu-1)U^*$ and substituting the first into the second we have,

$$[\omega\mu(2\omega-2-\omega\mu)+1]U^*L^*v_1 = (\mu-1)^2 v_1 \quad , \tag{83}$$

and

$$[\omega\mu(2\omega-2-\omega\mu)+1]L^*U^*v_2 = (\mu-1)^2 v_2 \quad . \tag{84}$$

Since from (73), the relations (83) and (84) become

$$\left.\begin{aligned} U^*L^*v_1 &= \frac{(\mu-1)^2}{\omega\mu(2\omega-2-\omega\mu)+1} v_1 \\ L^*U^*v_2 &= \frac{(\mu-1)^2}{\omega\mu(2\omega-2-\omega\mu)+1} v_2 \end{aligned}\right\} . \tag{85}$$

Thus, we obtain

$$B^2 v = \begin{bmatrix} U^*L^* & O \\ O & L^*U^* \end{bmatrix} \begin{bmatrix} v_1 \\ v_2 \end{bmatrix} = \left\{\frac{(\mu-1)^2}{\omega\mu(2\omega-2-\mu\omega)+1}\right\} \begin{bmatrix} v_1 \\ v_2 \end{bmatrix}, \tag{86}$$

from which we conclude that v is an eigenvector of B^2 and (74) is valid.

Following closely to a proof of Kahan [15] concerning the eigenvalues of the SSOR iteration matrix and those of B^2 we have the following proposition.

<u>Proposition 6.2</u>: Let A have the form (65). If η is an eigenvalue of the iteration matrix $H^{(2)}=I-M_2^{-1}A$ of (72) such that

$$\eta \neq \frac{1}{\omega} + \sqrt{\frac{1}{\omega^2} + \left(\frac{\omega-1}{\omega}\right)^2} \quad , \tag{87}$$

and if

$$\eta^2 = [(\omega-1)^2+\omega\eta(2-\omega\eta)]\delta \quad , \tag{88}$$

then δ is an eigenvalue of B^2

<u>Proof</u>:

The proof is similar to the one followed in Lemma 6.1.

We now proceed to the following theorem concerning the minimization of the P-condition number of the matrix $B_\omega^{(2)}$.

<u>Theorem 6.3</u>: If A possesses property A and σ_1-ordering then the P-condition number of the matrix $B_\omega^{(2)}$ is minimized if we let

$$\omega_1 = \begin{cases} \dfrac{1}{M(B)} = \omega_M \quad , & \text{if } M(B) \geq \frac{1}{2} \quad , \\ \dfrac{1}{1+\sqrt{1-4M(B)^2}} = \omega^* \quad , & \text{if } M(B) \leq \frac{1}{2} \quad , \end{cases} \tag{89}$$

where M(B) is the maximal eigenvalue of the matrix B and the corresponding value of $P(B_{\omega_1}^{(2)})$ is given by

$$P(B_{\omega_1}^{(2)}) \leq \begin{cases} \frac{1}{3}P(A) \quad , & \text{if } \omega_1 = \omega_M \\ 1 \quad , & \text{if } \omega_1 = \omega^* \end{cases} \quad . \tag{90}$$

<u>Proof</u>:

Let A have the form A=I-L-U, where (75) is valid for the matrices L and U. Let μ be an eigenvalue of $B_\omega^{(2)}$ and v an associated eigenvector.

Since now, $B_\omega^{(2)} v = \mu v$, we have that

$$(A-\mu M_2)v = 0 \ , \tag{91}$$

or assuming in v a partition similar to that of A we get from (70) and (91)

$$\begin{bmatrix} (1-\mu)I-\mu\omega^2 U^*L^* & (\mu\omega-1)U^* \\ (\mu\omega-1)L^* & (1-\mu)I-\mu\omega^2 U^*L^* \end{bmatrix} \begin{bmatrix} v_1 \\ v_2 \end{bmatrix} = 0 \ , \tag{92}$$

which simplifies to

$$\left.\begin{aligned} [(1-\mu)I-\mu\omega^2 U^*L^*]v_1 + (\mu\omega-1)U^*v_2 &= 0 \\ (\mu\omega-1)L^*v_1+[(1-\mu)I-\mu\omega^2 U^*L^*]v_2 &= 0 \end{aligned}\right\} . \tag{93}$$

Eliminating v_1 from the equations of (91) this results in

$$(1-\mu\omega)^2 U^*L^*v_2 - [(1-\mu)^2 I+\mu^2\omega^4(U^*L^*)^2-2(1-\mu)\mu\omega^2 U^*L^*]v_2 = 0. \tag{94}$$

Since the non-zero eigenvalues of (L+U) occur in pairs $\pm b_i$, (i=1,2,...,r) where r is less than or equal to the number of rows in L* or U*, the eigenvalues of U*L* are precisely b_i^2, (i=1,2,...,r) or zero.

Therefore, since $v_2 \neq 0$, we must have

$$(1-\mu\omega)^2 b_i^2-[(1-\mu)^2+\mu^2\omega^4 b_i^4-2(1-\mu)\mu\omega^2 b_i^2] = 0 \quad , \quad i=1,2,\ldots,r \ , \tag{95}$$

Now, the P-condition number of the matrix $B_\omega^{(2)}$ is given by the expression[(1)]

$$P(B_\omega^{(2)}) = \frac{M(B_\omega^{(2)})}{m(B_\omega^{(2)})}, \tag{96}$$

and is obtained by taking i=1 in both choices of the solution of (95) or in its transformed form

$$[(1+b_1)-\mu(\omega^2 b_1^2+\omega b_1+1)]\ [(b_1-1)+\mu(\omega^2 b_i^2-\omega b_1+1)] = 0 . \tag{97}$$

Thus,

$$P(B_\omega^{(2)}) = \left(\frac{1+b_1}{\omega^2 b_1^2+\omega b_1+1}\right)\left(\frac{\omega^2 b_1^2-\omega b_1+1}{1-b_1}\right), \tag{98}$$

$$= \left(\frac{1+b_1}{1-b_1}\right)\left(\frac{\omega^2 b_1^2-\omega b_1+1}{\omega^2 b_1^2+\omega b_1+1}\right). \tag{99}$$

Now, if $b_1 \geq \frac{1}{2}$, $P(B_\omega^{(2)})$ receives its minimum value with respect to ω at the point $\omega_M = \frac{1}{b_1}$, thus,

$$P(B_{\omega_M}^{(2)}) = \frac{1}{3}P(A), \tag{100}$$

where $P(A) = \frac{1+b_1}{1-b_1}$ is the P-condition number of A.

On the other hand, if $b_1 \leq \frac{1}{2}$, a minimum occurs at the point $\omega^* = \frac{2}{1+\sqrt{1-4b_i^2}}$. In that case $P(B_{\omega^*}^{(2)})=1$.

Finally, we conclude that (89) and (90) are valid, where $M(B)=b_1$.

Note that the above approach for determining a bound on $P(B_\omega^{(2)})$ is similar to that followed by Evans [2] concerning the classical scheme of preconditioning methods, when matrix A has property A.

The asymptotic rate of convergence when the matrix A possesses property A and conditioning matrix M_2 for $P(B_\omega^{(2)}) \gg 1$ is

---

(1) $M(B_\omega^{(2)})$, $m(B_\omega^{(2)})$ denote the maximal, minimal eigenvalue of $B_\omega^{(2)}$, respectively.

$$R_\infty(H^{(2)}_{\omega_M}) \doteq \frac{2}{P(B^{(2)}_{\omega_M})} \quad ,$$

or

$$R_\infty(H^{(2)}_{\omega_M}) \doteq \frac{6}{P(A)} \quad ,$$

whereas the reciprocal rate of convergence is given by

$$RR(H^{(2)}_{\omega_M}) \doteq \frac{P(A)}{6} \quad .$$

In the case of Property A we also have

$$R_\infty(B_{\bar\rho}) = R_\infty(B) \quad ,$$

where $B_{\bar\rho}$ is the iteration matrix of the Jacobi overrelaxation method [14] and

$$R_\infty(B_{\bar\rho}) \doteq \frac{1}{2P(A)} \quad .$$

Thus,

$$R_\infty(H^{(2)}_{\omega_M}) \doteq 12\ R_\infty(B_{\bar\rho}) \quad .$$

However, this improvement is really only of academic interest and does not convey the efficient use of this form of preconditioning for practical problems.

7. THE PDF-SI METHOD

Since the iteration matrix of the PDF method has positive eigenvalues, the semi-iterative technique ([8],[14]) when applied resulted in a $O(h^{\frac{1}{2}})$ acceleration of the convergence rate. The formula for the estimated PDF-SI method is

$$u^{(n)} = P_n(H_{\omega_1})u^{(0)} + k_n \quad , \tag{101}$$

where

$$P_n(H_{\omega_1}) = T_n\left[\frac{M(A_{\omega_1})+m(A_{\omega_1})-2A_{\omega_1}}{M(A_{\omega_1})-(A_{\omega_1})}\right] \Bigg/ T_n\left[\frac{M(A_{\omega_1})+m(A_{\omega_1})}{M(A_{\omega_1})-m(A_{\omega_1})}\right], \tag{102}$$

$$k_n = (I-P_n(H_{\omega_1}))A^{-1}b \ , \tag{103}$$

and T_n is the Chebyshev polynomial of degree n.

Taking $M(A_\omega) \gg m(A_\omega)$, we obtain the asymptotic rate of convergence of the PDF-SI method,

$$R_\infty(P_n(H_\omega)) \doteq \frac{2}{\sqrt{P(A_\omega)}} \quad .$$

The effectiveness of the PDF-SI method is tested by solving the six problems given in Table (1). The boundary values were taken to be zero on all sides of the unit square except for the side y=0, where they were taken to be unity. The starting vector $u^{(0)}$ was taken to be the vector with all its components equal to unity. The process terminated after n iterations where n satisfied the stopping procedure [14],

$$\frac{2r^{n/2}}{1+r^n} \leqslant 10^{-6} \quad , \tag{104}$$

where $r = \left(\frac{\sqrt{P(A_{\omega_1})}-1}{\sqrt{P(A_{\omega_1})}-1}\right)^2$.

The numerical results of Table (1) indicate an $O(h^{\frac{1}{2}})$ convergence even if the coefficient functions are not restricted to class $C^{(2)}$. This is an order of magnitude improvement of the PDF-SI over the PDF method. However, an $h^{-\frac{1}{2}}$ behaviour is attained by using the SSOR-SI. Also, both methods in the SI processes require approximately twice the work required by the SOR method.

As indicated in Table (1) the PDF-SI method is applicable for problems with certain kinds of discontinuities, as shown by problem IV and the one given by case V. The number of iterations required indicated $O(h^{\frac{1}{2}})$ acceleration of the convergence rate.

Further research is required to investigate the accuracy offered by the PDF-SI method as compared to the SSOR-SI method. The acceleration by SI techniques shows that the PDF model can be applied effectively to second degree and to non-stationary methods even in cases of differential equations containing discontinuous coefficients.

TABLE 1: Estimated parameters ω_1 and $P(B_{\omega_1})$

Prob.	Coefficients	h^{-1}	$\bar{b}$	$2\sqrt{\bar{b}}$	M	ω_1
I	$A=C=1$	20	.2500	1.0000	.9877	1.7288
		40	.2500	1.0000	.9969	1.8544
		60	.2500	1.0000	.9986	1.8992
II	$A=C=e^{10(x+y)}$	20	.2350	.9695	.9999	1.6065
		40	.2461	.9921	.9999	1.7779
		60	.2482	.9963	.9999	1.8436
III	$A=(1+2x^2+y^2)^{-1}$	20	.2499	.9997	.9967	1.8540
	$C=(1+x^2+2y^2)^{-1}$	40	.2499	.9992	.9992	1.9330
		60	.2500	1.0000	.9996	1.9422
IV	$A=C=\begin{cases}1+x, & 0\leq x\leq\frac{1}{2}\\ 2-x, & \frac{1}{2}\leq x\leq 1\end{cases}$	20	.2521	1.0041	.9914	1.7241
		40	.2501	1.0002	.9979	1.8790
		60	.2500	1.0000	.9990	1.9143
V	$A=1+4\lvert x-\frac{1}{2}\rvert^2$	20	.2499	.9997	.9977	1.8782
	$C=\begin{cases}1, & 0\leq x<\frac{1}{2}\\ 9, & \frac{1}{2}\leq x\leq 1\end{cases}$	40	.2500	1.0000	.9994	1.9357
		60	.2500	1.0000	.9997	1.9521
VI	$A=1+\sin\frac{\pi(x+y)}{2}$	20	.2366	.9728	.9999	1.6240
	$C=e^{10(x+y)}$	40	.2469	.9937	.9999	1.7996
		60	.2486	.9971	.9999	1.8607

continued

τ_1	$P(B_{\omega_1})$	$M(B_{\omega_1})$	n_{IE} ESTIMATED			OPTIMUM	
			PDF	SSOR	PDF-SI	h^{-1}	SOR
.8186	6.8760	2.1329	46	68	19	20	61
.5018	13.1661	3.7037	91	138	26	40	121
.3634	19.3686	5.2334	127	207	32	80	253
.9094	2.5377	1.5779	20	28	10	20	50
.6469	4.4997	2.5293	37	45	15	40	99
.4993	6.3872	3.4633	54	67	18	80	217
.5011	12.4429	3.6943	47	73	25	20	60
.2478	22.0863	7.7214	96	145	34	40	121
.2156	23.4865	8.8976	123	218	36	80	252
.8033	9.8038	2.1023	41	66	16	20	59
.4282	15.8543	4.3935	90	133	29	40	119
.3145	22.8430	6.0908	131	200	35	80	225
.4282	14.5976	4.3713	60	93	28	20	60
.2403	24.3183	8.0155	101	193	36	40	118
.1826	36.9602	10.6643	148	288	47	80	274
.8883	2.6561	1.6356	18	36	11	20	41
.6012	4.9865	2.7709	35	82	16	40	81
.4552	7.1754	3.8563	51	129	19	80	176

REFERENCES

1. Demetriou, I.C. - "Generalized Preconditioning Strategies", M.Phil. Thesis, University of Technology, Loughborough,U.K., 1980.

2. Evans, D.J. - "The Use of Preconditioning in Iterative Methods for Solving Linear Equations with Symmetric Positive Definite Matrices", J.I.M.A. 4, pp.295-314, 1968.

3. Evans, D.J. - "Comparison of the Convergence Rates of Iterative Methods for Solving Linear Equations with Preconditioning", Greek Mathematical Society, Carathéodory Symposium, pp.106-135, 1973.

4. Evans, D.J. - "An Algorithm for the Solution of Certain Tridiagonal Systems of Linear Equations", Comp.J. 15, pp.356-359, 1972.

5. Evans, D.J. - "On Preconditioned Iterative Methods for Elliptic Partial Differential Equations", Elliptic Problem Solvers Conference, Los Alamos Scientific Laboratory, pp. 261-269, edit. M. Schultz, Acad.Press, 1981.

6. Evans, D.J. and A. Hadjidimos - "On the Factorization of Special Symmetric Periodic and Non-Periodic Quindiagonal Matrices", Computing 21, pp.259-266, 1979.

7. Evans, D.J. and N.M. Missirlis - "The Preconditioned Simultaneous Displacement Method", Math. and Comp. in Simulation, 22, pp.256-263, 1980.

8. Golub, G.H. and R.S. Varga, - "Chebyshev Semi-iterative Methods, Successive Overrelaxation Iterative Methods and Second Order Richardson Iterative Methods", Num.Math. Part I and II, 3, pp.147-168, 1961.

9. Habetler, G.J. and E.L. Wachspress - "Symmetric Successive Overrelaxation in Solving Diffusion Difference Equations", Math.Comp. 15, pp.356-362, 1961.

10. Marchuk, G.I. - "Methods of Numerical Mathematics", (translated by J. Ruzicka from Russian), Springer-Verlag, New York, Heidelberg, Berlin, 1975.

11. Niethammer, W. - "Relaxation bei Komplexen Matrizen", Math. Zeitsch, 86, pp.34-40, 1964.

12. Wilkinson, J.H. - "The Algebraic Eigenvalue Problem", Clarendon Press, Oxford, 1965.

13. Young, D.M. - "Iterative Solution of Large Linear Systems", Academic Press, New York, 1971.

14. Young, D.M. - "On the Accelerated SSOR Method for Solving Large Linear Systems", Advances in Mathematics, 23, pp.215-271, 1977.

15. Kahan, W. - "Gauss Seidel Methods for Solving Large Systems of Linear Equations", Ph.D. Thesis, Univ. of Toronto, Canada, 1958.

PRECONDITIONED ITERATIVE METHODS FOR THE GENERALIZED EIGENVALUE PROBLEM WITH LARGE SPARSE MATRICES

D.J. Evans & J. Shanehchi

Department of Computer Studies, University of Technology, Loughborough, Leicestershire, U.K.

ABSTRACT

In this paper, a study of iterative methods suitable for the solution of the generalised eigenvalue problem for large sparse matrices is presented. The methods of interest, i.e. SOR and preconditioned methods are those in which no change in the structure of the original matrices occurs, and are suitable for the determination of the extreme eigenvalues and their corresponding eigenvectors. Finally, the advantage of the preconditioned approach becomes evident when coupled with the conjugate gradient algorithm whereby a powerful algorithm is revealed which copes adequately with this class of problems.

1. INTRODUCTION

In this paper we describe new preconditioned algorithms for the solution of large sparse eigenvalue problems by the minimisation of the Rayleigh quotient corresponding to the generalised eigenvalue problem,

$$Ax = \lambda Bx \quad , \tag{1}$$

where A and B are symmetric matrices with B positive definite.

The application of iterative methods, based on convergent splittings and in particular the S.O.R. method for the determination of the extreme eigenvalues of the above system has been studied by Ruhe [7-8]. In the ensuing work we shall consider the preconditioning method which was originally applied to determine the solution of linear systems, Evans [1], and determine its application for the eigenvalue problem (1).

Consequently, different splittings to that of the SOR method are proposed whereby the coefficient matrix is preconditioned to minimise the number of iterations required for one of the extreme eigenvalues and vectors to be obtained.

A general iterative procedure for the solution of an eigenvalue problem generates a sequence of vectors $x_1, x_2, \ldots, x_i, \ldots$ which in turn produces a sequence of eigenvalue approximations μ_i defined as,

$$\mu_i = \mu(x_i) = x_i^t A x_i / x_i^t B x_i \ . \tag{2}$$

In general, we consider the following splitting of the matrix C_i defined as,

$$C_i = A - \lambda_i B = V_i - H_i \ . \tag{3}$$

It can be seen that the sequence of vectors x_i satisfy the relation

$$x_{i+1} = x_i - p_i \ , \quad i=1,2,\ldots \ , \tag{4}$$

where,

$$p_i = V_i^{-1} C_i x_i \ , \tag{5}$$

which in the case of convergence, the vectors x_i after they have been normalised and their corresponding values μ_i will converge to an eigenvector and the corresponding eigenvalue of the system (1) respectively. It can be seen that in this case, the residual vector r_i defined as,

$$r_i = r(x_i) = (A - \mu_i B) x_i \ , \tag{6}$$

will converge to zero.

We shall now restrict ourselves to the study of the preconditioned technique which is to be used for the determination of the eigenvalues and to formulate the procedure.

2. FORMULATION OF THE ALGORITHM

The basic iteration formula used to evaluate the consecutive vectors x_i is defined as

$$x_{i+1} = V_i^{-1} H_i x_i \ , \tag{7}$$

where the matrices V_i and H_i are defined as,

$$V_i = (D_i - L) D_i^{-1} (D_i - U) \ , \tag{8}$$

and

$$H_i = LD_i^{-1}U \quad , \tag{9}$$

with L and U being the lower and upper triangular parts of the matrix $(A-\mu_i B)$ respectively and D_i being a diagonal matrix containing the diagonal entries of the matrix $(A-\mu_i B)$. It can be easily seen that the relation (3) is satisfied for the above splitting.

We now follow Ruhe [7-8] to prove that the above splitting produces a convergent method and also to give sufficient conditions under which the μ_i, $i=1,2,\ldots$ forms a monotone sequence.

Lemma 2.1:

Let x_i be the vectors evaluated by (7) and μ_i be the corresponding sequence computed by (2), then,

$$\mu_{i+1}-\mu_i = -p_i^t(V_i+H_i)p_i/x_{i+1}^t Bx_{i+1} \ . \tag{10}$$

Proof:

It can be shown (Ruhe [8]), that for any vector x_i evaluated by (7) and the corresponding μ_i obtained from (2) we have,

$$(\mu_{i+1}-\mu_i)x_{i+1}^t Bx_{i+1} = -2p_i^t r_i + p_i^t C_i p_i \ , \tag{11}$$

where p_i and r_i are defined in (5) and (6) respectively. However, the residual vector r_i can be expressed as,

$$\begin{aligned} r_i &= (A-\mu_i B)x_i = (V_i-H_i)x_i \\ &= V_i x_i - V_i V_i^{-1} H_i x_i = V_i x_i - V_i x_{i+1} = V_i p_i \ . \end{aligned} \tag{12}$$

Hence the expression (11) can be rewritten as,

$$(\mu_{i+1}-\mu_i)x_{i+1}^t Bx_{i+1} = -2p_i^t V_i p_i + p_i^t (V_i-H_i)p_i = -p_i^t(V_i+H_i)p_i \ , \tag{13}$$

from which the relation (10) can be derived.

From the above Lemma, we can extract the information and sufficient conditions for the μ_i's to form a monotone sequence. It can be easily justified that if the matrix (V_i+H_i) is a positive (negative) definite matrix for all $i=1,2,\ldots$, then the sequence μ_i will form a decreasing (increasing) sequence.

Lemma 2.2:

The matrices V_i and H_i are positive definite and positive semi-definite respectively, if D_i has no negative elements and are negative definite and negative semi-definite if D_i has no positive elements.

Proof: (Positive Definite Case)

Suppose that we have,

$$d_{i,i}>0 \ , \quad i=1,2,\ldots,n \ , \tag{14}$$

therefore D_i, can be written as

$$D_i = D_i^{\frac{1}{2}} D_i^{\frac{1}{2}} \ , \tag{15}$$

and we have,

$$V_i = (D_i-L)D_i^{-\frac{1}{2}}D_i^{-\frac{1}{2}}(D_i-U) = B\,B^t \ , \tag{16}$$

where

$$B = (D_i-L)D_i^{-\frac{1}{2}} \ . \tag{17}$$

The matrix H_i can also be written as,

$$H_i = LD_i^{-\frac{1}{2}}D_i^{-\frac{1}{2}}U = C\,C^t \ , \tag{18}$$

where

$$C = LD_i^{-\frac{1}{2}} \ . \tag{19}$$

The relations (16) show that V_i is a positive definite matrix whilst because of the structure of the matrix H_i in (18) we have the rank $(H_i)=n-1$ and therefore H_i is positive semi-definite.

From the above Lemmas we can deduce the following theorem.

Theorem 2.1:

If the starting vector x_1 in the iteration formula (7) is chosen such that the corresponding Rayleigh quotient μ_1 satisfies the inequality $\mu_1<\min\{a_{i,i}/b_{i,i}|i=1,2,\ldots,n\}$, then the μ_i's will form a decreasing sequence. Alternatively, if this vector is selected such that μ_1 satisfies $\mu_1>\max\{a_{i,i}/b_{i,i}|i=1,2,\ldots,n\}$, then the μ_i's will form an increasing sequence. In either case the method will converge in the following sense,

$$\mu_i \to \bar{\mu} = \lambda, \tag{20}$$

where λ is an eigenvalue of the system (7)

and

$$r_i \to 0 \ . \tag{21}$$

For the proof see Shanehchi [10].

Assuming that the eigenvalues of the system (1) are ordered as:

$$\lambda_1 \leq \lambda_2 \leq \ldots \leq \lambda_n \quad , \tag{22}$$

then in theory, to ensure the convergence of the method to one of the extreme eigenvalues we need to select the starting vector x_1 such that,

$$\mu_1 = \frac{x_1^t A x_1}{x_1^t B x_1} < \lambda_2 \quad , \text{ if } \lambda_1 \text{ is required,} \tag{23}$$

and

$$\mu_1 > \lambda_{n-1} \quad , \text{ if } \lambda_n \text{ is required.} \tag{24}$$

However, in practice, the conditions required by Theorem 2.1 seem to be sufficient for the convergency of the method to one of the extreme eigenvalues, (based on numerical experiments).

It can be shown, see Parlett [6], that using the Rayleigh quotient of a matrix in an iterative process for the determination of the eigenvalues, the accuracy of the eigenvalue obtained is the square of that of the eigenvectors. Hence, it seems reasonable to say that the rate of convergence of the system (7) is determined by the convergence rate of the limiting iteration,

$$x_{i+1} = V_\ell^{-1} H_\ell x_i \quad , \tag{25}$$

where V_ℓ and H_ℓ denote the limits of the matrices V_i and H_i respectively, i.e.

$$V_\ell = (D_\ell - L) D_\ell^{-1} (D_\ell - U) \quad , \tag{26}$$

and

$$H_\ell = L D_\ell^{-1} U \quad , \tag{27}$$

where

$$D_\ell = D(A) - \lambda D(B) \quad , \tag{28}$$

with λ being one of the extreme eigenvalues of A.

The numerical experiments carried out on this method showed that the efficiency of the algorithm is comparable with the Gauss-Seidel iteration method. However, there are two alternative strategies for accelerating the process:

i) a shift of origin on the iteration matrix,

ii) the introduction of a preconditioning parameter ω.

3. SHIFT OF ORIGIN OF THE ITERATION MATRIX

The iteration matrix for the above method

$$H_i = V_i^{-1} H_i \quad , \tag{29}$$

inherits some important properties which enable us to locate a suitable shift parameter to speed up the process. The first property is that the largest eigenvalue of the iteration matrix H_i converges to unity as the sequence μ_i converges to an eigenvalue. Hence, we have:

$$\lambda_{max}(H_i) \to 1. \tag{30}$$

Furthermore, it can be shown, see Shanehchi [10], that the matrix H_ℓ i.e. the limit of the iteration matrix H_i, is similar to a positive semi-definite matrix $\overline{H}_\ell$ defined as:

$$\overline{H}_\ell = QH_\ell Q^{-1} \quad , \tag{31}$$

where

$$\left.\begin{array}{l} Q = D_\ell^{-\frac{1}{2}}(D_\ell - U) \quad , \text{ if } \lambda_1 \text{ is evaluated} \\ Q = -(-D_\ell)^{-\frac{1}{2}}(D_\ell - U), \text{ if } \lambda_n \text{ is evaluated} \end{array}\right\} , \tag{32}$$

and since the matrix H_ℓ is singular we have,

$$0 \leq \lambda_{H_\ell} \leq 1 \quad , \tag{33}$$

with

$$\lambda_{min}(H_\ell) = 0 \quad \text{and} \quad \lambda_{max}(H_\ell) = 1 \; . \tag{34}$$

Now that the smallest and largest eigenvalues of the matrix H_ℓ are known, it is possible to apply the "shift of origin" technique related to the power method on the matrix H_ℓ and reduce the number of iterations required for the eigenvector to converge. That is, if γ_i, i=1,2,...,n denotes the eigenvalues of H_ℓ, then

$$0 = \gamma_1 \leq \gamma_2 \leq \gamma_3 \leq \ldots \leq \gamma_{n-1} \leq \gamma_n = 1 \; , \tag{35}$$

and the rate of convergence of the system (25) is momentarily governed by γ_{n-1}. To reduce this value we choose the shifting parameter p as:

$$p = \frac{\gamma_{n-1}}{2} - \varepsilon \; , \tag{36}$$

where ε=0.001 (say) and the iteration formula (25) is now replaced by,

$$x_{i+1} = H_\ell x_i - p x_i \; . \tag{37}$$

Although, in practice, the value of γ_{n-1} is generally unknown, good estimates of γ_{n-1} can be obtained when the size of the system increases forcing this value to become closer to 1. In the numerical experiments, it was observed that for n>100 the choice of p=0.45 to 0.49 can reduce the number of iterations required for the eigenvector to converge by 1/3.

4. INTRODUCTION OF THE PRECONDITIONING FACTOR ω

An alternative strategy for speeding up the process of evaluating the eigenvalues and eigenvectors of the system by the new splitting is to introduce an acceleration parameter into the iterative formula (7) whereby the new matrices V_i and H_i are defined as:

$$V_i = (D_i - \omega L) D_i^{-1} (D_i - \omega U) \quad , \tag{38}$$

and

$$H_i = \omega^2 L D_i^{-1} U + (1-\omega)(L+U) \quad . \tag{39}$$

It can be easily seen from the above two relations, that

$$C_i = A - \mu_i B = V_i - H_i \quad . \tag{40}$$

In general, with the above splitting of the matrix C_i, the matrix H_i is no longer a semidefinite matrix, although the matrix V_i remains a positive (negative) definite matrix provided that the conditions required for Theorem 2.1 are satisfied. We now attempt to select a suitable range for the preconditioning parameter ω such that for any value ω in that range, the matrix summation $V_i + H_i$ forms a definite matrix so that we will be able to apply Theorem 2.1 to prove the global convergence of the preconditioned method.

Lemma 4.1:

The matrix $V_i + H_i$, where V_i and H_i are defined in (38) and (39) is a positive definite matrix if ω is chosen to be in the range:

$$\frac{2-\sqrt{2}}{2} \leqslant \omega \leqslant \frac{2+\sqrt{2}}{2} \quad , \tag{41}$$

and that the initial vector x_1 is chosen such that

$$\mu_1 = \frac{x_1^t A x_1}{x_1^t B x_1} < \min\left\{\frac{a_{i,i}}{b_{i,i}} \;\middle|\; i=1,2,\ldots,n\right\}. \tag{42}$$

Proof:

The condition (42) ensures that:

$$d_{i,i} > 0, \quad i=1,2,\ldots,n,$$

therefore the matrix $D_i^{\frac{1}{2}}$ exists. However, the quadratic form of the matrix V_i+H_i for vector x_i can be expressed as follows:

$$x_i^t(V_i+H_i)x_i = x_i^t((D_i-\omega L)D_i^{-1}(D_i-\omega U)+\omega^2 LD_i^{-1}U+(1-\omega)(L+U))x_i$$

$$= x_i^t D_i x_i-(2\omega-1)x_i^t(L+U)x_i+2\omega^2 x_i^t LD_i^{-1}Ux_i . \quad (43)$$

Since ω is in the given range (41) and that matrix $LD_i^{-1}U$ is a positive semi-definite matrix, see Lemma (2.2), therefore, it is easy to see that:

$$(2\omega-1)^2 x_i^t LD_i^{-1}Ux_i \leq 2\omega^2 x_i^t LD_i^{-1}Ux_i . \quad (44)$$

Hence, we have,

$$x_i^t(V_i+H_i)x_i \geq x_i^t D_i x_i-(2\omega-1)x_i^t(L+U)x_i+(2\omega-1)^2 x_i^t LD_i^{-1}Ux_i$$

$$= x_i^t[(D_i-(2\omega-1)L)D_i^{-1}(D_i-(2\omega-1)U)]x_i$$

$$= x_i^t B_i^t B_i x_i > 0 \quad - \text{(see Lemma (2.2))} \quad ,$$

where

$$B_i = (D_i-(2\omega-1)L)D_i^{-\frac{1}{2}} ,$$

and the proof of the Lemma is complete.

The following Lemma can also be proved when the matrix V_i+H_i is negative definite.

Lemma 4.2:

The matrix V_i+H_i is a negative definite matrix if ω is in the range (41) and if the initial vector x_1 is chosen such that

$$\mu_1 > \max\{a_{i,i}/b_{i,i} \mid i=1,2,\ldots,n\}. \quad (45)$$

To analyse the rate of convergence of the preconditioned method we again consider the limiting iteration formula (25) where the matrices V_ℓ and H_ℓ are defined as.

$$V_\ell = (D_\ell-\omega L)D_\ell^{-1}(D_\ell-\omega U) , \quad (46)$$

and

$$H_\ell = \omega^2 LD_\ell^{-1}U+(1-\omega)(L+U) . \quad (47)$$

The limit of the iteration matrix $\mathcal{H}_i$ can be expressed as,

$$\mathcal{H}_\ell = ((D_\ell-\omega L)D_\ell^{-1}(D_\ell-\omega U))^{-1}(\omega^2 LD_\ell^{-1}U+(1-\omega)(L+U))$$

$$= ((D_\ell-\omega L)D_\ell^{-1}(D_\ell-\omega U))^{-1}((D_\ell-\omega L)D_\ell^{-1}(D_\ell-\omega U)-D_\ell+L+U)$$

$$= I-((D_\ell-\omega L)D_\ell^{-1}(D_\ell-\omega U))^{-1}C_\ell = I-B_\ell . \quad (48)$$

The matrix B_ω is defined as the preconditioned matrix of the method. It can be seen that the matrix B_ω is similar to a positive semi-definite matrix $\bar{B}_\omega$ defined as

$$\bar{B}_\omega = P B_\omega P^{-1} , \tag{49}$$

where

$$\left.\begin{array}{l} P = D_\ell^{\frac{1}{2}}(D_\ell - \omega L) \ , \quad \text{if } \lambda_1 \text{ is evaluated} \ , \\ P = -(-D_\ell)^{\frac{1}{2}}(D_\ell - \omega L) , \text{ if } \lambda_n \text{ is evaluated} \end{array}\right\} . \tag{50}$$

From the relation (48) we have that if γ_i and μ_i are the eigenvalues of the matrix H_ℓ and B_ω respectively, then they are real and are related through the relationship

$$\gamma_i = 1-\mu_j \quad , \mu_j>0 \ , \ i,j=1,2,\ldots,n \ . \tag{51}$$

As mentioned earlier, the rate of convergence of this method is governed by the second eigenvalue of the matrix H_ℓ in absolute value since its largest eigenvalue converges to unity as the method converges to an eigenvalue of A.

In order to be able to produce bounds for this eigenvalue of H_ℓ we first prove the following two Lemmas.

Lemma 4.3:

If the eigenvalues of the system $Ax=\lambda Bx$ are ordered as follows,

$$\lambda_1 \leqslant \lambda_2 \leqslant \cdots \leqslant \lambda_n \ , \tag{52}$$

then for any vector $\underline{x}$ we have

$$\lambda_1 \leqslant \rho(x) = \frac{x^t Ax}{x^t Bx} \leqslant \lambda_n \ . \tag{53}$$

Proof:

The system $Ax=\lambda Bx$ can be written as,

$$B^{-\frac{1}{2}}AB^{-\frac{1}{2}}y = \lambda y \quad \text{(since B is positive definite)} , \tag{54}$$

or

$$C^t ACy = \lambda y \ , \tag{55}$$

where

$$C = B^{-\frac{1}{2}} \ .$$

From (54) and (1) we have that the eigenvalues of the two systems are the same.

Since A is symmetric, we have

$$(C^t AC)^t = C^t A^t C = C^t AC \ , \tag{56}$$

therefore, the matrix $C^t AC$ is symmetric and hence for any vector $\underline{y}$ we have,

$$\lambda_1 \leq \frac{y^t C^t A C y}{y^t y} \leq \lambda_n \quad . \tag{57}$$

Now, to prove (53) we assume that $\underline{x}$ is an arbitrary vector. Suppose that $\underline{y}$ is a vector defined as,

$$y = B^{\frac{1}{2}} x \quad , \tag{58}$$

therefore, evaluating the Rayleigh quotient $\rho(x)$ for the vector x we have,

$$\rho(x) = \rho(B^{-\frac{1}{2}} y) = \frac{y^t B^{-\frac{1}{2}} A B^{-\frac{1}{2}} y}{y^t y} = \frac{y^t C^t A C y}{y^t y} \quad , \tag{59}$$

and according to (57),

$$\lambda_1 \leq \rho(x) = \frac{y^t C^t A C y}{y^t y} \leq \lambda_n \quad , \tag{60}$$

and the proof of the Lemma is complete.

Lemma 4.4:

The matrix $(A-\lambda_1 B)$ where λ_1 is the smallest eigenvalue of the system (1) is a positive semi-definite matrix.

Proof:

Let x be an arbitrary vector. We have to prove that for this vector,

$$x^t (A-\lambda_1 B) x \geq 0 \quad . \tag{61}$$

Suppose that the above inequality does not hold. Hence we have,

$$x^t (A-\lambda_1 B) x < 0 \quad , \tag{62}$$

or

$$x^t A x - \lambda_1 x^t B x < 0 \quad , \tag{63}$$

which implies

$$x^t A x < \lambda_1 x^t B x \quad . \tag{64}$$

Since B is positive definite, therefore $x^t B x \geq 0$ and we have,

$$\frac{x^t A x}{x^t B x} < \lambda_1 \quad . \tag{65}$$

The above relation says that there exists a vector x for which

$$\rho(x) = \frac{x^t A x}{x^t B x} < \lambda_1 , \tag{66}$$

and this is contradictory to Lemma (4.3). Hence, relation (62) can not be true and we have

$$x^t (A-\lambda_1 B) x \geq 0 \quad , \tag{67}$$

and the proof of the Lemma is complete.

The following Lemma can be similarly proved for the matrix $A-\lambda_n B$.

Lemma 4.5:

The matrix $A-\lambda_n B$ where λ_n is the largest eigenvalue of the system (1) is a negative semi-definite matrix.

Now that we have showed that the matrix C_ℓ is a semi-definite matrix, it can be seen, see Shanehchi [10] and Evans & Missirlis [2], that if the matrix C_ℓ is consistently ordered then the second smallest eigenvalue of the matrix B_ω i.e. μ_2 is bounded as follows,

$$\mu_2 \leq \frac{1}{\omega(2-\omega)} \qquad \omega\in\left[\frac{2-\sqrt{2}}{2}, \frac{2+\sqrt{2}}{2}\right], \tag{68}$$

and

or

$$\mu_2 \geq \begin{cases} \dfrac{1-\bar{p}}{1-\omega\bar{p}+\omega^2\bar{\beta}}, & \text{if } \bar{\beta}\geq 1/4 \text{ or if } \bar{\beta}\leq 1/4 \text{ and } \omega<\omega^* \\ \dfrac{2}{2+\omega^2\bar{\beta}}, & \text{if } \bar{\beta}\leq 1/4 \text{ and } \omega\geq\omega^* \end{cases}, \tag{69}$$

where $\bar{p}$ denotes the second largest eigenvalue of the Jacobi iteration matrix of C_ℓ i.e.

$$B = D_\ell^{-1}(L+U) \quad ,$$

and

$$\omega^* = \frac{2}{1+\sqrt{1-4\bar{\beta}}} \quad , \tag{70}$$

where

$$\bar{\beta} = S(\tilde{L}\tilde{U}) \quad , \tag{71}$$

with $\tilde{L}$ and $\tilde{U}$ defined as,

$$\tilde{L} = D_\ell^{-\frac{1}{2}} L D_\ell^{-\frac{1}{2}} \quad , \tag{72}$$

and

$$\tilde{U} = D_\ell^{-\frac{1}{2}} U D_\ell^{-\frac{1}{2}} \quad . \tag{73}$$

From the above bounds for μ_2 we can minimise the modulus of the second largest eigenvalue of H_ℓ denoted by $S_2(H_\ell)$ from which good estimates of the preconditioning parameter ω can be derived.

Thus, on equating (68) and (69) we have the quartic equation,

$$y(\omega) = 2\bar{\beta}\omega^4-2(\bar{p}+2\bar{\beta})\omega^3+(1+5\bar{p}+\bar{\beta})\omega^2-(3\bar{p}+2)\omega+1 = 0 \quad , \tag{74}$$

and assuming that ω_2 is a root of this equation which lies in the interval $(1,1+\frac{\sqrt{2}}{2})$, then if $\bar{\beta}\geq 1/4$ we have the value of ω which minimises $S_2(H_\ell)$ defined as,

$$\omega = \min\{\omega_2, \frac{\bar{p}}{2\bar{\beta}}\} \quad , \tag{75}$$

and the corresponding bounds can be seen to be,

$$S_2(H_\ell) \leq \begin{cases} \dfrac{1}{\omega_2(2-\omega_2)} - 1 & , \text{ if } \omega=\omega_2, \\ \dfrac{(4\bar{\beta}-\bar{p})\bar{p}}{(4\bar{\beta}-\bar{p}^2)} & , \text{ if } \omega= \dfrac{\bar{p}}{2\bar{\beta}}. \end{cases} \tag{76}$$

However, if $\bar{\beta}<1/4$ then we have two separate cases based on different values of $\bar{\beta}$, which are obtained from confining ω^* within the given range.

Thus, if $\bar{\beta}\geq 0.2426$ then we take ω as defined in (70) and if $\bar{\beta}<0.2426$ then ω is chosen to be

$$\omega = \min \{\frac{\bar{p}}{2\bar{\beta}} , \omega^*\} \quad . \tag{77}$$

Finally, the bounds for $S_2(H_\ell)$ when $\bar{\beta}<0.2426$ can be shown to be:

$$S_2(H_\ell) \leq \begin{cases} \dfrac{4\bar{\beta}\bar{p}-\bar{p}^2}{4\bar{\beta}-\bar{p}^2} & , \text{ if } \omega= \dfrac{\bar{p}}{2\bar{\beta}}, \\ \dfrac{\omega^*-1}{\omega^*} & , \text{ if } \omega=\omega^* \quad . \end{cases} \tag{78}$$

A thorough discussion on the above bounds can be found in Shanehchi [10] and Evans & Missirlis [2].

5. THE PRECONDITIONED CONJUGATE GRADIENT ALGORITHM

The determination of the eigenvalues and the corresponding eigenvectors of (large order sparse symmetric matrices) by the conjugate gradient algorithm has recently proved to be an efficient strategy for symmetric eigenvalue problems. Ruhe and Wiberg [9] have devised two practical algorithms on the application of the C-G method in conjunction with the inverse iteration,

$$(\lambda I-A)x_{i+1} = x_i \quad , \tag{79}$$

where λ is a fairly good approximation to an eigenvalue of the positive definite matrix A. The application of the C-G algorithm has also been considered by Nash [5] for the generalised eigenvalue problem where he discusses the use of alternative minimisation functions instead of the Rayleigh quotient as defined in (2).

The C-G method, Hestenes and Stiefel [3], evaluates the solution of the linear system

$$Cx = b \quad , \tag{80}$$

where C is a positive definite matrix, by the following algorithm,

$$x_0 = 0 ,$$
$$r_0 = p_0 = b ,$$
$$x_{i+1} = x_i + \alpha_i p_i , \quad \alpha_i = r_i^t r_i / p_i^t C p_i ,$$
$$r_{i+1} = r_i - \alpha_i C p_i , \tag{81}$$

and

$$p_{i+1} = r_{i+1} + \beta_i p_i , \quad \beta_i = r_{i+1}^t r_{i+1} / r_i^t r_i .$$

The above algorithm can be used to determine the eigensolution of the generalised eigenvalue problem expressed as

$$(\mu_i B - A) x_{i+1} = x_i , \tag{82}$$

by one of the following two practical algorithms.

The first algorithm of Ruhe and Wiberg [9] uses the above method (82) assuming that a reasonably good eigenvalue approximation λ of the generalised eigenvalue problem (1) is available. Using the following algorithm we can then improve the corresponding eigenvector x:

a) $b := x;\ c = \lambda B - A$

b) Perform one C-G step using (81)

c) $\varepsilon_c := \min(\varepsilon_0, \max(\gamma ||x_i||_2^{-1}, \varepsilon_1))$ for some $\gamma > 1$ (83)

d) If $||\hat{r}_i||_2 > \varepsilon_c$ then goto b) otherwise stop.

In the above algorithm $\hat{r}_i$ is defined as:

$$\hat{r}_i = r_i / ||x_i||_2 . \tag{84}$$

The second practical algorithm evaluates both the eigenvalue and the corresponding eigenvector as follows;

a) $b := x;\ \lambda = \mu(x) + ||s_1||_2^2 / d$

b) Perform one C-G step (81)

c) $\varepsilon_c := \min(\varepsilon_0, \max(\gamma ||x_i||_2^{-1}, \varepsilon_1))$ (85)

d) If $||\hat{r}_i||_2 > \varepsilon_c$ then goto b)

e) If $||\hat{s}_i||_2 > \varepsilon_1$ then goto a); otherwise stop.

where $s_i = (\mu_i B - A) x_i$, $\hat{s}_i = s_i / ||x_i||_2$ and d is some guess of the gap δ defined as,

$$\delta = \min_{2 \leq i \leq n} |\lambda - \lambda_i| . \tag{86}$$

For details of the above algorithms and the devices used to treat singular or indefinite systems, see Ruhe and Wiberg [9] and Luenberger [4].

We now apply the principle of the C-G algorithm on the generalised eigenvalue problem and attempt to solve the system (82) iteratively by the application of preconditioning techniques to the coefficient matrix $(\mu_i B-A)$. (Evans [1]).

The system (82) can be rewritten as:

$$x_{i+1} = V_i^{-1} H_i x_i \quad , \tag{87}$$

where V_i and H_i are the matrices defined in (38) and (39) respectively. Although the matrix $\mathcal{H}_i = V_i^{-1} H_i$ is a definite matrix (see Theorem 2.1) it is not in general symmetric and hence cannot be used directly as a substitute for the matrix C in (81). However, it can be easily shown, see Section 3, that the matrix $\mathcal{H}_i$ is similar to the symmetric matrix $\tilde{\mathcal{H}}_i$ defined as,

$$\tilde{\mathcal{H}}_i = Q_i \mathcal{H}_i Q_i^{-1} \quad , \tag{88}$$

where,

$$Q_i = (D_i - \omega U) D_i^{-\frac{1}{2}} \quad . \tag{89}$$

Hence we can now apply the C-G algorithm on the matrix $\tilde{\mathcal{H}}_i$ and solve the preconditioned system

$$y_{i+1} = \tilde{\mathcal{H}}_i y_i \quad , \tag{90}$$

for the vector y and after the solution has converged, the eigen-vector x corresponding to the original problem (1) can be obtained by solving the following system,

$$x = Q_i y \quad . \tag{91}$$

Note that the eigenvalues of the original system remain unchanged under the similarity transformation (88).

6. NUMERICAL EXPERIMENTS

In this section, we consider the proposed preconditioned method and also the preconditioned conjugate gradient algorithm described in the previous sections and investigate their behaviour in practice. For this we have selected 3 matrix examples for which one of the extreme eigenvalues and its corresponding eigenvector are evaluated with respect to different values of ω.

As shown in the previous section, in order to obtain the desired eigenvalue from the two extreme eigenvalues of the system (1), the starting vector x must be constructed such that one of the conditions stated in Theorem (2.1) is true. This initialisation of the eigenvector is not however as complicated as it may appear. Indeed, an efficient strategy, requiring only a few operations as described in Shanehchi [10] can be adopted.

Example 6.1:

Consider the following generalised eigenvalue problem,

$$Ax = \lambda Bx \ ,$$

where A and B are defined by (92) and (93) respectively. The above problem is obtained when finite element approximations are used for the solution of the boundary value problem.

$$-y''-\lambda q(x)y = 0 \ , \qquad q(x)\equiv 1,$$

with boundary conditions, $y(0) = y(1) = 0$,

in the form, $\omega(x) = \omega_\beta(x) = \sum_i \beta_i b_i(x)$,

where $b_i(x)$, $i=1,2,\ldots,n$ are linearly independent basis functions chosen to satisfy $b_i(0)=b_i(1)=0$, $i=1,2,\ldots$ and such that the matrices A and B are sparse and β is a vector of constants such that $||\beta||_2=1$, see Todd [11].

$$A=(n+1)\begin{bmatrix} 2 & -1 & & & 0 \\ -1 & 2 & -1 & & \\ & \ddots & \ddots & \ddots & \\ 0 & & -1 & 2 & -1 \\ & & & -1 & 2 \end{bmatrix} , \qquad (92)$$

$$B=\frac{1}{6(n+1)}\begin{bmatrix} 4 & -1 & & & 0 \\ -1 & 4 & -1 & & \\ & \ddots & \ddots & \ddots & \\ 0 & & -1 & 4 & -1 \\ & & & -1 & 4 \end{bmatrix} . \qquad (93)$$

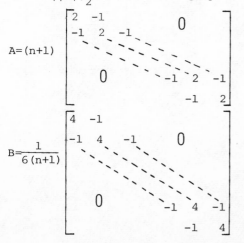

The extreme eigenvalues and the corresponding eigenvectors of the above problem of order 100 were obtained by the preconditioned algorithm and the SOR method and the results are tabulated in Table (1).

It is easily shown that the eigenvalues λ_i for the above generalised problem are determined theoretically from

$$\lambda_i = (12(n+1)^2 \sin^2(\frac{i\pi}{2(n+1)}))/(3-2\sin^2\frac{i\pi}{2(n+1)}) ,$$

$$i=1,2,\ldots,n.$$

The experimental results on the preconditioned method produced the smallest eigenvalue

$$\lambda_1 = 9.8704116,$$

which is accurate to 4 significant digits.

TABLE (1): The number of iterations and time units required to obtain the smallest eigenvalue and corresponding eigenvector of Example (6.1).

ω	SOR METHOD				PRECONDITIONED METHOD			
	EIGENVALUE		EIGENVECTOR		EIGENVALUE		EIGENVECTOR	
	No.of Iters.	Time Units	Total No.of Iters.	Total Time Units	No.of Iters.	Time Units	Total No.of Iters.	Total Time Units
1.0	2353	1347	2870	1422	1215	1096	1527	1218
1.1	1959	1122	2413	1188	993	903	1261	1008
1.2	1590	911	1984	968	712	645	937	734
1.3	1330	762	1669	811	618	560	802	632
1.4	1096	629	1386	671	462	417	608	475
1.5	883	507	1124	542	336	302	446	349
1.6	691	397	886	426	225	201	303	232
1.7	518	298	668	320	182	166	335	229
1.8	365	211	467	226	-	-	-	-
1.9	241	140	278	146	-	-	-	-

Example (6.2)

In this example, we shall consider the matrix A_2 derived from the finite difference discretisation of the Poisson partial differential equation

$$\nabla^2\phi = \frac{\partial^2\phi}{\partial x^2} + \frac{\partial^2\phi}{\partial y^2} = g(x,y) ,$$

on a square region with specified (Dirichlet) boundary conditions. A regular grid of mesh points of spacing h is chosen and the partial difference operator ∇^2 is replaced by an accurate finite difference approximation as given by the following computational molecule:

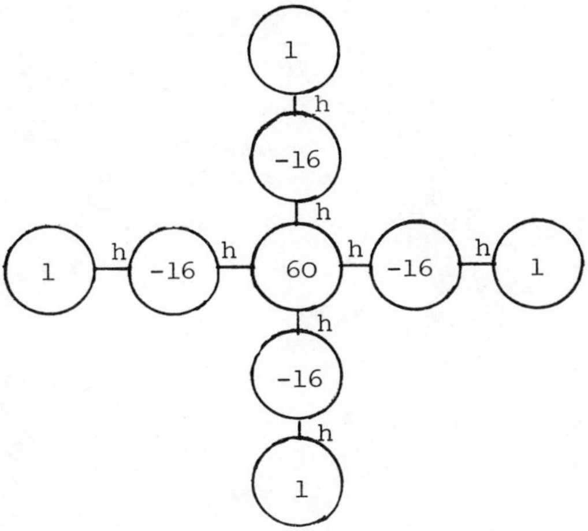

The eigenvalues of A_2 as defined in (94) are given by the following formula:

$$\lambda_{i,j} = 60-t_i-t_j \quad , \; i=1,2,\ldots,n, \quad j=1,2,\ldots,n,$$

where

$$t_k = 66-(8+2\cos\frac{k\pi}{n+1})^2 \quad , \; k=1,2,\ldots,n.$$

The eigenvalues of the matrix A_2 of order (225*225) evaluated for both the SOR method and the preconditioned method are tabulated together with the theoretical solutions in Table (2).

The number of iterations and the computational costs of the two methods for evaluating the smallest eigenvalue of A_2 are compared in Table (3) for different values of ω.

$$A_2=\begin{bmatrix} (B-I_n) & -16I_n & I_n & & & & \\ -16I_n & B & -16I_n & I_n & & 0 & \\ I_n & -16I_n & B & -16I_n & I_n & & \\ & \ddots & \ddots & \ddots & \ddots & \ddots & \\ & & I_n & -16I_n & B & -16I_n & I_n \\ & 0 & & I_n & -16I_n & B & -16I_n \\ & & & & I_n & -16I_n & (B-I_n) \end{bmatrix}_{n^2*n^2},$$

where

$$B=\begin{bmatrix} 59 & -16 & 1 & & & & \\ -16 & 60 & -16 & 1 & & 0 & \\ 1 & -16 & 60 & -16 & 1 & & \\ & \ddots & \ddots & \ddots & \ddots & \ddots & \\ & & 1 & -16 & 60 & -16 & 1 \\ & 0 & & 1 & -16 & 60 & -16 \\ & & & & 1 & -16 & 59 \end{bmatrix}. \tag{95}$$

(94)

TABLE (2): The Extreme Eigenvalues of A_2

SOR METHOD	PRECONDITIONED METHOD	THEORETICAL SOLUTION
0.925607 126.465775	0.9252610 126.465775	0.9252602 126.465776

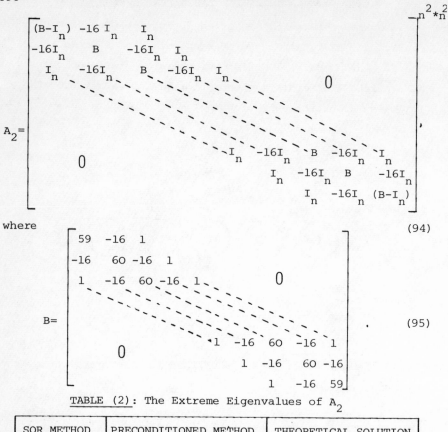

TABLE (3): The number of iterations and time units required to obtain the smallest eigenvalue and corresponding eigenvector of the matrix A_2

ω	SOR METHOD				PRECONDITIONED METHOD			
	EIGENVALUE		EIGENVECTOR		EIGENVALUE		EIGENVECTOR	
	No.of Iters.	Time Units	Total No.of Iters.	Total Time Units	No.of Iters.	Time Units	Total No.of Iters.	Total Time Units
1.0	161	238	211	262	70	183	97	219
1.1	138	206	180	226	62	162	86	195
1.2	118	176	155	195	48	125	69	154
1.3	102	152	132	167	43	112	60	136
1.4	88	132	113	145	36	94	50	114
1.5	77	115	97	126	30	79	41	95
1.6	69	103	83	111	27	70	36	84
1.7	65	98	75	105	78	204	230	401
1.8	72	108	87	117	-	-	-	-
1.9	113	170	130	179	-	-	-	-

Example (6.3):

Consider the tridiagonal matrix A_3 defined as:

$$\begin{bmatrix} 2 & -1 & & & & \\ -1 & 2 & -1 & & 0 & \\ & -1 & 2 & -1 & & \\ & & \ddots & \ddots & \ddots & \\ & 0 & & -1 & 2 & -1 \\ & & & & -1 & 2 \end{bmatrix} , \qquad (96)$$

which is derived from the finite difference dicretisation of the following 2-point boundary value problem,

$$-\frac{d^2\phi}{dx^2} = f(x) \quad , \quad a \leqslant x \leqslant b,$$

given the boundary conditions,

$$\phi = f(a) \quad \text{at } x=a,$$

$$\text{and} \qquad \phi = f(b) \quad \text{at } x=b.$$

The simplest central difference operator:

$$-\frac{d^2}{dx^2} = \frac{1}{h^2}\left\{ (-1) \overset{h}{—} (2) \overset{h}{—} (-1) \right\} + O(h^2),$$

is now chosen to replace the ordinary differential equation on a grid of evenly spaced points along the x-axis with spacing h and the resulting matrix A_3 is formed. The analytical solution for the eigenvalues of the above matrix is given by:

$$\lambda_k = 4\sin^2 \frac{k\pi}{2(n+1)} \quad , \; k=1,2,\ldots,n; \; n=200.$$

The smallest eigenvalue of the matrix (96) was evaluated theoretically from the above formula and then the corresponding eigenvector was determined using the preconditioned conjugate gradient algorithm (83). The number of iterations required for the method to determine the corresponding eigenvector for successive values of the preconditioning parameter ω together with the total number of units of time are given in Table (4). The results are to be compared with using the conjugate gradient algorithm (83) on the coefficient matrix A instead of $\tilde{H}_i$ which required 100 iterations and a total time unit of 364 to evaluate the eigenvector using the same initial starting value. It can be seen from Table (4) that an efficiency factor of 5 can be obtained from the preconditioned conjugate gradient method at near optimum values of ω when compared with the conjugate gradient algorithm using the inverse iteration (79).

Parameter ω	1.0	1.1	1.2	1.3	1.4	1.5	1.6	1.7	1.8	1.9
No. of Iterations	40	38	38	38	38	32	27	21	16	11
Total Time Units	251	238	238	238	238	201	170	133	103	71

TABLE (4)

The above example was also run on the preconditioned conjugate gradient algorithm (85) in order to obtain the smallest eigenvalue and the corresponding eigenvector. The total number of iterations and the time required for the algorithm to converge for different values of ω are tabulated in Table (5). When the conjugate gradient algorithm is used on the matrix (96) it can be seen that 1901 iterations are required to obtain the eigenvalue and the corresponding eigenvector in a total of 794 time units.

Parameter ω	1.0	1.1	1.2	1.3	1.4	1.5	1.6	1.7	1.8	1.9
No. of Iterations	75	67	65	65	57	48	39	29	22	17
Total Time Units	492	441	429	378	321	321	263	199	154	122

TABLE (5)

7. CONCLUSIONS

In the work presented, we have proposed two algorithms based on the preconditioning principle for the determination of the eigenvalues and eigenvectors of the generalised eigenvalue problem. The numerical experiments carried out on the algorithm described in Section 4 indicate the superiority of the algorithm over the SOR method for most cases, see Shanehchi [10]. Also it can be seen from Example (6.1) that the SOR method can give better results for some cases when ω is near its optimum value. However, the preconditioned method is more efficient than the SOR method for all other values of the parameter ω and is to be recommended for use in cases when the exact value of ω is unknown.

Finally, the application of the preconditioning technique to the conjugate gradient algorithm yields an extremely powerful and economical method (PCCG algorithm) which is confirmed by the results presented in Example (6.3). In the numerical experiments on this algorithm it was noted that a close starting vector was required for the PCCG algorithm (85) in order to converge to a particular eigenvalue. However, further numerical investigations are required in order to ascertain the pattern of behaviour of the optimum preconditioning parameter ω and to implement alternative minimisation functions to evaluate any intermediate eigenvalue before any specific recommendations can be given.

REFERENCES

1. Evans, D.J. - "The Use of Preconditioning in Iterative Methods for Solving Linear Equations with Symmetric Positive Definite Matrices", J.Inst.Math.Applics, 4, pp.295-314, 1968.

2. Evans, D.J. and N.M. Missirlis - "On the Preconditioned Jacobi Method for Solving Large Linear Systems", to be published.

3. Hestenes, M.R. and E. Stiefel - "Methods of Conjugate Gradients for Solving Linear Systems", Journal of Res., N.B.S., 49, pp.409-436, 1952.

4. Luenberger, D.G. - "Hyperbolic Pairs in the Method of Conjugate Gradients", SIAM J.Appl.Math., 17, pp.1263-1267, 1969.

5. Nash, J. and S.G. Nash - "Conjugate Gradient Methods for Solving Algebraic Eigenproblems", Proc. Symp. on Mini-computers and Large Scale Computation, Montreal, ed. P. Lylos (New York: American Chemical Society), pp.24-32, 1977.

6. Parlett, N. Beresford - "The Symmetric Eigenvalue Problem", Prentice Hall, 1980.

7. Ruhe, A. - "SOR Methods for the Eigenvalue Problem with Large Sparse Matrices", Mathematics of Computation, 28, 127, pp.697-710, 1974.

8. Ruhe, A. - "Iterative Eigenvalue Algorithms Based on Convergent Splittings", Journal of Computational Physics, 19, pp.110-120, 1975.

9. Ruhe, A. and T. Wiberg, - "The Method of Conjugate Gradients Used in Inverse Iteration", B.I.T., 12, pp.543-554, 1972.

10. Shanehchi, J. - "The Determination of Sparse Eigensystems and Parallel Linear System Solvers", Ph.D. Thesis, Loughborough University of Technology, 1980.

11. Todd, J. - "Basic Numerical Mathematics", Vol.2 Linear Algebra, pp.100-103, Acad.Press, 1977.

ON THE APPLICATION OF PRECONDITIONED DIRECT METHODS

D.J. Evans & M. Hatzopoulos[†]

Department of Computer Studies, University of Technology, Loughborough, Leicestershire, U.K.

ABSTRACT

Here, the preconditioning concept is shown to be applicable to direct methods of solution of ill-conditioned systems of linear equations. It is shown that in the solution of an equivalent preconditioned linear system by elimination methods the rounding errors incurred in the computational process are minimised, thus avoiding the need for higher precision arithmetic.

Finally, a comparison of the optimal scaling and preconditioning strategies is given for a number of well known ill-conditioned examples.

1. INTRODUCTION

In this paper we shall be concerned with the solution of ill-conditioned linear systems when the coefficient matrix is of small order, dense and positive definite.

Many authors (Bauer, 1963, etc.) have discussed ways of overcoming the problems of ill-conditioning and the main theme of all the methods is to transform the given system to one where the coefficient matrix is better conditioned.

Evans (1968) introduced one of these transformations for large sparse matrices and developed a class of iterative methods which possessed superior convergence rates.

Here we extend this preconditioning technique to ill-conditioned matrices for which the solution method is the Gaussian elimination scheme incorporating a pivoting strategy. It is considered that the proposed strategy could circumvent the need for double precision accumulation of inner products followed

[†] *Michigan Technological University, Michigan, U.S.A.*

by iterative refinement or even delay the onset of higher precision arithmetic when the coefficient matrix is ill-conditioned.

2. VECTOR AND MATRIX NORMS AND CONDITION NUMBERS

For the purpose of quantitatively discussing errors, it is convenient to be able to associate with any vector or matrix a non-negative scalar that in some sense measures its magnitude. Such measures which also satisfy some reasonable axioms are called norms. The most common vector norms are special cases of the family of L_p-norms.

$$||\underline{x}||_p = (|x_1|^p+|x_2|^p+\ldots+|x_n|^p)^{1/p}, \quad 1\leq p<\infty. \tag{2.1}$$

Two particular cases are often used:

p = 2: Euclidean norm:

$$||\underline{x}||_2 = (|x_1|^2+|x_2|^2+\ldots+|x_n|^2)^{1/2},$$

$p \to \infty$: Maximum norm:

$$||\underline{x}||_\infty = \max_{1\leq i\leq n} |x_i|. \tag{2.2}$$

Vector norms must have the following properties, which are analogous to properties of the usual concept of length:

1a. $||\underline{x}||>0$ if $\underline{x}\neq 0$, $||\underline{x}||=0$ if $x=0$.

2a. $||\alpha\underline{x}||=|\alpha|\,||\underline{x}||$, α a scalar.

3a. $||\underline{x}+\underline{y}||\leq||\underline{x}||+||\underline{y}||$.

For the maximum norm these are all easy to verify.

Matrix norms shall have similar properties:

1b. $||A||>0$ if $A\neq 0$, $||A||=0$ if $A=0$.

2b. $||\alpha A||=|\alpha|\,||A||$, α a scalar.

3b. $||A+B||\leq||A||+||B||$.

4b. $||AB||\leq||A||\;||B||$.

If a matrix norm and a vector norm are related in such a way that

5b. $||A\underline{x}||\leq||A||\;||\underline{x}||$,

is satisfied for any A and x, then the two norms are said to be consistent. For any vector norm, there exists a consistent matrix norm. In fact, such a norm is given by the matrix-bound norm subordinate to the vector norm

$$||A|| = \max_{\underline{x}\neq 0} \frac{||A\underline{x}||}{||\underline{x}||} . \qquad (2.3)$$

In the following we shall mainly use the maximum vector norm of equation (2.2) and the subordinate matrix-bound norm, which can be shown to be

$$||A||_{\infty} = \max_{1\leq i\leq n} \sum_{j=1}^{n} |a_{ij}| .$$

These have the advantage of being very simple to compute. Also, equation (2.3) has the attractive property that

$$||\ |A|\ || = ||A||, \quad |A|=(|a_{ij}|) .$$

This is not true for the matrix-bound norm subordinate to the Euclidean norm, for which the best we can say is

$$||\ |A|\ ||\leq n^{1/2}||A||.$$

We shall now investigate the condition of the problem of solving a non-singular linear system $A\underline{x}=\underline{b}$. We assume that the elements in A and $\underline{b}$ are the given data, and our goal is to estimate the effect of perturbations in $\underline{b}$ and A.

Suppose that A and $\underline{b}$ are perturbed, then the computed solution $\underline{x}+\delta\underline{x}$ satisfies the equation:

$$(A+\delta A)(\underline{x}+\delta\underline{x}) = \underline{b}+\delta\underline{b} . \qquad (2.4)$$

From (2.4) we have:

$$(A+\delta A)\delta\underline{x} = \delta\underline{b}-\delta A\underline{x} ,$$

or $$A(I+A^{-1}\delta A)\delta\underline{x} = \delta\underline{b}-\delta A\underline{x} ,$$

to give $$\delta\underline{x} = (I+A^{-1}\delta A)^{-1}A^{-1}(\delta\underline{b}-\delta A\underline{x}) ,$$

with $$||\delta\underline{x}||\leq||(I+A^{-1}\delta A)^{-1}||.||A^{-1}||.||\delta\underline{b}-\delta A\underline{x}||. \qquad (2.5)$$

If λ_i are the eigenvalues of $A^{-1}\delta A$ then:

$$||A^{-1}\delta A||\geq\lambda_i ,$$

and if we assume that

$$||A^{-1}\delta A||<1,$$

then $I+A^{-1}\delta A$ is non-singular and for any subordinate norm we have the result:

$$\frac{1}{(1+||A^{-1}\delta A||)}\leq||I+A^{-1}\delta A||\leq\frac{1}{(1-||A^{-1}\delta A||)} .$$

Substituting this in (2.5) gives:

$$||\delta\underline{x}|| \leq \frac{1}{(1-||A^{-1}\delta A||)} . ||A^{-1}|| . (||\delta\underline{b}|| + ||\delta A|| . ||\underline{x}||) , \qquad (2.6)$$

and for the relative error we have

$$\frac{||\delta\underline{x}||}{||\underline{x}||} \leq ||A|| . ||A^{-1}|| \left\{ \frac{||\delta\underline{b}||}{||\underline{b}||} + \frac{||\delta A||}{||A||} \right\} \frac{1}{(1-||A^{-1}|| . ||\delta A||)} . \qquad (2.7)$$

From (2.7) and $\delta A=0$ (i.e. we are considering perturbations only on the right hand side) we get:

$$\frac{||\delta\underline{x}||}{||\underline{x}||} \leq ||A|| . ||A^{-1}|| \frac{||\delta\underline{b}||}{||\underline{b}||} . \qquad (2.8)$$

For $\delta b=0$ (i.e. considering perturbations only on the left hand side) we get:

$$\frac{||\delta\underline{x}||}{||\underline{x}||} \leq ||A|| . ||A^{-1}|| \frac{||\delta A||}{||A||} . \frac{1}{(1-||A^{-1}|| . ||\delta A||)} . \qquad (2.9)$$

In deriving (2.7), we replaced $||A^{-1}\delta A||$ by $||A^{-1}|| . ||\delta A||$. We note that there are perturbations for which this is pessimistic. Since the result (2.7) is true for the $|| \; ||_2$ norm it is also true for the Euclidean norm.

If the quantity $||A|| . ||A^{-1}||$ is large, then small relative perturbation in A and $\underline{b}$ will produce large relative perturbations in $\underline{x}$ and the problem of solving $A\underline{x}=\underline{b}$ is then termed ill-conditioned.

We wish to assign a single "condition number" to the system $A\underline{x}=\underline{b}$ which in some way measures the difficulties associated with deriving a numerical solution.

The relations (2.7), (2.8) and (2.9) suggest such a number to be:

$$\text{cond}_n(A) = ||A||_n ||A^{-1}||_n ,$$

for a $||.||_n$ matrix norm.

Turing (1948) first recognised that the "condition number" should depend symmetrically on both A and A^{-1} or specifically on the product of their norms. He thus defined the M and N condition numbers as follows:

$$M(A) = n \max_{i,j} |a_{ij}| . \max|\alpha_{i,j}| . \tag{2.10}$$

$$N(A) = n^{-1} (\sum_{i,j} a_{ij}^2)^{\frac{1}{2}} (\sum_{i,j} \alpha_{ij}^2)^{\frac{1}{2}} , \tag{2.11}$$

where the terms (α_{ij}) are the inverse of the matrix $A=(a_{ij})$ of order n.

In the case of the spectral norm $(|| \; ||_2)$ we define the K-condition number (or spectral condition number) as:

$$K(C) = ||C||_2 . ||C^{-1}||_2 = \sqrt{\frac{\lambda_1}{\lambda_n}} , \tag{2.12}$$

where λ_1, λ_n are the largest and smallest eigenvalues of $C=A^H A$ respectively.

In the case of A being symmetric, (2.12) becomes,

$$K(A) = \frac{|\lambda_1|}{|\lambda_n|} , \tag{2.13}$$

where λ_1, λ_n the largest and smallest eigenvalues, respectively of A. The number (2.13) is called also the P-condition number and was first introduced by Von-Neumann and Goldstine, (1947) as

$$P(A) = \frac{|\lambda_1|}{|\lambda_n|} .$$

There, they proved that if A is a symmetric and positive definite matrix and Z is the computed inverse of A using the elimination method then:

$$|\mu| \leq 14.24 \; P(A) \; n^2 \varepsilon ,$$

where μ is the eigenvalue of largest modulus of I-AZ, n the order of A and ε the smallest number recognised by the particular machine in use.

For a time it was thought that ill-conditioning was related to the smallness of det(A) but from a knowledge of the Gaussian elimination method where the pivots are the principal leading minors indicates that difficulties may also arise when det(A) is also large. However, when A is normalised such that all the diagonal elements are unity then it is ill-conditioned when its inverse is large.

The following inequalities can be proved for the condition numbers defined above:

$$\frac{M(A)}{n^2} \leqslant N(A) \leqslant M(A) \ , \tag{2.14}$$

$$\frac{M(A)}{n} \leqslant PA() \leqslant mN(A) \ . \tag{2.15}$$

3. THE PRECONDITIONING METHOD

Let the system

$$A\underline{x} = \underline{b} \ , \tag{3.1}$$

where A is symmetric, positive definite and has the form,

$$A = I+L+L^T \ , \tag{3.2}$$

where I is the identity matrix and L is a strictly lower triangular matrix.

Consider now the non-singular lower and upper triangular matrices $(I+\omega L)$ and $(I+\omega L^T)$ where ω is a parameter to be defined later. Then the system (3.1) can be transformed into the equivalent system

$$(I+\omega L)^{-1}A(I+\omega L^T)^{-1}[(I+\omega L^T)\underline{x}] = (I+\omega L)^{-1}\underline{b} \ , \tag{3.3}$$

or

$$B_\omega \underline{y} = \underline{d}_\omega \ . \tag{3.4}$$

where

$$B_\omega = (I+\omega L)^{-1}A(I+\omega L^T)^{-1} \ , \tag{3.5}$$

$$\underline{d}_\omega = (I+\omega L)^{-1}\underline{b} \ , \tag{3.6}$$

and

$$\underline{y} = (I+\omega L^T)\underline{x} \ . \tag{3.7}$$

For ω=0 it can be seen that the new system reverts back to its original form (3.1). Thus, by introducing the transformation above we allow ω to play the role of a preconditioning parameter such that as ω varies in a restricted range $0<\omega<w$ (say) a minimum value of the condition number of the matrix B_ω is obtained (Evans, 1968). Thus, at the optimum ω we solve the system (3.4) and from (3.7) we can calculate the solution $\underline{x}$ to the original system.

In fact B_ω can be written in the form,

$$B_\omega = S_\omega + S_\omega^T \ , \tag{3.8}$$

where $$S_\omega = (I+\omega L)^{-1}(\tfrac{1}{2}I+L)(I+\omega L^T)^{-1}\ , \tag{3.9}$$

and S_ω can be accurately computed by using simple algorithms of backward and forward substitution on the rows and columns of A. (Evans, 1973).

Let $\underline{x}$ be an eigenvector of B_ω corresponding to the eigenvalue μ. Then, by definition, we have,

$$(I+\omega L)^{-1}A(I+\omega L^T)^{-1}\underline{x} = \mu\underline{x}\ . \tag{3.10}$$

If we write $$\underline{y} = (I+\omega L^T)^{-1}\underline{x}\ , \tag{3.11}$$

then (3.10) can be expressed in the form

$$A\underline{y} = \mu(I+\omega L)(I+\omega L^T)\underline{y}\ , \tag{3.12}$$

which on expanding becomes

$$A\underline{y} = \mu((1-\omega)I+\omega A+\omega^2 LL^T)\underline{y}\ , \tag{3.13}$$

and finally, in a more compact form

$$A\underline{y} = \mu((1-\tfrac{\omega}{2})^2 I+\omega A+\omega^2(LL^T-\tfrac{1}{4}I))\underline{y}\ , \tag{3.14}$$

as given by Andersson (1975). If we now premultiply (3.14) by $\underline{y}^T$ we have the final result,

$$\underline{y}^T A\underline{y} = \mu\underline{y}^T((1-\tfrac{\omega}{2})^2 I+\omega A+\omega^2(LL^T-\tfrac{1}{4}I))\underline{y}\ . \tag{3.15}$$

Since matrix A is symmetric and positive definite, then it can be shown that there always exists a number $a \neq 0$ such that

$$a = \sup_{\underline{y}} \frac{\underline{y}^T\underline{y}}{\underline{y}^T A\underline{y}}\ . \tag{3.16}$$

If we now assume that there exists a number b for which the relationship

$$b = \sup_{\underline{y}} \frac{\underline{y}^T(LL^T-\tfrac{1}{4}I)\underline{y}}{\underline{y}^T A\underline{y}}\ , \tag{3.17}$$

holds, then from (3.15), (3.16) and (3.17) we have the result

$$\frac{1}{\mu} \lesssim (1-\tfrac{\omega}{2})^2 a+\omega+\omega^2 b. \tag{3.18}$$

The previous relationship holds for every eigenvalue of B_ω and for the minimum eigenvalue, i.e. μ_{min} we have the result

$$\frac{1}{\mu_{min}} \lesssim (1-\tfrac{\omega}{2})^2 a+\omega+\omega^2 b. \tag{3.19}$$

Again from (3.10) it can be shown with further analysis that the following result is valid,

$$(I+\omega L)(I+\omega L^T) = \omega(2-\omega)A+((\omega-1)I+\omega L)((\omega-1)I+\omega L)^T , \quad (3.20)$$

which when substituted in (3.12) yields

$$A\underline{y} = \mu(\omega(2-\omega)A+((\omega-1)I+\omega L)((\omega-1)I+\omega L)^T)\underline{y} . \quad (3.21)$$

After premultiplication by $\underline{y}^T$ we obtain,

$$\underline{y}^TA\underline{y} = \mu\underline{y}^T(\omega(2-\omega)A+((\omega-1)I+\omega L)((\omega-1)I+\omega L)^T)\underline{y}. \quad (3.22)$$

However, since by definition, the matrix

$$((\omega-1)I+\omega L)((\omega-1)I+\omega L)^T ,$$

is positive semi-definite, i.e.,

$$\underline{y}^T((\omega-1)I+\omega L)((\omega-1)I+\omega L)^T\underline{y}\geq 0, \quad (3.23)$$

then we have the result,

$$\underline{y}^TA\underline{y}\geq\mu\underline{y}^T(\omega(2-\omega)A)\underline{y} , \quad (3.24)$$

which is valid for all eigenvalues of the matrix B_ω and especially so for the largest one, μ_{max}. Thus we have the result,

$$\mu_{max} \leq \frac{1}{\omega(2-\omega)} , \quad \omega\in(0,2) . \quad (3.25)$$

Then, by definition, the P-condition number of B_ω can be expressed as

$$P(B_\omega) = \frac{\mu_{max}}{\mu_{min}} \leq \frac{(2-\omega)a}{4\omega} + \frac{1}{(2-\omega)} + \frac{\omega b}{(2-\omega)} . \quad (3.26)$$

The minimum value of the right hand side of (3.26) can be shown to be

$$\sqrt{\tfrac{1}{2}(2b+1)a} + \tfrac{1}{2} , \quad (3.27)$$

and it occurs for an optimum value of ω given by

$$\omega_{opt} = 2/(1+\sqrt{2(2b+1)/a}) . \quad (3.28)$$

Thus, we have shown that for the given matrix B_ω, as given in (3.5) the P condition number is bounded and satisfies the relationship

$$P(B_{\omega_{opt}}) \leq \sqrt{\tfrac{1}{2}(2b+1)a} + \tfrac{1}{2}. \quad (3.29)$$

Theorem 3.1: For the symmetric and positive definite matrix $A=I+L+L^T$, then, if

$$a = \sup_{\underline{y}} \frac{\underline{y}^T\underline{y}}{\underline{y}^TA\underline{y}} \quad \forall \quad \underline{y}\in R^n, \ \underline{y}\neq 0 , \quad (3.30)$$

and

$$b = \sup_{\underline{y}} \frac{\underline{y}^T(LL^T-\frac{1}{4}I)\underline{y}}{\underline{y}^TA\underline{y}} , \quad (3.31)$$

then the P condition number of the preconditioned matrix B_ω is bounded as in (3.29) for an optimum value of ω given by (3.28).

For the more general case when $A=D+L+L^T$ we first use a pre-transformation of the form,

$$D^{-\frac{1}{2}}AD^{-\frac{1}{2}} \quad , \tag{3.32}$$

which can be regarded as a form of pre-scaling, (Forsythe & Wasow, 1967) to recover the given matrix form as in (3.2) although in this case the system does not revert back to its original form for $\omega=0$. This is confirmed from the numerical results presented in Sections 4 and 5. The preconditioned matrix now has the form,

$$B_\omega = D^{\frac{1}{2}}(D+\omega L)^{-1}A(D+\omega L^T)^{-1}D^{\frac{1}{2}} \quad , \tag{3.33}$$

and similar to the above theorem we can prove that,

$$P(B_{\omega_{opt}}) \leq \sqrt{\tfrac{1}{2}(2b+1)a} + \tfrac{1}{2} \quad , \tag{3.34}$$

for

$$\omega_{opt} = 2/(1+\sqrt{2(2b+1)/a} \ . \tag{3.35}$$

If instead of (3.30) and (3.31) we use the alternative upper bounds

$$a = \sup_{\underline{y}} \frac{\underline{y}^T D\underline{y}}{\underline{y}^T A\underline{y}} \quad , \quad \underline{y}\in R^n \quad , \tag{3.36}$$

and

$$b = \sup_{\underline{y}} \frac{\underline{y}^T(LD^{-1}L^T - \frac{1}{4}D)\underline{y}}{\underline{y}^T A\underline{y}} \quad . \tag{3.37}$$

In fact when,

$$|LD^{-1}L^T| \leq \frac{1}{4}|D| \quad , \tag{3.38}$$

we have $\quad b = 0 \ ,$

and we have the alternative sharper bound,

$$P(B_{\omega_{opt}}) \leq \sqrt{\frac{a}{2}} + \tfrac{1}{2} \ . \tag{3.39}$$

4. NUMERICAL EXPERIMENTS

In order to test the theoretical analysis derived earlier we now consider the following test example.

Let the (n×n) matrix:

$$A_n = (a_{ij}) = \frac{1}{i+j-1} \quad , \ i,j=1,2,\ldots,n.$$

or

$$A_n = \begin{bmatrix} 1 & \frac{1}{2} & \frac{1}{3} & \cdots & \frac{1}{n} \\ \frac{1}{2} & \frac{1}{3} & \frac{1}{4} & \cdots & \frac{1}{n+1} \\ \vdots & \vdots & \vdots & & \vdots \\ \frac{1}{n} & \frac{1}{n+1} & \cdots & & \frac{1}{2n-1} \end{bmatrix} . \qquad (4.1)$$

The Hilbert matrices enjoy the reputation of being very ill-conditioned matrices with respect to inversion and occur in the solution of boundary value problems by Rayleigh Ritz variational methods. It is well known that the condition number increases exponentially with the order of the matrix. The P and M condition numbers are given (approximately) by:

$$P(A_n) \simeq e^{3.5n} ,$$

$$M(A_n) \simeq k\, e^{3.525n} ,$$

where k is a constant.

From (4.1) it can be seen that as the order of the matrix increases each row of the matrix becomes almost equal to the next row (i.e. the rows become 'nearly' linearly dependent.

Consider the systems,

$$A_n \underline{x} = \underline{b}_n , \qquad (4.2)$$

where A_n is the (n×n) Hilbert matrix and b_n is chosen such that

$$b_n(i) = \sum_{j=1}^{n} a_{ij} .$$

This ensures that the system then has the solution,

$$x^T = (1,1,\ldots,1).$$

For n=4(1)9 we solve (4.2) and its corresponding pre-conditioned form (3.4) and in the following tables we give the results a) without preconditioning and b) with optimal pre-conditioning.

a) Without preconditioning

Order of Matrix	P-Condition Number	N-Condition Number	M-Condition Number	Relative Euclidean Error
4	$1.551.10^{4}$	$5.841.10^{3}$	$2.592.10^{3}$	$0.21822567.10^{-8}$
5	$4.766.10^{5}$	$9.616.10^{4}$	$8.960.10^{5}$	$0.32079296.10^{-6}$
6	$1.495.10^{7}$	$2.519.10^{6}$	$2.646.10^{7}$	$0.34054993.10^{-5}$
7	$4.748.10^{8}$	$6.882.10^{7}$	$9.338.10^{8}$	$0.41603360.10^{-3}$
8	$1.675.10^{10}$	$1.940.10^{9}$	$3.406.10^{10}$	$0.11399340.10^{-1}$
9	$1.456.10^{12}$	$6.271.10^{10}$	$1.239.10^{12}$	$0.10302179.10^{0}$

TABLE 4.1

b) Using preconditioning with optimum ω.

Order of Matrix	P-Condition Number	N-Condition Number	M-Condition Number	Relative Euclidean Error
4	$5.621.10^{2}$	$1.440.10^{2}$	$4.732.10^{2}$	$0.17172306.10^{-9}$
5	$1.070.10^{4}$	$2.179.10^{3}$	$2.449.10^{4}$	$0.14056164.10^{-8}$
6	$2.339.10^{5}$	$3.970.10^{3}$	$3.890.10^{5}$	$0.25911078.10^{-7}$
7	$5.542.10^{6}$	$8.022.10^{5}$	$1.005.10^{7}$	$0.20001276.10^{-6}$
8	$1.388.10^{8}$	$1.757.10^{7}$	$3.346.10^{8}$	$0.14416701.10^{-4}$
9	$3.667.10^{9}$	$4.106.10^{8}$	$9.192.10^{9}$	$0.41562611.10^{-3}$

TABLE 4.2

In the accompanying Figure 1 the average decimal places of accuracy obtained is plotted against the order of the Hilbert matrix for a) without preconditioning and b) with optimal preconditioning.

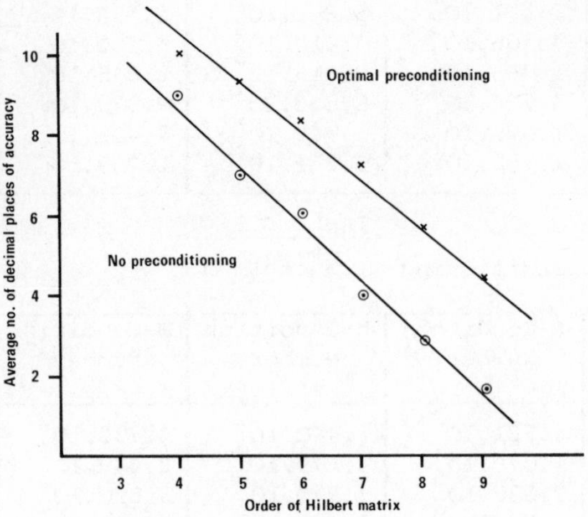

Figure 1. No. of digits accuracy vs. order of Hilbert matrix

The results clearly indicate an improvement of 10^2-10^3 in accuracy which can be derived from the application of the preconditioning strategy.

5. COMPARISON OF OPTIMAL SCALING AND PRECONDITIONING STRATEGIES

Let the (n×n) ill-conditioned system be represented as,

$$A\underline{x} = \underline{b} \; , \tag{5.1}$$

and let P and Q be two (n×n) non-singular matrices. Then (5.1) can be shown to be equivalent to the systems

$$PAQ\underline{y} = P\underline{b} \; , \tag{5.2}$$

and

$$\underline{x} = Q\underline{y} \; . \tag{5.3}$$

So we can solve the system

$$B\underline{y} = \underline{b}' \; ,$$

where B=PAQ, $\underline{b}'=P\underline{b}$ and then from (5.3) we calculate the solution $\underline{x}$.

The problem now is to find suitable preconditioning matrices

P and Q such that matrix B is better conditioned than A.

a) Bauer's Method

The most well-known method of this type is Bauer's method (1963) in which diagonal matrices are used for the preconditioning matrices P and Q.

The following results relating to this approach have been proved for the norm subordinate to the maximum norm.

1) Let D_1 and D_2 be two diagonal matrices, then:

$$\min_{D_1,D_2} \operatorname{cond}(D_1AD_2) = \pi(|A|.|A^{-1}|) , \tag{5.4}$$

where π is the Perron root of $|A|.|A^{-1}|$.

2) The expression for the minimum (5.4) is achieved for both

$$D_1 = X_1^{-1} ,$$

and
$$D_2 = X_2 ,$$

where $X_1\underline{e}$ is the right Perron vector of $|A|.|A^{-1}|$ and $X_2\underline{e}$ is a right Perron vector of $|A^{-1}|.|A|$.

3) The value of X_2 can be determined from:

$$|A^{-1}|X_1\underline{e} = X_2\underline{e} .$$

4) The minimum conditions

$$\min_{D_1} \operatorname{cond}(D_1A) = \operatorname{lub}\,(|A^{-1}|.|A|) , \tag{5.5}$$

$$\min_{D_2} \operatorname{cond}(AD_2) = \operatorname{lub}\,(|A|.|A^{-1}|) , \tag{5.6}$$

are achieved for both $D_2=X_2$ and $D_1=X_1^{-1}$ where X_1 and X_2 are determined by:

$$|A^{-1}|\underline{e} = X_2\underline{e} , \tag{5.7}$$

$$|A|\underline{e} = X_1\underline{e} . \tag{5.8}$$

The above method is termed 'optimal scaling'.

The following results can be proved for optimal scaling:

a) Optimal scaling for the maximum norm as in (5.4) is characterised by the matrices D_1AD_2 and $D_2^{-1}A^{-1}D_1^{-1}$ each having equal row sums of absolute values.

b) Optimal scaling for the maximum norm as in (5.5),(5.6) is characterised by the matrices D_1A or $(AD_2)^{-1}$ respectively having equal row sums of absolute values.

To assess the amount of computational work required, we see that for the optimal scaling strategy we need first to calculate A^{-1} and it is known that this calculation requires $O(n^3)$ multiplicative operations. Then, we have the following work to carry out:

1. the computation of $|A|.|A^{-1}|$,
2. the computation of the Perron root π and Perron vector $X_1\underline{e}$ of $|A|.|A^{-1}|$ from
$$|A|.|A^{-1}|X_1\underline{e} = X_1\underline{e} \ ,$$
3. the calculation of X_2 from:
$$A^{-1}X_1\underline{e} = X_2\underline{e} \ ,$$
4. finally, we have to calculate the matrix,
$$B = D_1AD_2 \ ,$$
where $D_1=X_1^{-1}$ and $D_2=X_2$.

The major disadvantage of the optimal scaling strategy is the á fortiori computation of A^{-1}, a task which in fact we are trying to avoid when A is ill-conditioned.

Whilst in comparison the preconditioning strategy for a given value of ω, requires arithmetic operations (multiplications and divisions) of order $O(n^3)$, in order to form the preconditioned matrix B_ω before proceeding to the solution stage so one can infer that the two strategies are comparable in orders of computational work requirements.

The preconditioning strategy as shown earlier, also, clearly does not change the structure of the matrix, i.e. a positive definite matrix remains so, whereas, in optimal scaling a symmetric positive definite matrix can change to a non-positive definite and even to a non-symmetric matrix. This is a disadvantage since then the use of more efficient direct methods of solution (i.e. Cholesky) for the solution of the transformed system is not available. This is illustrated in the following example.

Consider the matrix:

$$A = \begin{bmatrix} 1 & 1 & 1 \\ 1 & 2 & 3 \\ 1 & 3 & 6 \end{bmatrix},$$

from which it is easy to check that:

$$A = \begin{bmatrix} 1 & & \\ 1 & 1 & \\ 1 & 2 & 1 \end{bmatrix} \begin{bmatrix} 1 & 1 & 1 \\ & 1 & 2 \\ & & 1 \end{bmatrix},$$

i.e., A is symmetric and positive definite matrix from which we can infer that the preconditioned matrix B_ω is also symmetric and positive definite.

Optimal scaling for the maximum norm transforms A into

$$B = D_1 A D_2 = \begin{bmatrix} \frac{3}{5}\sqrt{10} & 3 & \frac{2}{5}\sqrt{10} \\ 1 & 10 & 2 \\ \frac{1}{5}\sqrt{10} & 3 & \frac{4}{5}\sqrt{10} \end{bmatrix},$$

with

$$B^{-1} = \begin{bmatrix} \frac{1}{2}\sqrt{10} & -3 & \frac{1}{2}\sqrt{10} \\ -1 & 10 & -2 \\ \frac{1}{4}\sqrt{10} & -3 & \frac{3}{4}\sqrt{10} \end{bmatrix}.$$

Thus, although the original matrix A is symmetric and positive definite, the transformed matrix B is non-symmetric. Of course, the advantage of optimal scaling is that we know the exact minimum condition number (over the set of all diagonal matrices) and that the computation of B (after the computation of D_1 and D_2) is both easy and accurate.

In the following, some results are given which compares the effect of optimal scaling and preconditioning on the condition of the coefficient matrix and on the error in the solution of the associated linear system.

Example 5.1:

From the quindiagonal matrix defined as,

$$A = (a_{ij}) = \begin{cases} 6, & \text{if } i=j, \\ -4, & \text{if } i=j+1 \text{ or } i=j-1, \\ 1, & \text{if } i=j+2 \text{ or } i=j-2, \\ 0, & \text{elsewhere,} \end{cases}$$

we consider the 20×20 linear system:

$$\begin{bmatrix} 6 & -4 & 1 & & & & & \\ -4 & 6 & -4 & 1 & & & 0 & \\ 1 & -4 & 6 & -4 & 1 & & & \\ & 1 & -4 & 6 & -4 & 1 & & \\ & & \ddots & \ddots & \ddots & \ddots & \ddots & \\ & 0 & & \ddots & \ddots & \ddots & \ddots & 1 \\ & & & & \ddots & \ddots & \ddots & -4 \\ & & & & & 1 & -4 & 6 \end{bmatrix} \begin{bmatrix} x_1 \\ x_2 \\ x_3 \\ \vdots \\ \vdots \\ \vdots \\ \vdots \\ x_{20} \end{bmatrix} = \begin{bmatrix} 3 \\ -1 \\ 0 \\ \vdots \\ \vdots \\ \vdots \\ -1 \\ 3 \end{bmatrix}. \quad (5.9)$$

This system is known to have the solution $x_i=1$, $(i=1,2,\ldots,20)$. The condition numbers of the coefficient matrix of (5.9) are given by,

P-condition number = 7390.92,

N-condition number = 870.52,

and M-condition number = 6640.99.

When the system (5.9) is solved by the elimination method the relative Euclidean Error norm is:

$$0.11555393.10^{-7}.$$

In the following Table 5.1 we give the P,N,M condition numbers, trace and the relative Euclidean error norm versus the preconditioning parameter ω for the preconditioning linear system derived from (5.9).

ω	P-condition number	Trace	N-condition number	M-condition number	Relative Euclidean Error Norm
0	7390.92	20	870.52	6640.99	$0.4193.10^{-7}$
0.2	4418.20	16.91	569.06	5343.99	$0.32382.10^{-7}$
0.4	2682.34	14.57	373.02	4278.48	$0.17306.10^{-7}$
0.6	1638.57	12.73	242.19	3295.95	$0.66207.10^{-8}$
0.8	1056.85	11.26	153.87	2444.92	$0.21688.10^{-8}$
1	709.43	10.07	94.39	1725.87	$0.21688.10^{-8}$
1.2	480.15	9.10	55.03	1138.81	$0.16258.10^{-8}$
1.4	320.43	8.36	30.03	700.42	$0.11604.10^{-8}$
1.6	209.51	7.94	15.48	390.45	$0.26888.10^{-9}$
1.7	172.28	7.97	13.39	284.48	$0.16187.10^{-9}$
1.8	155.13	8.43	9.53	210.26	$0.51826.10^{-10}$
1.9	193.10	10.36	11.86	163.21	$0.88863.10^{-10}$
2	522.64	20	32.70	135.74	$0.69357.10^{-10}$

TABLE 5.1

In the following, Table 5.2, the results of optimal scaling and optimal preconditioning are compared.

	Original System	Optimally Scaled System	Optimal Preconditioning
max norm cond.no. $\|\|A\|\|_\infty.\|\|A^{-1}\|\|$	9680.0	7376.89	361.21
N-condition no.	870.52	116.12	9.53
M-condition no.	6640.99	860.84	210.26
P-condition no.	7390.92	965.56	155.13
Relative Euclidean Error	$0.115553.10^{-7}$	$0.474362.10^{-8}$	$0.518269.10^{-10}$

TABLE 5.2

Example 5.2:

Consider the non-symmetric system,

$$\begin{bmatrix} 8 & 7 & & & & & & \\ 7 & 7 & 6 & & & & 0 & \\ 6 & 6 & 5 & & & & & \\ 5 & 5 & 5 & 4 & & & & \\ 4 & 4 & 4 & 4 & 3 & & & \\ 3 & 3 & 3 & 3 & 3 & 2 & & \\ 2 & 2 & 2 & 2 & 2 & 2 & 1 & \\ 1 & 1 & 1 & 1 & 1 & 1 & 1 & \end{bmatrix} \begin{bmatrix} x_1 \\ x_2 \\ x_3 \\ x_4 \\ x_5 \\ x_6 \\ x_7 \\ x_8 \end{bmatrix} = \begin{bmatrix} 15 \\ 20 \\ 23 \\ 24 \\ 23 \\ 20 \\ 15 \\ 8 \end{bmatrix}, \qquad (5.10)$$

having the solution: $x^T=(1,1,1,1,1,1,1,1)$.

The matrix A can be seen to be of lower Hessenberg form and its condition numbers are:

$$K(A) = 2.201.10^5,$$
$$N(A) = 3.790.10^4,$$
$$M(A) = 3.680.10^5.$$

The relative Euclidean error norm obtained in solving (5.10) by Gaussian elimination is:

$$0.18765.10^{-6}\ .$$

In the following Table 5.3 we give the K, N and M condition numbers and the Euclidean error norm of (5.10) versus the preconditioning parameter ω. We also give the smallest and largest eigenvalues of $B_\omega^T B_\omega$.

ω	Smallest Eigenvalue	Largest Eigenvalue	K-condition Number	N-condition Number	M-condition Number	Relative Euclidean Error Norm
0	$0.6001.10^{-9}$	$0.2042.10^{2}$	$1.844.10^{5}$	$2.693.10^{4}$	$1.724.10^{5}$	$0.1898.10^{-6}$
0.4	$0.1217.10^{-8}$	$0.2728.10$	$4.734.10^{4}$	$9.825.10^{3}$	$1.310.10^{5}$	$0.7460.10^{-7}$
0.6	$0.1877.10^{-8}$	$0.1470.10$	$2.798.10^{4}$	$7.263.10^{3}$	$1.127.10^{5}$	$0.2721.10^{-7}$
0.7	$0.2211.10^{-8}$	$\underline{0.1463.10}$	$\underline{2.572.10^{4}}$	$6.566.10^{3}$	$1.122.10^{5}$	$0.1479.10^{-7}$
0.8	$0.2575.10^{-8}$	$0.2141.10$	$2.883.10^{4}$	$6.175.10^{3}$	$1.097.10^{5}$	$0.8848.10^{-8}$
0.9	$0.3019.10^{-8}$	$0.3237.10$	$3.274.10^{4}$	$\underline{6.076.10^{3}}$	$1.082.10^{5}$	$0.3977.10^{-8}$
1	$0.3435.10^{-8}$	$0.5024.10$	$3.824.10^{4}$	$6.289.10^{3}$	$1.066.10^{5}$	$0.5987.10^{-8}$
1.2	$0.4104.10^{-8}$	$0.1289.10^{2}$	$5.505.10^{4}$	$7.872.10^{3}$	$\underline{1.049.10^{5}}$	$0.5171.10^{-8}$
1.3	$\underline{0.4362.10^{-8}}$	$0.2113.10^{2}$	$6.959.10^{4}$	$9.436.10^{3}$	$1.372.10^{5}$	$\underline{0.6662.10^{-8}}$
1.4	$0.4317.10^{-8}$	$0.3487.10^{2}$	$8.988.10^{4}$	$1.171.10^{4}$	$1.755.10^{5}$	$0.1545.10^{-8}$
1.6	$0.4179.10^{-8}$	$0.9533.10^{2}$	$4.776.10^{5}$	$1.948.10^{4}$	$3.471.10^{5}$	$0.2323.10^{-7}$
2	$0.2635.10^{-8}$	$0.1290.10^{4}$	$6.998.10^{5}$	$8.717.10^{4}$	$2.580.10^{6}$	$0.7953.10^{-7}$

cont.

TABLE 5.3

Similarly in Table 5.4 the comparison of the results for optimal scaling and optimal preconditioning are given.

	Original System	Optimally Scaled System	Optimal Pre-conditioning
max norm cond.no. $\|\|A\|\|_\infty . \|\|A^{-1}\|\|_\infty$	$3.755.10^5$	$6.425.10^4$	$8.830.10^4$
N-condition no.	$3.790.10^5$	$2.373.10^4$	$6.076.10^3$
M-condition no.	$3.686.10^5$	$1.285.10^5$	$1.081.10^5$
K-condition no.	$2.201.10^6$	$1.811.10^5$	$3.274.10^4$
Relative Euclidean Error	$0.187656.10^{-6}$	$0.105563.10^{-6}$	$0.397726.10^{-8}$

TABLE 5.4

In this example it can be seen that although the two methods decrease the condition numbers by approximately the same factor there is a factor of 10^{-2} between the error norms. Thus, we can infer that preconditioning has more effect on the minimisation of the error norm than possibly optimal scaling.

In another example, the non-symmetric Cauchy matrix given by

$$a_{ij} = 1/(1/i)+(1/j+6), \quad i,j=1,2,...,6, \qquad (5.11)$$

was considered and we compare preconditioning results with optimal scaling in Table 5.5.

	Original System	Optimally Scaled System	Optimal Pre-conditioning
max norm cond.no. $\|\|A\|\|_\infty . \|\|A^{-1}\|\|_\infty$	$2.8409.10^{10}$	$9.7841.10^9$	$7.891.10^8$
N-condition no.	$2.3756.10^9$	$2.1719.10^9$	$8.752.10^7$
M-condition no.	$9.8684.10^9$	$7.1704.10^9$	$9.7451.10^8$
Relative Euclidean Error	$0.104109.10^{-1}$	$0.805565.10^{-2}$	$0.894377.10^{-5}$

TABLE 5.5

In the above table the minimum eigenvalue of A^TA and $B_\omega^TB_\omega$ was too small for the available routine to evaluate the K-condition number.

Finally, the following (6×6) Vandermonde matrix was considered:

$$A = \begin{bmatrix} 1 & 1 & 1 & 1 & 1 & 1 \\ 1 & 2 & 3 & 4 & 5 & 6 \\ 1 & 2^2 & 3^2 & 4^2 & 5^2 & 6^2 \\ 1 & 2^3 & 3^3 & 4^3 & 5^3 & 6^3 \\ 1 & 2^4 & 3^4 & 4^4 & 5^4 & 6^4 \\ 1 & 2^5 & 3^5 & 4^5 & 5^5 & 6^5 \end{bmatrix} . \tag{5.12}$$

In the following Table 5.6 we compare the condition numbers (the N,M,K and max.norm condition numbers) of (5.12) and those of the optimally scaled and the optimal preconditioned forms.

	Original Matrix	Optimally Scaled	Optimal Preconditioning
max.norm cond. no.	$1.281.10^6$	$4.412.10^3$	$6.538.10^2$
K-condition no.	$7.311.10^5$	$6.604.10^3$	$3.756.10^2$
N-condition no.	$1.220.10^5$	$1.160.10^3$	$6.245.10$
M-condition no.	$1.975.10^6$	$3.863.10^3$	$8.267.10^2$

TABLE 5.6

6. COMPARISON OF OPTIMAL SCALING AND PRECONDITIONING ON A 2×2 MATRIX

Consider the (2×2) matrix,

$$A = \begin{bmatrix} a & b \\ c & d \end{bmatrix} , \tag{6.1}$$

with real integers $a,b,c,d \in R^*$ and $ad-bc>0$.

I. For optimal scaling we can derive from the previous section the following results,

$$A^{-1} = \frac{1}{\det(A)} \begin{bmatrix} d & -b \\ -c & a \end{bmatrix} , \tag{6.2}$$

with

$$|A|.|A^{-1}| = \frac{1}{\det(A)} \begin{bmatrix} (ad+bc) & 2ab \\ 2cd & (ad+bc) \end{bmatrix} , \tag{6.3}$$

and

$$|A^{-1}|.|A| = \frac{1}{\det(A)} \begin{bmatrix} (ad+bc) & 2bd \\ 2ac & (ad+bc) \end{bmatrix} . \tag{6.4}$$

The common Perron eigenvalue of (6.3) and (6.4) is:

$$\pi(|A|.|A^{-1}|) = \frac{(ad+bc+2\sqrt{abcd})}{\det(A)} , \tag{6.5}$$

whilst the right Perron eigenvector of (6.3) is:

$$x_1 = \begin{bmatrix} \sqrt{abcd} \\ cd \end{bmatrix} , \tag{6.6}$$

and the right Perron eigenvector of (6.4) is:

$$x_2 = \begin{bmatrix} \sqrt{abcd} \\ ac \end{bmatrix} , \tag{6.7}$$

Finally, we obtain,

$$D_1 = \begin{bmatrix} \sqrt{abcd} & 0 \\ 0 & cd \end{bmatrix} , \quad D_2 = \begin{bmatrix} \sqrt{abcd} & 0 \\ 0 & ac \end{bmatrix} ,$$

and

$$D_1^{-1}AD_2 = \begin{bmatrix} a & \sqrt{\frac{abc}{d}} \\ \sqrt{\frac{abc}{d}} & a \end{bmatrix} ,$$

with

$$\text{cond}_\infty(D_1^{-1}AD_2) = \frac{(ad+cb+2\sqrt{abcd})}{\det(A)} . \tag{6.8}$$

II. Now applying preconditioning after normalising A, we obtain,

$$A' = \begin{bmatrix} 1 & \frac{b}{\sqrt{ad}} \\ \frac{c}{\sqrt{ad}} & 1 \end{bmatrix} ,$$

from which we can establish,

$$(I+\omega L)^{-1} = \begin{bmatrix} 1 & 0 \\ -\frac{\omega c}{\sqrt{ad}} & 1 \end{bmatrix} , \text{ and } (I+\omega U)^{-1} = \begin{bmatrix} 1 & -\frac{b}{\sqrt{ad}} \\ 0 & 1 \end{bmatrix} .$$

Thus, the preconditioned system,

$$B_\omega = (I+\omega L)^{-1}A'(I+\omega U)^{-1} = \begin{bmatrix} 1 & \frac{(b-b\omega)}{\sqrt{ad}} \\ \frac{(c-\omega c)}{\sqrt{ad}} & (1-\frac{2bc\omega}{ad} + \frac{cb\omega^2}{ad}) \end{bmatrix} , \tag{6.9}$$

and

$$B_\omega^{-1} = \frac{ad}{(ad-bc)} \begin{bmatrix} (1-\frac{2bc}{ad}\omega+\frac{cb\omega^2}{ad}) & \frac{(b\omega-b)}{\sqrt{ad}} \\ \frac{(c\omega-c)}{\sqrt{ad}} & 1 \end{bmatrix} . \tag{6.10}$$

It is easy to check that for $\omega=1$ we have the minimum value of $||B_\omega||_\infty$ and the minimum value for $||B_\omega^{-1}||$ (Evans, 1968). In this case, we have:

$$B_{\omega=1} = \begin{bmatrix} 1 & 0 \\ 0 & (1-\frac{bc}{ad}) \end{bmatrix}, \quad B_{\omega=1}^{-1} = \begin{bmatrix} 1 & 0 \\ 0 & \frac{ad}{(ad-bc)} \end{bmatrix},$$

from which we can determine the following results,

$$||B_{\omega=1}||_\infty = 1, \text{ and } ||B_{\omega=1}^{-1}|| = \frac{ad}{(ad-bc)} = \frac{ad}{\det(A)} .$$

Hence,

$$\text{cond}_\infty(B_{\omega=1}) = \frac{ad}{\det(A)}, \tag{6.11}$$

and comparing (6.11) to (6.8) we have the result,

$$\text{cond}_\infty(B_{\omega=1}) \leq \text{cond}_\infty(D_1^{-1}AD_2) . \tag{6.12}$$

III. For a numerical example we consider the ill-conditioned matrix,

$$A = \begin{bmatrix} 100 & 50 \\ 59.9 & 30 \end{bmatrix} .$$

We can easily determine,

$$A^{-1} = \begin{bmatrix} 6 & -10 \\ -\frac{59.9}{5} & 20 \end{bmatrix},$$

and the result,

$$\text{cond}_\infty(A) = 150.\frac{(159.9)}{5} = 4797.$$

From (6.8) we have that,

$$\text{cond}_\infty(D_1^{-1}AD_2) = 2390.4,$$

whilst from (6.12) we have,

$$\text{cond}_\infty(B_{\omega=1}) = 600.$$

Thus, it can be seen the preconditioning strategy can yield lower condition numbers than optimal scaling.

7. REFERENCES

1. L. Andersson - "S.S.O.R. Preconditioning of Toeplitz Matrices with Applications to Elliptic Equations", Computer Science Report, Chalmers Univ. of Technology, Goteburg, Sweden, (1975).

2. F.L. Bauer - "Optimally Scaled Matrices", Num.Math. 5, pp.73-87, (1963).

3. D.J. Evans - "The Use of Preconditioning in Iterative Methods for Solving Linear Equations with Symmetric Positive Definite Matrices", J.I.M.A. 4, pp.295-314, (1968).

4. D.J. Evans - "Comparison of the Convergence Rates of Iterative Methods for Solving Linear Equations with Preconditioning", Greek Math.Soc., C. Carathéodory Symp., Sept. 3-7th, 1973, pp.106-135.

5. D.J. Evans -"Iterative Sparse Matrix Algorithms", in 'Software for Numerical Mathematics', ed. D.J. Evans, Acad.Press, pp.49-83, (1974).

6. D.J. Evans and M. Hatzopoulos - "A Comparison of Optimal Scaling and Preconditioning", A.C.M. Signum Newsletter, 4, 2, pp.20-22, (1979).

7. G.E. Forsythe and W. Wasow - "Finite Difference Methods for Partial Differential Equations", J. Wiley & Sons (1960).

8. M. Hatzopoulos and D.J. Evans - "On the Direct Solution of Preconditioned Linear Equations", Bull. Greek Math.Soc., 16, pp.30-44, (1975).

9. M. Hatzopoulos - "Preconditioning and Other Computational Techniques for the Direct Solution of Linear Equations", Ph.D. Thesis, Loughborough University of Technology (1974).

10. J. Von Neumann and H.H. Goldstine - "Numerical Inverting of Matrices of Higher Order", Bull.Amer.Math.Soc., 53, pp.1021-1099, (1947).

11. A.M. Turing - "Round Off Errors in Matrix Processes", Quart.J.Mech.Appl.Math., 1, pp.287-308, (1948).

12. J.H. Wilkinson - "The Solution of Ill-Conditioned Linear Equations", in 'Math. Methods for Digital Computers', Vol. 2, edits. A. Ralston & M. Wilf, J. Wiley & Sons, pp.65-93 (1967).

DEVELOPMENT OF AN ICCG ALGORITHM FOR LARGE SPARSE SYSTEMS

ALAN JENNINGS

Civil Engineering Department, Queens University, Belfast

Classical iterative methods often exhibit very poor convergence characteristics when applied to structural analysis problems. This paper outlines how an incomplete Choleski conjugate gradient algorithm has been developed for general purpose use with large order sparse symmetric positive definite equations including those arising from structural analysis.

1. A HISTORICAL NOTE

One of the key landmarks in the history of structural computing was the publication in 1930 by Hardy Cross of the moment distribution method for the iterative solution of rigidly-jointed plane frame problems. If the method is analysed mathematically it is found to be equivalent to Gauss-Seidel iteration of the relevant stiffness equations involving joint rotation variables although no explicit record is kept of the joint rotations. Instead the residual joint moments are recorded and updated. Moment distribution seems to be the first iterative method to be extensively applied to structural problems. The immediate impact it had can be judged by the fact that the 10 page paper attracted 147 pages of discussion, most of which was by people recording how they had successfully utilised the method [1]. Until computers became available moment distribution was the normal method of solving rigidly-jointed plane frame problems [2], and is still taught today. Moment distribution was the forerunner of hand relaxation methods as applied to finite difference methods [3] of solving elliptic partial differential equations (first paper in 1935).

When digital computers became available it was soon established that Livesley's method was the most effective way of solving general frame problems [4]. For a rigidly-jointed plane frame this meant using three variables per joint - two orthogonal displacements and rotation, thus permitting programs to be written which could analyse any geometry of frame by inputting the joint coordinates and information on the member incidences and properties. The resulting equations are ill-conditioned and hence exhibit extremely bad convergence characteristics.

Whereas not all structural problems will yield such poor conditioning as rigidly-jointed frames, it is generally true to say that all large order problems will have poor or very poor conditioning. Although iterative methods gain advantage over elimination methods because of the lack of fill-in, in cases where this benefit is most to be sought (where the number of equations is large), one finds that poor convergence rates destroy their potential usefulness. This is the most important reason why they have not been used very much for solution of structural problems by computer, despite the earlier successes of the moment distribution method for hand solution.

2. A GLANCE AT WHY CONVERGENCE IS OFTEN VERY POOR

Consider a simple portal frame (Figure 1) in which the area and second moment of area of the cross-sections of the columns and beam are A_1, I_1 and A_2, I_2 respectively. The Livesley displacement coordinates will be $\{x_B\ y_B\ \theta_B\ x_C\ y_C\ \theta_C\}$ so that the stiffness equations relating these variables to the joint forces $\{H_B\ V_B\ M_B\ H_C\ V_C\ M_C\}$ will be as shown in Figure 2.

Since the axial stiffness terms EA_1/ℓ_1 and EA_2/ℓ_2 tend to be larger than the sway stiffness terms $12EI_1/\ell_1{}^3$ and $12EI_2/\ell_2{}^3$ by a factor of more than 100, the convergence of all point iterative methods is destroyed by the one off-diagonal coefficient in position (4,1) which is almost as large as the diagonal coefficients in rows 1 and 4. The mathematical effect of this coefficient is to make the equations ill-conditioned. The physical effect is to couple the displacements x_B and x_C. Indeed, if the frame is

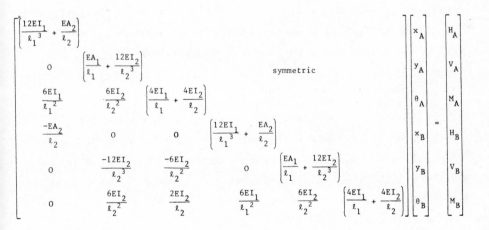

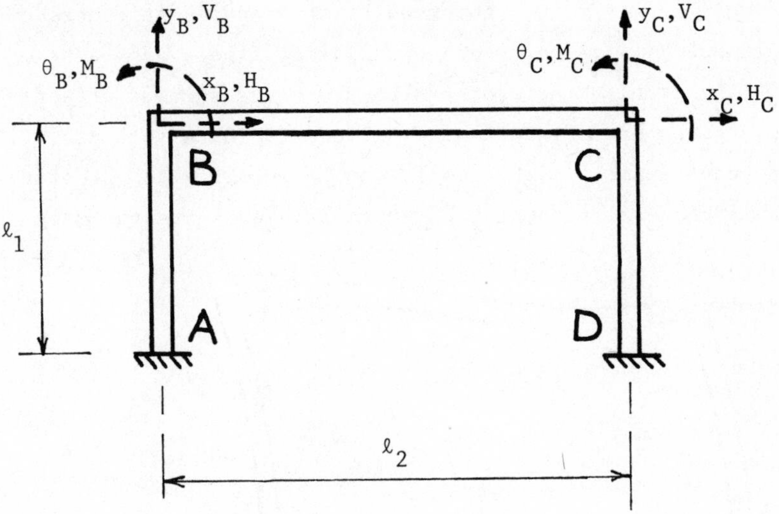

FIGURE 1 A portal frame

$$
\begin{bmatrix}
\left(\frac{12EI_1}{\ell_1^3}+\frac{EA_2}{\ell_2}\right) & & & & & \\
0 & \left(\frac{EA_1}{\ell_1}+\frac{12EI_2}{\ell_2^3}\right) & & \text{symmetric} & & \\
\frac{6EI_1}{\ell_1^2} & \frac{6EI_2}{\ell_2^2} & \left(\frac{4EI_1}{\ell_1}+\frac{4EI_2}{\ell_2}\right) & & & \\
\frac{-EA_2}{\ell_2} & 0 & 0 & \left(\frac{12EI_1}{\ell_1^3}+\frac{EA_2}{\ell_2}\right) & & \\
0 & \frac{-12EI_2}{\ell_2^3} & \frac{-6EI_2}{\ell_2^2} & 0 & \left(\frac{EA_1}{\ell_1}+\frac{12EI_2}{\ell_2^3}\right) & \\
0 & \frac{6EI_2}{\ell_2^2} & \frac{2EI_2}{\ell_2} & \frac{6EI_1}{\ell_1^2} & \frac{6EI_2}{\ell_2^2} & \left(\frac{4EI_1}{\ell_1}+\frac{4EI_2}{\ell_2}\right)
\end{bmatrix}
\begin{bmatrix} x_A \\ y_A \\ \theta_A \\ x_B \\ y_B \\ \theta_B \end{bmatrix}
=
\begin{bmatrix} H_A \\ V_A \\ M_A \\ H_B \\ V_B \\ M_B \end{bmatrix}
$$

FIGURE 2 Stiffness equations for portal frame

loaded as shown in Figure 3, the sway displacement induced has a magnitude which is mainly governed by the bending stiffness of the columns (EI_1) rather than the axial stiffness of the beam (EA_2). The problem therefore is to avoid the serious effects on convergence of large off-diagonal coefficients such as (4,1) in the above example. Whereas it may be possible to predetermine by structural reasoning which coefficients are the ones

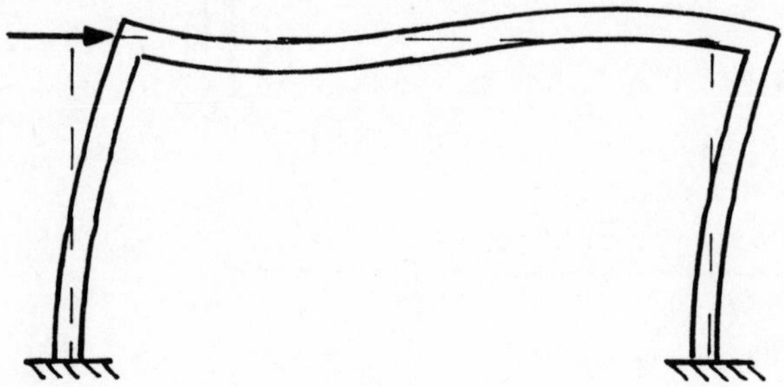

FIGURE 3 Sway action of a portal frame

which will cause trouble, evidence is available in the equations as assembled and therefore it should be possible to devise a solution process which sorts out its own salvation.

3. INCOMPLETE FACTORIZATION

A possible way of achieving this objective is to perform a factorization of the coefficient matrix which only takes account of the strong off-diagonal elements. Although this factorization will be incorrect, an iterative method based on it should have the adverse effect of these elements removed. If the equations to be solved are

$$Ax = b \tag{1}$$

the incomplete factorization may be expressed as

$$A = LL^T - C \tag{2}$$

where C is the matrix of elements omitted from the factorization. An iterative refinement procedure for solving equations (1) based on this incomplete factorization is

$$LL^T x^{(k+1)} = b + Cx^{(k)} \tag{3}$$

For the method to be much more efficient than point iterative methods it is necessary for the convergence rates to be substantially improved when only a few off-diagonal elements are contained in L. Some criterion is required to decide which elements can be omitted during the elimination phase and this needs to be done by comparing the magnitude of the off-diagonal element with the appropriate diagonal elements, either the original ones or their current values at the appropriate time during elimination.

A procedure of this type was tested by Tuff and Jennings [5] and was found to yield good convergence rates for a large number of structural problems. However one difficulty encountered was that with some values of rejection parameter the factorization failed due to loss of positive definiteness. A simple example of this is the matrix

$$\begin{bmatrix} 1 & 1 & 0.1 \\ 1 & 2 & 2.1 \\ 0.1 & 2.1 & 4.2 \end{bmatrix}$$

which loses its positive definite property if the small elements in positions (3,1) and (1,3) are omitted.

4. A ROBUST TECHNIQUE

From minimax properties of eigenvalues [6] it is known that the eigenvalues of a symmetric matrix will not be reduced if a symmetric positive semi-definite matrix is added to it. Hence a satisfactory modification to the above matrix would be

$$\begin{bmatrix} 1 & 1 & 0.1 \\ 1 & 2 & 2.1 \\ 0.1 & 2.1 & 4.2 \end{bmatrix} + \begin{bmatrix} 0.1 & 0 & -0.1 \\ 0 & 0 & 0 \\ -0.1 & 0 & 0.1 \end{bmatrix} = \begin{bmatrix} 1.1 & 1 & 0 \\ 1 & 2 & 2.1 \\ 0 & 2.1 & 4.3 \end{bmatrix} \quad (4)$$

The method of Jennings and Malik [7] makes use of this principal but the relative magnitude of the additions to the diagonal are adjusted to be appropriate to the magnitude of those elements. Thus if element a^*_{ij} is to be rejected $\left(\frac{a^*_{jj}}{a^*_{ii}}\right)^{\frac{1}{2}}|a^*_{ij}|$ is added to a^*_{jj} and $\left(\frac{a^*_{ii}}{a^*_{jj}}\right)^{\frac{1}{2}}|a^*_{ij}|$ is added to a^*_{ii} where a^*_{ij}, a^*_{ii} and a^*_{jj} are modified elements of the matrix at the stage of the reduction at which the rejection of element a^*_{ij} is considered. A further modification is to adopt the conjugate gradient technique for solving the iterative equations rather than equation (3) which can be shown to be a form of iterative refinement. In order to use the conjugate gradient method it is necessary to apply it to the preconditioned equations

$$\overline{A}y = \overline{b} \quad (5)$$

where $y = L^T x$, $\overline{b} = L^{-1}b$ and

$$\overline{A} = L^{-1}AL^{-T} \quad (6)$$

The matrix $\overline{A}$ should not be obtained explicitly because the only operation required of $\overline{A}$ is the matrix premultiplication

$$u^{(k)} = \overline{A}p^{(k)} \quad (7)$$

where $u^{(k)}$ and $p^{(k)}$ are vectors involved in the iteration process. This operation can be performed implicitly either using equation (6) or

$$\overline{A} = I - L^{-1}CL^{-T} \quad (8)$$

the latter being obtained by substituting the value of A from equation (2).

This technique is robust in so far as:

(a) The factorization can never fail due to loss of positive definiteness nor can the conditioning be made worse by the lowest eigenvalue of LL^T being closer to zero than the lowest eigenvalue of A.

(b) Since the conjugate gradient method is universally convergent for all symmetric positive definite matrices no problems can arise through lack of convergence.

Preliminary tests [7,8] using $\bar{A}$ as in equation (8) showed that the method could achieve significant reductions in the number of non-zero elements to be stored as compared with standard factorization techniques while at the same time obtaining solutions with less total number of non-zero multiplications than other iterative methods including preconditioned and block conjugate gradient methods for a set of various problems generating equations with orders up to n = 432.

5. RELATIONSHIP TO OTHER ICCG METHODS

Although the above method has been called 'Partial Elimination', it does bear a strong relationship with the methods of Meijerink and van der Vorst [9] and Kershaw [10] which are called 'incomplete Choleski-conjugate gradients' (ICCG) and indeed could be classified under the same title. Meijerink and van der Vorst applied ICCG to M-matrices whose properties permit stable factorizations with no diagonal correction terms. Kershaw considers general application to symmetric positive definite equations but the method differs in two ways from the Jennings-Malik algorithm.

(i) The factorization is carried out within a pre-specified storage area. Elements within this area take part in the elimination and any fill-in elements occurring outside this area are omitted. This technique is more suitable where the equations are of regular form such as for a set of finite difference equations specified over a rectangular grid.

However, where the sparsity pattern of the equations is irregular and where some individual elements may give much stronger coupling between the variables than others, there would be problems defining what is a satisfactory storage area.

(ii) No modifications to the diagonal are introduced unless a zero or negative pivot is encountered. At that stage the pivot concerned is modified to be positive and factorization is continued. However no modifications are made to pivots which are positive but very small.

The occurrence of a small pivot (maybe much smaller than any pivot in the complete factorization) would certainly be expected to cause problems with accuracy and conditioning. Despite this possible disadvantage of Kershaw's method he has reported some impressive results when applied to different types of finite difference equations, particularly where the original equations are ill-conditioned.

Munksgaard [11] has developed an ICCG method which does allow for fill-in. Apart from using a different form of diagonal modification, the method differs from the technique presented here by adopting a minimal degree pivoting strategy. The program organisation is therefore very different to the one given below.

6. AN ALGORITHM TO TAKE FULL ACCOUNT OF SPARSITY

One of the advantages of the ICCG method in comparison with elimination is the reduction in storage space required if full advantage is taken of sparsity. Examination of the previous test results by Jennings and Malik revealed that the storage requirements when using equation (6) would often be less than when using equation (8). Since also the coefficient matrix A is preserved this method was chosen for efficient implementation using a sparse storage scheme having no restrictions on the pattern of sparsity [12]. The two main stages of the algorithm are therefore:

(a) Incomplete factorization: Determine L from A, preserving A

(b) Conjugate gradients: Use L and A in CG iteration.

The least amount of storage space required for a low density sparse matrix is a real store for each non-zero element, an integer store for the column identifies of each element, and an integer store to give a pointer to the starting address of each row [13]. Thus the upper triangle of the matrix

$$A = \begin{bmatrix} 8.3 & -3.2 & 0 & -2.7 \\ -3.2 & 13.7 & -5.6 & 0 \\ 0 & -5.6 & 10.5 & -2.7 \\ -2.7 & 0 & -2.7 & 9.2 \end{bmatrix} \qquad (9)$$

would be stored in a sequential packed store as

REAL ARRAY A[8.3 -3.2 -2.7 13.7 -5.6 10.5 -2.7 9.2]
INTEGER ARRAY AC[1 2 4 2 3 3 4 4]
INTEGER ARRAY ADDRESS [1 4 6 8]

Since there is no undue difficulty performing the iterative phase of the algorithm with L^T and the upper triangle of A stored in this way, the main problem is to find how to implement the factorization phase taking account of the following difficulties:

(a) Some fill-in may take place.
(b) The matrix A still needs to be available after elimination.
(c) All modifications to a diagonal element arising through element rejection must be carried out before it is used as a pivot.

The operations required to determine row i of L^T may be expressed mathematically by

$$a^*_{ii} = a_{ii} - \sum_{k=1}^{i-1} \ell^2_{ik} + \sum_{k=1}^{i-1} g_{ik} + \sum_{j=i+1}^{n} h_{ij} \qquad (10)$$

$$a^*_{ij} = a_{ij} - \sum_{k=1}^{i-1} \ell_{ik}\ell_{jk} \qquad (j>i) \qquad (11)$$

$$\ell_{ii} = \sqrt{a^*_{ii}} \qquad (12)$$

$$\ell_{ij} = a^*_{ij}/\ell_{ii} \qquad (j>i) \qquad (13)$$

where a^*_{ii} and a^*_{ij} are the Gaussian coefficients at the start of the i-th pivoting step with the g_{ik} and h_{ij} terms present to represent the diagonal modifications obtained when weak elements on column i or row i are omitted. Because there is no facility in the storage of L^T which enables easy column scanning, the method adopted is to generate the modifications $-\ell^2_{ik}$, $-\ell_{ik}\ell_{jk}$ and g_{ik} when row k of L^T is being formed and store them until they are needed for row i. Two storage facilities are employed, one is an array of order n to store the updated diagonal elements and the other is a randomly packed store to contain the off-diagonal modifications $-\Sigma\ell_{ik}\ell_{jk}$.

As a simple illustration of how the incomplete factorization is implemented consider the matrix

$$A = \begin{bmatrix} 3.9 & 0.1 & 2 & 0 & 2 & 0 & 2 \\ & 3.9 & 0 & -4 & 0 & 2 & 0 \\ & & 4.6 & 0.2 & 1 & 0 & 5 \\ & & & 4.9 & 0 & -3 & 0 \\ & & & & 9 & -3 & 0 \\ & & & & & 7 & 0 \\ & & & & & & 13 \end{bmatrix} \tag{14}$$

The first operation is to inspect row 1 to see which elements may be omitted. With ψ as the rejection parameter, the general rule is that an element $a^{(k)}_{ik}$ is rejected if

$$(a^{(k)}_{ik})^2 < \psi^2 a^{(k)}_{kk}\, a^{(k)}_{ii} \tag{15}$$

With $\psi = 0.2$ element (1,2) will be omitted and the diagonal elements (1,1) and (2,2) both increased by 0.1 (according to the rule given in §4). The remaining elements in row 1 take part in the factorization giving $\ell_{11} = 2$ and $\ell_{13} = \ell_{15} = \ell_{17} = 1$. On account of the Gaussian modification of the rest of the upper triangle, elements in positions (3,3), (3,5), (3,7), (5,5), (5,7) and (7,7) are all reduced by 1. The diagonal element modifications are performed immediately and the off-diagonal modifications are entered into the random store whether or not there are corresponding non-zero elements in the matrix A. Row 2 is examined next and the first operation is to combine any row 2 modifications stored in the random store with row 2 of A (in this case no modifications are required). Since no elements pass the rejection test the factorization of row 2 can take place and the resulting modifications to elements (4,4), (4,6) and (6,6) recorded. Continuing the process yields eventually

$$L^T = \begin{bmatrix} 2 & 0 & 1 & 0 & 1 & 0 & 1 \\ & 2 & 0 & -2 & 0 & 1 & 0 \\ & & 2 & 0 & 0 & 0 & 2 \\ & & & 1 & 0 & -1 & 0 \\ & & & & 3 & -1 & 0 \\ & & & & & 2 & 0 \\ & & & & & & 3 \end{bmatrix}$$

The randomly packed temporary store for the Gaussian modifications requires two integers to be stored per element (the column number and a pointer for the next element) and also an address for the first stored element on each row. When elements are no longer required the space they vacate is made available for new elements. Hence the total storage requirement is not very large, particularly when ψ is large.

7. FURTHER CONSIDERATIONS

The above algorithm is completely automatic apart from the need to choose the rejection parameter ψ. With $\psi = 1$, no off-diagonal elements are included in the factorization and so the method is equivalent to the conjugate gradient method with row and column scaling (However the presence of diagonal modifications make the scaling different to that necessary to give unit diagonal elements). With $\psi = 0$, all elements are included in the factorization making it complete. In this case the conjugate gradient process halts after one iteration provided that rounding error does not affect the tolerance test. Hence the method may be considered to act as a bridge between iterative techniques and elimination, and it would be expected that a decrease in ψ would achieve a faster convergence rate at the expense of having more elements in L (and also the temporary random working store). In general this is true although there is no mathematical proof that changes in convergence rate or number of elements in L vary monotonically with ψ nor are there ways of obtaining realistic bounds on the convergence rate.

Numerical tests suggest that with ψ chosen in the interval $0.01 < \psi < 0.1$ (approximately) the method yields much better convergence characteristics than standard iterative methods including the conjugate gradient method without incurring the use of much additional storage space. A program written with an ICCG solution for the equations is therefore capable of solving larger problems within the available storage space than if an elimination solution had been adopted. For larger problems (average bandwidth > 50 approximately), the method is likely to be more efficient than both elimination and iterative methods. If a set of equations have multiple right-hand sides it will be more advantageous to use a low value of ψ so that more computation is carried out in the unrepeated factorization phase and less in the repeated iterative phase. Since the factorization phase usually takes much less time than the iterative phase, it is not too costly in machine time to repeat the analysis with a larger value of ψ if, at the first attempt, the chosen value of ψ yielded an L matrix which overran its allocated storage space during the first phase.

8. CONCLUSION

The ICCG method with diagonal modification according to the Jennings-Malik algorithm is robust in that it can accept any symmetric positive definite set of equations without the possibility of pivoting problems or deterioration of convergence characteristics. It is also possible to be programmed in such a way that full use is made of sparsity. The ICCG method appears to be the first iterative technique having good (but not proven) convergence characteristics which is available to structural engineers and others wishing to solve symmetric positive definite equations with poor conditioning on a computer.

REFERENCES

1. CROSS, H. - Analysis of Continuous Frames by Distributing Fixed-end Moments. Transactions ASCE, Vol. 96, pp. 1-156, 1932.

2. LIGHTFOOT, E. - Moment Distribution, Spon, London, 1961.

3. SOUTHWELL, R.V. - Relaxation Methods in Theoretical Physics, Oxford University Press, 1946.

4. LIVESLEY, R.K. - Analysis of Rigidly-Jointed Frames by an Electronic Digital Computer. Engineering, London, Vol. 176, pp. 230-233, 1953.

5. TUFF, A.D., and JENNINGS, A. - An Iterative Method for Large Systems of Linear Structural Equations. Int. J. for Num. Meth. in Engng., Vol. 7, pp. 175-183, 1973.

6. WILKINSON, J.H. - The Algebraic Eigenvalue Problem, Clarendon Press, Oxford, 1965.

7. JENNINGS, A., and MALIK, G.M. - Partial Elimination. J. Inst. Maths. Applics., Vol. 20, pp. 307-316, 1977.

8. JENNINGS, A., and MALIK, G.M. - The Solution of Sparse Linear Equations by the Conjugate Gradient Method. Int. J. for Num. Meth. in Engng., Vol. 12, pp. 141-158, 1978.

9. MEIJERINK, J.A., and VAN DER VORST - An Iterative Solution Method for Linear Systems of Which the Coefficient Matrix is a Symmetric M-Matrix. Math. Comp. Vol. 31, pp. 148-162, 1977.

10. KERSHAW, D.S. - The Incomplete Cholesky-Conjugate Gradient Method for the Iterative Solution of Systems of Linear Equations. J. Comp. Phys., Vol. 26, pp. 43-65, 1978.

11. MUNKSGAARD, N. - Solving Sparse Symmetric Sets of Linear Equations by Preconditioned Conjugate Gradients. Report CSS67, AERE Harwell, Oxfordshire, 1979.

12. AJIZ, M.A., and JENNINGS, A. - A Robust ICCG Algorithm, Civil Engineering Dept. Report, Queens University, Belfast, 1981 (Fortran or Algol versions are available).

13. JENNINGS, A. - Matrix Computation for Engineers and Scientists, Wiley, London 1977.

PRECONDITIONED CONJUGATE GRADIENT METHODS FOR TRANSONIC FLOW CALCULATIONS

YAU SHU WONG
Institute for Computer Applications in Science and Engineering

MOHAMED M. HAFEZ
Joint Institute for Advancement of Flight Sciences
The George Washington University

ABSTRACT

The preconditioning technique has generally been accepted as an efficient procedure for accelerating the rate of convergence of an iterative method. One of the well-known examples is the preconditioned version of the conjugate gradient method for the solution of systems of linear equations. In this paper we study the application of the preconditioning technique for transonic flow problems, in which the governing equations are nonlinear and of mixed elliptic-hyperbolic type. Two iterative methods are presented, which are based on using a preconditioned conjugate gradient algorithm. Numerical results are reported which show that the present methods are computationally more efficient than the standard iterative procedure based on the successive line over-relaxation method.

1. INTRODUCTION

The study of transonic aerodynamics has received considerable attention in the recent past, this is due to the fact that modern transport aircrafts operate at transonic speeds. The mathematical formulation of the transonic flow problem is well known, but its solution is not straightforward to obtain, because the governing partial differential equations are nonlinear and of mixed type. The standard numerical procedure for transonic flow calculations

---

Research of the first author was supported by NASA Contract NAS1-15810 while he was in residence at ICASE, NASA-Langley Research Center, Hampton, VA 23665. The second author's work was supported by NASA-Langley Research Center Cooperative Agreement NCC1-24.

is based on the successive line overrelaxation method. However, not only does that method require an estimation of a relaxation parameter, but it also suffers from slow convergence as well. Alternative computational procedures such as the approximate factorization methods have been suggested recently [Holst and Ballhaus, 1979]. These methods have been shown to provide faster convergence rates than the successive line overrelaxation method if a set of iteration parameters is properly determined. Motivated by the difficulties in choosing optimal parameters a priori, it is therefore of strong interest to develop an efficient and reliable method which does not depend upon iteration parameters. Two such methods are presented in this paper based on using a preconditioning technique with the conjugate gradient method.

We discuss the basic mathematical formulation for the transonic flow problem in section 2, the standard relaxation method and the approximate factorization methods in section 3, two iterative procedures based on the preconditioned conjugate gradient algorithms in section 4, computational results in section 5, and finally, concluding remarks are given in section 6.

The purpose of this paper is to study the application of preconditioning techniques to accelerate the rate of convergence of iterative methods for transonic flow problems. Because of space limitations, many important aspects of the numerical solution for transonic flow calculations are not included.

2. TRANSONIC FLOW CALCULATIONS

2.1 Mathematical Formulation

The basic differential equation governing the flow of an inviscid, isentropic fluid is given by a kinematical relation representing the conservation of mass

$$\nabla \cdot \rho \vec{q} = 0 \qquad [1]$$

where

$$\vec{q} = \nabla \phi ,$$

$$\rho = (M_\infty^2 a^2)^{\frac{1}{\gamma-1}} ,$$

$$a^2 = \frac{1}{M_\infty^2} - \frac{\gamma-1}{2} (q^2-1) .$$

Here ϕ is the velocity potential, ρ, the density, M_∞, the Mach number at infinity, a, the local speed of sound and γ, the ratio of specific heats.

The tangential and wake boundary conditions, and the requirement that the velocity vanishes at infinity, complete the formulation of the governing equation.

For a two-dimensional flow in cartesian coordinates, Equation [1] can be expressed in the form

$$(\rho\phi_x)_x + (\rho\phi_y)_y = 0 \qquad [2]$$

This is known as the transonic full potential equation. In this paper we shall concentrate on the iterative solution of Equation [2]. From the mathematical point of view, the main difficulties associated with Equation [2] are as follows:

(a) the equation is nonlinear, where ρ is a function of ϕ_x and ϕ_y

(b) the equation is of mixed type: it changes from elliptic in subsonic regions to hyperbolic in supersonic regions, and the boundary between these regions is not known.

(c) the equation admits discontinuous solutions such as shocks which may exist in the flow.

(d) both compression and expansion shocks are admitted by the equation, and an additional condition must be introduced in order to eliminate the expansion shocks since they are physically meaningless.

For low Mach numbers ($M<1$) the flow is completely subsonic, and for high Mach numbers ($M>1$) the flow is completely supersonic (for airfoils with sharp leading and trailing edges). Figure 1 shows a typical transonic flow around an airfoil.

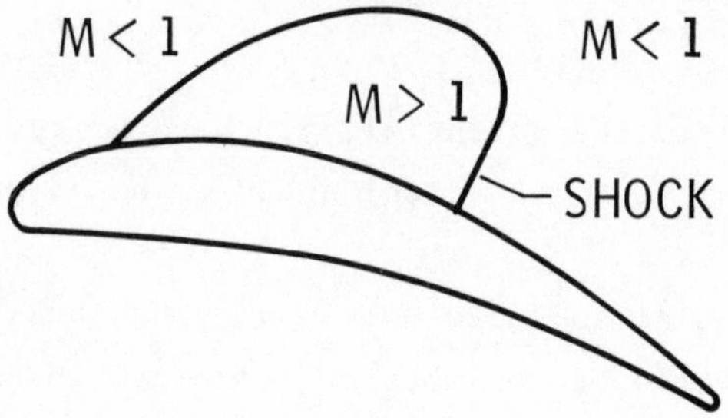

FIGURE 1. Transonic flow around an airfoil.

2.2 Numerical Procedures

It was first suggested by Murman and Cole [1971] that a type-dependent finite difference scheme could be used for transonic flow calculations. In their method central differences are used in subsonic regions, and upwind differences are used in the supersonic regions. It should be noted that an artificial viscosity is effectively introduced by using the upwind biased scheme in the supersonic regions, which in turn, is needed to eliminate the expansion shocks.

Another method, namely the artificial density method [Hafez, South and Murman, 1978], has recently been proposed where an artificial viscosity is easily implemented. In this method the transonic potential equation [2] is rewritten in the form

$$(\bar{\rho}\phi_x)_x + (\bar{\rho}\phi_y)_y = 0 \qquad [3]$$

where

$$\bar{\rho} = \rho - \mu\rho_s \Delta S \ ,$$

$$\mu = \max\ (0,\ 1 - a^2/q^2)\ .$$

Here ρ, a and q are defined as in Equation [1]. The term $\rho_s \Delta S$ is the product of the streamwise density gradient and the step length along a streamline. The use of $\bar{\rho}$ in Equation [3] produces a dissipative term when correct differencing is applied. It has been shown that retarding the density produces the same effect as the artificial viscosity introduced by the type-dependent schemes [Hafez, South and Murman, 1978].

Using the artificial density method, a central difference approximation can be applied to the modified equation [3] regardless of whether the region is subsonic or supersonic. An immediate consequence of this is that the regular structure of the matrix equation which results from the discretization of the linearized transonic equation is preserved. It should be noted that this property will be destroyed when an artificial viscosity is explicitly used. Because of this advantage the artificial density method is used in this paper.

3. ITERATIVE PROCEDURES

Assuming the density $\bar{\rho}$ is known from the previous iteration, a central difference approximation to Equation [3] is given by

$$(D_x^- \bar{\rho}_{i+\frac{1}{2},j} D_x^+ + D_y^- \bar{\rho}_{i,j+\frac{1}{2}} D_y^+)\ \phi_{i,j} = 0. \qquad [4]$$

where D_x^- and D_x^+ are the standard backward and forward difference operators in the x-direction, they are defined as follows.

$$D_x^- \phi_{i,j} = (\phi_{i,j} - \phi_{i-1,j})/\Delta x$$

$$D_x^+\phi_{i,j} = (\phi_{i+1,j} - \phi_{i,j})/\Delta x$$

Similar definitions are given for D_y^- and D_y^+.

The solution of the continuous problem is thus reduced to the solution of the following matrix equation

$$A(\phi^n)\ \phi^{n+1} = b \qquad [5]$$

where A is a symmetric positive definite matrix, whose elements are calculated from the previous solution ϕ^n.

A simple way to solve Equation [5] is by a first degree iterative scheme

$$\delta\phi^n = -\ \tau r^n \qquad [6]$$

where $\delta\phi^n = \phi^{n+1} - \phi^n$, $r^n = b - A(\phi^n)\ \phi^n$, and τ is an iteration parameter.

The simple iterative scheme [6] can be regarded as an iteration in pseudo-time, where the term $\delta\phi^n$ produces a time-step level of the scheme. If $\tau = \Delta t/\alpha$ then Equation [6] is a discretization for the equation

$$\alpha\phi_t = (\bar{\rho}\phi_x)_x + (\bar{\rho}\phi_y)_y \qquad [7]$$

Note that this scheme will converge for subsonic flow only. In order to obtain convergence for a transonic flow a ϕ_{xt} term must be included in the left hand side of Equation [7]. [Hafez and South, 1979].

In order to accelerate the convergence of the iterative scheme, a preconditioning matrix M, can be introduced. The preconditioned version of Equation [6] is

$$\delta\phi = -\ \tau M^{-1}\ r^n$$

or

$$M\delta\phi = -\tau r^n \quad [8]$$

It is not hard to see that the matrix M determines the rate at which the iterative scheme converges. Generally speaking, if M is a good approximation of A, then a fast convergence rate can be expected.

To implement the preconditioned scheme [8] effectively, the matrix M must be easily invertible. A common approach is to factor M such that

$$M = M_1 M_2 \quad [9]$$

and the product of the factors is an approximation to A. Furthermore, M_1 and M_2 should have a simple structure, such as a triangular matrix or a tridiagonal matrix.

We show that the successive line overrelaxation method and the approximate factorization methods can be described by the iterative scheme [8]. The main difference between these methods is in the choice of the preconditioning matrix M, and consequently, different rates of convergence result.

3.1 Successive Line Overrelaxation (SLOR) Method.

The SLOR method has been regarded as the standard iterative procedure for transonic flow calculations. The SLOR algorithm can be expressed by Equation [8], where

$$M\delta\phi_{i,j} = \left[\frac{1}{\Delta x}\left(-\frac{\overline{\rho}_{i+\frac{1}{2},j}}{\Delta x} - \overline{\rho}_{i-\frac{1}{2},j} D_x^-\right) + \left(D_y^- \overline{\rho}_{i,j+\frac{1}{2}} D_y^+\right)\right]\delta\phi_{i,j} \quad [10]$$

It should be noted that the scheme is semi-implicit, in the sense that it is explicit in the x-direction, and is implicit in the y-direction where an inversion of a tridiagonal matrix equation for a given value of i is required. Notice that $D_x^-\delta\phi_{i,j}$ in Equation [10] generates a ϕ_{xt} term, which is needed for the convergence in the supersonic regions.

Although the SLOR algorithm is reliable for transonic flow calculations, its convergence rate is slow for many practical problems. It is known that the rate of convergence of a fully implicit scheme can be faster than that of a semi-implicit or explicit iteration scheme. In the following, different pre-conditioning matrices M are studied, their choice will make the iterative scheme described in [8] become fully implicit.

3.2 Alternating Direction Implicit (ADI) Method.

Using the standard ADI scheme the matrix M can be constructed as follows

$$\alpha M\delta\phi_{i,j} = -(\alpha - D_x^- \bar{\rho}_{i+\frac{1}{2},j} D_x^+)(\alpha - D_y^- \bar{\rho}_{i,j+\frac{1}{2}} D_y^+)\ \delta\phi_{i,j} \qquad [11]$$

where α is an iteration parameter.

It is easily seen that the scheme is fully implicit. Multiplying the factors in Equation [11], it gives

$$\alpha M\delta\alpha_{i,j} = \alpha\left(-\alpha - \frac{D_x^- \bar{\rho}_{i+\frac{1}{2},j} D_x^+ D_y^- \bar{\rho}_{i,j+\frac{1}{2}} D_y^+}{\alpha} + D_x^- \bar{\rho}_{i+\frac{1}{2},j} D_x^+ + D_y^- \bar{\rho}_{i,j+\frac{1}{2}} D_y^+\right)\delta\phi_{i,j} \qquad [12]$$

Thus M is an approximation to the original problem, and the first two terms in the bracket represent the errors associated with the ADI scheme. It should be noted that there is no ϕ_{xt} term in this scheme, instead the first error term $\alpha\delta\phi_{i,j}$ produces a ϕ_t term in the iteration.

3.3 Approximate Factorization (AF) Method.

The AF method was first studied by Ballhaus and Steger [1975] for the transonic small disturbance equation, and it subsequently was applied to the transonic full potential equation by Holst and Ballhaus [1979]. In the AF scheme the matrix M is chosen as follows

$$\alpha M\delta\phi_{i,j} = -(\alpha D_x^- - D_y^- \bar{\rho}_{i,j+\frac{1}{2}} D_y^+)(\alpha - \bar{\rho}_{i+\frac{1}{2},j} D_x^+)\ \delta\phi_{i,j} \qquad [13]$$

Multiplying the factors of M gives

$$\alpha M\delta\phi_{i,j} = \alpha\left(-\alpha D_x^- - \frac{D_y^- \ \bar{\rho}_{i,j+\frac{1}{2}} D_y^+ \ \bar{\rho}_{i+\frac{1}{2},j} D_x^+}{\alpha}\right.$$

$$\left. + D_x^- \bar{\rho}_{i+\frac{1}{2},j} D_x^+ + D_y^- \ \bar{\rho}_{i,j+\frac{1}{2}} \ D_y^+\right) \delta\phi_{i,j} \qquad [14]$$

It is important to note that unlike the ADI scheme the first error term $\alpha D_x^- \delta\phi_{i,j}$ in Equation [14] does produce a ϕ_{xt} term in the iteration.

In addition to the relaxation parameter τ associated with the iterative scheme [8], both ADI and AF methods require an estimation of the parameter α. It has been found that the high-frequency and low-frequency errors can be reduced effectively by choosing a suitable sequence for α.

Computational experiments show that the ADI scheme provides an excellent convergence rate for subsonic flow calculations. However, it is difficult to obtain convergence for cases of transonic flows. Divergence of the ADI scheme on nonuniform grids has been observed [South, 1981]. On the other hand, the AF scheme gives a good convergence rate for transonic flows provided the values of τ and α are properly chosen. The key success of the AF scheme is associated with the fact that a ϕ_{xt} term which is needed for convergence in the supersonic regions is included in this scheme.

It should be mentioned that another implicit approximate factorization scheme has recently been investigated by Sankar et al [1981], in which the factors of M are based on using the strongly implicit procedure [Stone, 1968]. Like the AF method the strongly implicit procedure also requires an estimation of a sequence of parameters α.

4. PRECONDITIONED CONJUGATE GRADIENT METHOD

In the previous section we have described several iterative procedures for transonic flow calculations. All of these methods can be regarded as using a preconditioning technique for a first-degree iterative scheme. The main difficulty in the application of these schemes efficiently is the requirement of choosing optimal parameters for τ and α. It should be pointed out that improper values for these parameters would lead to a slow convergence or even a divergence for the iterative scheme. To overcome this difficulty we shall now consider the method of conjugate gradients (CG).

The CG method was first proposed by Hestenes and Stiefel [1952] for solving a symmetric and positive definite system of linear equation. The basic CG algorithm for the solution of Equation [5] is as follows.

Let ϕ^o be an arbitrary vector, compute

$$r^o = b - A\phi^o \quad , \; p^o = r^o$$

Then for n=0, 1, 2, ..., compute

$$\phi^{n+1} = \phi^n + \alpha_n p^n$$

$$r^{n+1} = r^n - \alpha_n Ap^n \qquad [15]$$

where

$$\alpha_n = (r^n,r^n)/(p^n,Ap^n)$$

$$p^{n+1} = r^{n+1} + \beta_n p^n$$

where

$$\beta_n = (r^{n+1},r^{n+1})/(r^n,r^n)$$

Note that no estimation of iteration parameters is required in the CG algorithm. Other important properties and features of this method can be found in Reid [1971].

The CG algorithm given in Equations [15] can be rewritten in the following three-terms recurrence relation

$$(\phi^{n+2}-2\phi^{n+1}+\phi^{n})+(1-\frac{\alpha_{n+1}}{\alpha_n}\beta_n)(\phi^{n+1}-\phi^{n}) = \alpha_{n+1}r^{n+1}$$

or

$$\delta^2\phi + \omega\delta\phi = -\eta r^{n+1} \qquad [16]$$

Clearly it is a second-degree iterative scheme. If $\beta_n=0$ for all n, Equation [16] then reduces to a first degree scheme.

Over the last few years the preconditioning technique has been successfully applied with the CG algorithm for solving large sparse systems of linear equations [Evans, 1968; Axelsson, 1974; Concus, Golub and O'Leary, 1976; Meijerink and van der Vorst, 1976 Wong, 1979, etc.]. Let M be the preconditioning matrix, and consider the system

$$M^{-1}A\phi = M^{-1}b \qquad [17]$$

The preconditioned CG (PCG) algorithm for Equation [17] is as follows.

Let ϕ^o be an arbitrary vector, compute

$$r^o = b - A\phi^o ,$$

Solve $Mz^o = r_o$, Set $p^o = z^o$

Then for n = 0, 1, 2, ... compute

$$\phi^{n+1} = \phi^n + \alpha_n p^n$$

$$r^{n+1} = r^n - \alpha_n Ap^n$$

where

$$\alpha_n = (r^n, z^n)/(p^n, Ap^n) \qquad [18]$$

Solve $\qquad Mz^{n+1} = r^{n+1}$

$$p^{n+1} = z^{n+1} + \beta_n p^n$$

where $\qquad \beta_n = (r^{n+1}, z^{n+1})/(r^n, z^n).$

The PCG algorithm can be viewed as a preconditioning version of a second-degree iterative scheme, namely

$$M\,\delta^2\phi + \omega M\delta\phi = -\,\eta r^{n+1} \qquad [19]$$

Note that at each step in the PCG algorithm, the solution of the linear system Mz=r is required. In order to solve this matrix equation efficiently, M is usually chosen to be the product of M_1M_2 as defined in [9]. Different factorization for the matrix M lead to different PCG algorithms.

In this paper we shall consider two types of preconditioning matrices, in which M is an approximate factorization of A and $M=M_1M_2$.

(a) Row-sum agreement factorization [Wong, 1979]. In this factorization the following conditions are satisfied:

I) $(M_1)_{i,j}=(A_L)_{i,j}$ and $(M_2)_{i,j}=(A_U)_{i,j}$ where M_1 and M_2 are a lower and upper triangular matrices respectively, and M_1 and M_2 have nonzero elements only in those positions which correspond to the nonzero elements in the lower or upper triangular part of A.

II) For $A_{i,j}\neq 0$ and $i\neq j$, $M_{i,j}=A_{i,j}$ i.e. the off-diagonal elements of M whose locations correspond to the nonzero off-diagonal elements of A are set to those values.

III) The row-sums of M are the same as those of A.

(b) Symmetric successive overrelaxation (SSOR) preconditioning [Axelsson,1974].

Let A be written in the form

$$A = D - L - U$$

where D is a diagonal matrix, and L and U are strictly lower and upper triangular matrix respectively, then the preconditioning matrix M can be expressed by

$$M = \frac{1}{\omega(2-\omega)} (D-\omega L)\ D^{-1}\ (D-\omega U)$$

where ω is a relaxation parameter. The rate of convergence of the PCG method is not as sensitive as to choice of ω in the SOR iterative scheme.

Since the transonic potential equation [2] is nonlinear in nature, the solution of a nonlinear system of algebraic equations is required. The solution of this nonlinear problem can be obtained by solving a sequence of linearized systems of equations, however, the matrices generated by these linearized systems are different for each iteration step. In the problem considered here, the matrix equation is modified according to the solution obtained from the previous iteration. Consequently, the preconditioning matrix M will vary from iteration to iteration. The requirement for calculating the factorizations of M at each iteration step will increase the computational work considerably; thus, it would be very advantageous to choose a preconditioning matrix independent of the iteration step.

Recall that the matrix A in Equation [5] corresponds to the finite-difference approximation to

$$\partial_x(\bar{\rho}\partial_x) + \partial_y(\bar{\rho}\partial_y) \qquad [20]$$

Now consider another matrix B which results from the discretization for the following operator

$$\bar{\rho}\ (\partial_{xx} + \partial_{yy}) \qquad [21]$$

Let the preconditioning matrix M be an approximate factorization to B instead of A, with

$$M = M_1 M_2 M_3 \ , \qquad [22]$$

where M_1 is a diagonal matrix whose elements are given by $\bar{\rho}_{i,j}$, and the product of $M_2 M_3$ is an approximation to the operator $(\partial_{xx}+\partial_{yy})$. The elements of M_2 and M_3 can be computed by either the row-sum agreement factorization or the SSOR preconditioning, and these values remain unchanged throughout the calculations.

The preconditioning based on [22] and [21] does not include a ϕ_{xt} term in the iterative scheme. However, if B is modified so that it is a difference approximation to

$$\bar{\rho}(\partial_{xx} + \partial_{yy} + \varepsilon\partial_x) \ , \qquad [23]$$

where ε is a parameter to be discussed shortly. Then a ϕ_{xt} term is introduced. The preconditioning matrix M is defined in Equation [22], where M_1 is the diagonal matrix denoting the density elements at each grid point, the product $M_2 M_3$ is an approximate factorization to $(\partial_{xx} + \partial_{yy} + \varepsilon\partial_x)$.

In this paper two iterative procedures based on PCG methods are investigated.

(a) SLOR + CG scheme

This is a combination of the SLOR scheme and the PCG algorithm, the preconditioning matrix is based on [21] and [22]. The relaxation step is needed for convergence in cases of transonic flows because the PCG algorithm with the factorization given in [21] and [22] does not include a ϕ_{xt} term.

The present combined iterative scheme is similar to that proposed by Jameson [1976], in which a relaxation scheme and a fast solver are used. However, there are many advantages in using the CG method over the fast solver routine:

I) There is no restriction on the grid in any direction in the computational domain.

II) Since only the overall convergence for the nonlinear problem is of interest, it is sufficient to use a small number of iterations in the PCG algorithm for the solution of the linearized system. This in turn provides a considerable saving in the computational work.

III) The PCG algorithm can be easily extended for three-dimensional problems.

(b) PCG scheme

With the modified approximate factorization [22] and [23], a ϕ_{xt} term is explicitly included. Consequently a convergence for this purely PCG scheme for subsonic and transonic flow calculations can be ensured.

Note that if the parameter ε in [23] is a non-zero constant, then a ϕ_{xt} term will be introduced in both subsonic and supersonic regions. The ϕ_{xt} term is needed for convergence in the supersonic region only; however, its presence in the subsonic regions may decrease the convergence rate of the iterative scheme. To overcome this problem, ε is chosen as follows

$$\varepsilon = \frac{-1}{\Delta x}\,\mu \qquad [24]$$

where

$$\mu = \max(0,\ 1 - \frac{1}{M^2}) \quad ,$$

$$M^2 = \frac{a^2}{q^2} \quad .$$

a^2 and q^2 are defined in Equations [1] and [2]. The parameter μ can be regarded as a switching function, such that ε is zero in subsonic regions but nonzero in supersonic regions. With this definition for ε, the factorization of M must be modified according to the development of the supersonic regions. However, it is sufficient to update the approximate factorization for every k step, and k=5 will be used in the numerical experiments reported in the next section.

5. COMPUTATIONAL RESULTS

In order to compare the performance of the PCG methods and the standard iterative procedure based on the SLOR method, two test problems for transonic potential calculations around a circular cylinder and NACA 0012 airfoil are examined. Particular attention is focused on the comparison of rates of convergence for subsonic and transonic flow problems. The CPU times in seconds on the CYBER-175 computer are also reported. In all cases the maximum residual, R_{max}, is used to measure the convergence of the iterative process.

Preconditioning techniques based on the SSOR method and the row-sum agreement factorization have been compared, and the convergence rate of the former is slower than that of the latter for all problems tested. Thus the results for the PCG method presented here will be based on the row-sum agreement factorization to determine the preconditioning matrix M.

5.1 Flow Calculations Around a Circular Cylinder

The SLOR and the combined SLOR + CG methods are compared. A 61x31 grid is used in all cases with a uniform mesh in the θ-direction and a 15 percent stretching in the r-direction. Figures 2 and 3 show the rates of convergence of the two methods for subsonic and transonic cases respectively. The pressure distributions around the circular cylinder are plotted in Figure 4. Figure 5 shows the rates of convergence for different values of ω, the relaxation parameter in the SLOR method. Note that the convergence for the combined SLOR + CG method is almost the same for a wide range of ω.

5.2 Flow Calculations Around NACA 0012 Airfoil at Zero Angle of Attack

A 61x31 grid is used in all calculations, and the Mach number

M_∞=0.6 corresponds to a subsonic flow, while M_∞=0.85 corresponds to a transonic flow. Figures 6 and 7 show a comparison of the SLOR and SLOR + CG methods for two different grid systems. In Figure 6 a highly stretched grid is used which corresponds to stretching the physical domain from $-\infty$ to $+\infty$ in the x-direction and from 0 to $+\infty$ in the y-direction. In contrast, Figure 7 corresponds to a finite physical domain, in which uniform grids are used in both x- and y-directions. It is clear that the combined SLOR + CG method is not sensitive to the stretched grids. Figure 8 shows the results for the purely PCG method. The corresponding pressure distributions are shown in Figure 9.

6. CONCLUSION

Iterative procedures for subsonic and transonic flow calculations have been studied in this paper. It appears that current existing procedures, which include the SLOR, ADI, AF methods, can be regarded as the application of a preconditioning technique to a first degree iterative scheme. All these methods require an estimation of iteration parameters in order to obtain an optimal convergence rate. On the other hand, we show that the PCG algorithm provides a second degree iterative scheme, in which no estimation of any parameter is needed. This particular feature provides the basis for an efficient and reliable iterative procedure. Even for the case of the SLOR + CG scheme, the convergence of this combined method is not sensitive to the relaxation parameter as is the case for the SLOR scheme. For all problems tested here, the PCG methods provide faster rates of convergence than the SLOR method. They give excellent results for subsonic cases, and gives a modest improvement for transonic cases as well. It should be mentioned that the PCG methods have also been tested for transonic finite element calculations with similar success [Wong and Hafez, 1981]. It is important to note that the present PCG methods are not sensitive to the stretched grids, and their power increases when a finer grid is used and when a higher accuracy is required. It holds promise for 3D calculations as well.

REFERENCES

1. AXELSSON, O. - On Preconditioning and Convergence Acceleration in Sparse Matrix Problems, Report CERN 74-10, Geneva, Switzerland, 1974.

2. BALLHAUS, W. F., and STEGER, J. L. - Implicit Approximate Factorization for the Low-Frequency Transonic Equation. NASA TMX-73, 082, 1975.

3. EVANS, D. J. - The Use of Pre-conditioning in Iterative Methods for Solving Linear Equations with Symmetric Positive Definite Matrices, J. Inst. Maths. Applics., Vol. 4, pp. 295, 1968.

4. HAFEZ, M. M., SOUTH, J. C., and MURMAN, E. M. - Artificial Compressibility Methods for Numerical Solution of Transonic Full Potential Equation. AIAA Paper No. 78-1148, 1978.

5. HAFEZ, M. M., and SOUTH, J. C. - Vectorization of Relaxation Methods for Solving Transonic Full Potential Equation, GAMM Workshop on Numerical Methods for the Computation of Inviscid Transonic Flow with Shock Waves, Stockholm, Sweden, 1979.

6. HESTENES, M. R., and STIEFEL, E. - Methods of Conjugate Gradients for Solving Linear Systems, J. Res. Nat. Bur. Stand., Vol. 49, pp. 409, 1952.

7. HOLST, T. L., and BALLHAUS, W. F. - Fast Conservative Schemes for the Full Potential Equation Applied to Transonic Flow, AIAA J., Vol. 17, pp. 1038-1045, 1979.

8. JAMESON, A. - Transonic Flow Calculations, VKI Lecture Series 87, March 1976.

9. MEIJERINK, J. A., and VAN DER VORST, H. A. - An Iterative Solution Method for Linear Systems of Which the Coefficient Matrix is a Symmetric M-Matrix, Math. Comput., Vol. 31, pp. 148-162, 1977.

10. MURMAN, E. M., and COLE, J. D. - Calculation of Plane Steady Transonic Flow, AIAA J., Vol. 9, pp. 114-121, 1971.

11. REID, J. K. - On the Method of Conjugate Gradients for the Solution of Large Sparse Systems of Linear Equations, Large Sparse Sets of Linear Equations, Ed. Reid, J. K., Academic Press, 1971.

12. SANKAR, N. L., MALONE, J. B., and TASSA Y. - An Implicit Conservative Algorithm for Steady and Unsteady Three Dimensional Transonic Potential Flows, AIAA 5th Computational Fluid Dynamics Conference Proceedings, 1981.

13. SOUTH, J. C. - Private Communication, 1981.

14. STONE, H. L. - Iterative Solution of Implicit Approximations of Multi-Dimensional Partial Differential Equations, SIAM J. Numer. Anal., Vol. 5, pp. 530, 1960.

15. WONG, Y. S. - Pre-conditioned Conjugate Gradient Methods for Large Sparse Matrix Problems, Numerical Methods in Thermal Problems, Ed. Lewis, R. W., and Morgan, K., Pineridge Press, 1979.

16. WONG, Y. S. and HAFEZ, M. M. - Application of Conjugate Gradient Methods to Transonic Finite Difference and Finite Element Calculations, AIAA 5th Computational Fluid Dynamics Conference Proceedings, 1981.

17. CONCUS, P., GOLUB, G. H., O'LEARY, D. P. - A Generalized Conjugate Method for the Numerical Solution of Elliptic Partial Differential Equations, Sparse Matrix Computations, Ed. Bunch, J. R. and Rose, D. J., Academic Press, 1976.

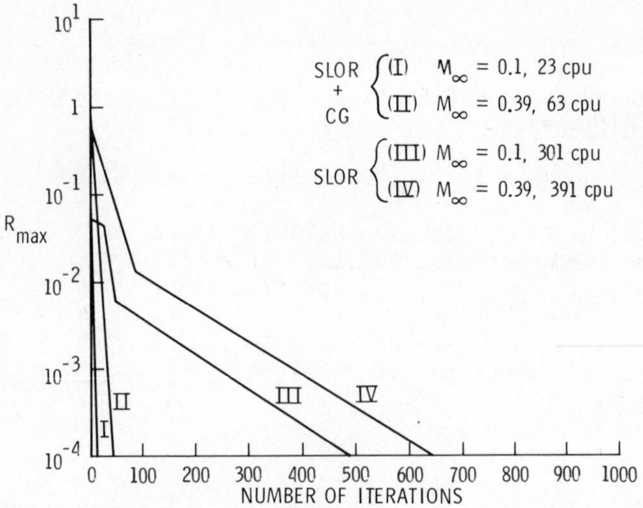

FIGURE 2. Rates of convergence for subsonic calculations around a circular cylinder.

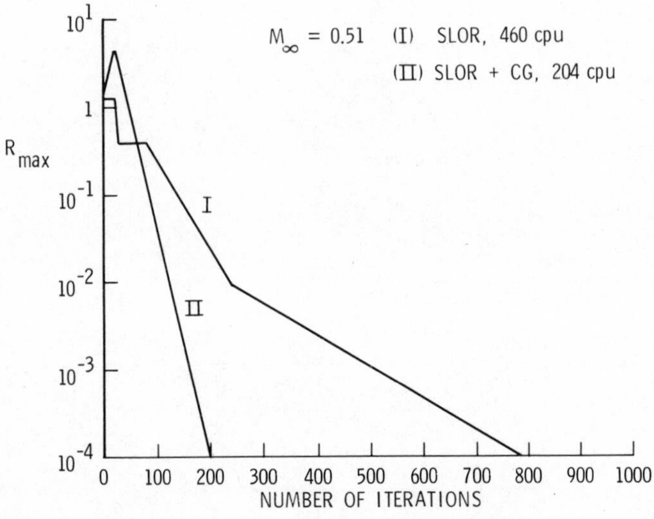

FIGURE 3. Rates of convergence for transonic calculations around a circular cylinder.

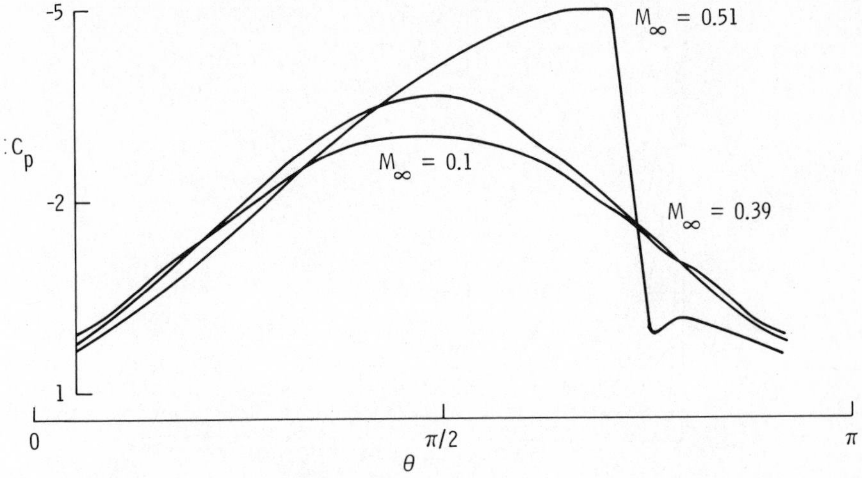

FIGURE 4. Pressure distributions around a circular cylinder at different Mach numbers.

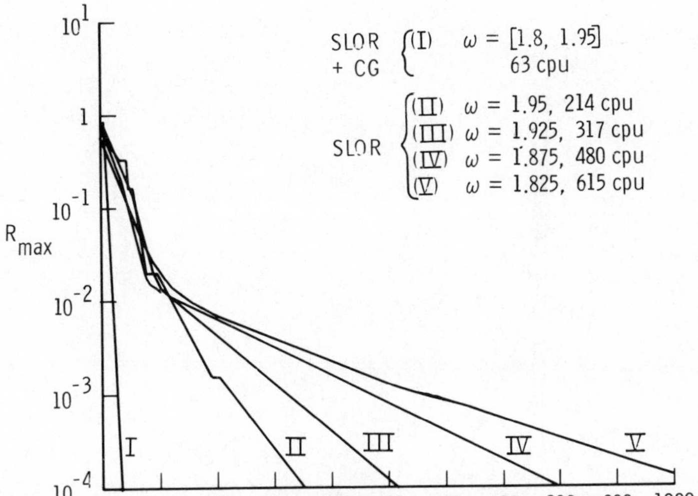

FIGURE 5. Dependence of convergence rates on relaxation parameter ω at $M_\infty = 0.39$.

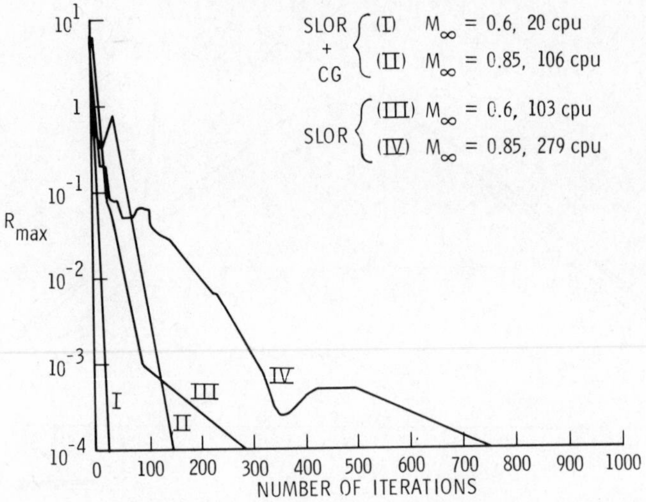

FIGURE 6. Rates of convergence for flow calculations around NACA 0012 airfoil with highly stretched grid.

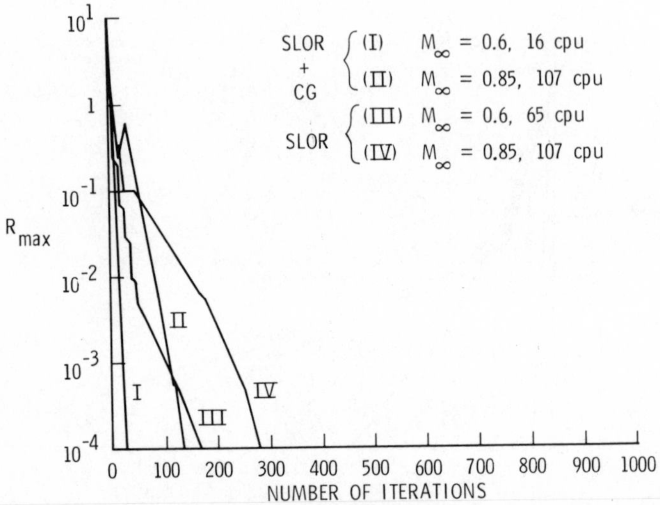

FIGURE 7. Rates of convergence for flow calculations around NACA 0012 airfoil with uniform grid.

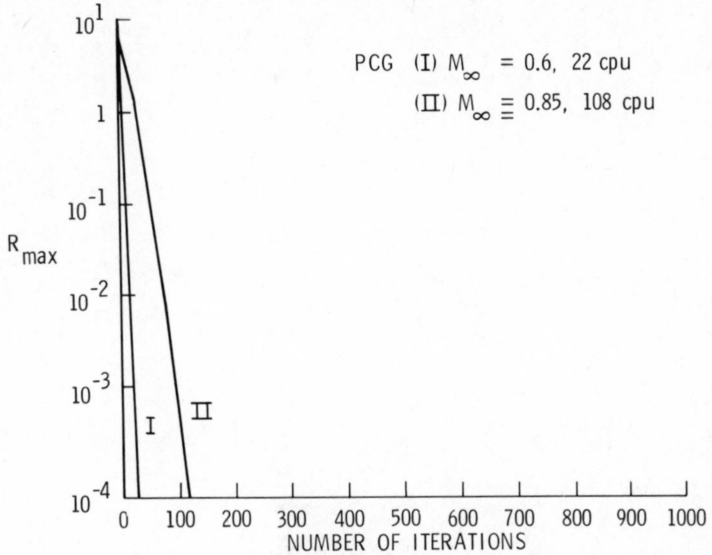

FIGURE 8. Rates of convergence for flow calculations around NACA 0012 airfoil.

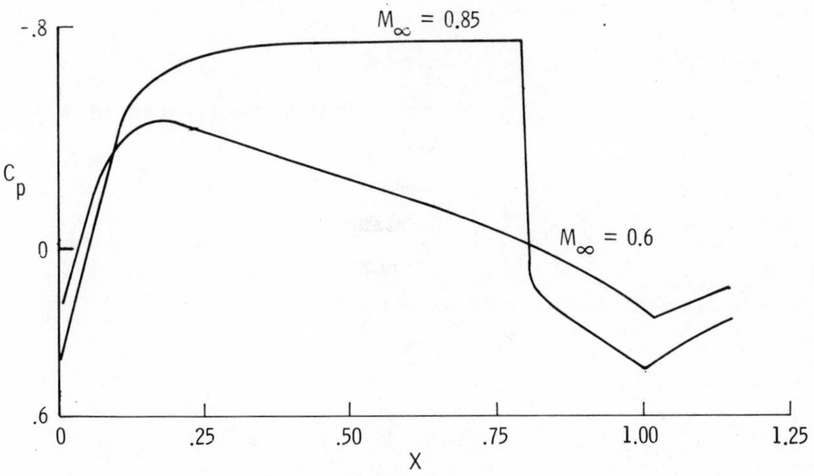

FIGURE 9. Pressure distributions around NACA 0012 airfoil at different Mach numbers.

AN EFFICIENT PRECONDITIONED CONJUGATE GRADIENT METHOD. APPLICATION TO THE SOLUTION OF NONLINEAR PROBLEMS IN FLUID DYNAMICS

Roland GLOWINSKI*, Jacques PERIAUX**, Olivier PIRONNEAU***
* Paris VI University, LA 189, Place Jussieu, 75230 Paris Cedex 05
** AMD/BA, B.P. 300, 92214 St-Cloud, France
*** Paris-Nord University, Place du 8 Mai 1945, 93200 St-Denis, France

ABSTRACT

We describe in this paper a new preconditioning technique, which has very good performances, compared to more classical methods, when applied to problems in which one has to solve a large number of linear systems associated to the same matrix. Applications to nonlinear problems in Fluid Dynamics illustrate the possibilities of the method.

1. INTRODUCTION.

The numerical solution of large industrial problems modelled by nonlinear partial differential equations leads, after convenient *finite element* approximations, to very large (10^4 unknowns and more) finite dimensional nonlinear systems. These last years conjugate gradient methods with scaling have become an efficient tool (cf. BARTELS-DANIEL [1], DOUGLAS-DUPONT [2], CONCUS-GOLUB-O'LEARY [3]), assuming that iterations are done with a convenient euclidian metric related to the so-called scaling or preconditioning matrix.

The size and complexity of industrial applications have motivated several authors to use incomplete Cholesky factorizations in order to avoid out of core solution methods. Let us mention among these authors AXELSSON [4], MEIJERINK-VAN DER VORST [5], KERSHAW [6], MANTEUFEL [7], WESSELING-SONNEFELD [8].

The main goal of this paper is to present an incomplete Cholesky factorization method of a new kind and apply it to the *in core* solution of nonlinear problems arising from Fluid Mechanics applications ; using this method we shall avoid time consuming transfers between externals disks and the core memory of the computer.

Numerical experiments will show the efficiency of this new technique ; details of comparisons to more classical incomplete factorization methods can be found in GLOWINSKI-PERIAUX-PIRONNEAU [9] (see also GLOWINSKI-MANTEL-PERIAUX-PIRONNEAU-POIRIER [10], GLOWINSKI-MANTEL-PERIAUX-PERRIER-PIRONNEAU [11]).

2. GENERALITIES ON THE PRECONDITIONED SOLUTION OF BOUNDARY VALUE PROBLEMS.

We shall discuss in this section the numerical solution of boundary value problems by iterative methods with preconditioning ; to avoid abstraction we have chosen to discuss the solution of two families of test problems.

2.1. A first test problem. The linear heat equations.

Let Ω be a bounded domain of $\mathbb{R}^N$ (N=1,2,3 in practice) and let Γ be its boundary ; using a convenient system of physical units, *heat conduction* phenomenon are described ($]0,T[$ is a *time* interval) by

$$\frac{\partial\phi}{\partial t} - \Delta\phi = f \text{ in } \Omega\times]0,T[, \tag{2.1}$$

$$\phi = g \text{ on } \Gamma\times]0,T[, \tag{2.2}$$

$$\phi(x,0) = \phi_o(x) \text{ in } \Omega. \tag{2.3}$$

Using the standard implicit Euler time discretization scheme we derive from (2.1)-(2.3) the following semi-discretization scheme where $\phi^n(x) \simeq \phi(x,n\Delta t)$,

$$\phi^o = \phi_o, \tag{2.4}$$

and for $n \geq 0$,

$$(\frac{I}{\Delta t} - \Delta)\phi^{n+1} = f^{n+1} + \frac{\phi^n}{\Delta t} \text{ on } \Omega, \tag{2.5}$$

$$\phi^{n+1} = g^{n+1} \quad \textit{on } \Gamma \tag{2.6}$$

(with $f^n(x) = f(x,n\Delta t)$, $g^n(x) = g(x,n\Delta t)$).

Using then a convenient *space discretization* (by finite elements or finite differences) we obtain a *fully discrete* approximation of (2.1)-(2.3) by

$$\phi_h^o = \phi_{oh} \tag{2.7}$$

and for $n \geq 0$

$$A_h\phi_h^{n+1} = F_h^{n+1} , \tag{2.8}$$

where (2.8) approximates simultaneously (2.5),(2.6).

The above matrix A_h is *symmetric, positive definite, sparse,* and in the present case *independent* of n ; in practice this N×N matrix A_h may be such that $N > 10^4$; from these properties it is therefore very tempting to use a *Cholesky factorization* $A_h = L_h L_h^t$, done once and for all, to solve the sequence of problems (2.8). A most important reason to use Cholesky factorization is that it preserves the bandwidth. In fact even if L_h has the same bandwidth than A_h it contains in general much more non zero elements. We have represented on Figure 2.1 the matrices A_h and L_h^t with a visualization of the zero and non zero elements ; this figure illustrates very well the structure of A_h and L_h^t (therefore L_h) concerning the distribution of the zero and non zero elements.

From the above example it appears that there are situations in which it is possible to store A_h *in core* (in fact its lower triangular part, for example, if A_h is symmetric) but not L_h ; therefore solving the two triangular systems associated with the solution of (2.8), namely

$$L_h z_h^{n+1} = F_h^{n+1}, \tag{2.9}$$

$$L_h^t \phi_h^{n+1} = z_h^{n+1}, \tag{2.10}$$

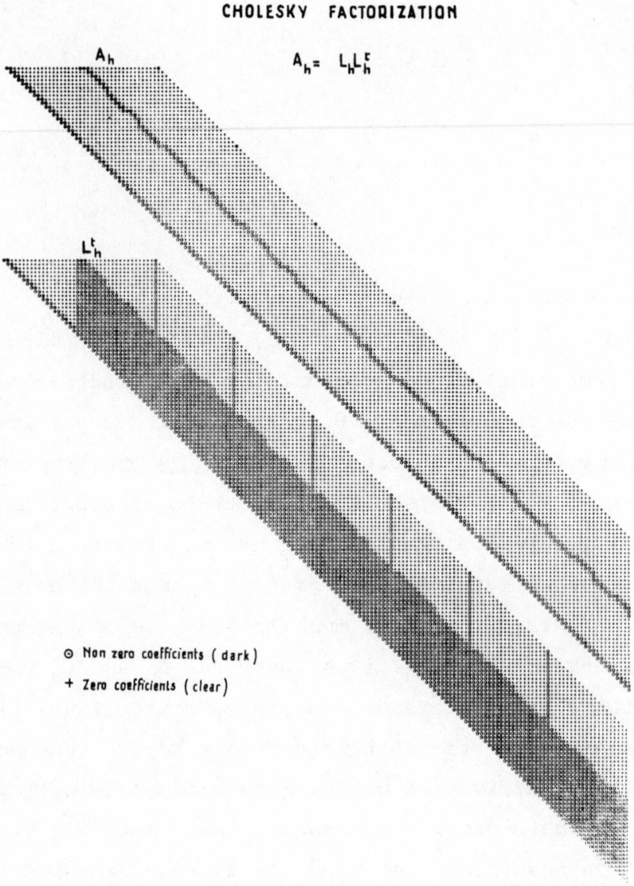
CHOLESKY FACTORIZATION
A_h
$A_h = L_h L_h^t$
L_h^t
⊙ Non zero coefficients (dark)
+ Zero coefficients (clear)

FIGURE 2.1

will require tape or disk transfers, resulting in a possibly prohibitive computation time for very large problems. A classical remedy to the above difficulty is to introduce a matrix $\tilde{L}_h$, *sparser than* L_h, which is such that $\tilde{A}_h$ defined by

$$\tilde{A}_h = \tilde{L}_h \tilde{L}_h^t , \tag{2.11}$$

is spectrally as close as possible to A_h ; then (omitting indices n and h) one may solve (2.8) by

$$\phi_o \textit{ given}, \tag{2.12}$$

then for $m \geq 0$,

$$\tilde{A}\phi_{m+1} = (\tilde{A}-A)\phi_m + F, \tag{2.13}$$

or by conjugate gradient variants of (2.12),(2.13).

In order to construct $\tilde{L}_h$ we consider two types of situations :

(i) For a given matrix A_h the number of problems (2.8) that we have to solve (differing one to each other by their right hand sides, only) *is not very large* ; in that case we highly recommend *approximate factorization methods* like those discussed in, e.g., [5],[6],[7].

(ii) The number of problems that we have to solve for the same A_h is very large ; in that latter case it may be interesting to construct a "better" $\tilde{L}_h$ via a more costly process, including for example as a preliminary step the construction of L_h. This point of view will be discussed in Sec. 3.

2.2. Description of a preconditioned conjugate gradient algorithm for solving linear equations.

Before discussing a second family of examples, related to nonlinear problems, we shall describe a conjugate gradient algorithm with preconditioning to solve *linear systems* whose matrix is *symmetric* and *positive definite*. Suppose that this system is

$$\underset{\sim}{A}\underset{\sim}{\phi} = \underset{\sim}{F} \tag{2.14}$$

where $\underset{\sim}{A}$ is an $N \times N$ *symmetric matrix, positive definite,* and where $\underset{\sim}{\phi}, \underset{\sim}{F} \in \mathbb{R}^N$; we denote by $(\cdot,\cdot)$ the usual scalar product of $\mathbb{R}^N$. Then (2.14) is equivalent to the minimization problem

$$\begin{cases} \textit{Find } \underset{\sim}{\phi} \in \mathbb{R}^N \textit{ such that} \\ J(\underset{\sim}{\phi}) \le J(\underset{\sim}{\psi}) \quad \forall \underset{\sim}{\psi} \in \mathbb{R}^N \end{cases} \tag{2.15}$$

where $J : \mathbb{R}^N \to \mathbb{R}$ is defined by

$$J(\underset{\sim}{\psi}) = \frac{1}{2}(\underset{\sim}{A}\underset{\sim}{\psi},\underset{\sim}{\psi}) - (\underset{\sim}{b},\underset{\sim}{\psi}). \tag{2.16}$$

Let $\underset{\sim}{S}$ be a second matrix, which is also symmetric and *positive definite* ; using $\underset{\sim}{S}$ as a preconditioner we can solve (2.14), (2.15) using the following *conjugate gradient* algorithm

<u>Step 0</u> : <u>Initialization</u>

$$\underset{\sim}{\phi}^o \textit{ arbitrarily given in } \mathbb{R}^N, \tag{2.17}$$

$$\underset{\sim}{g}^o = \underset{\sim}{S}^{-1}(\underset{\sim}{A}\underset{\sim}{\phi}^o - \underset{\sim}{F}), \tag{2.18}$$

$$\underset{\sim}{w}^o = \underset{\sim}{g}^o, \tag{2.19}$$

then for $n \ge 0$, *assuming that* $\underset{\sim}{\phi}^n, \underset{\sim}{g}^n, \underset{\sim}{w}^n$ *are known we compute* $\underset{\sim}{\phi}^{n+1}$, $\underset{\sim}{g}^{n+1}, \underset{\sim}{w}^{n+1}$ *using*

<u>Step 1</u> : <u>Descent</u>

$$\underset{\sim}{\phi}^{n+1} = \underset{\sim}{\phi}^n - \lambda_n \underset{\sim}{w}^n, \tag{2.20}$$

where

$$\lambda_n = \operatorname*{Arg\,Min}_{\lambda \in \mathbb{R}} J(\underset{\sim}{\phi}^n - \lambda \underset{\sim}{w}^n)$$

i.e.

$$\lambda_n = \frac{(\underset{\sim}{S}\underset{\sim}{g}^n,\underset{\sim}{w}^n)}{(\underset{\sim}{A}\underset{\sim}{w}^n,\underset{\sim}{w}^n)} \quad \left(= \frac{(\underset{\sim}{S}\underset{\sim}{g}^n,\underset{\sim}{g}^n)}{(\underset{\sim}{A}\underset{\sim}{w}^n,\underset{\sim}{w}^n)}\right) . \tag{2.21}$$

Step 2 : Calculation of the new descent direction

$$\underset{\sim}{g}^{n+1} = \underset{\sim}{g}^n - \lambda_n \underset{\sim}{S}^{-1} \underset{\sim}{A}\underset{\sim}{w}^n, \tag{2.22}$$

$$\gamma_{n+1} = \frac{(\underset{\sim}{S}\underset{\sim}{g}^{n+1},\underset{\sim}{g}^{n+1})}{(\underset{\sim}{S}\underset{\sim}{g}^n,\underset{\sim}{g}^n)} , \tag{2.23}$$

$$\underset{\sim}{w}^{n+1} = \underset{\sim}{g}^{n+1} + \gamma_{n+1}\underset{\sim}{w}^n , \tag{2.24}$$

$n = n+1$, *go to* (2.20).

The convergence of (2.17)-(2.24) in at most N iterations (if round-off errors are neglected) is a well-known result.

Remark 2.1 : The closer is S to A the faster is the convergence of (2.17)-(2.24) with regard to the number of iterations (if $\underset{\sim}{S} = \underset{\sim}{A}$ we need only *one* iteration). Actually, in practice, $\underset{\sim}{S}$ has to be much easier to factorize than $\underset{\sim}{A}$, but has to be such that the spectrum of $\underset{\sim}{S}^{-1}\underset{\sim}{A}$ is a narrow one in order to give good convergence properties to algorithm (2.17)-(2.24).

Remark 2.2 : The above preconditioned conjugate gradient algorithm appears as a compromise between a purely iterative method, and a direct one, for solving the linear problem (2.14).

2.3. A second family of test problems. Least square solution of nonlinear Poisson problems.

2.3.1. Formulation of the test problems.

We follow [12, Sec. 2], [13, Chap. 7] ; let $\Omega \in \mathbb{R}^N$ be a bounded domain with a smooth boundary. We consider the nonlinear Dirichlet problem

$$
\begin{aligned}
& -\Delta\phi - T(\phi) = 0 \\
& \phi = 0 \text{ on } \Gamma.
\end{aligned}
\tag{2.25}
$$

We do not discuss here the *existence* and *uniqueness* properties of the solutions of (2.25) since we do not want to be specific about operator T. Since various iterative methods for solving (2.25) are described in the two above references we shall concentrate on *conjugate gradient methods with scaling* via convenient *least square formulations*.

2.3.2. A H^{-1} least square method for solving (2.25).

Let us introduce first some *Sobolev's functional spaces* which seem optimally suited for the following study

$$H^1(\Omega) = \{v \in L^2(\Omega),\ \frac{\partial v}{\partial x_i} \in L^2(\Omega) \quad \forall i=1,\dots N\}, \tag{2.26}$$

$$H_o^1(\Omega) = \{v \in H^1(\Omega),\ v=0 \text{ on } \Gamma\}\ . \tag{2.27}$$

Then $H^1(\Omega)$ is a *Hilbert space* with the scalar product

$$(v,w)_{H^1(\Omega)} = \int_\Omega (vw+\nabla v\cdot\nabla w)\,dx$$

and the corresponding norm. Moreover $H_o^1(\Omega)$ is a *closed subspace* of $H^1(\Omega)$. Since Ω is bounded, then $H_o^1(\Omega)$ is a Hilbert space with the scalar product

$$(v,w)_{H_o^1(\Omega)} = \int_\Omega \nabla v\cdot\nabla w\ dx, \tag{2.28}$$

and the corresponding norm $||v||_{H_o^1(\Omega)} = (\int_\Omega |\nabla v|^2\ dx)^{1/2}$ is equivalent to the norm induced by $H^1(\Omega)$. Let $H^{-1}(\Omega) = (H_o^1(\Omega))'$ be the topological dual space of $H_o^1(\Omega)$. If $L^2(\Omega)$ has been identified to its dual space, then

$$H_o^1(\Omega) \subset L^2(\Omega) \subset H^{-1}(\Omega),$$

moreover $\Delta = \underset{\sim}{\nabla}^2$ is an isomorphism from $H_o^1(\Omega)$ onto $H^{-1}(\Omega)$. In the following the duality pairing between $H^{-1}(\Omega)$ and $H_o^1(\Omega)$ will be denoted by $<\cdot,\cdot>$, where $<\cdot,\cdot>$ is such that

$$<f,v> = \int_\Omega fv\,dx \quad \forall f \in L^2(\Omega), \quad \forall v \in H_o^1(\Omega). \tag{2.29}$$

The topology of $H^{-1}(\Omega)$ is defined by $||\cdot||_{-1}$, where

$$||f||_{-1} = \sup_{v\in H_o^1(\Omega)-\{0\}} \frac{|<f,v>|}{||v||_{H_o^1(\Omega)}} . \tag{2.30}$$

From the above considerations a most natural *least square formulation* for solving the model problem (2.25), is

$$\underset{v\in H_o^1(\Omega)}{\mathrm{Min}} \quad ||\Delta v+T(v)||_{-1} . \tag{2.31}$$

It is clear that if (2.25) has a solution, then this solution will be a solution of (2.31) for which the cost function will vanish. Let us introduce $\xi \in H_o^1(\Omega)$ by

$$\begin{aligned} \Delta\xi &= \Delta v+T(v) \ \textit{in } \Omega, \\ \xi &= 0 \ \textit{on } \Gamma, \end{aligned} \tag{2.32}$$

then (2.31) reduces to

$$\underset{v\in H_o^1(\Omega)}{\mathrm{Min}} \quad ||\Delta\xi||_{-1} , \tag{2.33}$$

where, in (2.33), ξ is a function of v via (2.32) ; actually it can be proved that if $||\cdot||_{-1}$ is defined by (2.30) with $<\cdot,\cdot>$ obeying (2.29), then

$$||\Delta v||_{-1} = ||v||_{H_o^1(\Omega)} \qquad \forall v \in H_o^1(\Omega). \tag{2.34}$$

It follows then from (2.34) that (2.33) may be formulated also by

$$\underset{v\in H^1_o(\Omega)}{\text{Min}} \int_\Omega |\nabla \underset{\sim}{\xi}|^2 \, dx, \tag{2.35}$$

where ξ is still a function of v through (2.32).

Remark 2.3 : Nonlinear boundary value problems have been treated by CEA-GEYMONAT [14], LOZI [15] using formulations very close to (2.32),(2.35).

2.3.3. Conjugate gradient solution of the least square problem (2.32),(2.35).

Let us define $J : H^1_o(\Omega) \to \mathbb{R}$ by

$$J(v) = \frac{1}{2}\int_\Omega |\nabla \underset{\sim}{v}|^2 \, dx, \tag{2.36}$$

where ξ is a function of v in accordance with (2.32) ; then (2.35) may also be written as

$$\underset{v\in H^1_o(\Omega)}{\text{Min}} \quad J(v). \tag{2.37}$$

To solve (2.37) we shall use a *conjugate gradient* algorithm. Among the possible conjugate gradient algorithms we have selected the *Polak-Ribière* version (cf. POLAK [16]) since this algorithm produced the best performances (compared to other variants) in the preliminary numerical tests we did (the good performances of the Polak-Ribière algorithm are discussed in POWELL [16]). Let us denote by J' the differential of J ; then the Polak-Ribière version of the conjugate gradient method, applied to the solution of (2.37) is :

Step 0 : Initialization

$$\phi^o \in H^1_o(\Omega) \text{ given}, \tag{2.38}$$

then compute $g^o \in H^1_o(\Omega)$ *from*

$$-\Delta g^o = J'(\phi^o) \text{ in } \Omega, \qquad g^o = 0 \text{ on } \Gamma, \tag{2.39}$$

and set

$$w^o = g^o. \tag{2.40}$$

Then for $n \geq 0$*, assuming* ϕ^n, w^n, g^n *known, compute* $\phi^{n+1}, g^{n+1}, w^{n+1}$ *by*

Step 1 : Descent

$$\lambda^n = \underset{\lambda \in \mathbb{R}}{\text{Arg Min}}\, J(\phi^n - \lambda w^n), \tag{2.41}$$

$$\phi^{n+1} = \phi^n - \lambda^n w^n. \tag{2.42}$$

Step 2 : Construction of the new descent direction

Define $g^{n+1} \in H_o^1(\Omega)$ *by*

$$-\Delta g^{n+1} = J'(\phi^{n+1}) \text{ in } \Omega \text{ , } g^{n+1} = 0 \text{ on } \Gamma, \tag{2.43}$$

then

$$\gamma^{n+1} = \frac{\int_\Omega \underset{\sim}{\nabla} g^{n+1} \cdot \underset{\sim}{\nabla}(g^{n+1} - g^n)\, dx}{\int_\Omega |\underset{\sim}{\nabla} g^n|^2\, dx}, \tag{2.44}$$

$$w^{n+1} = g^{n+1} + \gamma^{n+1} w^n \tag{2.45}$$

$n = n+1$, *go to* (2.41).

The two non trivial steps of algorithm (2.38)-(2.45) are

(i) The solution of the *single variable* minimization problem (2.41) ; the corresponding line search can be achieved by *dichotomy* or *Fibonacci* methods (see for example [16],[18], [19]). We observe that each evaluation of $J(v)$, for a given argument v, requires the solution of the linear Dirichlet problem (2.32) to obtain the corresponding ξ.

(ii) The calculation of g^{n+1} from ϕ^{n+1} which requires the solution of two linear Dirichlet problems (namely (2.32) with $v=\phi^{n+1}$ and (2.43)).

Calculation of $J'(\phi^n)$ and g^n : We refer to [13, Chap. 7] where we prove that $\forall v,w \in H_0^1(\Omega)$ we have

$$<J'(v),w> = \int_{\tilde{\Omega}} \underset{\sim}{\nabla}\xi \cdot \underset{\sim}{\nabla} w \, dx - <T'(v)\cdot w,\xi> \tag{2.46}$$

where, in (2.46), T' is the differential of T at v, and ξ the solution of (2.32) corresponding to v. Therefore $J'(v) \in H^{-1}(\Omega)$ may be identified with the linear functional defined on $H_0^1(\Omega)$ by

$$w \to \int_{\Omega} \underset{\sim}{\nabla}\xi \cdot \underset{\sim}{\nabla} w \, dx - <T'(v)\cdot w,\xi> . \tag{2.47}$$

It follows then from (2.43),(2.46),(2.47) that g^n is the solution of the following *linear variational problem*

$$\text{Find } g^n \in H_0^1(\Omega) \text{ such that } \forall w \in H_0^1(\Omega)$$

$$\int_{\Omega} \underset{\sim}{\nabla} g^n \cdot \underset{\sim}{\nabla} w \, dx = \int_{\tilde{\Omega}} \underset{\sim}{\nabla}\xi^n \cdot \underset{\sim}{\nabla} w \, dx - <T'(\phi^n)\cdot w,\xi^n> , \tag{2.48}$$

where ξ^n is the solution of (2.32) corresponding to $v = \phi^n$.

Remark 2.4 : The fact that $J'(v)$ is known through (2.46) is not at all a drawback if a *Galerkin* or a *finite element* method is used to approximate (2.25). Indeed we only need to know the values of $<J'(v),w>$ for w belonging to a *basis* of the *finite dimensional subspace* of $H_0^1(\Omega)$ corresponding to the approximation under consideration.

Remark 2.5 : The least square method described in the present section can be combined to *arc length continuation* techniques for solving nonlinear problems with multiple solutions (nonlinear eigenvalue problems for example) ; see for more details [10, Sec. 2.3.4], [13, Chap. 7], [20].

2.3.4. <u>On the practical implementation of the preconditioning operation in algorithm</u> (2.38)-(2.45).

We shall use in practice a *finite dimensional* variant of (2.38)-(2.45) to solve a *finite dimensional approximation* of (2.25) ; from this observation we shall discuss the numerical solution of the following problem in $\mathbb{R}^N$

$$\tilde{F}(\tilde{\phi}) = \tilde{0} \qquad (2.49)$$

(where $\tilde{F} : \mathbb{R}^N \to \mathbb{R}^N$, $\tilde{\phi} \in \mathbb{R}^N$) by a preconditioned conjugate algorithm via a nonlinear least square formulation.

Let $\tilde{B}$ be a *symmetric, positive definite* N×N matrix ; to $\tilde{v} \in \mathbb{R}^N$ we associate $\tilde{\xi}(\tilde{v}) = \tilde{\xi} \in \mathbb{R}^N$ by

$$\tilde{B}\tilde{\xi} = \tilde{F}(\tilde{v}). \qquad (2.50)$$

A nonlinear least square formulation of (2.50) is then

$$\operatorname*{Min}_{\tilde{v} \in \mathbb{R}^N} \frac{1}{2} (\tilde{B}\tilde{\xi}, \tilde{\xi}), \qquad (2.51)$$

where in (2.51), $\tilde{\xi}$ is a function of $\tilde{v}$ via (2.50) and where $(\cdot,\cdot)$ denotes the usual scalar product of $\tilde{\mathbb{R}}^N$. We define $J : \mathbb{R}^N \to \mathbb{R}$ by

$$J(\tilde{v}) = \frac{1}{2} (\tilde{B}\tilde{\xi}(\tilde{v}), \tilde{\xi}(\tilde{v})) \ ;$$

let us detail now the discrete analogue of (2.38)-(2.45) for solving (2.49) via (2.50),(2.51) :

<u>Step 0</u> : <u>Initialization</u>

$$\tilde{\phi}^o \in \mathbb{R}^N, \ \textit{given}, \qquad (2.52)$$

$$\tilde{g}^o = \tilde{B}^{-1} J'(\tilde{\phi}^o), \qquad (2.53)$$

$$\tilde{w}^o = \tilde{g}^o. \qquad (2.54)$$

Then for $n \geq 0$, *assuming that* ϕ^n, g^n, w^n *are known, we define* ϕ^{n+1}, g^{n+1}, w^{n+1} *by*

Step 1 : Descent

$$\lambda^n = \operatorname{Arg}\operatorname*{Min}_{\lambda \in \mathbb{R}} J(\phi^n - \lambda w^n), \tag{2.55}$$

$$\phi^{n+1} = \phi^n - \lambda^n w^n. \tag{2.56}$$

Step 2 : Construction of the new descent direction

$$g^{n+1} = B^{-1} J'(\phi^{n+1}), \tag{2.57}$$

$$\gamma^{n+1} = \frac{(Bg^{n+1}, g^{n+1} - g^n)}{(Bg^n, g^n)}, \tag{2.58}$$

$$w^{n+1} = g^{n+1} + \gamma^{n+1} w^n, \tag{2.59}$$

$n = n+1$, *go to* (2.55).

We have concerning $J'(v)$

$$J'(v) = (F'(v))^t \xi = (F'(v))^t B^{-1} F(v) \quad \forall v \in \mathbb{R}^N. \tag{2.60}$$

From the above relations a most important tool for solving (2.49), via (2.50),(2.51) and algorithm (2.52)-(2.59), will be an efficient solver for linear systems whose matrix is B ; consider a *Cholesky's factorization* of B such as

$$B = LL^t \; ; \tag{2.61}$$

if L is not sparse enough compared to B, it may be necessary to solve also these linear systems by an *iterative method* with preconditioning. Let

$$Bx = c \quad (x, c \in \mathbb{R}^N) \tag{2.62}$$

be such a system ; we may use in particular to solve (2.62) a

conjugate gradient method with a preconditioning associated to
the matrix

$$\tilde{\underset{\sim}{B}} = \tilde{\underset{\sim}{L}}\tilde{\underset{\sim}{L}}^t \tag{2.63}$$

where $\tilde{\underset{\sim}{L}}$ is precisely obtained from $\underset{\sim}{B}$ via an incomplete Cholesky
factorization process. One may find the above approach a bit
complicated since it involves the *conjugate gradient outer loop*
(2.52)-(2.59) and *conjugate gradient inner loops* for solving
various internal systems like (2.62) ; at that point it is very
tempting to use $\tilde{\underset{\sim}{B}}$ instead of $\underset{\sim}{B}$ in (2.50),(2.51) and also in
(2.52)-(2.62). If $\tilde{\underset{\sim}{B}}$ is well chosen, *using* $\tilde{\underset{\sim}{B}}$ *instead of* $\underset{\sim}{B}$ *will
increase the number of outer iterations*, but since the internal
systems (2.62) are replaced by linear systems like

$$\tilde{\underset{\sim}{B}}\underset{\sim}{x} \ (= \tilde{\underset{\sim}{L}}\tilde{\underset{\sim}{L}}^t\underset{\sim}{x}) = \underset{\sim}{c} \ (\underset{\sim}{x},\underset{\sim}{c} \in \mathbb{R}^N) \tag{2.64}$$

much easier to solve, the resulting global algorithm may be faster.

3. ON A NEW INCOMPLETE CHOLESKY FACTORIZATION METHOD.

Let $\underset{\sim}{A} = (a_{ij})_{1 \leq i,j \leq N}$ be a N×N *symmetric, positive definite, banded* matrix of bandwidth 2m+1 (one says that $\underset{\sim}{A}$ is 2m+1-*diagonal*) ;
we clearly have

$$a_{ij} = 0 \ \textit{if} \ |i-j| > m. \tag{3.1}$$

Consider the *linear system*

$$\underset{\sim}{A}\underset{\sim}{x} = \underset{\sim}{b} \ ; \tag{3.2}$$

to solve (3.2) we may use a *Cholesky's factorization*

$$\underset{\sim}{A} = \underset{\sim}{L}\underset{\sim}{L}^t \tag{3.3}$$

of $\underset{\sim}{A}$, with $\underset{\sim}{L}$ *lower triangular*. It is well-known that (3.1) implies

$$\ell_{ij} = 0 \ \textit{if} \ i-j > m, \tag{3.4}$$

but as mentioned above (see Figure 2.1), some *zero elements* in
the band of $\underset{\sim}{A}$ may be such that the corresponding elements of $\underset{\sim}{L}$

are $\neq 0$.

We shall use $\underset{\sim}{L}$ to construct an *incomplete Cholesky's factorization* of $\underset{\sim}{A}$; to define and discuss the properties of the following methods it is convenient to introduce

$$K = \{\{i,j\} \mid 0 \leq i-j \leq m,\ a_{ij} = 0\}\ , \tag{3.5}$$

$$K^* = \{\{i,j\} \mid 0 \leq i-j \leq m,\ \ell_{ij} \neq 0\}\ , \tag{3.6}$$

$$n(K) = \text{Card }(K). \tag{3.7}$$

We give now a *positive constant* C, and from $\underset{\sim}{L}$ and C we define the two following *lower triangular matrices*

$$\begin{aligned} &\tilde{\underset{\sim}{L}}_C = (\tilde{\ell}_{ij})\ ,\ \textit{with} \\ &\tilde{\ell}_{ij} = 0 \textit{ if } \{i,j\} \in K \textit{ and } |\ell_{ij}| < C, \\ &\tilde{\ell}_{ij} = \ell_{ij} \textit{ if the contrary holds,} \end{aligned} \tag{3.8}$$

$$\begin{aligned} &\tilde{\underset{\sim}{L}}{}'_C = (\tilde{\ell}'_{ij}),\ \textit{with} \\ &\tilde{\ell}'_{ij} = 0 \textit{ if } \{i,j\} \in K \textit{ and } (\ell_{ij}) \leq C \min(\ell_{ii},\ell_{jj}), \\ &\tilde{\ell}'_{ij} = \ell_{ij} \textit{ if the contrary holds.} \end{aligned} \tag{3.9}$$

<u>Remark 3.1</u> : We observe that

(i) *If* $C < \min\limits_{\{i,j\}\in K} |\ell_{ij}|$ *then* $\tilde{\underset{\sim}{L}}_C = \underset{\sim}{L}$,

(ii) *If* $C < \min\limits_{\{i,j\}\in K} \left(\dfrac{|\ell_{ij}|}{\min(|\ell_{ii}|,|\ell_{jj}|)}\right)$ *then* $\tilde{\underset{\sim}{L}}{}'_C = \underset{\sim}{L}$,

(iii) *If* $C > \max\limits_{i,j} |\ell_{ij}|$ *then* $\{\{i,j\} \mid \tilde{\ell}_{ij} = 0\} = K$,

(iv) *If* $C > \max\limits_{i,j} \left(\dfrac{|\ell_{ij}|}{\min(|\ell_{ii}|,|\ell_{jj}|)}\right)$ *then* $\{\{i,j\} \mid \tilde{\ell}'_{ij} = 0\} = K$.

If (iii) and (iv) hold, then $\tilde{\underset{\sim}{L}}_C$ and $\tilde{\underset{\sim}{L}}{}'_C$ have their non zero elements in the same position than those of $\underset{\sim}{A}$, and are therefore close to the incomplete Cholesky operators discussed in [5],[6].

However the "approximate Cholesky's factors in [5],[6] are constructed *during* the Cholesky's factorization of $\underset{\sim}{A}$, and not after like in our methods ; from this fundamental difference it follows that

- the approximate factors in [5], [6] are more economical to construct than those defined by (3.8),(3.9) and therefore may be interesting to use if the number of linear systems associated to $\underset{\sim}{A}$ is small.
- The matrices $\tilde{\underset{\sim}{A}}_c$ and $\tilde{\underset{\sim}{A}}'_c$, defined from $\tilde{\underset{\sim}{L}}_c$ and $\tilde{\underset{\sim}{L}}'_c$ by

$$\tilde{\underset{\sim}{A}}_c = \tilde{\underset{\sim}{L}}_c\tilde{\underset{\sim}{L}}_c^t , \tag{3.10}$$

$$\tilde{\underset{\sim}{A}}'_c = \tilde{\underset{\sim}{L}}'_c\tilde{\underset{\sim}{L}}'^t_c \tag{3.11}$$

are *spectrally closer* to $\underset{\sim}{A}$ than the corresponding matrices in [5], [6] ; moreover, *once constructed,* their factors have basically the same storage requirement than those in [5],[6]. Therefore in industrial environment where the number of linear systems associated to a given $\underset{\sim}{A}$ may be very large, our approach appears to lead to more efficient algorithms than those in [5], [6], even if the construction of $\underset{\sim}{L}$ requires an *out of core* process.

Practical construction of $\tilde{\underset{\sim}{L}}_c$ and $\tilde{\underset{\sim}{L}}'_c$:

For practical applications it is feasible to choose C such that a given percentage of non zero elements of $\tilde{\underset{\sim}{L}}_c$ and $\tilde{\underset{\sim}{L}}'_c$ is kept. Thus, giving d, $0 < d < 100$, we may define $\tilde{\underset{\sim}{L}}_{(d/100)}$ and $\tilde{\underset{\sim}{L}}'_{(d/100)}$ as follows :

We define from C *and* $\tilde{\underset{\sim}{L}}_c$, $\tilde{\underset{\sim}{L}}'_c$

$$\tilde{K}_c = \{\{i,j\} | \{i,j\} \in K,\ \tilde{\ell}_{ij} \neq 0\}, \tag{3.12}$$

$$\tilde{K}'_c = \{\{i,j\} | \{i,j\} \in K,\ \tilde{\ell}'_{ij} \neq 0\}, \tag{3.13}$$

$$n(\tilde{K}_c) = \text{Card}\ (\tilde{K}_c),\ n(\tilde{K}'_c) = \text{Card}\ (\tilde{K}'_c), \tag{3.14}$$

and then we adjust C *in order to obtain either*

$$\tilde{\underset{\sim}{L}}_{(d/100)} = \tilde{\underset{\sim}{L}}_c \text{ with C such that } n(\tilde{K}_c)/n(K)=d/100, \tag{3.15}$$

or

$$\tilde{\underset{\sim}{L}}'_{(d/100)} = \tilde{\underset{\sim}{L}}'_c \text{ with C such that } n(\tilde{K}'_c)/n(K) = d/100 \tag{3.16}$$

We observe that d=0 (resp. d=100) leads to the same $\tilde{\underset{\sim}{L}}_c$ and $\tilde{\underset{\sim}{L}}'_c$ than (iii) and (iv) in Remark 3.1 (resp. leads to matrix $\underset{\sim}{L}$ of (3.3)).

Remark 3.2 : If $\underset{\sim}{A}$ comes from a *finite element approximation* one may also proceed as follows to obtain an approximate Cholesky's factorization of $\underset{\sim}{A}$:

Let $\mathcal{T}_h$ be a finite element partition of a given domain Ω (that we suppose *polyhedral* for simplicity) ; we have

$$\bar{\Omega} = \bigcup_{T \in \mathcal{T}_h} T \tag{3.17}$$

where in (3.17) the finite elements T are adjacent polyhedras ; let

$$\{P_i\}_{i=1}^N \tag{3.18}$$

be the set of the nodes of $\mathcal{T}_h$ associated to the finite element method under consideration. The *complementary* set $\bar{K}$ of K (K is defined by (3.5)) is defined by

$$\bar{K} = \{\{i,j\} \mid 0 \le i-j \le m,\ \exists\, T \in \mathcal{T}_h \text{ such that } P_i, P_j \in T\}\ ; \tag{3.19}$$

we introduce also

$$\bar{\bar{K}} = \{\{i,j\} \mid 0 \le i-j \le m, \exists k \textit{ such that } P_i, P_k \in T,\ P_j, P_k \in T' \textit{ with } T, T' \in \mathcal{C}_h\} \tag{3.20}$$

We define finally from $\bar{\bar{K}}$ and $\underset{\sim}{L}$ an approximate Cholesky factorization of $\underset{\sim}{A}$ by

$$\begin{aligned} &\bar{\bar{\underset{\sim}{L}}} = (\bar{\bar{\ell}}_{ij}), \\ &\bar{\bar{\ell}}_{ij} = \ell_{ij} \textit{ if } \{i,j\} \in \bar{\bar{K}}, \quad \bar{\bar{\ell}}_{ij} = 0 \textit{ otherwise.} \end{aligned} \tag{3.21}$$

With such a construction $\bar{\bar{\underset{\sim}{L}}}$ is independent of the numbering of $\{P_i\}_{i=1}^N$; in two-dimension the number of non zero elements in the band of $\bar{\bar{\underset{\sim}{L}}}$ is approximately 20%, but in three-dimension it is around 50% which is too large in view of the problems that we want to solve.

Remark 3.3 : The introduction of $\underset{\sim}{\tilde{L}}{}'_c$ defined by (3.9) was also motivated by finite element approximations. Indeed a quick and easy analysis shows that if $\Omega \subset R^3$, then $\ell_{ij} = O(h)$; on the other hand $\ell_{ij} = O(1)$ if $\Omega \subset \mathbb{R}^2$. Thus it is necessary to discard the small elements by a test on their relative magnitude instead of their absolute value ; observe however that this problem does not arise for $\underset{\sim}{\tilde{L}}_{(d/100)}$ and $\underset{\sim}{\tilde{L}}'_{(d/100)}$.

4. NUMERICAL TESTS. (I) SOLUTION OF AN ELLIPTIC LINEAR PROBLEM. We consider the application of the above algorithms to the solution of *linear problems in finite dimension* such as

$$\underset{\sim}{A}\underset{\sim}{\phi} = \underset{\sim}{F}. \tag{4.1}$$

In this test, (4.1) comes from a *finite element* approximation of the *elliptic boundary value problem* whose boundary conditions are partly *Neumann* and partly *Dirichlet*

$$-\Delta\phi = f \textit{ in } \Omega,$$

$$\phi = g_1 \textit{ on } \Gamma_1, \frac{\partial\phi}{\partial n} = g_2 \textit{ on } \Gamma_2 \tag{4.2}$$

where, in (4.2), $\Gamma_1 \cup \Gamma_2 = \partial\Omega$, $\Gamma_1 \cup \Gamma_2 = \emptyset$; the following results correspond to *piecewise linear, globally continuous finite element approximations*.

We have used algorithm (2.17)-(2.24) of Sec. 2.2, with various preconditioning matrices $\underset{\sim}{S}$, and also with different numbering of the nodes. The first example is obtained by the discretization of a two-dimensional problem (4.2) ; the corresponding number of unknowns is 600 and the Cholesky's factor $\underset{\sim}{L}$ in

$$\underset{\sim}{A} = \underset{\sim}{L}\underset{\sim}{L}^t \tag{4.3}$$

has about 1.2×10^4 non zero elements ; we have shown on Figure 4.1 the number of iterations required to solve (4.1) by algorithm (2.17)-(2.24) using various incomplete Cholesky's factorizations of $\underset{\sim}{A}$ to construct the preconditioning matrix $\underset{\sim}{S}$; the stopping test is

$$||\underset{\sim}{A}\underset{\sim}{\phi}^n - \underset{\sim}{F}||^2 \leq \varepsilon \tag{4.4}$$

(here $\varepsilon = 10^{-7}$) ; the norm used in (4.4) is the usual euclidian norm of $\mathbb{R}^N$.

We observe the very good performances obtained with $d/100 \simeq$ 5%. We also have indicated the results corresponding to the method in [5]. Other tests have shown that up to 5% the number of iterations with $\tilde{\underset{\sim}{L}}_{d/100}$ is not very sensitive to mesh refinement (of course computer time increases). We have shown on Figure 4.2 the results of similar experiments for a 3-dimensional problem

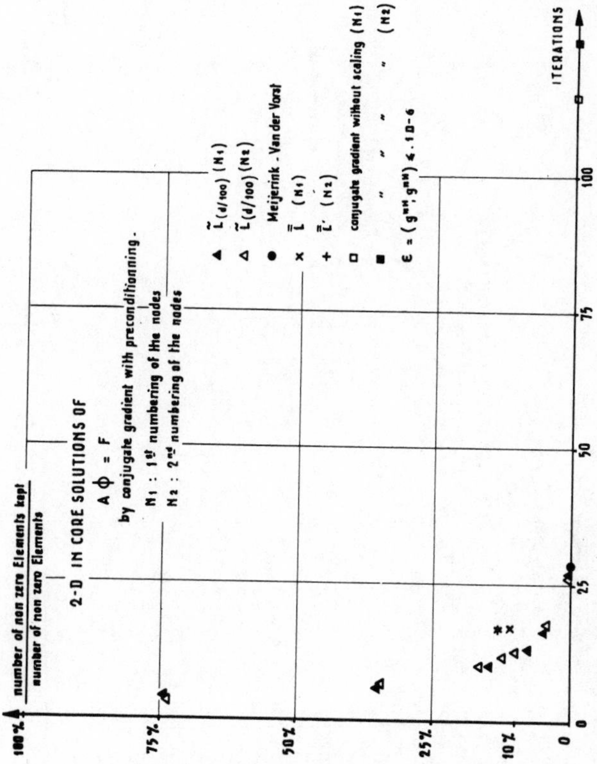
2-D IN CORE SOLUTIONS OF
A Φ = F
by conjugate gradient with preconditionning -
N1 : 1st numbering of the nodes
N2 : 2nd numbering of the nodes
number of non zero Elements kept
number of non zero Elements
Meijerink - Van der Vorst
conjugate gradient without scaling (N1)
ITERATIONS
100 %
75 %
50 %
25 %
10 %
0
25
50
75
100

FIGURE 4.1

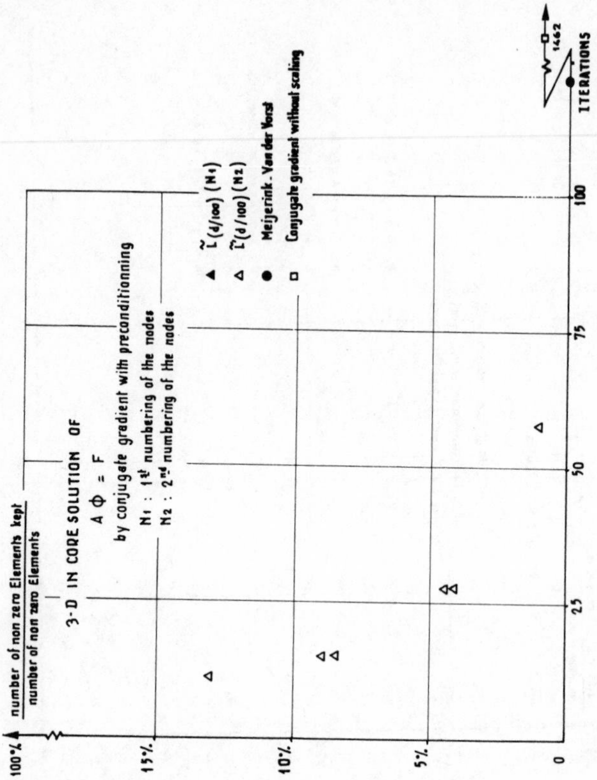
number of non zero Elements kept / number of non zero Elements
3-D IN CORE SOLUTION OF
A Φ = F
by conjugate gradient with preconditionning
N1 : 1st numbering of the nodes
N2 : 2nd numbering of the nodes
L̃(d/100) (N1)
L̃(d/100) (N2)
Meijerink . Van der Vorst
Conjugate gradient without scaling
100%
15%
10%
5%
0
25
50
75
100
1462
ITERATIONS

FIGURE 4.2

(4.2) ; this time the number of unknowns is 5328 and the Cholesky's factor $\underset{\sim}{L}$ in (4.3) has about 1.5×10^6 non zero elements.

5. NUMERICAL TESTS. (II) APPLICATION TO TRANSONIC POTENTIAL FLOWS FOR COMPRESSIBLE INVISCID FLUIDS.

For more details on this important problem see [12], [13, Chap. 7], [21], [22].

5.1. Mathematical formulation.

Under simplifying assumptions for which we refer to, e.g. [12], [13],[21],[22] the flows under consideration are modelled by

$$\underset{\sim}{\nabla}\cdot\rho(\phi)\underset{\sim}{\nabla}\phi = 0 \textit{ in } \Omega, \tag{5.1}$$

where

$$\rho(\phi) = \rho_o\left(1 - \frac{|\underset{\sim}{\nabla}\phi|^2}{\frac{\gamma+1}{\gamma-1}C_*^2}\right)^{\frac{1}{\gamma-1}} . \tag{5.2}$$

In (5.1),(5.2) : Ω is the domain of the flow, ϕ a velocity potential ($\underset{\sim}{v} = \underset{\sim}{\nabla}\phi$ where $\underset{\sim}{v}$ is the *flow velocity*), ρ the fluid density, γ (= 1.4 in *air*) the ratio of specific heats, C_* the critical velocity.

In addition to (5.1),(5.2) we have to prescribe *boundary conditions* ; a typical set of boundary condition is

$$\begin{aligned} &\phi = g_1 \textit{ on } \Gamma_1,\ \rho\frac{\partial\phi}{\partial n} = g_2 \textit{ on } \Gamma_2, \\ &\textit{where } \Gamma_1\cup\Gamma_2 = \partial\Omega,\ \Gamma_1\cup\Gamma_2 = \emptyset, \end{aligned} \tag{5.3}$$

and also Kutta-Joukowsky conditions if Ω is a *multi-connected domain* ; we have finally to include an *entropy condition* to avoid *rarefaction shocks* since such shocks are *unphysical*.

5.2. Finite element approximation and iterative solution. The numerical solution of the above transonic flow problems, by methods using *finite elements* approximations, *least square* formulations and *conjugate gradient* algorithms, is discussed in [12],[13],[21]-[23].

Suppose that Ω has been approximated by a polyhedral domain - still denoted by Ω - and denote by $\mathcal{T}_h$ a finite element partition of Ω, with (as usual) $h = \max_{T\in\mathcal{T}_h} d(T)$, with $d(T)$ = diameter of T. Assuming that the T are triangles (resp. tetrahedrons) in 2-dimension (resp. in 3-dimension), we approximate $H^1(\Omega)$ by

$$H_h^1 = \{v_h | v_h \in C^o(\bar{\Omega}),\ v_h|_T \in P_1 \quad \forall T \in \mathcal{T}_h\},$$

with P_1 = space of polynomials of degree ≤ 1. We look for approximate solutions in

$$V_h = \{v_h | v_h \in H_h^1,\ v_h|_{\Gamma_1} = g_{1h}\}$$

where g_{1h} is an approximation of g_1 ; we associate to V_h

$$V_{oh} = \{v_h | v_h \in H_h^1 ,\ v_h|_{\Gamma_1} = 0\}.$$

For simplicity we do not discuss here the Kutta-Joukowsky conditions ; then an approximate transonic flow problem can be formulated as follows

$$\underset{\phi_h \in V_h}{\text{Min}} \{\frac{1}{2}\int_\Omega |\underset{\sim}{\nabla}\xi_h|^2\, dx + \mathcal{E}_h(\phi_h)\} , \tag{5.4}$$

where, in (5.4), ξ_h is a function of ϕ_h via the *state problem*

$$\begin{aligned} &\xi_h \in V_{oh}, \\ &\int_\Omega \underset{\sim}{\nabla}\xi_h \cdot \underset{\sim}{\nabla} v_h\, dx = \int_\Omega \rho(\phi_h)\underset{\sim}{\nabla}\phi_h \cdot \underset{\sim}{\nabla} v_h dx - \int_{\Gamma_2} g_{2h} v_h d\Gamma \ \forall v_h \in V_{oh}, \end{aligned} \tag{5.5}$$

and where $\mathcal{S}_h$ is a *discrete entropy functional* taking very large values if the velocity field $\nabla\phi_h$ contains rarefactions shocks (implementation of the entropy condition using an *upwinding of the density* is discussed in [11], [13]).

The discrete problem (5.4),(5.5) which is very close to (2.50),(2.51), can be solved by conjugate gradient methods with scaling very close to (2.52)-(2.59). It is very convenient to define $\underset{\sim}{B}$, in these variants of (2.52)-(2.59), from the bilinear form

$$\{v_h, w_h\} \to \int_\Omega \underset{\sim}{\nabla} v_h \cdot \underset{\sim}{\nabla} w_h \, dx,$$

thus $\underset{\sim}{B}$ can be viewed as a discrete Laplace operator and therefore each iteration of the above conjugate gradient algorithm will require the solution of several discrete Poisson problems, and the various comments done in Sec. 2.3.4 hold for this discrete transonic flow problem.

5.3. Numerical experiments.

We consider as first test problem, the flow around a NACA 0012 airfoil with $M_\infty = .8$ and a zero angle of attack (thus the Kutta-Joukowsky condition is automatically satisfied) ; from these values there exists a *supersonic* pocket in the flow and a subsonic-supersonic transition with *shock* (see Figure 5.3 for the mach distribution on the skin of the airfoil ; we observe a very sharp shock).

Following Sec. 5.2 the above flow has been approximated by a finite element method using about 1400 triangles and 600 nodes ; the Cholesky's factors associated to the corresponding discrete Laplace operator has about 1.1×10^4 non sero elements ; the discrete transonic flow problem (5.4),(5.5) is then solved by a conjugate gradient method with a preconditioning operator $\underset{\sim}{B}$ which is a discrete variant of $-\Delta$ with appropriate *boundary conditions* ; the above preconditioning method implies the solution at each iteration of several discrete Poisson problems; these problems can also be solved by conjugate gradient methods with preconditioning, namely those methods discussed in Sec. 2.2, Sec. 3 and also Sec. 4.

We have indicated on Figure 5.1 the C.P.U. time required to solve to a given accuracy the above flow problem, according to the incomplete Cholesky methods used to solve at each iteration the discrete Poisson problems required by the nonlinear least squares-conjugate gradient algorithm ; we observe again the very good performances of the method using $\tilde{\underset{\sim}{L}}_{(d/100)}$, even with small values of d ($\simeq 10$).

If $\tilde{\underset{\sim}{L}}$ is an incomplete Cholesky's factor of $\underset{\sim}{B}$, it is very tempting, as mentioned in Sec. 2.3.4, to use $\tilde{\underset{\sim}{B}} = \tilde{\underset{\sim}{L}}\tilde{\underset{\sim}{L}}^t$ to replace $\underset{\sim}{B}$ introduced in Secs. 2.3.4 and 5.2 ; thus instead of the state equation (5.5) we shall introduce a state equation whose matrix in the left hand side will be $\tilde{\underset{\sim}{B}}$; similarly, instead of $\int_\Omega |\underset{\sim}{\nabla}\xi_h|^2\, dx$ we shall use in (5.4) the quadratic form associated to $\tilde{\underset{\sim}{B}}$. Such a choice leads to some simplification of the iterative process (depending upon the degree of *incompleteness*), at the possible expense however of a greater number of iterations. We have used the above strategy to solve the transonic flow problem, taking $\tilde{\underset{\sim}{B}}$ $(= \tilde{\underset{\sim}{L}}\tilde{\underset{\sim}{L}}^t)$. The corresponding results are summarized on Fig. 5.2 which shows the behavior of a *reference residual* as a function of the nomber of iteration ; using the notation of Sec. 2.3.4, we have taken as residual $(\underset{\sim}{B}\hat{\underset{\sim}{\xi}}^n, \hat{\underset{\sim}{\xi}}^n)$ where $\hat{\underset{\sim}{\xi}}^n$ is defined from $\underset{\sim}{\phi}^n$ by

$$\tilde{\underset{\sim}{B}}\hat{\underset{\sim}{\xi}}^n = \underset{\sim}{F}(\underset{\sim}{\phi}^n),$$

where $\{\underset{\sim}{\phi}^n\}_n$ is the sequence of approximate solutions generated by the nonlinear least squares-conjugate gradient method (using $\tilde{\underset{\sim}{B}} = \tilde{\underset{\sim}{L}}\tilde{\underset{\sim}{L}}^t$ instead of $\underset{\sim}{B}$) ; in addition to this behavior of the residual we have shown on Fig. 5.3 the velocity distribution on the airfoil at iteration 80 for different choices of the Cholesky's factors. We observe that the shock approximation is rather poor

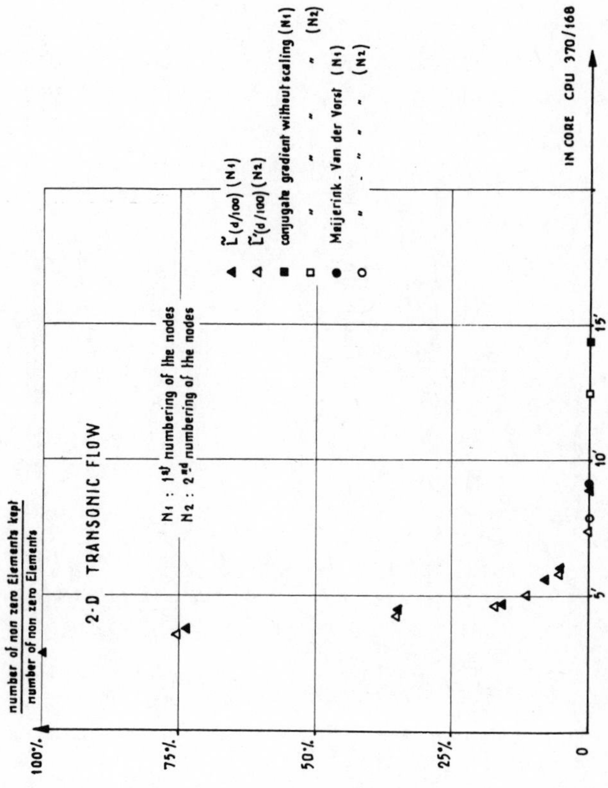
number of non zero Elements kept / number of non zero Elements
2-D TRANSONIC FLOW
N1 : 1st numbering of the nodes
N2 : 2nd numbering of the nodes
L̃(d/100) (N1)
L̃'(d/100) (N2)
conjugate gradient without scaling (N1)
" " " (N2)
Meijerink . Van der Vorst (N1)
" " " (N2)
IN CORE CPU 370/168
100%
75%
50%
25%
0
5'
10'
15'

FIGURE 5.1

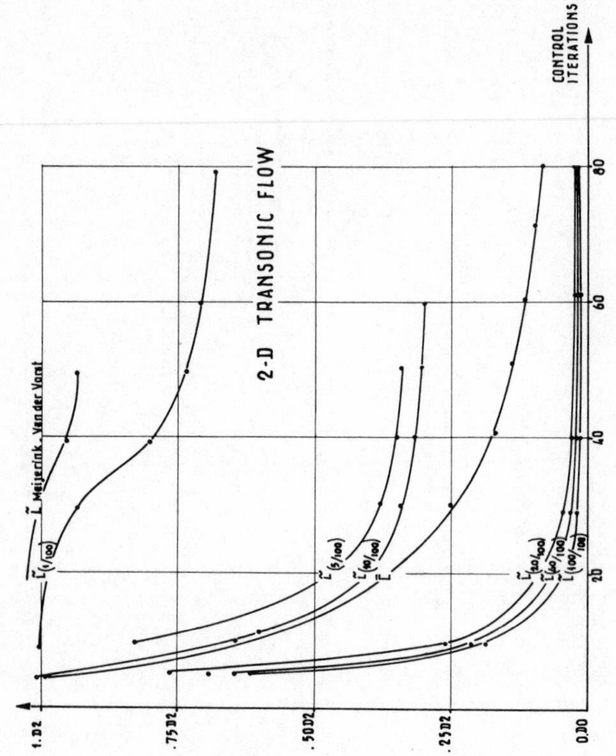
2-D TRANSONIC FLOW
CONTROL
ITERATIONS
1.D2
.75D2
.50D2
.25D2
0.D0
20
40
60
80
Meijerink, Van der Vorst

FIGURE 5.2

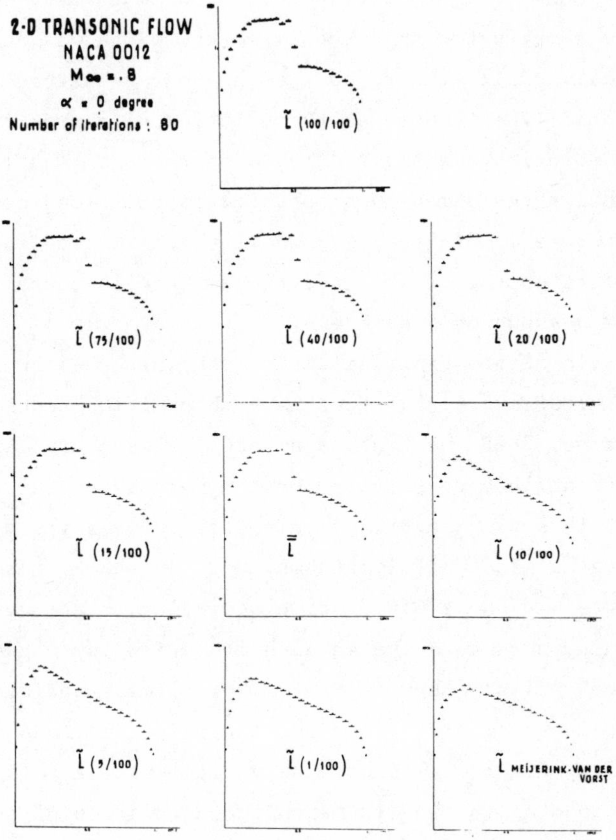
2-D TRANSONIC FLOW
NACA 0012
M∞ = .8
α = 0 degree
Number of iterations : 80
L̃ (100/100)
L̃ (75/100)
L̃ (40/100)
L̃ (20/100)
L̃ (15/100)
L̿
L̃ (10/100)
L̃ (5/100)
L̃ (1/100)
L̃ MEIJERINK-VAN DER VORST

FIGURE 5.3

if the degree of incompleteness is too large ; this is not surprising since discontinuities mean large spectrum, and clearly for small d, or for the Meijerink-Van de Vorst method, the high frequencies have been too much damped.

The second transonic flow test problem that we consider is much more complicated than the first one which was actually of sufficiently small size to be treated in core without difficulty ; now we would like to discuss the solution of a problem whose standard solution needs *out of core* operations (mostly disk transfers). This three-dimensional problem is the flow in and around the air intake of Figure 5.4 ; we want to simulate in fact the effect of an oblique wind ; additional details concerning the problem are shown on Figure 5.4.

Details of the finite element mesh which has been used are shown on Figures 5.4, 5.5 ; we have about 6×10^3 nodes, 2.5×10^4 tetrahedrons, 7×10^4 non zero elements ; these very large numbers imply that an in core solution by the methods of Sec. 5.2 is not possible. This fact justifies incomplete Cholesky's procedure for our conjugate gradient algorithms.

Figure 5.6 shows C.P.U. time according to the type of preconditioning used to solve at each iteration (by conjugate gradient) the various discrete Poisson problems associated to the nonlinear least squares - conjugate gradient methods discussed in Sec. 5.2. Figure 5.7 shows the variation of a *reference residual* (defined as in two-dimension) as a function of the iteration number, and according to the type of matrices $\underset{\sim}{\tilde{B}} = \underset{\sim}{\tilde{L}}\underset{\sim}{\tilde{L}}^t$ replacing the discrete variant $\underset{\sim}{B}$ of $-\Delta$ in the cost function (5.4) and the state equation (5.5).

The *supersonic* region on the surface of the air intake is shown on Figure 5.4.

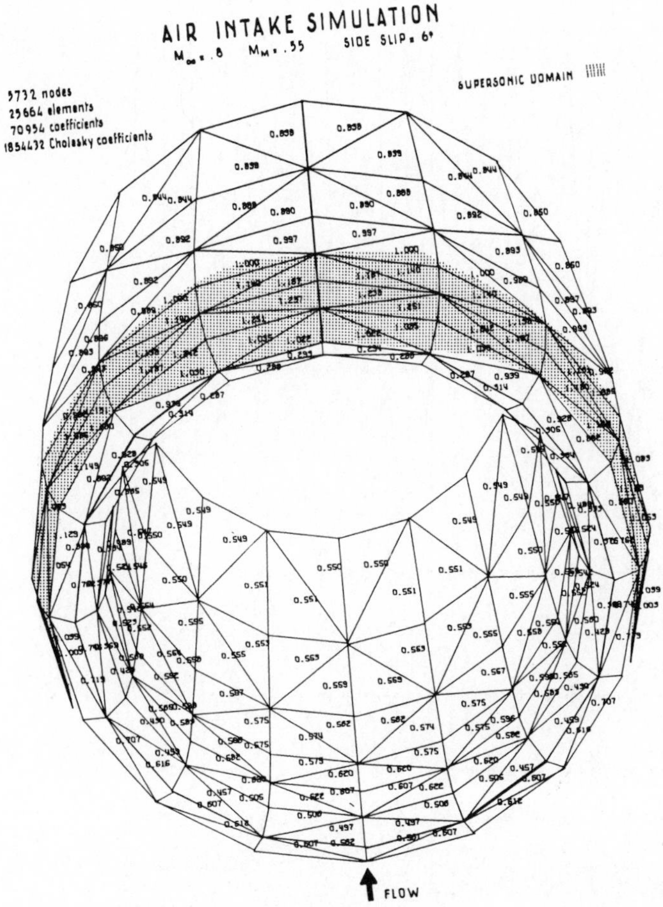
AIR INTAKE SIMULATION
M∞ = .8 MM = .55 SIDE SLIP = 6°
SUPERSONIC DOMAIN
5732 nodes
25664 elements
70954 coefficients
1854432 Cholesky coefficients
FLOW

FIGURE 5.4

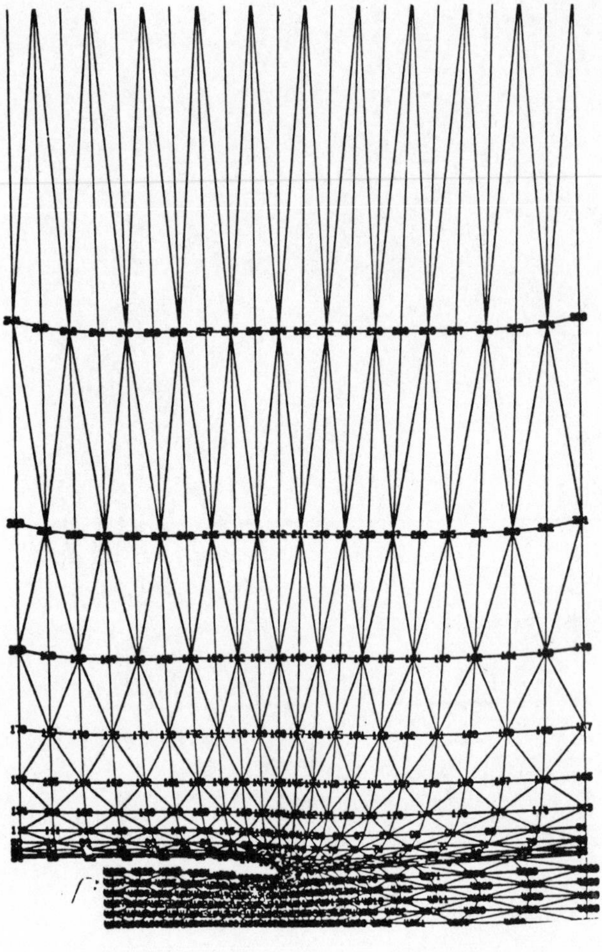

FIGURE 5.5

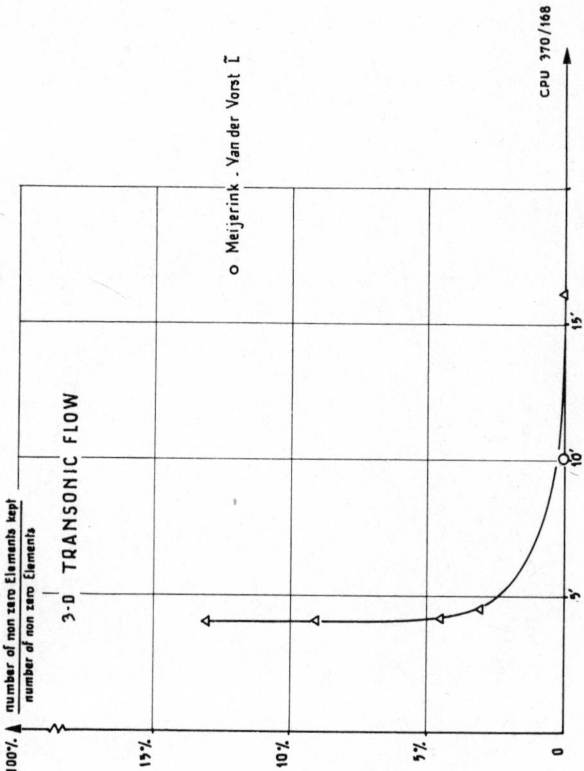
number of non zero Elements kept / number of non zero Elements
3-D TRANSONIC FLOW
o Meijerink . Van der Vorst Ĩ
CPU 370/168
100%
15%
10%
5%
0
5'
10'
15'

FIGURE 5.6

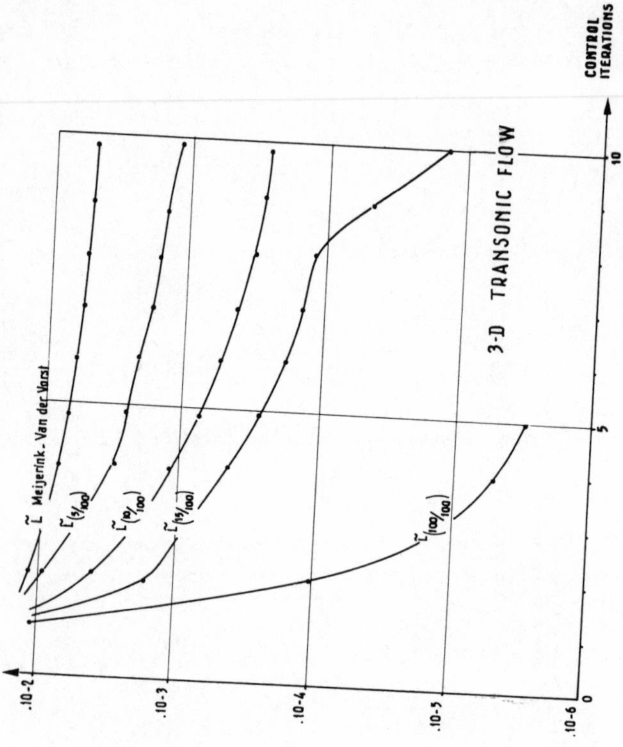
3-D TRANSONIC FLOW
CONTROL ITERATIONS
Meijerink. Van der Varst
.10-2
.10-3
.10-4
.10-5
.10-6
0
5
10

FIGURE 5.7

6. - NUMERICAL TESTS. (III) APPLICATION TO THE NAVIER-STOKES EQUATIONS FOR INCOMPRESSIBLE VISCOUS FLOWS.

6.1. Synopsis.

The preconditioned solution of the Navier-Stokes equations for incompressible viscous flows is discussed in a more detailed fashion in [11]. In this section we shall briefly describe how the general methods of Secs. 2, 3 apply to those Navier-Stokes problems.

6.2. Mathematical formulation.

We consider a *Newtonian incompressible viscous fluid*. If Ω and Γ denote respectively the region of the flow and its boundary, then this flow is governed by the *Navier-Stokes equations*

$$\frac{\partial \underset{\sim}{u}}{\partial t} - \nu\Delta\underset{\sim}{u} + (\underset{\sim}{u}\cdot\underset{\sim}{\nabla})\underset{\sim}{u} + \underset{\sim}{\nabla}p = \underset{\sim}{f} \text{ in } \Omega, \tag{6.1}$$

$$\underset{\sim}{\nabla}\cdot\underset{\sim}{u} = 0 \text{ in } \Omega \text{ (incompressibility condition)} \tag{6.2}$$

which in the steady case reduce to

$$-\nu\Delta\underset{\sim}{u} + (\underset{\sim}{u}\cdot\underset{\sim}{\nabla})\underset{\sim}{u} + \underset{\sim}{\nabla}p = \underset{\sim}{f} \text{ in } \Omega, \tag{6.3}$$

$$\underset{\sim}{\nabla}\cdot\underset{\sim}{u} = 0. \tag{6.4}$$

In (6.1)-(6.4) :

- $\underset{\sim}{u}$ is the *flow velocity*,
- p is the *pressure*,
- ν is the *viscosity* of the fluid,
- $\underset{\sim}{f}$ is a density of *external* forces.

Boundary conditions have to be added ; typical boundary conditions are for example

$$\underset{\sim}{u} = \underset{\sim}{g} \text{ on } \Gamma. \tag{6.5}$$

From (6.2),(6.4), $\underset{\sim}{g}$ has to obey $\int_\Gamma \underset{\sim}{g}\cdot\underset{\sim}{n}\, d\Gamma = 0$ where $\underset{\sim}{n}$ is the unit vector of the outward normal at Γ. Finally for the time dependent problem (6.1),(6.2) an initial condition such as

$$\underset{\sim}{u}(x,0) = \underset{\sim}{u}_o(x) \ a.e. \ in \ \Omega, \tag{6.6}$$

where $\underset{\sim}{u}_o$ is given, is prescribed. A *linear* problem of particular interest, namely the *Stokes problem* is obtained by cancelling the nonlinear terms in (6.1),(6.3).

6.3. Finite element approximation and iterative solution.
Using *implicit time discretization schemes,* like those discussed in [11],[12],[13],[21] we reduce the solution of the time dependent problem (6.1),(6.2),(6.5),(6.6) to that of a sequence of time independent problems very close to the steady Navier-Stokes problem (6.3)-(6.5).

Using then a convenient nonlinear least square conjugate gradient method (see [12] for more details) we reduce the solution of the Navier-Stokes problem (6.3)-(6.5) to a sequence of Stokes problems. We can say to summarize *that solving the steady and unsteady Navier-Stokes equations can be reduced to the solution of a sequence of Stokes problems*. Using in turn the results of [11],[12],[13],[21],[24], it appears that *the solution of these Stokes problems can be reduced to the solution of a sequence of Poisson problems.*

If we suppose that the *finite element approximations* to be used are those described in [11],[12],[13],[21],[24], then the solution of the steady and unsteady Navier-Stokes equations is eventually reduced to the solution of a sequence of discrete Poisson problems. From these reduction properties it is clear that a most important tool in order to solve the Navier-Stokes equations is again *efficient discrete Poisson solvers*. Thus the methods for solving linear problems discussed in Secs. 3 and 4 are still the basic tool for solving the above Navier-Stokes problems. Moreover most of the comments of Sec. 5, related to the solution of another nonlinear fluid dynamics problem still hold for the present problem.

The numerical results of Sec. 6.4 have been obtained using the preconditioning methods described with more details in [11]; actually using alternating direction methods like those described in the present paper we can greatly improve the performances of the Navier-Stokes solvers in [11] ; such an alternating direction method to solve the time dependent problem, but also quite efficient to solve steady flow problems, is the following (we consider the time discretization only) :

$$\underset{\sim}{u}^{o} = \underset{\sim}{u}_{o}, \tag{6.7}$$

then for $n \geq 0$ *compute* $\{\underset{\sim}{u}^{n+1/2}, p^{n+1/2}\}$ *and* $\underset{\sim}{u}^{n+1}$ *from* $\underset{\sim}{u}^{n}$, *by solving*

$$\frac{\underset{\sim}{u}^{n+1/2}-\underset{\sim}{u}^{n}}{(\Delta t/2)} - \frac{\nu}{2}\Delta\underset{\sim}{u}^{n+1/2} + \underset{\sim}{\nabla}p^{n+1/2} = \underset{\sim}{f}^{n+1/2} + \frac{\nu}{2}\Delta\underset{\sim}{u}^{n} - (\underset{\sim}{u}^{n}\cdot\underset{\sim}{\nabla})\underset{\sim}{u}^{n} \quad \textit{in } \Omega,$$

$$\underset{\sim}{\nabla}\cdot\underset{\sim}{u}^{n+1/2} = 0 \textit{ in } \Omega, \tag{6.8}$$

$$\underset{\sim}{u}^{n+1/2} = \underset{\sim}{g}^{n+1/2} \textit{ on } \Gamma,$$

and

$$\frac{\underset{\sim}{u}^{n+1}-\underset{\sim}{u}^{n+1/2}}{(\Delta t/2)} - \frac{\nu}{2}\Delta\underset{\sim}{u}^{n+1} + (\underset{\sim}{u}^{n+1}\cdot\underset{\sim}{\nabla})\underset{\sim}{u}^{n+1} = \underset{\sim}{f}^{n+1} + \frac{\nu}{2}\Delta\underset{\sim}{u}^{n+1/2} -$$

$$- \underset{\sim}{\nabla}p^{n+1/2} \textit{ in } \Omega, \tag{6.9}$$

$$\underset{\sim}{u}^{n+1} = \underset{\sim}{g}^{n+1} \textit{ on } \Gamma,$$

respectively ; we have used the notation $f^{j}(x) = f(x,j\Delta t), g^{j}(x) = g(x,j\Delta t)$. The finite element approximation of (6.7)-(6.9) is discussed in [13, Chap. 7] and [25]. We observe however that (6.8) is very close to a steady Stokes problem and that (6.9) is a *nonlinear Dirichlet system,* very close to the nonlinear Dirichlet model problem (2.25) ; actually (6.9) can be solved by least-square - conjugate gradient methods very close to those used to solve (2.25).

6.4. Numerical experiments.

We consider a fairly complicated time dependent problem since it is the simulation at Re = 750 of a two-dimensional unsteady incompressible Navier-Stokes flow *in* and *around* an *idealized air intake* (the front part of which is shown on Figures 6.1, 6.2), at a very large angle of attack (40 degrees) ; more details on this problem are given in [11] and [26] (and also [21],[24] for a similar problem at Re = 250). Using the finite element method in [11],[12],[13, Chap. 7], [21],[24],[25],[26], we introduce two finite element triangulations, the *coarser* one - $\mathcal{T}_h$ - to approximate the *pressure* (see Fig. 6.1), and $\tilde{\mathcal{T}}_h$ (see Fig. 6.2) to approximate the velocity. Using those grids we have defined *globally continuous, piecewise linear approximations* for both pressure and velocity. The corresponding number of nodes and triangles are respectively :

for the *pressure*, 1555 *nodes* , 2921 *triangles,*

for the *velocity*, 6023 *nodes* , 11684 *triangles.*

The Cholesky's factors of the discrete analogue of $-\Delta$ associated to $\mathcal{T}_h$ (resp. $\tilde{\mathcal{T}}_h$) contain about 1.65×10^5 (resp. 1.25×10^6) non zero elements ; thus from these large numbers a standard solution will require an *out of core* procedure.

Using incomplete Cholesky's factorization methods for solving by preconditioned conjugate gradient the discrete Poisson problems, we have been able to solve *in core, once* the approximate Cholesky's factor $\tilde{\underset{\sim}{L}}_h$ is constructed, the approximate Navier-Stokes problem ; actually since the solution of each discrete Stokes problem can be reduced (cf.the above references) to the solution of a finite number of discrete Poisson problems (7 in

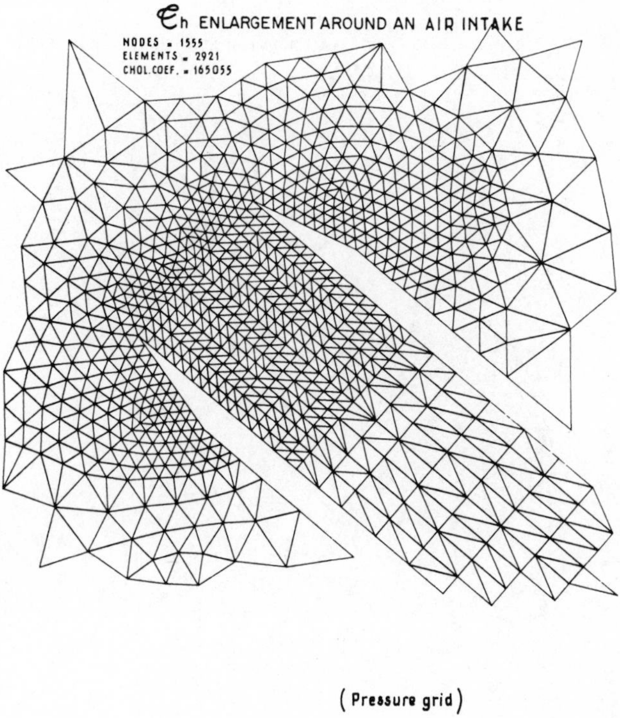
$\mathcal{C}_h$ ENLARGEMENT AROUND AN AIR INTAKE
NODES = 1555
ELEMENTS = 2921
CHOL.COEF. = 165055
(Pressure grid)

FIGURE 6.1

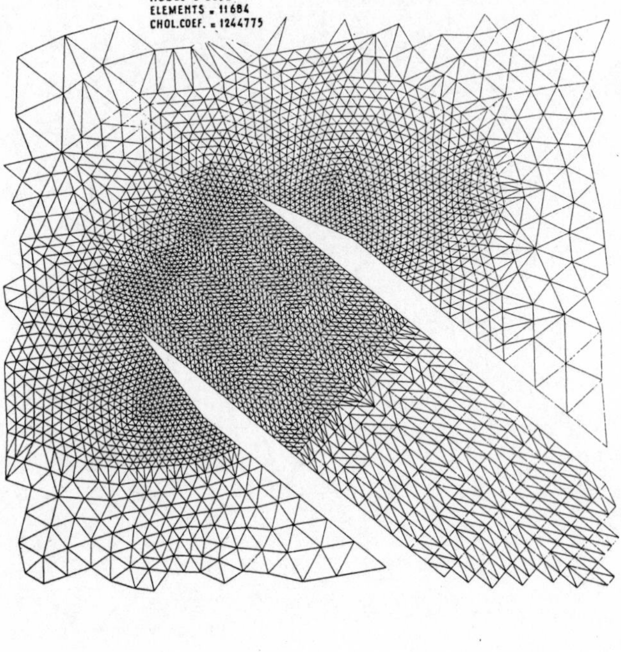
$\tilde{\mathcal{C}}_h$ ENLARGEMENT AROUND AN AIR INTAKE
NODES = 6032
ELEMENTS = 11684
CHOL.COEF. = 1244775

FIGURE 6.2

2-D, 9 in 3-D) *plus* a discrete boundary integral problem, whose matrix is *symmetric, positive definite,* and *full*, we have experimented the iterative solution of that last problem by preconditioned conjugate gradient also. Let us denote by $\underset{\sim}{S}_h$ the above boundary matrix ; since it is symmetric and positive definite we have

$$\underset{\sim}{S}_h = \underset{\sim}{R}_h \underset{\sim}{R}_h^t \tag{6.10}$$

where $\underset{\sim}{R}_h$ is a regular, lower triangular full matrix. We can define from $\underset{\sim}{R}_h$ various incomplete factors of $\underset{\sim}{S}_h$, using for example the methods described in Sec. 3 ; doing so we obtain an incomplete Cholesky's factor $\underset{\sim}{R}_h$ from which we define $\underset{\sim}{S}_h = \underset{\sim}{R}_h \underset{\sim}{R}_h^t$ which is still symmetric and positive definite and which can be used, therefore, as a preconditioner.

We have shown on Figure 6.3 the cost of 10 iterations according to the preconditioning strategy for the discrete Poisson problems and the above discrete boundary integral problem ; from Fig. 6.3 it is clear than as important saving of computer resources is obtained using $\underset{\sim}{\tilde{L}}_{(d/100)}$ and $\underset{\sim}{\tilde{R}}_{(d'/100)}$ with reasonably small values of d and d' (particularly of d).

We have indicated on Fig. 6.4 the velocity distribution at t=6 ; we observe a rather complicated *vortex* configuration. More results on this 2-D simulation are given in [11].

7. CONCLUSION.

From the above study it appears that conjugate gradient methods with preconditioning are well-suited to the solution of large problems, like those coming from industrial applications ; indeed up to 95% of core memory storage can be saved when compared to a direct method, and the computer time can be divided by 100 when compared to a non-preconditioned conjugate gradient method.

In conclusion preconditioned conjugate gradient is no more time consuming than direct methods but if a convenient precondi-

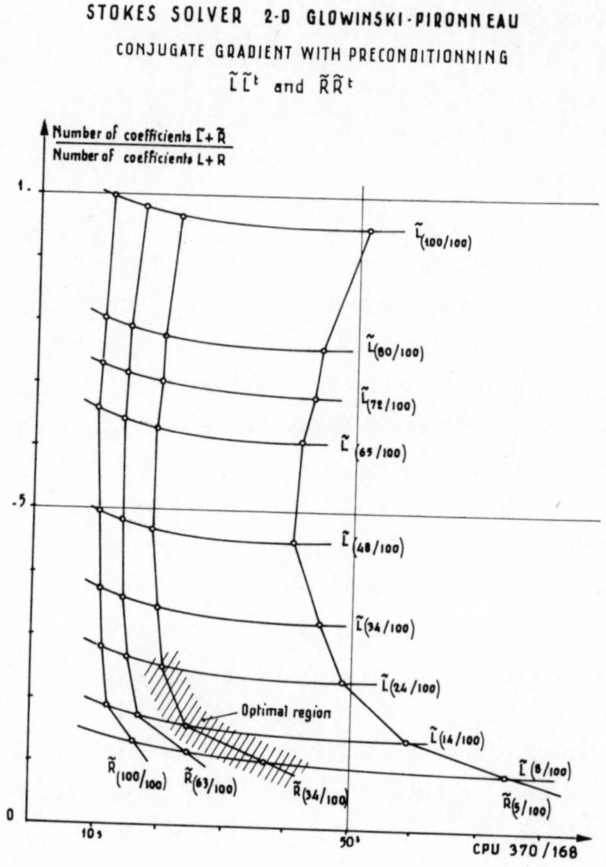
STOKES SOLVER 2-D GLOWINSKI-PIRONNEAU
CONJUGATE GRADIENT WITH PRECONDITIONNING
$\tilde{L}\tilde{L}^t$ and $\tilde{R}\tilde{R}^t$
Number of coefficients $\tilde{L}+\tilde{R}$ / Number of coefficients L+R
1.
.5
0
10s
50s
CPU 370/168
$\tilde{L}$(100/100)
$\tilde{L}$(80/100)
$\tilde{L}$(72/100)
$\tilde{L}$(65/100)
$\tilde{L}$(48/100)
$\tilde{L}$(34/100)
$\tilde{L}$(24/100)
$\tilde{L}$(14/100)
$\tilde{L}$(8/100)
Optimal region
$\tilde{R}$(100/100)
$\tilde{R}$(63/100)
$\tilde{R}$(34/100)
$\tilde{R}$(5/100)

FIGURE 6.3

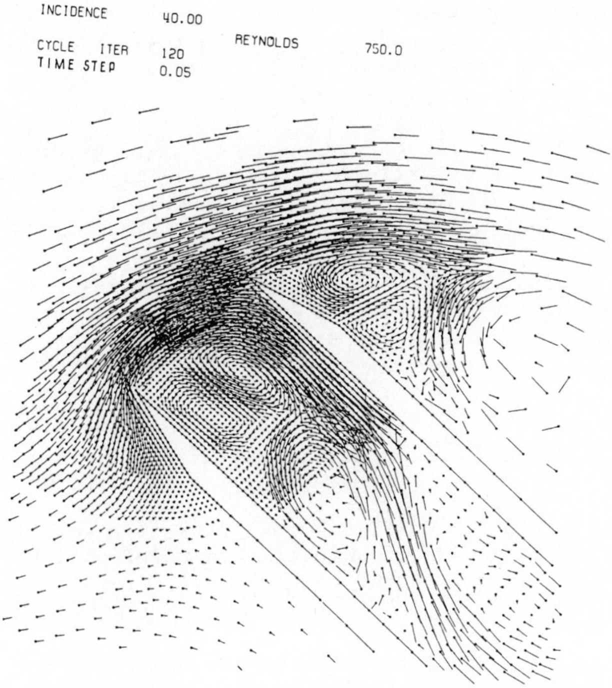

FIGURE 6.4

tioning is done it can use only 5% of the memory required by a direct method. From such properties this method is a very valuable tool for solving complicated Fluid Dyanmics like those considered in the above sections.

REFERENCES

[1] BARTELS R., DANIEL J.W., A conjugate gradient approach to nonlinear boundary value problems in irregular region, in G.A. Watson (ed.), Conference on the numerical solution of differential equations, Dundee, 1973, Lecture Notes in Math. Vol. 363, Springer-Verlag, Berlin, 1973.

[2] DOUGLAS J., DUPONT T., Preconditioned conjugate gradient iteration applied to Galerkin methods for mildly nonlinear Dirichlet problem, in J.R. Bunch and D.J. Rose (eds.), Sparse Matrix Computations, Acad. Press, New-York, 1976,

[3] CONCUS P., GOLUB G.H., O'LEARY D.P., Numerical solution of nonlinear partial differential equations by a generalized conjugate gradient method, Computing, 19, (1977), pp. 321-340.

[4] AXELSSON O., A class of iterative methods for finite element equations, Comp. Meth. Appl. Mech. Eng., 9, (1976), 2, pp. 123-138.

[5] MEIJERINK J.A., VAN DER VORST H.A., An iterative solution method for linear systems of which the coefficient matrix is a symmetric M-matrix, Math. of Comp., 31, (1977), pp. 148-162.

[6] KERSHAW D.S., The incomplete Cholesky-conjugate gradient method for the iterative solution of systems of linear equations, J. of Comp. Physics, 26, (1978), pp. 43-65.

[7] MANTEUFFEL J.A., Solving structures problems iteratively with a shifted incomplete Cholesky preconditioning, in Computing Methods in Applied Sciences and Engineering, R. Glowinski, J.L. Lions (eds.), North-Holland, Amsterdam, 1980, pp. 427-444.

[8] WESSELING P., SONNEVELD P., Numerical experiments with a multiple grid and a preconditioned Lanczos type method, in Approximation Methods for Navier-Stokes problems, Springer-Verlag, Berlin, R. Rautman (ed.), Lecture Notes in Math., Vol. 771, 1980, pp. 543-562.

[9] GLOWINSKI R., PERIAUX J., PIRONNEAU O., An efficient preconditioning scheme for iterative numerical solution of

partial differential equations, Applied Mathematical Modelling, 4, (1980), pp. 187-192.

[10] GLOWINSKI R., MANTEL B., PERIAUX J., PIRONNEAU O., POIRIER G., An efficient preconditioned conjugate gradient method applied to nonlinear problems in fluid dynamics via least square formulation, in Computing Methods in Applied Sciences and Engineering, R. Glowinki, J.L. Lions (eds.), North-Holland, Amsterdam, 1980, pp. 445-487.

[11] GLOWINSKI R., MANTEL B. PERIAUX J., PERRIER P., PIRONNEAU O., On an efficient new preconditioned conjugate gradient method, application to the in core solution of the Navier-Stokes equations via nonlinear least squares and finite element methods, in Vol. 2 of the Proceedings of the Third International Conference on Finite Elements in Flow Problems, Banff, Alberta, Canada, 10-13 June, 1980, pp. 229-255.

[12] BRISTEAU M.O., GLOWINSKI R., PERIAUX J., PERRIER P., PIRONNEAU O., POIRIER G., Application of optimal control and finite element methods to the calculation of transonic flows and incompressible viscous flows, in Numerical Methods in Applied Fluid Dynamics, B. Hunt (ed.), Academic Press, London, 1980, pp. 203-312.

[13] GLOWINSKI R., Numerical Methods for Nonlinear Variational Problems, 2nd edition, Springer-Verlag (to appear).

[14] CEA J., GEYMONAT G., Une méthode de linéarisation via l'optimisation, Bol. Istituto Nazionale Alta Mat., Symp. Math., 10, (Bologna, 1972), pp. 431-451.

[15] LOZI R., Analyse Numérique de certains problèmes de bifurcation, Thèse de 3ème Cycle, Université de Nice, 1975.

[16] POLAK E., Computational Methods in Optimization, Acad. Press, New-York, 1971.

[17] POWEL M.J.D., Restart procedure for the conjugate gradient method, Math. Programming, 12, (1977), pp. 148-162.

[18] WILDE D.J., BEIGHTLER C.S., Foundations of Optimization, Prentice Hall, Englewood Cliffs, N.J., 1967.

[19] BRENT R., Algorithms for minimization without derivatives, Prentice Hall, Englewood Cliffs, N.J., 1973.

[20] REINHART L., Sur la résolution numérique de problèmes aux limites non linéaires par des méthodes de continuation, Thèse de 3ème cycle, Université Pierre et Marie Curie,

Paris, June 1980.

[21] PERIAUX J., Résolution de quelques problèmes non linéaires en aérodynamique par des méthodes d'éléments finis et de moindres carrés fonctionnels, Thèse de 3ème cycle, Université Pierre et Marie Curie, Juin 1979.

[22] BRISTEAU M.O., GLOWINSKI R., PERIAUX J., PERRIER P., PIRONNEAU O., POIRIER G., Transonic flow simulations by finite elements and least squares methods, in Vol. 1 of the Proceedings of the Third International Conference on Finite Elements in Flow Problems, Banff, Alberta, Canada, 10-13 June, 1980, pp. 11-29.

[23] BRISTEAU M.O., Application of optimal control theory to transonic flow computations by finite element methods, in Computing Methods in Applied Sciences and Engineering, 1977, (II), Proceedings, IRIA, Paris, R. Glowinski, J.L. Lions (Ed.), Lecture Notes in Physics, 91, Springer, Berlin, 1979, pp. 103-124.

[24] BRISTEAU M.O., GLOWINSKI R., MANTEL B., PERIAUX J., PERRIER P., PIRONNEAU O., A finite element approximation of Navier-Stokes equations for incompressible viscous fluids. Iterative methods of solution, in Approximation methods for Navier-Stokes problems, R. Rautman (ed.), Lecture Notes in Math., Vol. 771, Springer-Verlag, Berlin, 1980, pp. 78-128.

[25] GLOWINSKI R., MANTEL B., PERIAUX J., Numerical solution of the time dependent Navier-Stokes equations for incompressible viscous fluids by finite element and alternating direction methods, to appear in the Proceedings of the Conference on Numerical Methods in Aeronautical Fluid Dynamics, Reading, 30th March - 1st April, 1981.

[26] GLOWINSKI R., MANTEL B., PERIAUX J., PIRONNEAU O., A finite element approximation of Navier-Stokes equations for incompressible viscous fluids. Functional least-square methods of solution, in : Computer Methods in Fluids, C. Taylor, K. Morgan, C.A. Brebbia (eds.), Pentech Press, 1980, pp. 84-133.

PRECONDITIONED CONJUGATE GRADIENT ALGORITHMS FOR SOLVING FINITE DIFFERENCE SYSTEMS

D.A.H. Jacobs
Central Electricity Generating Board, Leatherhead,
Surrey, England.

ABSTRACT

The solution of large systems of algebraic equations is a major part of many finite difference and finite element models. The increasing size of these models in terms of the number of nodal values or degrees of freedom and the complexity of the problems being considered makes solution difficult and expensive.
The paper describes research aimed at obtaining methods which can be implemented as general purpose algorithms for problems with coefficient matrices which need not be symmetric nor positive definite nor non-singular. The results of a series of tests on a range of problems are described and from these conclusions on the optimal choice of method for general purpose algorithms are drawn.

1. INTRODUCTION

We consider the solution of large systems of algebraic equations arising from finite difference models of elliptical partial differential equations, although the methods described are more widely applicable. Such systems of equations are often large which generally implies that they are expensive to solve. However they possess considerable structure which can be exploited to save computation. The equations may be linear. Alternatively, they may be equations which have been linearized or are being solved by Newton iteration; in either case the repeated solution of sequences of linear systems is required.

The conjugate-gradient method was developed by Hestenes and Stiefel [1] and in theory is a direct method for solving symmetric, positive definite systems of algebraic equations. By 'direct' we mean that in a finite and predetermined number of computations, and also a predetermined amount of storage, it theoretically obtains the 'exact' solution to the system. However, when the calculations are executed on a digital computer working with finite precision arithmetic, number storage and representation, the method behaves much more like an iterative procedure [2]. For a system of algebraic equations of order N, the conjugate gradient method using exact arithmetic would involve at most N steps. What is of considerable importance is that at each step the solution is progressively refined. With finite precision arithmetic the finite termination property no longer holds, and the solution attainable to the machine precision is obtained only after, at best, a very large number of steps. However it is generally found that each step does progressively improve the solution.

Two points in real applications are of significance. First, generally a solution is required to only a finite precision, typically three or four decimal places. Thus for a linear problem, if a reasonable rate of convergence can be achieved, only a small number of iterations of an iterative method need be used. For a non-linear problem in which the solution is obtained by successive steps of a linearized model, the solution of each 'linear' step generally starts from a good approximation, namely the last approximate solution. In addition, for such problems, only an approximate solution is required at each step since the equations being solved are linearized approximations of the real set. Secondly when using the conjugate gradient method on a symmetric, positive definite system it can be shown [3] that each of the steps successively refines the solution, namely the residual is strictly monotonically decreasing. Combining these features means that the conjugate gradient method can be viewed and indeed used, as an iterative method. Clearly one would be aiming to terminate the iterations long before N steps had been used, and so it is the convergence property that is most important.

The rate at which the solution is refined when using the conjugate gradient method depends on the matrix of coefficients -

to be precise it depends on the number of distinct eigenvalues and the spread of their values. In the extreme case of a unit matrix, convergence is obtained in only one step. The fewer distinct eigenvalues, the fewer steps are needed to obtain the exact solution. When a solution is only required to a finite precision, it is not only the number of distinct eigenvalues that is relevant, but also the clustering of the eigenvalues into groups. Reid [2] and Jennings [4] detail some illuminating experiments . The theoretical result, of which these are manifestations, is that the conjugate gradient method requires, using exact arithmetic, one step for each distinct eigenvalue. A number of theoretical results about the properties and behaviour of the conjugate gradient method can be deduced from analogous results derived for the Lanczos [5] algorithm for finding the eigenvalues of a matrix. In practice if the eigenvalues are clustered around a small number of distinct values, then the use of conjugate gradients will significantly reduce the residual in a number of steps equal to the number of clusters.

However for the system of algebraic equations derived from finite difference or finite element models of partial differential equations, the eigenvalues of the coefficient matrix are generally well distributed. Indeed in the case of Poisson's equation on a uniform Cartesian square mesh, using finite differences, the eigenvalues are uniformly distributed. Thus the conjugate gradient method is not suitable for such systems, at least not without some form of modification.

Meijerink and van der Vorst [6] produced the breakthrough in the use of conjugate gradient methods for such problems by employing a preconditioning which transfomed the system of equations into a form for which a conjugate gradient method 'converges' faster, in fact one in which the eigenvalues are clustered. Such a procedure turns out to be highly efficient. Their particular method is applicable to symmetric matrices which have a special form, termed M matrices.

This paper details some further developments of this method, namely variations and extensions of preconditioned conjugate gradient methods to enable it to be used on a very much wider class of problems. In Section 2 various types of preconditionings are considered and compared, including one based on the approximate factorization used in the Strongly Implicit Procedure [7]. The results of a large number of tests are described in Section 3 and 'the' best choice of preconditioning is made.

2. TYPES OF PRECONDITIONED CONJUGATE GRADIENT METHODS

2.1 Introduction

The method described by Meijerink and van der Vorst [6] was restricted to systems characterized by symmetric 'M' matrices. The very rapid rates of convergence achieved have directed efforts towards extending or adapting the method to a wider range of systems. In this section a number of these extensions are described. In Section 3 the results of the use of some of them on a variety of problems are described.

2.2 Symmetric 'M' Matrices

The original methods of Meijerink and van der Vorst [6] were restricted to symmetric 'M' matrices. The sparsity pattern of the approximate Cholesky factors was forced to be identical to that of the triangular parts of the coefficient matrix A (the sparsity being defined in terms of a diagonal structure) in the case of 1CCG(0), in the case of ICCG(3) three extra diagonals were allowed to have non-zero elements in the factor matrices over and above those in the corresponding positions of the matrix A. In both cases the values of the non-zero elements of the approximate factors were determined by forcing the equality of the elements of the preconditioning matrix LL^T and the matrix A on those diagonals in positions defined by the non-zero diagonals of the matrix A.

For this special case of symmetric 'M' matrices, Meijerink and van der Vorst [6] proved that these approximate factorizations produced non-singular factor matrices L and that the factorization process was at least as stable as a complete Cholesky factorization.

In two later papers (Meijerink and van der Vorst [8]. [9]) the ICCG(0) and ICCG(3) methods are shown to be two particular members of a whole class of methods for symmetric 'M' matrices for which the incomplete Cholesky factors are derived by either ignoring fill-ins altogether or by accepting fill-ins if they meet a particular condition. The method of symmetric successive over-relaxation of Axelsson [10] is also shown to belong to this case. In [8] a comparison of various strategies on the selection of the extra diagonals in the factor matrix L is given based on a number of trial solutions of real problems.

For symmetric 'M' matrices which do not have a diagonally structured sparsity, Meijerink and van der Vorst [9] mention the analogous ICCG method, using a sparsity pattern in the factors identical to that in the original matrix. They present no results for this case.

2.3 General Symmetric Positive Definite Matrices

The method described by Meijerink and van der Vorst [6] was restricted to systems characterized by 'M' matrices. Kershaw [11] very successfully used the same basic method with a minor modification for solving a wider class of systems for which the matrix is symmetric positive definite, but not necessarily an 'M'-matrix. With an 'M'-matrix, Meijerink and van der Vorst had proved that the main diagonal terms of the approximate factors are always positive for the inexact factorization they used, as they are for the exact factorization. A similar result does not hold for non-'M' matrices. It is therefore necessary to ensure that the matrix factor is non-singular in order that the preconditioning is non-singular. This criterion is satisfied if the main diagonal elements of the approximate factors are all non-zero. Moreover if the minimization property for the measure h^2 of the residual is to carry over to the preconditioned systems, the preconditioning matrix LL^T must also be positive definite. To achieve these requirements, Kershaw [11] recommended forcing all values of the main diagonal elements to be positive. If any main diagonal element was calculated to be non-positive during the approximate factorization, he set it to a 'suitable' positive value. Such a modification, he argued, would only have a very limited effect on the eignevalue spectrum of the preconditioning matrix, provided the number of such 'corrections' was small. The basis of the calculation for the other elements of the approximate factors is otherwise the same as that of Meijerink and van der Vorst [6], namely the equality of non-zero elements of the matrix A with the corresponding elements of the preconditioning matrix LL^T, apart from the main diagonal modification which may be needed, as described above.

Kershaw described very favourable experience with this method, and this has been echoed by other users.

Munksgaard [12] developed an analogous algorithm and implemented it in code which made no assumptions about a regular sparsity structure for the matrix A. His method also incorporates a very flexible treatment of fill-ins whereby the user specifies a threshold factor which is used to gauge the importance of fill-ins compared to the main diagonal element of that row and column. The fill-in is ignored if it is smaller in magnitude than the tolerance times the main diagonal element. Like Kershaw [11], Munksgaard modified the diagonal coefficients of the approximate factor if they become non-positive to ensure the positive definiteness of the preconditioning matrix.

There is another difference between the preconditionings used by Munksgaard and those of Kershaw and Meijerink and van der Vorst. The latter calculated the elements of the preconditioning matrix from relationships obtained by forcing, wherever possible, the equality of the non-zero elements of the matrix A with the corresponding elements of the preconditioning matrix LL^T. Munksgaard based his calculation for the off-main diagonal elements of the factor matrices on the same relationship; however for the main diagonal terms of the factors he used the method proposed by Gustafsson [13] which equated each row sum of the matrix A with the corresponding row sum of the preconditioning matrix LL^T. The desirability of maintaining the row sum can be estimated from an examination of the Rayleigh quotients [14] of the coefficient matrix and of the preconditioning matrix, taking Poisson's equation as the example differential equation.

Munksgaard [12] reports results for his incomplete threshold factorization used to precondition a system prior to solution by conjugate gradients on a wide range of problems. The method is certainly an interesting flexible approach which probably has considerable potential for more general sparse systems than those derived from finite element and finite difference models using reasonably regular meshes.

Wong [15] compared three different forms of preconditioning for a five diagonal symmetric, positive definite system. Each enforced the equality of row sums of the product LL^T and the original matrix A, but they differed in the other equalities used to prescribe the elements of the factors. He found the 'best', based on a series of tests, was to employ:

(i) row sums equality of LL^T and A

and (ii) equality of non-zero elements of A which are off the main diagonal with the corresponding elements of LL^T.

From the work summarized here we can conclude that at present the best algorithm for preconditioning a system of symmetric positive definite equations prior to solution by conjugate gradients forces the equality of row sums of the original matrix and of the preconditioning matrix. For many applications restricting the sparsity of the approximate Cholesky factors to be the same as in the respective triangular part of the original matrix provides a reasonable approximation with which to precondition the original system. A suitable condition for the specification of the elements of the approximate factors (in addition to forcing the equality of row sums) is the equality of the off-main diagonal elements of the preconditioning matrix with those of the original matrix.

For diagonally structured matrices such as those derived when using finite differences, this algorithm is particularly economic on storage requirements for the approximate Cholesky factors since only the main diagonal elements need be stored (or calculated), the off-main diagonal elements being simply related to those of the original matrix. The threshold preconditioning method of Munksgaard provides a extension to this preferred alrogithm for the case when the fill-ins that are ignored in this algorithm are relatively large. For general applications it is not considered that the added complexity and overheads would warrant using such a method.

2.4 General Asymmetric Positive Definite Matrices

The bi-conjugate gradient algorithm [16] provides a conjugate gradient type of algorithm for solving non-symmetric matrices. Wong [15] describes the generalization of the algorithm described in the previous sub-section for preconditioning non-symmetric matrices. An approximate LU factorization is used in which the sparsity of the approximate factor matrices L and U is the same as the respective parts of the original matrix A. The elements of L and U are determined from analogous constraints to those used in the above algorithm, namely the equality of row sums of the matrix A and the preconditioning matrix LU, and the equality of non-zero off-main diagonal elements of A with the corresponding elements of the matrix LU. Wong reports results on a number of test problems with orders up to 8000 for which the rates of convergence per step are of comparable 'quality' to those obtained from similar sized and conditioned symmetric systems. Note that the bi-conjugate gradient method does involve more computation per iteration than the basic conjugate gradient algorithm.

2.5 SIP Preconditioning

In this sub-section we describe an alternative preconditioning which we conceived for use with both symmetric and asymmetric systems of equations. The approximate factorization is based on that used by Stone [7] in the Strongly Implicit Procedure (SIP) which has been used with considerable success on a wide variety of problems.

Given a general matrix A (which need not be symmetric, nor positive definite), there exists an exact LU factorization provided the leading sub-matrices are non-singular. In general if the matrix A is a banded structure, the exact factors L and U will have non-zero elements in almost every location within this band.

Stone devised an iterative scheme (SIP) based on an approximate LU factorization of a five diagonal matrix such as that obtained from a finite difference model of Poisson's equation in a rectangle.

In place of using the exact LU factorization in which a large amount of fill-in occurs, consider the product of two matrices L and U in which L has only three non-zero diagonals taken in the same positions as those in the low triangular part of the matrix A, and the matrix U has only three non-zero diagonals taken in the same positions as those in the upper triangular part of the matrix A. The product of two such matrices is a seven diagonal matrix which approximates a five diagonal matrix.

In this manner we obtain an approximate LU factorization of the coefficient matrix A with

$$\Delta T = LUT + O(\Delta x^3, \Delta y^3, \Delta x \Delta y).$$

We can use this to precondition the system of equations which will then be solved using a suitable conjugate gradient method.

Since $B = L^{-1} AU^{-1}$ is approximately equal to the unit matrix, we would expect this matrix to have only a few distinct eigenvalues. The preconditioned system should therefore be solved more efficiently by conjugate gradients than the original system.

The objective in deriving this preconditioning was to employ a 'complete' cancellation of the extra unwanted elements in the approximate factorization. The preconditioning matrix then not only preserves row sums but can be shown to differ from the original matrix by a small amount quantifiable in terms of the mesh size in the finite difference scheme. It was anticipated, because of this very close relationship, that the basic conjugate gradient method could be used to solve the preconditioned system irrespective of symmetry, positive definiteness, etc.

A large number of test problems have been used to evaluate this method and to compare and contrast it with other preconditioning algorithms. The results, which will be described in the next section, indicate that for symmetric positive definite systems, the SIP preconditioning method and the algorithm described in the previous sub-section are equivalently good. However, surprisingly, for non-symmetric and singular systems, for which the bi-conjugate gradient method is used to solve the preconditioned system, the SIP preconditioning is totally unsatisfactory.

2.6 Preconditioning Singular Systems

In seeking a preconditioning algorithm suitable for general and wide application, the problem of solving singular systems of algebraic equations arises. Such systems are obtained, for example, when modelling Poisson's equation in a region on the boundary of which conditions on the derivative alone are prescribed [17]. We have tried several ways of treating such problems.

First in seeking the approximate factors of the matrix, one must ensure that these factors are not themselves singular. To ensure this we simply use an extension of the pragmatic approach of Kershaw [11] which he recommended for positive definite systems which are not characterized by an 'M' matrix. This ensures that the preconditioning matrix is non-singular, but the preconditioned systems to be solved by a conjugate gradient method is, of course, still singular, this being carried over from the original matrix. The basic conjugate gradient method has proved, as shown in the next section, unsuitable for such systems. The bi-conjugate gradient algorithm which was developed for both non-symmetric and indefinite systems has worked very well in these cases. Indeed it will be seen in the next section that for such problems, the rate of convergence is only slightly less fast than for problems with the same number of algebraic equations but which are non-singular.

When solving singular systems of equations with any gradient method, the possibility arises of choosing a direction of search which is parallel to the eigenvector corresponding to the zero eigenvalue. For a conjugate gradient algorithm this is most unlikely unless the initial solution exactly satisfies the equations.

2.7 Summary

The separate parts of this section have described a range of different preconditionings which can be applied prior to using a conjugate gradient method. Taken together they cover the whole spectrum of possible matrix types. For some systems, a number of different possible preconditions have been detailed. In the next section these are compared and contrasted (where this has not been done previously) by using the various combinations of preconditionings and conjugate gradient methods to solve a number of different problems.

3. COMPARISON OF ALGORITHMS

3.1 Introduction

In this section a number of the methods of preconditioning a system of equations together with solution by a conjugate gradient method are compared and contrasted on a range of different problems, both in terms of the size of system and the complexity, which could be quantified by the condition number of the matrix.

The objective is to identify a suitable algorithm to form the basis of general purpose solution programs. The solution methods are briefly detailed in the next sub-section and the test problems described in Section 3.3. In Section 3.4 results are tabulated and discussed.

3.2 The Solution Algorithms

All the programs employed double precision real arithmetic on the IBM 370/168 computer. Conclusions were difficult to make from earlier comparisons of tests for which single precision was used, confirming the need to use a reasonable precision of computation with conjugate gradient algorithms.

The Strongly Implicit Procedure is used as a measure against which to compare the conjugate gradient based methods. It was devised by Stone [7] and has been developed and implemented in a number of packages [18].

Algorithm 1: Basic Conjugate Gradient Method

This used the basic conjugate gradient algorithm, with no preconditioning. The residual calculation employed was the recursive form, as with all the other methods. The calculation of α_i used the cross formula $(p_i, r_i)/(p_i, Ap_i)$ and for β_i the formula $-(r_{i+1}, Ap_i)/(p_i, Ap_i)$. These same formulae were used in all the other algorithms which were based on conjugate gradients. A measure of the utility of the various preconditionings can be obtained by comparing the efficiencies of the methods in which they are employed with that of this algorithm.

Algorithm 2: Diagonal Scaling With the Basic Conjugate Gradient Method

This algorithm employed the simplest form of preconditioning for which the approximate factorization is restricted to be a main diagonal matrix only whose non-zero elements are simply the main diagonal elements of the original matrix. The resulting algorithm is then the same as applying the conjugate gradient method to the original system after it has been scaled by its diagonal. This scaling is, in one measure, optimal since it approximately minimizes the condition number amongst all diagonal scalings [19]. Note also that no additional storage is required for the preconditioning factors.

Algorithm 3: ICCG(0)

This algorithm is that of Meijerink and van der Vorst [6]. The preconditioning factors have the same diagonal sparsity structure as that of the respective triangles of the original matrix. The elements of the factors are chosen so that the elements on the non-zero diagonals of the original matrix are equal to the corresponding elements of the preconditioning matrix. Note that row sums are not preserved in this preconditioning.

Algorithm 4: 'Row Sums Equality' Preconditioning With Basic Conjugate Gradients

This algorithm is the chosen algorithm of Wong [15]. The preconditioning factors have the same diagonal sparsity structure as that of the respective triangles of the original matrix. The elements of the factors are chosen to ensure the equality of row sums of the preconditioning matrix and of the original matrix, and of the corresponding elements on the non-zero off-main diagonals of these two matrixes.

Algorithm 5: SIP Preconditioning With Basic Conjugate Gradients

For this algorithm the preconditioning was based on the approximate LU factorization together with complete cancellation of the extra terms using the physical representation of the preconditioning matrix. As described in Section 2.5, this preconditioning, like that of algorithm 4, preserved row sums.

Algorithms 6 to 10

These are the analogous algorithms to algorithms (1) to (5) which use the bi-conjugate gradient method instead of the basic conjugate gradient method. The calculations of both the residual and bi-residual are based on the use of the recursive formulae, and the calculation of the α_i uses the expression $\frac{(r_i, p_i)}{(\bar{p}_i, Ap_i)}$ and of β_i the expression $-\frac{(r_{i+1}, Ap_i)}{(\bar{p}_i, Ap_i)}$.

Note on Implementation

The basic conjugate gradient method is only applicable to symmetric positive definite systems. However in deriving finite difference approximations to elliptic partial differential equations, the systems obtained are not always symmetric, although by a scaling and some substitution of known values of the elements of the vector x, they may be capable of being made symmetric. The basic conjugate gradient method, and indeed the preconditionings, which are directed at symmetric systems, appear to work quite satisfactorily for such 'modified' systems, although it is of course necessary to use an LU factorization rather than a Cholesky one for the preconditioning. For this reason all the algorithms are written with sufficient generality to be applicable to asymmetric systems [20].

Some of the test problems used produce systems of equations which are either singular or are highly ill-conditioned. The determination of approximate matrix factors for use in subsequent preconditioning has already been discussed in Section 2.6. All the implementations of the algorithms have built-in checks to detect a singular or near singular approximate factors, namely a zero or very 'small' element on the main diagonal compared to that of the original matrix. The replacement procedure described in Section 2.6 is then used. The tables of results indicate when such a procedure was invoked.

3.3 The Test Examples

Laplace's Equation With Dirichlet Boundary Conditions

The first test problems involved the solution of the systems of algebraic equations resulting from a finite difference model of Laplace's equation on a uniform square mesh with equal numbers of nodes in each direction. The boundary conditions specify the function value to be unity at all points of the boundary and an initial approximation is used which is generated by a uniform distribution random number generator. The boundary nodes are included in the mesh, as would normally be done for a real problem.

For the first phase of tests mesh sizes of 11 and 31 nodes in each direction were used. For the second phase, ten different mesh sizes were used, with 11,21,31,41,51,61,71,81,101 and 201 nodes in each of the two directions, i.e. producing systems with orders from 121 to 40401. The systems are positive definite and could be transformed to be symmetric.

Laplace's Equation With Neumann Boundary Conditions

The second test problems involved the solution of a similar problem to that described above, the difference being the use of zero derivative (Neumann) conditions on the whole of the boundary. The solution to the elliptic partial differential equation is not unique, indeed an arbitrary constant can be added to the solution. The algebraic system of equations is singular, with a rank one deficiency. For the first phase of tests, two different sizes of mesh were used, namely with 11 and 31 nodes in each direction. For the second phase of tests the same problems with 101 and 201 nodes were added. The initial approximation to the solution was the same as that used for the first set of test problems.

The systems of equations obtained from this model therefore range in order from 121 to 40401. They are singular with rank-one deficiency. The equations for the nodes on the boundary contribute a non-symmetric part to the matrix.

Laplace's Equation With Mixed Conditions

This particular problem has been used because it provides an example for which the condition number of the matrix of coefficients can be varied across a wide range by changing only one parameter without altering the size of the system. The problem arose from a gross simplification of a model of galvanic corrosion [21].

The solution to Laplace's equation is sought in unit square region. The boundary conditions are as shown below

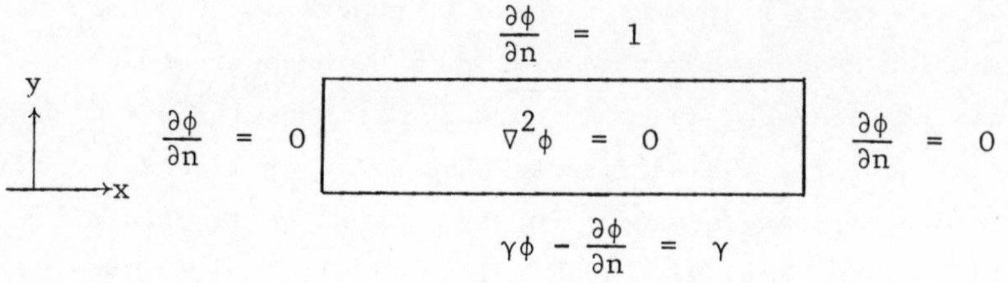

For non-zero values of the parameter γ, the analytic solution for suitably chosen reference axes, is

$$\phi = y .$$

However if $\gamma = 0$, the boundary conditions are all Neumann, and the system of algebraic equations is singular. The solution is then $\phi = y + k$ where k is an undetermined constant. By taking values of γ from zero upwards, systems of equations can be obtained with a wide range of condition numbers. The mesh used had 15 nodes in the x direction and 21 in the y direction with uniform mesh increments in each of these directions. Two values of γ were used, namely 1.0 and 10^{-5}.

The system of equations obtained are of order 315 and are non-symmetric. The condition number ranges from about 10^2 to about 10^8.

3.4 Tables of Results and Commentary

Table 1 summarizes the results obtained using all the algorithms described in Section 3.2 to solve the first phase of problems described in Section 3.3, that is two Dirichlet problems, two pure Neumann problems and two mixed boundary condition problems. This first phase of problems was used to narrow the number of algorithms to be condidates for further tests, particularly on large problems which are expensive in computational time to solve, with the aim of identifying an all-round general purpose algorithm.

The main result given in the tables for each test is the number of iterations used to reduce the maximum absolute element of both the residual vector and the change vector to less than 10^{-4}. Note that the residual and change which are used are those of the original system, namely the system Ax = b, and not of the preconditioned system. Although some extra computation is involved in the algorithm to obtain these quantities, it does mean that direct comparisons can be made.

As noted in Section 3.2, a check is made in all the algorithms on the singularity or near-singularity of the approximate factor matrices. The tolerance factor used is 10^{-2}. A note is made in the tables of those test cases in which this modification was invoked.

In comparing and contrasting the various algorithms on the different problems, the relative computational speeds of the algorithms should be borne in mind, but not to the exclusion of seeking a robust and reliable scheme. As a guide, the times for a conjugate gradient step of the preconditioning algorithms are all about equal, and they are about twice the time for an iteration of the SIP method. The algorithms employing bi-conjugate gradients use about twice the computation as that of the conjugate gradient algorithms.

The following conclusions can be drawn from the first table of results.

For the symmetric positive definite systems considered, i.e. the first test problem, the basic conjugate gradient method is only marginally improved by diagonal scaling - though note that the original system is probably not very far from being optimally scaled anyway. All the other preconditioning methods produce significant gains in convergence efficiency. The most effective scheme emerges as algorithm (4) (and indeed algorithm (9) which is equivalent in this case). Note that the algorithm (5) which employs a fully corrected SIP type preconditioning introduces a non-symmetry into the system of preconditioned equations which is

TABLE 1 Number of Steps of the Conjugate Gradient Algorithm to Achieve Convergence

Algorithm		Laplace-Dirichlet 11×11	Laplace-Dirichlet 31×31	Laplace-Neumann 11×11	Laplace-Neumann 31×31	Laplace-Mixed 15×21 $\gamma=1.0$	Laplace-Mixed 15×21 $\gamma=10^{-5}$
SIP		9	17	15	24	D	NC
CG	1	36	83	34	73	79	100*
CG	2	30	69	32	78	NC	NC
CG	3	10	23	13	26	NC	D
CG	4	8	14	23†	48†	D	NC†
CG	5	12	27	15	32	NC	NC
Bi-CG	6	25	58	34	73	79	100*
Bi-CG	7	26	57	29	68	83	100*
Bi-CG	8	9	20	12	26	28	37
Bi-CG	9	8	14	14†	29†	39†††	41†
Bi-CG	10	10	24	15	33	33	43

<u>Key</u> D : Divergent

NC : Not Converging

* : Failed to achieve convergence test

† : Modification in preconditioning used because of small main diagonal

TABLE 2 Number of steps of the Conjugate Algorithms to Achieve Convergence

Problem	Mesh	SIP	CG 3	CG 4	Bi-CG 8	Bi-CG 10
Laplace - Dirichlet	11 × 11	9	10	8	9	8
	21 × 21	13	17	12	15	12
	31 × 31	17	23	14	20	14
	41 × 41	29	25	17	25	17
	51 × 51	35	34	20	30	19
	61 × 61	53	35	22	34	20
	71 × 71	NC	41	23	36	22
	81 × 81	NC	51	25	34	24
	101 × 101	NC	50*	28	50*	27
	201 × 201	NC	50*	41	50*	40
Laplace - Neumann	11 × 11	15	13	23†	12	14†
	31 × 31	24	26	48	26	29
	101 × 101	NC	50*	50*†	50*	50*†
	201 × 201	NC	50*	50*†	46	50*†

Key NC : Not Converging

* : Failed to achieve convergence test

† : Modification in preconditioning used because of small main diagonal

evident from the slightly improved iteration performance when a bi-conjugate gradient algorithm (algorithm (10) is used contrasted with algorithm (5)). For the other preconditionings applied to these symmetric systems, the bi-conjugate gradient based algorithm closely replicates the results for the conjugate gradient algorithm.

For the non-symmetric and singular systems obtained from the first phase of the second class of test problems, namely Laplace's equation with Neumann boundary conditions, the first point of note is that all the algorithms required more iterations than on the problem of the same size but from the first class of test problems. Although the algebraic system is singular and non-symmetric, the algorithm (3) (ICCG(0)) is the most effective of the algorithms which use the basic conjugate gradient method. The rate of convergence is only marginally slower than that obtained from the comparable Dirichlet problem. This contrasts with the results for algorithm (4) for which the number of steps taken for these problems is significantly larger than for the symmetric system of comparable size, unless the bi-conjugate gradient method is used (algorithm (9)) in which case the number of steps is only approximately doubled. Indeed the results for algorithms (3) and (9) are very similar in terms of steps used. The preconditioning based on SIP factorization (algorithm (4)) provides very similar results to those of algorithm (3), and like that algorithm, it does not appear to demand the use of bi-conjugate gradients for these systems, supporting the conjecture of Section 2.5 that the preconditioned system should be very close to the identity system.

The third problem considered produces a non-symmetric system of equations, which although being non-singular for non-zero values of γ, can have a very large condition number. None of the preconditioning algorithms which used the basic conjugate gradient method provided satisfactory solution procedures. The algorithms (6) to (10) which use the biconjugate gradient method, all produce convergent schemes. The best amongst them are the algorithms (8) and (9), with algorithm (10) being slightly less effective.

From this first phase of results, the number of methods to be tested further can be considerably reduced. The results of Table 1 show that the preconditionings based on that used in ICCG(0) (as in algorithms (3) and (8)), the equality of row sums and on off-main diagonal elements (as in algorithms (4) and (9)), and on the corrected SIP factorization (as in algorithms (5) and (10)), provide the 'best' preconditioning for subsequent solution by an appropriate conjugate gradient method. Unlike the first two of these preconditionings, the one based on the SIP factorization generally produces a non-symmetric preconditioned system, even in the case when the original system is symmetric. A bi-conjugate gradient method might well have to be used for all systems in a general prupose algorithm. For non-symmetric systems and for singular systems, the algorithms ((5) and (10)) based on this SIP preconditioning are never 'better' than the comparable method using a preconditioning of one of the other two types. In addition the requirement to store the off-main diagonal elements of the approximate factors is a further disadvantage of the SIP form of preconditioning. The generalization to systems derived from models other than finite differences or finite elements on regular meshes, and to general sparsity patterns is also not readily derivable. For these reasons, we conclude that despite its relatively good performance on a number of problems, the SIP form of preconditioning is unsuitable as the basis of a general algorithm, since other algorithms provide comparable performance without the features and restrictions mentioned above.

The selection of the 'best' algorithm for general purpose application is, at least using the results given in this paper as a basis for comparison, between the ICCG(0) preconditioning of algorithms (3) and (8), and the row sums equality preconditioning of algorithms (4) and (9). To provide further evidence on which to base a selection of one of these preconditionings, and to extend the sizes of the problems being considered, the algorithms (3), (4), (8) and (9) were all used to solve the additional test problems in the second phase of tests. The additional tests comprised the additional eight mesh sizes for the first test problem, namely Laplace's equation with Dirichlet boundary conditions (mesh sizes of 21,41,51,61,71,81,101 and 201 in each coordinate direction), and two additional mesh sizes for the Laplace problem with Neumann boundary conditions (mesh sizes of 101 and 201 in each coordinate direction). Table 2 gives the results of solving these problems with the four algorithms under test, together with the results when the SIP solution procedure is used. Note that the results for mesh sizes of 11 × 11 and 31 × 31 are included from the previous table for completeness.

The ability of the algorithms to solve problems with very large numbers of equations (over 5000) is of particular importance since SIP works satisfactorily up to about this size, but fails for larger systems (at least in two dimensions). The results for the first problem given in Table 2 show that the preconditioned conjugate gradient methods, and in particular algorithm (4) (and its bi-conjugate gradient form - algorithm (9)) continue to be very effective as the problem size increases. The results for the first test problem on the different meshes for algorithm (4) and SIP illustrate that for a system with a small number of equations, SIP is more efficient in terms of computer time to achieve a particular level of residual. However as the number of equations increases, the number of iterations increases faster than $\sqrt{\text{number of equations}}$. In contrast the algorithm (4) based on the row-sums equality preconditioning with the conjugate gradient method uses more computation for small problems than SIP, but the number of iterations increases less fast than $\sqrt{\text{number of equations}}$, and continues to be very effective even for very large problems (over 40 000 equations). The changeover point from SIP to algorithm (4) in terms of the most efficient algorithm measured by computational effort is at a problem size of about 55 × 55 for this particular problem, namely Laplace's equation with Dirichlet boundary conditions. (Note that had we selected the mixed boundary condition problem, the results for which are given in Table 1, the changeover point would have been much lower since SIP fails to converge on the 15 × 21 mesh, i.e. 315 equations.)

For the non-symmetric and singular systems obtained from the second test problem, the test cases with 101 and 201 nodes in each direction proved too difficult to obtain 'converged' solutions within the 50 iteration limit. The levels to which convergence was achieved were examined and compared, they indicated a slight, but general superiority of algorithms (3) and (8) over the algorithms (4) and (9) respectively. From these results it is interesting to note that the system, which has a non-symmetric system of equations, is 'better' solved by using the basic conjugate gradient method if simple preconditioning is used (algorithm (3) in preference to algorithm (8)), whereas if row sums equality preconditioning is used, the algorithm (9) which uses the bi-conjugate gradient method is more effective than algorithm (4) which uses the conjugate gradient algorithm.

The selection of the most appropriate algorithm for a general purpose solution package is therefore not easily made between the algorithm (3) (and its bi-conjugate form (8)) and algorithm (4) (and its biconjugate form (9)), at least not from the data presented in the tables. One algorithm is better for some problems but not for others. Although it would be possible to provide both preconditionings, graphs of the convergence rates of the test cases showed that the algorithm (4) (and its bi-conjugate form (9)) produced smoother (and indeed generally monotonic) reductions in the residual and change, compared with those of algorithm (3) (and (8)). Thus we have selected algorithm (4) to be used if the original system is symmetric and non-singular, and algorithm (9) otherwise.

4. CONCLUSIONS

The preconditioned conjugate gradient methods are the most exciting development in the area of iterative solvers for large systems of algebraic equations. They provide faster convergence than other iterative methods on a very wide range of problems, and their convergence rates appear to be less affected by the order of the system than other methods. They are therefore ideally suited to solving very large systems for which direct methods are often too demanding on computation and/or storage to be viable, and for which other iterative methods do not provide acceptable convergence rates. Fast convergence contributes not only to the economy with which a solution is obtained, but also, at least as important in some applications, it means that the error bounds on the solution can be made tighter; indeed the user pays only a small 'price' to do a few extra iterations which will provide a much greater increase in the accuracy of the solution than would be the case if the convergence rate was much slower.

A number of different forms of preconditioning have been proposed by various authors. The 'best' preconditioning strategy based on the comparisons described in this paper and on the work of others is based on forcing the equality of each row sum of the preconditioning matrix with the comparable row sum of the original matrix. Such a requirement is not sufficient to uniquely determine the elements of the preconditioning matrix, but the work described in this paper and that of other researchers as reviewed here indicates that the various additional requirements are in general of less significance than the row-sums equality condition. For a preconditioning based on an approximate Cholesky or LU factorization, dependent on the symmetry or otherwise of the system, enforcing the equality of the non-zero off-main diagonal elements of the original matrix with those of the preconditioning matrix, in addition to the equality of rwo sums, provides an algorithm which is amongst the best, if not 'the' best, from the comparisons we have made. In addition this preconditioning has considerable attractions for its simplicity, the need, at least for a range of applications to certain diagonally structured matrices, to store only the main diagonal elements of the approximate factors used for preconditioning and its straightforward generalization to and implementation for systems with random sparsity rather than diagonally structured sparsity. Such a preconditioning can be applied to both symmetric and non-symmetric positive definite systems. For indefinite and/or singular systems it is necessary to ensure that the preconditioning matrix is not singular. It has been found adequate to check whilst calculating the approximate Cholesky or LU factors to be used for preconditioning that the factors are non-singular. If a main diagonal element of these factors becomes zero or very small, the equality of row sums on that row is relaxed in order to obtain a non-singular factor, and an 'appropriate' main diagonal value is used (e.g. that of the original matrix). Such a modification prevents the use of a singular preconditioning without, it appears, detrimentally affecting the 'quality' of the preconditioning.

The basic conjugate gradient method was derived for solving symmetric positive definite systems. For such systems it can be proved that each step of the algorithm reduces the chosen norm of the residual. For non-symmetric or indefinite systems, the biconjugate gradient algorithm provides a means of solution which in practice appears to have similar convergence rates to the conjugate gradient method on 'comparable' symmetric positive definite, it systems, without the deficiencies of the method based on solving the normal equations. However, the monotonic decay of the norm of the residual cannot be proved for this algorithm. Although in practice, as stated, this does not seem to be a limitation, it does mean that greater care on the convergence criteria and on iteration termination must be exercised.

5. ACKNOWLEDGEMENTS

The work described in this paper has been influenced by stimulating discussions with many people, John Ried (of UKAEA, Harwell), Geoffrey Markham and John Pritchard in particular. John Pritchard in addition designed and developed the test bed referred to in the text which provided a very easy to use and easy to modify facility to compare and contrast the various algorithms.

The work was carried out at the Central Electricity Research Laboratories and is published by permission of the Central Electricity Generating Board.

6. REFERENCES

1. HESTENES, M.R. and STIEFEL, E. - Methods of Conjugate Gradients for Solving Linear Systems. J. Res. Nat. Bur. Standards, Vol. 49, pp. 409-436, 1952

2. REID, J.K. - On the Method of Conjugate Gradients for the Solution of Large Sparse Systems of Equations, in 'Large Sparse Sets of Linear Equations', Edited by J.K. Reid, Academic Press, London, 1971

3. JENNINGS, A. - Matrix Computation for Engineers and Scientists, John Wiley, Chichester, 1977

4. JENNINGS, A. - Influence of the Eigenvalue Spectrum on the Convergence Rate of the Conjugate Gradient Method, J. Inst. Maths. Applics., 20, 61-72, 1977

5. LANCZOS, C. - An Iteration Method for the Solution of the Eigenvalue Problem of Linear Differential and Integral Operators, J. Res. Nat. Bureau Stand., 45, 255-282, 1950

6. MEIJERINK, J.A. and van der Vorst, H.A. - An Iterative Solution Method for Systems of Which the Coefficient Matrix is a Symmetric M-matrix, Maths. Comp., 31, 148-162, 1977

7. STONE, H.L., Iterative Solution of Implicit Approximations of Multi-dimensional Partial Differential Equations, SIAM J. Numer. Anal., 5, 530-558, 1968

8. MEIJERINK, J.A. and van der Vorst, H.A., 1978, Guidelines for the Usage of Incomplete Decompositions in Solving Sets of Linear Equations as Occur in Practical Problems, Private Communication, 1978

9. MEIJERINK, J.A. and van der Vorst, H.A. - Incomplete Decompositions as Preconditioning for the Conjugate Gradient Algorithm, in 'Conjugate Gradient Methods and Similar Techniques', Edited by I.S. Duff, UKAEA, Harwell Report No. R9636, 1979

10. AXELSSON, O. - A generalized SSOR Method, Nordisk Tidskrift for Information Behandling, 13, 443-467, 1972

11. KERSHAW, D.S. - The Incomplete Cholesky-Conjugate Gradient Method for the Iterative Solution of Systems of Linear Equations, J. Comp. Phys., 26, 43-65, 1978

12. MUNKSGAARD, N. - Solving Sparse Symmetric Sets of Linear Equations by Preconditioned Conjugate Gradients, UKAEA Harwell Report No. CSS67, 1979

13. GUSTAFSSON, I. - A Class of First Order Factorization Methods, Chalmers University of Technology Research Report No. 77.04R, 1977

14. WILKINSON, J.H. - The Algebraic Eigenvalue Problems, Clarendon Press, Oxford, 1965

15. WONG, Y.S. - Iterative Methods for Problems in Numerical Analysis, D. Phil. Thesis, Oxford, 1978

16. FLETCHER, R. - Conjugate Gradient Methods for Indefinite Systems, in Proc. of Dundee Conference on Numerical Analysis - 1975, Edited by G.A. Watson, Springer Verlag, 1976

17. JACOBS, D.A.H. - Solving Poisson's Equation Subject to Neumann Boundary Conditions, CERL Note No. RD/L/N 116/80, 1980

18. JACOBS, D.A.H. - A Summary of Subroutines and Packages for Solving Elliptic and Parabolic Partial Differential Equations, CERL Note No. RD/L/N 70/80, 1980

19. VAN DER SLUIS, A. - Condition Numbers and Equilibration of Matrices, Numer. Math., 14, 14-23, 1969

20. JACOBS, D.A.H. - Generalizations of the Conjugate Gradient Method for Solving Non-symmetric and Complex Systems of Algebraic Equations, CERL Note No. RD/L/N 70/80, 1980

21. ELSON, G.D. - GALVAN - A Program to Numerically Solve the Equations Pertaining to a Galvanic Corrosion Model, CEGB HQCC Program No. CC/P556, 1978

AN APPROXIMATE QIF METHOD FOR PARALLEL COMPUTERS

D.J. Evans & E.A. Lipitakis

Department of Computer Studies, University of Technology, Loughborough, Leicestershire, U.K.

ABSTRACT

A parallel linear solver based on a sparse factorization scheme of a general banded coefficient matrix is introduced. The use of sparsity in the factorization process leads to a significant reduction on the number of processors required for the solution of such banded linear systems over the corresponding number for other parallel solvers.

1. INTRODUCTION

Recently the increased use of parallel computation with its established superiority in speed over the serial computation has led to the development of new parallel algorithms as well as to the re-structuring and re-design of well-known serial numerical algorithms [19],[16],[11],[2].

An efficient parallel algorithm for the solution of the non-singular linear system

$$A\underline{u} = \underline{s} \,, \tag{1.1}$$

on a parallel computer of the "Single Instruction Stream-Multiple Data" type (SIMD) [10],[18] using the Quadrant Interlocking Factorization (QIF) method has been presented in [3]. It should be noted that in the SIMD computers there are N-independent processors capable of performing N-tasks on different sets of data, while the computational speed on each processor (with N-fold parallelism) is N-times faster than the computation on a serial processor. After the introduction of the QIF method, which has been originated from the Folding Algorithm [2] and possesses a large degree of parallelism, i.e., the solution process can be obtained in O(N) operations, several modifications and extensions of the method have been presented [4],[5],[6],[7].

In this paper we investigate the adaptability of both direct and iterative methods for solving the linear system (1.1), with the coefficient matrix A being a non-singular banded diagonally-dominant matrix, by introducing a sparse version of the QIF method originated from the Extendable Sparse Factorization techniques [14],[12],[9]. The proposed parallel linear solver then can be used either as a direct method by considering the exact factorization of A (the "limit" case) or as an iterative method, where the sparse factorization of A is used in conjunction with implicit schemes, for an efficient solution of system (1.1). The new method requires a maximum number of $\max\{2(r_1+r_2)^2, 2(r_1+r_2+1)\}$ processors working in parallel for the sparse factorization (in which only the r_1 and r_2 outermost off-diagonal terms have been retained), while the corresponding numbers of processors for the complete factorization of a full (dense) and a banded (of semi-bandwidth p) (N×N) matrix are respectively given by $2(N-2)^2$ and $\max\{2p^2, 2(p+1)\}$ [3],[5].

2. STATEMENT OF THE PROBLEM AND THE QIF SOLUTION METHOD

We consider the linear system (1.1) where the coefficient matrix A is an (N×N) general non-singular compact dense matrix and $\underline{u}, \underline{s}$ are two N-dimensional vectors given respectively by

$$\underline{u} = [u_1, u_2, \ldots, u_N]^T; \quad \underline{s} = [s_1, s_2, \ldots, s_N]^T . \qquad (2.1)$$

The solution of the system (1.1) can be obtained serially in a time proportional to N by a large number of related methods and between them the LU factorization methods have been very successful providing efficient linear system solvers.

A parallel algorithm, applicable to SIMD machines and based on an alternative LU factorization of the coefficient matrix A (i.e. the factors are interlocking matrix quadrants, cf.[3]), had led to the derivation of the Quadrant Interlocking

Factorization (QIF) method for the solution of system (1.1). The QIF method, possessing a large degree of parallelism by obtaining the solution in O(N) operations, compares favourably with similar existing methods [16],[11],[2] and a review of the method based on the Choleski root free factorization is given below.

Let A be a non-singular matrix as defined earlier with elements $a_{i,j}|i,j=1(1)N$, and assume that there exist matrices W,D and Z such that

$$A = W.D.Z \ , \tag{2.2}$$

where W,Z have the general forms,

$$W \equiv \begin{bmatrix} 1 & & & & & & & 0 \\ w_{21} & 1 & & 0 & & & 0 & w_{2,N} \\ w_{31} & w_{32} & 1 & \cdot & \cdot & 0 & w_{3,N-1} & w_{3,N} \\ \vdots & \vdots & \vdots & \cdot & & \vdots & \vdots & \vdots \\ w_{N-2,1} & w_{N-2,2} & w_{N-2,3} & & \cdot & 1 & w_{N-2,N-1} & w_{N-2,N} \\ w_{N-1,1} & w_{N-1,2} & & 0 & & & 1 & w_{N-1,N} \\ w_{N,1} & & & & & & & 1 \end{bmatrix} = [W_1, W_2, \ldots, W_N], \tag{2.3}$$

$$Z \equiv \begin{bmatrix} z_{11} & z_{12} & z_{13} & \cdot & \cdot & \cdot & z_{1,N-2} & z_{1,N-1} & z_{1,N} \\ & z_{22} & z_{23} & \cdot & \cdot & \cdot & z_{2,N-2} & z_{2,N-1} & \\ & & z_{33} & \cdot & \cdot & \cdot & z_{3,N-2} & & \\ & & & \cdot & & \cdot & & & \\ & 0 & & & \cdot & & 0 & & \\ & & & \cdot & & \cdot & & & \\ & & 0 & \cdot & \cdot & \cdot & z_{N-2,N-2} & & \\ & 0 & z_{N-1,3} & \cdot & \cdot & \cdot & z_{N-1,N-2} & z_{N-1,N-1} & \\ 0 & z_{N,2} & z_{N,3} & \cdot & \cdot & \cdot & z_{N,N-2} & z_{N,N-1} & z_{N,N} \end{bmatrix} = [Z_1, Z_2, \ldots, Z_N], \tag{2.4}$$

and D is a diagonal matrix with positive diagonal entries, viz.

$$D \equiv \text{diag}\{\tilde{d}_1, \tilde{d}_2, \ldots, \tilde{d}_N\} \ . \tag{2.5}$$

The components $W_i, Z_i | i=1(1,N$ in the right hand side of (2.3), (2.4) are respectively the ith column vectors of the matrices W, Z^T and by $a_{i,j}^{(k)} | i,j=1(1)N$ we denote the elements of the matrices A_k, $k=1(1)[\frac{N-1}{2}]$ (where [t] denotes the largest integer not exceeding the real number t) which are defined such that

$$A_1 = A = \sum_{i=1}^{N} \tilde{d}_i W_i Z_i^T , \tag{2.6a}$$

$$A_k = A_1 - \sum_{i=1}^{k-1} \tilde{d}_i W_i Z_i^T - \sum_{i=N-k+2}^{N} \tilde{d}_i W_i Z_i^T \Big| k=2(1)[\tfrac{N-1}{2}], \tag{2.6b}$$

$$A_{k+\frac{1}{2}} = A_1 - \sum_{i=1}^{k} \tilde{d}_i W_i Z_i^T - \sum_{i=N-k+2}^{N} \tilde{d}_i W_i Z_i^T \Big| k=1(1)[\tfrac{N-1}{2}]. \tag{2.6c}$$

Then, by adapting the fundamental assumptions of parallelism, i.e., a parallel replacement statement requires negligible time while any other parallel arithmetic operation needs the same time step to be performed, the elements of the W,D,Z matrices can be obtained by applying the following parallel algorithm (known as the NQIF2 method [6]).

For $k=1(1)[\frac{N-1}{2}]$

$$\tilde{d}_k = a_{k,k}^{(k)} , \tag{2.7}$$

$$z_{k,j} = a_{k,j}^{(k)}/\tilde{d}_k \; ; \; w_{j,k} = a_{j,k}/\tilde{d}_k, \text{ for } j=k+1(1)N-k+1, \tag{2.8}$$

$$A_{k+\frac{1}{2}} = A_k - A_k^{(k)} Z_k^T , \tag{2.9}$$

$$\tilde{d}_{N-k+1} = a_{N-k+1,N-k+1}^{(k+\frac{1}{2})} , \tag{2.10}$$

$$z_{N-k+1,j} = a_{N-k+1,j}^{(k+\frac{1}{2})}/\tilde{d}_{N-k+1} ; \; w_{j,N-k+1} = a_{j,N-k+1}^{(k+\frac{1}{2})}/\tilde{d}_{N-k+1} ,$$

$$\text{for } j=k+1(1)N-k, \tag{2.11}$$

$$A_{k+1} = A_{k+\frac{1}{2}} - A_{N-k+1}^{(k+\frac{1}{2})} Z_{N-k+1}^T , \tag{2.12}$$

(where $A_\ell^{(k)}$ denotes the ℓth column of the matrix A_k).

Finally, in order to complete the factorization process for $k=[\frac{N+1}{2}]$, if N is even, the steps (2.7),(2.10) are carried out while if N is odd only the step (2.7) is executed.

Due to the use of the three-term factorization (2.2) and the choice of the half step in (2.7)-(2.12) this parallel algorithm requires a considerably smaller number of time steps as well as a maximum number of processors than earlier versions of QIF method.

The solution of the linear system (1.1) then can be obtained equivalently by solving the systems:

$$W\underline{y} = \underline{s}\ ; \quad D\underline{v} = \underline{y}\ ; \quad Z\underline{u} = \underline{v}\ , \tag{2.13}$$

which require a total number of (7N-6) time steps while the maximum number of processors working in parallel is $(N-1)^2$, cf. [6]).

3. THE SPARSE QUADRANT INTERLOCKING FACTORIZATION METHOD

In this section we extend the applicability of the QIF methods to the numerical solution of banded linear systems, which occur in the finite difference or finite element discretization of P.D.E.'s in mathematical physics and engineering, on parallel SIMD computers by introducing the Sparse Quadrant Interlocking Factorization (SQIF) method.

The SQIF method is based on the well-established idea of the Extendable Sparse Factorization methods for serial computers [13],[14],[12].

Let us consider the linear system (1.1), where A is now a general non-singular banded real (N×N) matrix of semibandwidth $p \ll N$, and assume that there exist matrices R, W_s, D_s, Z_s such that,

$$A + R = W_s D_s Z_s\ , \tag{3.1}$$

where W_s, D_s, Z_s are sparse forms of W,D,Z [cf. (2.3)-(2.5)] and (A+R) is a simple replacement of A leading to "easily solvable" systems $W_s D_s Z_s \underline{u} = \underline{s}$.

In the following we outline a strategy whereby W_s, D_s, Z_s are easily determined and the approximate factorization (3.1) can be as accurate as we require. We define the matrices W_s, D_s, Z_s such that:

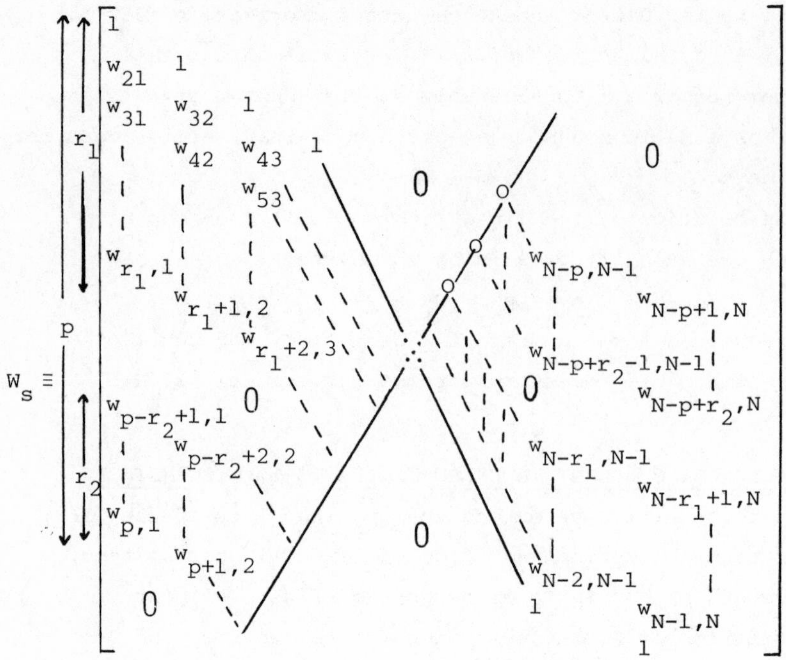

(3.2)

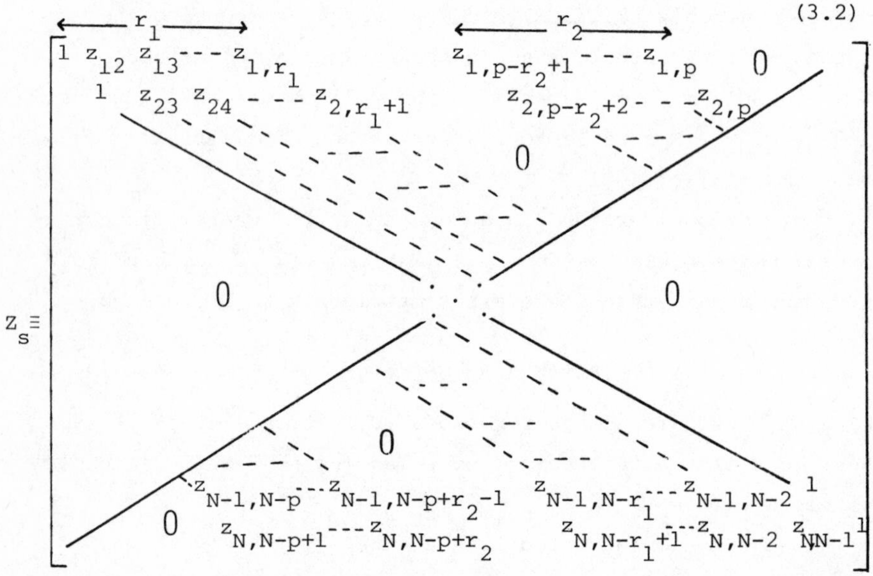

(3.3)

and $$D_s \equiv \text{diag}\{d_1, d_2, \ldots, d_N\}. \qquad (3.4)$$

It can be easily proved by induction that both W_s and Z_s are banded matrices of the same semibandwidth p and zero-structure as the coefficient matrix A which additionally is assumed to possess the diagonally dominance property, a sufficient condition for numerical stability, cf. [5].

Our approach for the derivation of the SQIF method is based on the fact that only r_1 and r_2 outermost off-diagonal entries are kept in the matrix quadrants, cf. (3.2),(3.3). Consequently the choice of the parameters $r_1,r_2\in[1,p]$ (with $r_1+r_2\leq p$) enable us to obtain an (r_1,r_2)-dependent approximate solution to the original linear system (1.1) by solving the system $W_sD_sZ_s\underline{u}=\underline{s}$. The sparse approximate factorization then can be shown to yield an iterative procedure for the solution of (1.1).

Previous research work [9],[12] indicates that for 1D and 2D Laplace-like problems the values of parameters r_1 and r_2 are given by $r_1\in[1,2]$ and $r_2\in[0,3]$ respectively, while for more complicated types of problems, i.e. Biharmonic and 3D-problems, the value of r_2 is expected to be greater than 3 (specifically, for the Biharmonic and the 3D-seven-point Laplace operator the value of r_2 depends on the linear dimensionality of the problem).

It should be noted that if $(r_1+r_2)=p$ outermost off-diagonal entries have been retained in W_s,Z_s, then $W_s\equiv\tilde{W},D_s\equiv\tilde{D}$ and $Z_s\equiv\tilde{Z}$ (the "limit" case), where $\tilde{W}$ and $\tilde{Z}$ are interlocking matrix quadrants retaining p outermost off-diagonal entries. The coefficient matrix A is then factorized exactly, i.e., $A=\hat{W}\tilde{D}\tilde{Z}$ to yield a parallel algorithmic procedure (the QIF2 technique) for the solution of banded linear systems by determining the elements of $\tilde{W},\tilde{D},\tilde{Z}$ [5]. Since the latter factorization can be obtained from its approximate counterpart when $(r_1+r_2)\to p$ (i.e. $W_s^{-1}\tilde{W}=I+O(\varepsilon)$ and $Z_s^{-1}\tilde{Z}=I+O(\varepsilon)$) the SQIF method is to be considered as the general case.

Since the original matrix A is banded i.e.,

$$a_{i,j}^{(\ell)} = 0, \quad \text{for } i,j=1(1)N \text{ and } |i-j|>p \text{ with } \ell=1(1)[\tfrac{N-1}{2}],$$

the elements of W_s, D_s, Z_s can be obtained by applying the following parallel algorithm:

<u>For $k=1(1)[\frac{N-p}{2}]$</u>

$$d_k = a_{k,k}^{(k)}, \quad d_{N-k+1} = a_{N-k+1,N-k+1}^{(k)}, \tag{3.5}$$

$$\left.\begin{aligned} z_{k,j} &= a_{k,j}^{(k)}/d_k, \quad w_{j,k} = a_{j,k}^{(k)}/d_k, && (3.6)\\ z_{N-k+1,N-j+1} &= a_{N-k+1,N-j+1}^{(k)}/d_{N-k+1}, && (3.7)\\ w_{N-j+1,N-k+1} &= a_{N-j+1,N-k+1}^{(k)}/d_{N-k+1}, && (3.8) \end{aligned}\right\} \begin{array}{l}\text{for } j=k+1(1)k+p \text{ and } |j-k|<r_1,\\ \text{or } j=k+1(1)k+p\\ \text{and } |j-k|>p-r_2-1,\end{array}$$

$$A_{k+1} = A_k - A_k^{(k)} Z_k^T - A_{N-k+1} Z_{N-k+1}^T. \tag{3.9}$$

<u>Then, for $k=[\frac{N-p}{2}]+1(1)[\frac{N-1}{2}]$</u>

$$d_k = a_{k,k}^{(k)}, \tag{3.10}$$

$$\left.\begin{aligned} z_{k,j} &= a_{k,j}^{(k)}/d, \\ w_{j,k} &= a_{j,k}^{(k)}/d, \end{aligned}\right\} \begin{array}{l}\text{for } j=k+1(1)N-k+1 \text{ and } |j-k|<r_1,\\ \text{or } j=k+1(1)N-k+1 \text{ and } |j-k|>p-r_2-1,\end{array} \tag{3.11}$$

$$A_{k+\frac{1}{2}} = A_k - A_k^{(k)} Z_k^T, \tag{3.12}$$

$$d_{N-k+1} = a_{N-k+1,N-k+1}^{(k+\frac{1}{2})}, \tag{3.13}$$

$$\left.\begin{aligned} z_{N-k+1,j} &= a_{N-k+1,j}^{(k+\frac{1}{2})}/d_{N-k+1}, && (3.14)\\ w_{j,N-k+1} &= a_{j,N-k+1}^{(k+\frac{1}{2})}/d_{N-k+1}, && (3.15) \end{aligned}\right\} \begin{array}{l}\text{for } j=k+1(1)N-k \text{ and } |N-k+1-j|<r_1,\\ \text{or } j=k+1(1)N-k \text{ and } |N-k+1-j|>p-r_2-1,\end{array}$$

$$A_{k+1} = A_{k+\frac{1}{2}} - A_{N-k+1}^{(k+\frac{1}{2})} Z_{N-k+1}^T. \tag{3.16}$$

Finally, for $k=[\frac{N+1}{2}]$, if N is even, the steps (3.5),(3.6) are carried out to find $d_k, d_{k+1}, z_{k,k+1}$ and $w_{k+1,k}$, while if N is odd only the step (3.10) is executed to find d_k. It should be pointed out that when an interval of the form $j=\alpha(1)\beta$, (or $|j|<\gamma$) appears after a statement in the parallel algorithm (3.5)-(3.16) (henceforth called the SQIF algorithm) the statement is assumed to be executed simultaneously for all indices in the interval $[\alpha,\beta]$ (or $[1,\gamma)$ respectively).

An approximate solution of the linear system (1.1) can then be obtained from $W_s D_s Z_s \underline{u}=\underline{s}$ and setting

$$Z_s\underline{u} = \underline{v}\ ,\quad D_s\underline{v} = \underline{y}\ , \tag{3.17}$$

the problem reduces to solve the system,

$$W_s\underline{y} = \underline{s}\ . \tag{3.18}$$

The system (3.18) can be equivalently re-written as

$$\sum_{i=1}^{N} y_i W_i = \underline{s}=\underline{s}^{(1)}\ , \tag{3.19}$$

and by introducing the result vectors $\underline{s}^{(k+\frac{1}{2})}$, $\underline{s}^{(k)}\,|k=1(1)[\frac{N+1}{2}]$ such that

$$\underline{s}^{(1)} = \underline{s}\ ;\ \underline{s}^{(k+\frac{1}{2})} = \underline{s}^{(k)}-\underline{y}_k W_k\ ;\ \underline{s}^{(k+1)} = \underline{s}^{(k+\frac{1}{2})}- \underline{y}_{N-k+1}W_{N-k+1}, \tag{3.20}$$

it can be solved by applying the SQIF algorithm as follows:

<u>For $k=1(1)[\frac{N-p}{2}]$</u>

$$y_k = s_k^{(k)}\ ,\quad y_{N-k+1} = s_{N-k+1}^{(k)}\ , \tag{3.21}$$

$$s^{(k+1)} = s^{(k)}-y_k W_k-y_{N-k+1}W_{N-k+1}\ , \tag{3.22}$$

while, <u>for $k=[\frac{N-p}{2}]+1(1)[\frac{N-1}{2}]$</u>

$$y_k = s_k^{(k)}\ ,\quad s^{(k+\frac{1}{2})} = s^{(k)}-y_k W_k\ , \tag{3.23}$$

$$y_{N-k+1} = s_{N-k+1}^{(k+\frac{1}{2})}\ ,\quad s_{k+1} = s^{(k+\frac{1}{2})}-y_{N-k+1}W_{N-k+1}. \tag{3.24}$$

and, for $k=[\frac{N+1}{2}]$, if N is even, the first equations of (3.23) and (3.24) are respectively carried out to determine the values of $\underline{y}_k$ and $\underline{y}_{k+1}$, while, if N is odd, only the first equation of (3.23) is executed giving $\underline{y}_k$.

The second system of (3.17) can be simply solved as

$$v_i = y_i/d_i \mid i=1(1)N. \tag{3.25}$$

The first system of (3.17) can be equivalently re-written as

$$\sum_{i=1}^{N} u_i Z_i^* = \underline{v}=\underline{v}^{(1)}, \tag{3.26}$$

where $Z_i^* | i=1(1)N$ are the column vectors of Z.

Then, by introducing the result vectors $\underline{v}^{(i)}, \underline{v}^{(i+\frac{1}{2})}$ $|i=1(1)[\frac{N+1}{2}]$ such that

$$v^{(1)} = v\ ;\quad v^{(\ell-k+3/2)} = v^{(\ell-k+1)} - u_{N-k+1} Z^*_{N-k+1}, \tag{3.27}$$

$$v^{(\ell-k+2)} = v^{(\ell-k+3/2)} - u_k Z_k^*, \tag{3.28}$$

where $k=[\frac{N+1}{2}](-1)1$ and $\ell=[\frac{N+1}{2}]$ (except for N odd and $k=\ell$ when we set $v^{(3/2)}=v^{(1)}=v$ and $v^{(2)}=v^{(1)}-u_\ell Z_\ell^*$) the system (3.26) can then be solved as follows:

If N is even, set

$$u_{\ell+1} = v_{\ell+1}^{(1)}\ ;\quad v^{(3/2)} = v^{(1)} - u_{\ell+1} Z^*_{\ell+1}\ ; \tag{3.29}$$

$$u_\ell = v_\ell^{(3/2)}\ ;\quad v^{(2)} = v^{(3/2)} - u_\ell Z_\ell^*, \tag{3.30}$$

otherwise

$$u_\ell = v_\ell^{(1)}\ ;\quad v^{(2)} = v^{(1)} - u_\ell Z_\ell^*, \tag{3.31}$$

Then, <u>for $k=[\frac{N-1}{2}](-1)[\frac{N-p}{2}]+1$</u>

$$u_{N-k+1} = u_{N-k+1}^{(\ell-k+1)};\quad v^{(\ell-k+3/2)} = v^{(\ell-k+1)} - u_{N-k+1} Z^*_{N-k+1}, \tag{3.32}$$

$$u_k = v_k^{(\ell-k+3/2)}\ ;\quad v^{(\ell-k+2)} = v^{(\ell-k+3/2)} - u_k Z_k^*. \tag{3.33}$$

while, <u>for $k=[\frac{N-p}{2}](-1)1$</u>

$$u_k = v_k^{(\ell-k+1)}\ ;\quad u_{N-k+1} = v_{N-k+1}^{(\ell-k+1)}, \tag{3.34}$$

$$v^{(\ell-k+2)} = v^{(\ell-k+1)} - u_k Z_k^* - u_{N-k+1} Z^*_{N-k+1}. \tag{3.35}$$

Note that the last step (3.35) is not executable for k=1.

Under the fundamental assumptions for parallelism mentioned in Section 2 and by observing that when $k=1(1)[\frac{N}{2}]-p$

(assuming that $p[\frac{N}{2}]$) the non-zero elements of the matrix products in (3.9) do not overlap, i.e., for the above values of k, the two subtractions in two time steps in (3.9) can be done simultaneously in one time step, from which it can be readily found that the sparse factorization of matrix A requires a total number of time steps equal to

$$3(N-1)-2[\frac{N-p}{2}]-0.5|[\frac{N}{2}]-p|-0.5([\frac{N}{2}]-p),$$

by using a maximum number of $\max\{2(r_1+r_2)^2,4(r_1+r_2)\}$ processors working in parallel to perform all the operations involved.

The first system of (3.17) requires a total number of time steps

$$2(N-1)-[\frac{N-p}{2}]-0.5|[\frac{N}{2}]-p|-0.5([\frac{N}{2}]-p)\ ,$$

with a maximum number of $2(r_1+r_2)$ processors, while the second system of (3.17) is solved in one time step requiring a number of N processors. The system (3.18) requires a total number of $2(N-1)-2[\frac{N-p}{2}]$ time steps by using a maximum number of $2(r_1+r_2)$ processors. Consequently, the total number of time steps for both factorization and solution of the system under consideration is

$$7(N-1)-5[\frac{N-p}{2}]-|[\frac{N}{2}]-p|-([\frac{N}{2}]-p)+1,$$

while the corresponding maximum number of processors working in parallel at the same time is $\max\{2(r_1+r_2)^2,4(r_1+r_2),N\}$.

For comparative reasons the total number of time steps as well as the maximum number of processors required for the factorization, the solution of auxiliary systems (3.17),(3.18) and the complete solution of the original system are given respectively in Tables 1,2 and 3 for several versions of the QIF method (i.e., the QIF1 [5],[6], NQIF2 [7] and SQIF methods).

METHOD	TOTAL NUMBER OF TIME STEPS	MAX.NO. OF PROCESSORS
QIF1	$3(N-1)-1.5(1+(-1)^N)-2[\frac{N-p}{2}]-0.5\|[\frac{N}{2}]-p\|-0.5([\frac{N}{2}]-p)$	$\max\{2p^2, 2(p+1)\}$
NQIF2	$3(N-1)-2[\frac{N-p}{2}]-0.5\|[\frac{N}{2}]-p\|-0.5([\frac{N}{2}]-p)$	$\max\{2p^2,4p\}$
SQIF	$3(N-1)-2[\frac{N-p}{2}]-0.5\|[\frac{N}{2}]-p\|-0.5([\frac{N}{2}]-p)$	$\max\{2(r_1+r_2)^2, 4(r_1+r_2)\}$

TABLE 1: The factorization of the Coefficient Matrix A

METHOD	TOTAL NUMBER OF TIME STEPS	MAX.NO. OF PROCESSORS
QIF1	$4.5(N-1)-0.75(1+(-1)^N)-3[\frac{N-p}{2}]-0.5\|[\frac{N}{2}]-p\|-0.5([\frac{N}{2}]-p)+1$	$2p$
NQIF2	$4(N-1)-3[\frac{N-p}{2}]-0.5\|[\frac{N}{2}]-p\|-0.5([\frac{N}{2}]-p)+1$	$\max\{2p,N\}$
SQIF	$4(N-1)-3[\frac{N-p}{2}]-0.5\|[\frac{N}{2}]-p\|-0.5([\frac{N}{2}]-p)+1$	$\max\{2(r_1+r_2),N\}$

TABLE 2: The Solution of the Auxiliary Systems (3.17),(3.18)

METHOD	TOTAL NUMBER OF TIME STEPS	MAX.NO. OF PROCESSORS
QIFI	$7.5(N-1)-2.25(1+(-1)^N)-5[\frac{N-p}{2}]-\|[\frac{N}{2}]-p\|-([\frac{N}{2}]-p)+1$	$\max\{2p^2, 2(p+1)\}$
NQIF2	$7(N-1)-5[\frac{N-p}{2}]-\|[\frac{N}{2}]-p\|-([\frac{N}{2}]-p)+1$	$\max\{2p^2,4p,N\}$
SQIF	$7(N-1)-5[\frac{N-p}{2}]-\|[\frac{N}{2}]-p\|-([\frac{N}{2}]-p)+1$	$\max\{2(r_1+r_2)^2, 4(r_1+r_2),N\}$

TABLE 3: The Complete Solution of the Original System (including the factorization process)

The SQIF technique can then be combined into semi-direct methods of the form:

$$W_s D_s Z_s \delta\underline{u}_{i+1} = \alpha\underline{r}_i \ , \tag{3.36}$$

$$W_s D_s Z_s \delta\underline{u}_{i+1} = \alpha\underline{r}_i + \beta\delta\underline{u}_i \ , \tag{3.37}$$

and $$W_s D_s Z_s \delta\underline{u}_{i+1} = \alpha_\nu\underline{r}_i + \beta_\nu\delta\underline{u}_i \ , \tag{3.38}$$

where $$\delta\underline{u}_{i+1} = \underline{u}_{i+1} - \underline{u}_i \ , \quad \underline{r}_i = \underline{s} - A\underline{u}_i \ , \tag{3.39}$$

with α,β and α_ν,β_ν being preconditioned acceleration parameters and Chebychev sequences of parameters respectively defined in [17].

Furthermore, the new method can be successfully used in conjunction with the Conjugate Gradient method [15] leading to accelerated parallel solution schemes.

Until recently the experimental and numerical approach to verifying the theoretical improvements of the method particularly on large scale problems was not really feasible due to the lack of an available sufficiently large number of processors, say 500 or more. It is expected that in the near future computing technology will be sufficiently advanced to provide a multi-microprocessor system with the required number of processors in a much more compact way [1].

CONCLUSION

It can be readily seen from [3],[4],[5],[7] that the number of rows and columns of the coefficient matrix under consideration is proportional to the maximum number of processors and the total number of time steps required for the solution of the linear system respectively.

The new SQIF method, using the same total number of time steps for the solution of the original linear system as the NQIF2 method, requires only a number of $\max\{2(r_1+r_2)^2, 4(r_1+r_2), N\}$ processors. By adjusting the parameters r_1, r_2 we are able to obtain both the complete factorization (i.e. the

NQIF2 method) and the sparse factorization schemes. Finally, the sparse factorization approach in parallel computation leads to significant savings in the number of processors over existing parallel solvers especially for large problems.

REFERENCES

1. Barker, D.C. - "A Distributed Array Processor for the Simulation of Heat Transfer in Buildings", Proc. of the first Intern.Conf. on Numer.Meth. in Thermal Problems, Univ. College Swansea, pp.943-953, 1979.

2. Evans, D.J., Hatzopoulos, M. - "The Solution of Certain Banded Systems of Linear Equations Using the Folding Algorithm", Comp.J. 19, pp.184-187, 1976.

3. Evans, D.J., Hatzopoulos, M. - "A Parallel Linear System Solver", J.Comp.Maths. 7, pp.227-238, 1979.

4. Evans, D.J., Hadjidimos, A. - "A Modification of the Quadrant Interlocking Factorization Parallel Method", Int.J.Comp.Maths. 8, pp.149-166, 1980.

5. Evans, D.J., Hadjidimos, A. - "Parallel Solution of Certain Banded, Symmetric and Centrosymmetric Systems by Using the Quadrant Interlocking Factorization Method", Math.Comp. Sim. 23, pp.180-187, 1981.

6. Evans, D.J., Hadjidimos, A., Noutsos, D. - "Parallel Solution of Linear Systems by Quadrant Interlocking Factorization Methods", Comp.Meth.Appl.Mech. & Eng. 29, pp.97-107, 1981.

7. Evans, D.J., Hadjidimos, A., Noutsos, D. - "The Parallel Solution of Banded Linear Equations by the New Quadrant Interlocking Factorization (QIF) Method", Int.J.Comp.Math. 9, pp.151-162, 1981.

8. Evans, D.J., Lipitakis, E.A. - "On Sparse LU Factorization Procedures for the Solution of Parabolic Differential Equations in Three-Space Dimensions", Int.J.Comp.Math. 7, pp.315-338, 1978.

9. Evans, D.J., Lipitakis, E.A. - "A Normalized Implicit Conjugate Gradient Method for the Solution of Large Sparse Systems of Linear Equations", Comp.Meth.Appl.Mech. & Engng. 23, pp.1-20, 1980.

10. Flynn, M.J. - "Some Computer Organizations and Their Effectiveness", IEEE Trans. on Computers C-21, pp.948-960, 1972.

11. Lambiotte, J.J. - "The Solution of Linear Systems of Equations on a Vector Computer", Doctoral Thesis, Univ. of Virginia, U.S.A., 1975.

12. Lipitakis, E.A., Evans, D.J. - "On the Numerical Solution of Large Linear Finite Element Systems by Normalized Factorization Procedures", to be published.

13. Lipitakis, E.A., Evans, D.J. - "Normalized Factorization Procedures for the Solution of Self-Adjoint Elliptic Partial Differential Equations in Three-Space Dimensions" Math. & Comp. in Sim. XXI, pp.189-196, 1979.

14. Lipitakis, E.A. Evans, D.J. - "Solving Non-Linear Elliptic Difference Equations by Extendable Sparse Factorization Procedures, Computing, 24, pp.325-334,1980.

15. Sameh, A.H. - "Block Conjugate Gradient Methods and Parallel Computation", Proc. of Conf. on Elliptic Problem Solvers, Los Alamos, pp.405-412, edit. M. Schultz, Acad. Press, 1981.

16. Sameh, A.H., Kuck, D.J. - "Linear System Solvers for Parallel Computers", T.R. No. UIUCDCS-R-75-701, Dept. Comp.Science, Univ. of Illinois, USA, 1975.

17. Stiefel, E.L. - "Kernel Polynomials in Linear Algebra and their Numerical Applications", Nat.Bur. Standards in Appl.Math. Ser 49, pp.1-22, 1958.

18. Stone, H.S. - "Problems of Parallel Computation", in 'Complexity of Sequential and Parallel Numerical Algorithms', (Edit. J.F. Traub), pp.1-16, New York, Acad. Press, 1973.

19. Stone, H.S. - "An Efficient Parallel Algorithm for the Solution of a Tridiagonal Linear System of Equations", J.A.C.M. 20, pp.27-38, 1973.

INDEX

A

A optimum, 302

Accelerated Overrelaxation (AOR),49

Acceleration factor, 49

Acceleration parameters, 2,5,6,7,12, 17,18,37, 40,43

Adaptive parameter, 138,317

Alternating Direction Implicit (ADI) method, 54,81,94,109, 117,446

Alternating Direction Preconditioning (ADP) method, 57,84,101, 117
  Simultaneous Displacement, 90
  Richardson, 90

Applications
  Airfoil, 442,454,487
  Asymmetric systems, 516
  Biharmonic problem, 19,57,99,161, 173,233,286
  Boundary value problem, 19
  Constrained minimisation problem, 296
  Elliptic problem, 19,208,251, 265,481
  Eigenvalue problem, 394
  Fluid dynamics, 463,485,497
  Finite elements, 283,463
  Heat conduction problem, 464
  Laplace problem, 19,54,92,136, 206,233,522
  Navier Stokes flow, 500
  Nonlinear elliptic problem, 211
  Non-self adjoint problem, 340
  Parabolic problem, 28,54
  Plane frame problems, 425
  Portal frame, 426
  Self adjoint problem, 82,150, 191,364
  Singular systems, 518
  Step motor, 296
  Transonic flow problem, 440,454, 485
  Two point B.V. problem, 72,366, 397

Approximate elimination algorithm, 34

Approximate factorisation, 10,191, 446,467

Approximate Q.I.F. method, 539

Artificial density method, 442

B

Back substitution process, 26,72, 194

Bauer's method, 413

Bi-conjugate gradient method, 528

Block factorisation, 284

Bordering technique, 192

C

Chebyshev polynomials, 4,18,61,198, 243,300,329

Choleski method, 435,465

Column scaling, 435

Compact preconditioning, 189

Condition numbers: K, 405
  " " M, 405
  " " N, 405
  " " P, 3,71,404

Conditioning matrix, 10,356

Convex hull, 335

Conjugate Gradient method, 207

Conjugate Gradient Richardson method, 309

Conjugate Gradient Tchebyshev method, 302

Cyclic Chebyshev iterative method, 6

D

Diagonal correction terms, 431

Diagonal scaling, 409,520

Direct methods
  Choleski, 10
  Direct factorisation, 356
  Gaussian elimination, 10,34
  Q.I.F., 538
  LDL^T, 192
  LDU, 10
  LU, 10
  Normalised symmetric factorisation, 10

Drop criterion, 228

E

Ellipsoids, 39,59,243,330

Eigenvalue spectrum, 3,11,204

Extended to the Limit (EL) LDL^T factorisation, 192,207

Extrapolated factor, 47,62

Extrapolated method
  A.D.I., 54,81,94,109,117,446
  Gauss Seidel, 50,117,425
  Jacobi, 50,70,117
  Non-stationery, 47,61
  Stationery, 47,59
  S.O.R., 50,117

F

Fibonacci method, 146,473

Fill in, 9,33,192,220

Fill in parameter, 192

Fixed space factorisation, 227

Fixed storage allocation, 219

Forward substitution process, 26,72

Fractional step scheme, 356

G

Gauss-Seidel method, 49,117,425

Generated matrix, 25

Generalised eigenvalue problem, 393

Generalised preconditioning, 74

Gradient methods, 39

H

Horizontal row elimination, 91,108

I

I.C.C.G. method, 390

In core solution, 464

Incomplete
  Choleski method, 249,463,477
  Crout decomposition, 245
  Elimination method, 10,28
  Factorisation, 219,222,267,312, 428
  LU decomposition, 243,272

Inverse iteration, 390

Iterative methods
  A.D.I., 54,81,94,109,117,446
  Conjugate Gradient, 42,472,520
  Chebyshev S.I., 5,6,243
  Extrapolated Aitken, 8
  Gauss Seidel, 49,117,425
  Implicit, 190
  Jacobi, 2,49,70,117
  Jacobi overrelaxation, 50
  Non-stationery, 47,61
  Richardson, 3,4,5,81,298,305
  Simultaneous Displacement, 3,81, 117
  Stationery, 47,59
  Steepest descent, 40
  S.I.P., 520
  S.L.O.R., 445
  S.O.R., 7,49,117
  S.S.O.R., 8,117
  Variational, 40

Iterative refinement, 430

J

Jacobi method, 2,49,70,117

Jacobi overrelaxation method,50

K

K condition number, 405

L

Least squares solution, 469

M

M condition number, 200,405

Matrix
  Biharmonic, 19,57,173
  Cauchy, 419
  Cyclic, 73
  Hessenberg, 418
  Hilbert, 410
  Indefinite, 295
  Jacobi, 53
  L matrix, 272
  M matrix, 52,244,272,511
  Quindiagonal, 30,107,191,415
  Toeplitz, 180
  Tridiagonal, 180
  Unsymmetric, 321
  Vandermonde, 420

Modified A.D.P. method, 151,163

Modified Incomplete Choleski (MIC) method, 224,265

Modified Incomplete Choleski C.G. (MICCG), 275

Modified P.S.D. method, 119

Monotone sequence, 381

N

N condition number, 405

Nested dissection, 219

Norms
  matrix, 402
  vector, 402

O

Off-diagonal modification, 225

Optimal scaling, 412

P

P condition number, 3,14,78,81,93, 105,125

Parallel computer, 537

Partial elimination, 431

Perron root/vector, 413

Perturbed system, 403

Picard/Newton iteration, 212

Preconditioned acceleration parameters, 195

Preconditioned iterative methods
  Simultaneous Displacement, 11,17, 37,70,117,119,322
  Chebyshev method, 17,37,70,327
  Conjugate Gradient, 42,390,448, 467
  Jacobi, 117
  Richardson, 17,37,70
  Steepest descent, 40
  Variational method, 39

P.S.D. - C.G. method, 134

P.S.D. - S.D. method, 132

P.S.D. - S.I. method, 129

P.S.D. - V.E. method, 133

Preconditioned direct method, 401

Preconditioned residual, 39,70

Preconditioning by direct factorisation (PDF), 356

P.D.F. - S.I. method, 374

Preconditioning parameter, 17,76, 385,408

Preconditioning polynomial, 296, 303

Prescaling, 409

Q

Quadrant Interlocking Factorisation (Q.I.F.) method, 538

R

Rayleigh Quotient, 380

Regular splitting, 265

Rejection parameter, 429

Residual polynomial, 298

Relaxation methods, 425

Row scaling, 435

Row-sum agree factorisation, 450

Row-sums equality, 224

S

S.I.M.D. parallel computer, 537

Strongly implicit
  Method, 520
  Preconditioning, 521

S.L.O.R. method, 445

S.O.R. method, 7,49,117

S.S.O.R. method, 8,117,179,513

S.S.O.R. preconditioning, 179,450, 517

Shift parameter, 383

Sparse elimination iterative method, 36

Sparse preconditioning methods, 16

Sparse Q.I.F. method, 541

Spectral condition number, 2,116, 179

Splitting, 2,47,74,117,265,356,381

Symmetric implicit C.G. method, 207

T

Triangular preconditioning (T.P.), 69,96

V

Vertical row elimination 92,108

W

W.D.Z factorisation, 539

Z

Z-matrix property, 224